EMERGENCY RESPONSE AND
EMERGENCY MANAGEMENT LAW

Second Edition

EMERGENCY RESPONSE AND EMERGENCY MANAGEMENT LAW

Cases and Materials

By

WILLIAM C. NICHOLSON, ESQ.

Assistant Professor of Criminal Justice (Retired)
North Carolina Central University
Durham, North Carolina

CHARLES C THOMAS • PUBLISHER, LTD.
Springfield • Illinois • U.S.A.

Published and Distributed Throughout the World by

CHARLES C THOMAS • PUBLISHER, LTD.
2600 South First Street
Springfield, Illinois 62704

© 2012 by CHARLES C THOMAS • PUBLISHER, LTD.

ISBN 978-0-398-08831-6 (hard)
ISBN 978-0-398-08832-3 (paper)
ISBN 978-0-398-08833-0 (ebook)

Library of Congress Catalog Card Number: 2012019683

With THOMAS BOOKS *careful attention is given to all details of manufacturing
and design. It is the Publisher's desire to present books that are satisfactory as to their
physical qualities and artistic possibilities and appropriate for their particular use.*
THOMAS BOOKS *will be true to those laws of quality that assure a good name
and good will.*

Printed in the United States of America
UBC-R-3

Library of Congress Cataloging-in-Publication Data

Nicholson, William C.
    Emergency response and emergency managment law: cases and materials
/ William C. Nicholson – 2nd ed.
        p. cm.
    Includes bibliographical references and index.
    ISBN 978-0-398-08831-6 (hard) – ISBN 978-0-398-08832-3 (pbk.) – ISBN
978-0-398-08833-0 (ebook)
    1. Emergency management–Law and legislation–United States. I. Title.

KF3750.N53 2012
344.7305'34–dc23
                                                    2012019683

Thomas Arthur Gowling III

August 17, 1952 – September 5, 2011

Thomas A. Gowling III. Photo courtesy of Carol Gowling.

*T*ommy *and I have been friends since high school. He served for many years as a member of several Volunteer Fire Departments (VFDs) in the Washington, DC area, including Cabin John, Bethesda-Chevy Chase, Rockville and Potomac.*

I recall a number of times in our youth – while the rest of our friends and I were partying – when Tom was either absent for training or called to a scene. Tommy left, often as the fun was starting, much to our puzzlement. When he came back, smelling of smoke, he had that satisfied, quizzical little smile displayed by many volunteers.

Tommy was an adopted, only child whose Dad was diagnosed with Multiple Sclerosis only 3 weeks after the adoption was finalized. He had to grow up quickly. Tommy never hesitated to help around the house when called to do so, no matter what else was happening with his peers. Likewise, he jumped immediately when the call came to join his brother firefighters.

Tommy and I had many talks over the years about the relationship between law and emergency response as well as his experiences as a volunteer. The "boots on the ground" viewpoints of Tommy and several other serving emergency responders and emergency managers helped make the first edition a better text. Their criticisms of the first edition helped guide me to improve the second edition. Adding the section on Post-Traumatic Stress Disorder (PTSD) springs in part from a discussion Tommy and I had about how continued exposure to pain and violence affects emergency responders.

As an only child, Tommy wanted siblings very much. We who count ourselves among Tommy's long-time friends were very blessed in having him – and one another – as a kind of surrogate family. Nothing in his life gave Tommy greater happiness than his loving wife Carol and their close and caring family, including his three children, Megan Gowling, Katelyn Gowling, Ryan Gowling, and two stepdaughters, Brittany Karakostas and Kristina Karakostas. Never have I seen a blended family – or any family – closer than they are.

Among our discussions one late and soggy night, Tommy and I agreed that, like all emergency responders, he could retire to Fiddler's Green and that, as a sailor, I am eligible to end up there as well. He grudgingly admitted that even being a lawyer doesn't disqualify me. So, Tommy, save me a space and let's have a cold one when my time comes. Once again, we'll jaw over the pains and glories of the world with the emergency responders and sailors who have gone before.

PREFACE

This book is the second edition of a text that was well received by both academics and practitioners in the emergency response and emergency management world. While some counseled resting on the laurels garnered by that book, in the eight years since its publication there have been enough changes in law that major elements needed updating.

As the first book to be published on emergency response and emergency management law, the earlier text filled a substantial gap in the legal literature. The additional material covered by the second edition addresses historic expansions in the law that have taken place since the first edition's publication.

This book's structure carries forward from that of the first edition. The text begins with emergency response law, to assist in understanding the daily legal challenges faced by the dedicated public servants who serve as frontline troops in emergent situations. These contents are arranged in a tiered manner, beginning with the duty to respond and proceeding through the wide range of legal issues that arise during response. Broader issues of emergency management law follow. That discussion begins with the responsibilities of local and state governments, after which federal emergency management law is considered.

To set the stage for response law discussion, the text examines the origin and end of the duty to act. Legal issues arise even before response, including planning, training and exercising. Training accidents sometimes happen despite preventative measures. The text examines case law on the topic.

Case law explicates the major issues that arise from the use of emergency response vehicles. These topics include law enforcement's varying responses to "hot pursuit," vehicle accidents, and the potential for criminal charges against the driver of an emergency vehicle involved in an accident.

The first step in emergency response is usually dispatch. Even before an emergency responder arrives on the scene, legal issues may arise involving delayed dispatch, prearrival instructions, or failure to meet dispatch standards.

The text evaluates topics of particular interest to Emergency Medical Services (EMS). Matter explored includes the ability of proper documentation to assist in protecting from liability, patient consent, treatment of minor patients, "Good Samaritan" acts, and delayed response to emergency scenes.

After an accident occurs, the response must be properly managed. Safety for responders flows from the legal requirements for use of the incident management system, well-written mutual aid agreements, and proper standard operating procedures. These elements create a "structure of safety." Failure to use them properly may result in both legal liability and death or injury to responders. This section contains a great deal of material added to the second edition on the National Incident Management System (NIMS), National Response Plan (NRP), and National Response Framework (NRF), as well as the National Fire Protection Association Standard 1600 Standard on Disaster/Emergency Management and Business Continuity Programs (NFPA 1600).

The law imposes specific requirements on a hazardous materials response. The Occupational Safety and Health Administration (OSHA) Hazardous Waste Operations and Emergency Response (HAZWOPER) standard and Section 472 of the Life Safety Code of the National Fire Protection Association "Standard for Professional Competence of Responders to Hazardous Materials Incidents" provide a structure for such incidents. Case law discusses OSHA penalties for violations.

Volunteers may be either a vital resource or a legal liability. Competent volunteer organizations provide needed infrastructure for important support functions. At the other end of the spectrum, emergent volunteers may interfere in the smooth functioning of incidents. The federal Volunteer Protection Act of 1997 is a tort reform law providing a small measure of shelter for unpaid assistants.

The law contains important privileges and immunities for emergency responders. The common law "Fireman's Rule" prevents responders from suing victims for injuries sustained while on duty, while the rescue doctrine carves out some protection for responders from the acts of others. Varying legal approaches by different states to the very real challenges posed by Post-Traumatic Stress Disorder illustrate that emergency responders are not equally protected everywhere. The World Trade Center Site Litigation provides important lessons for emergency responders who worked the site following the attack. Choices made by managers affected the number of responder casualties resulting from this catastrophic event. The lessons learned apply to everyday events as well as to large-scale occurrences such as the New York attacks.

Emergency management is an all-hazards discipline that provides an invaluable tool for mitigation, preparedness, response, and recovery. It provides a structure for unifying all resources available to units of government. Mitigation lessens or eliminates the effect of potential emergencies. Preparedness includes planning, training to the plan, and exercising the plan, followed by revision of the plan to reflect lessons learned. Response is the actual reaction to emergencies and disasters, when the benefits of mitigation and preparedness result in increased safety for responders and more rapid control of events. Recovery involves restoring matters at least to their status before the event. Recovery blends into mitigation to help assure that future events are either avoided entirely or that their potential effects are lessened.

At the state level, gubernatorial emergency powers vary significantly. Both strong and weak governors have challenges created by the statutes that create and regulate their emergency powers.

Both state and local units of government have specific emergency planning requirements. Federal law requires that planning for release of Extremely Hazardous Substances be undertaken on the local level by the Local Emergency Planning Committee (LEPC). The responsibilities of emergency management, however, include planning for all hazards. The LEPC and local emergency management must, therefore, closely coordinate their plans in order to ensure that they provide a seamless approach to all hazards. Different states approach this federal mandate in varying ways.

On the federal level, the Federal Emergency Management Agency (FEMA) coordinates emergency management efforts. The Stafford Act and the National Response Framework (NRF) created pursuant thereto provide a skeleton upon which the various levels of government build in order to counter calamities. A variety of other federal plans work together with the NRF to assure complete federal support for responses to all types of disaster. One of FEMA's major responsibilities entails issuing, administering, and managing grants. Grantees and subgrantees use many sources to assure that they utilize grants in a lawful manner. FEMA emphasizes mitigation grants in order to lessen costs of subsequent disaster relief. When the first edition of this text was published, mitigation grants focused on natural hazards.

The national focus on terrorism led to FEMA's incorporation into the cabinet-level Department of Homeland Security (DHS). The subordination of tiny FEMA into the gargantuan DHS has brought with it many challenges, including losses of funding and some historic missions. FEMA's response to Hurricane Katrina in 2005 was largely seen as ineffectual and amateurish. An outcry for FEMA's renewed status as an independent agency resulted in legislation designed to ensure that its protected status within DHS would be restored, similar to that of the Secret Service and Coast Guard.

Attorneys and emergency management are important partners for one another. Attorneys cannot continue to tell themselves that "it doesn't happen here." Emergency managers cannot continue to say, "This is an emergency. I don't have time for legalities." Both parties must learn the law beforehand, and take appropriate mitigation and planning steps to lessen the likelihood of litigation. They must also work together through the emergency management process in order to facilitate a response that is both safe and least likely to result in liability. The failure of an attorney advising the leader of a unit of government to understand emergency response and emergency management law may arguably be malpractice. The text includes a recent article discussing the barriers facing attorneys and emergency managers as they do their best to protect their local jurisdictions from liability and other hazards.

Mitigation of legal risk is a difficult task, made even tougher in an economic environment that demands justification for every expenditure. All too often, the attitude of the fiscal body may be, "It hasn't happened. Wait for an event before we authorize expensive legal fees." The text posits a metric for calculating the risk of going without proactive legal advice for emergency management. The goal of this section is to give emergency managers a more concrete measurement to help those with the purse strings understand the risks of going without mitigation-level legal advice to emergency management.

An important part of mitigation revolves around wise land-use policies. Sometimes, however, wise land use may result in individuals who own property which loses all or part of its value due to these policies losing substantial value. The text includes a case examining the issues that arise when these interests conflict.

The lessons of emergent events often point to ways in which preparedness can be enhanced. One such road to improvement is paved with a number of laws and FEMA's Comprehensive Preparedness Guide (CPG) 101, Version 2.0 (November, 2010) which, among other matters, provide strong guidance and support for language services for persons with limited English proficiency

(LEP). The text examines legal requirements for language services to LEP populations in all phases of emergency management and provides practical guidance for finding competent interpreters and making them a part of the emergency management team.

Preparedness cases discuss aspects of planning issues. Another case delves into the Fifth U.S. Circuit Court of Appeals' advance preparedness plans for operations during a disaster.

An included law review article is a good short guide to emergency management law. One must recall, however, that every state has its own unique set of laws covering this area. So, while the article is a good general guide, it is a snapshot of the law in one jurisdiction – Florida – at one time. It may be of help to someone who has purchased the book, then put it on the shelf meaning to get to it later – only to find that later arrives along with a major disaster. (If this is the case – after the immediate trouble is over – read the recovery section, and consider the resources referenced there. Make sure to read the section on mitigation before recovery is too far along.)

The two cases resulting from Hurricane Katrina illustrate challenges during recovery. The first examines the remedies available to applicants for federal assistance for home rebuilding funds. Another case discusses FEMA's options when providing debris removal from a private canal following application for assistance by the local unit of government.

The Moving Forward Section examines the requirements that ethical law imposes on government employees and how those remain in place despite some ethical lapses during the period after 9/11. The book closes with a broadly viewed consideration of how our nation and its emergency management laws have changed in the wake of the September 11, 2001 attacks.

Emergency response and emergency management law are constantly evolving to meet the challenges of an ever-changing world. *Emergency Response and Emergency Management Law: Cases and Materials, Second Edition* surveys the law regulating response by the fire service, hazardous materials teams, emergency medical services, law enforcement, and volunteer groups. The text also examines the varying authorities underlying emergency management as well as its ever-increasing legal obligations. Law school classes as well as practicing attorneys will find the text to be a vital resource for learning emergency response and emergency management law. As with the First Edition, this book should prove to be popular with students of emergency management at all levels in both emergency management-specific as well as public administration programs. The book's potential audience also includes practicing emergency responders and emergency managers who will find its straightforward style to be both comprehensible and useful in their preparedness efforts. The work provides a firm base of legal knowledge for a partnership composed of emergency responders, emergency management professionals, and their attorneys. One of the book's major goals is passing on relevant, useful knowledge to another key attorney group: those who will be at the side of business and government chief executives in all four phases of emergency management – not only in the aftermath of emergencies and disasters.

W.C.N.

INTRODUCTION TO THE SECOND EDITION

America has changed significantly since the appearance of the first edition of this text in 2003. The first edition came out almost two years after the terrorist attacks of September 11, 2001. By that time, the law regarding the book's subject matter had begun to alter markedly, including such landmark enactments as the USA PATRIOT Act of 2001 and the Homeland Security Act of 2002. These far-reaching laws transformed the relationships between emergency management and emergency response at all levels of government. Even more importantly, though beyond the scope of this book, such laws, as well as perceived failures of government during this period (relevant examples of which are described herein), reinforced feelings of cynicism and disconnection from government among many of the populace. Ten years after the 9-11 attacks, the result is a lack of trust in government, marked by symptoms such as the rise of the "tea party" and limits on the ability of the Federal government to provide funding for the Keynesian solution of "priming the pump" to spend our way out of the recession in which we find ourselves.

The 2012 publication date for the second edition is not mere coincidence. This anniversary marks over a decade since the terrorist attacks of September 11, 2001 – events that significantly altered American perceptions of the nature and importance of the various threats facing the nation. Major reorderings of national priorities regarding emergency response and emergency management flowed from those changed perceptions. These reorderings were enshrined in the form of laws at the federal, state, and local levels.

The new legal developments have by and large flowed from the federal level downwards to the state and local units of government. They have resulted in numerous additional requirements for state and local responders. (Many of them n the form of the dreaded "unfunded mandates" of which the federal government is so fond.) The reader can decide for him- or herself the extent to which these have been valuable. Certainly, many have been endorsed by relevant, knowledgeable people at all levels (including the actual responders), whose voice is generally considered to be the most authoritative. Interestingly, several of the most basic legal requirements for emergency responders have remained unchanged.

Other aspects of the law have changed in huge ways. FEMA is no longer a small but independent agency. Instead, its concern with natural hazards is a small voice compared to the law enforcement focus in the Department of Homeland Security.

The addition of other, extensive legal mandates binding every emergency responder and emergency manager meant that the 2003 edition of this book was simply becoming more dated, and less helpful, as every year went by. Given these new developments in the field, I had no option but to bring out an updated version of the text.

This truly is a new edition of the book. The majority of "new editions" have changed perhaps 10 percent of their content. As the reader can tell merely by hefting the text, this manuscript is significantly larger than the prior version, with about one-half its contents being totally new material. The reason for this is that the author has spent considerable time and effort in updating relevant portions of the text itself as well as enlisting the assistance of additional scholars with specific expertise. The goal is to assist in providing a tool that will help the emergency responder, emergency manager, and the lawyer – who MUST be a part of the team – to work closely with both of their organizations.

Responding to an emergency event in real time is difficult enough without simultaneously needing to internalize an unfamiliar area of law. Those who learn and understand the legal underpinnings of all phases of emergency management will be rewarded with a complete understanding of the discipline's capabilities.

The reader should not be mistaken – understanding and complying with the law is not a free ticket to avoid involvement with the court system. In all emergency situations, something goes wrong, often catastrophically so. In our litigious society, for many people the first reaction when injury or property damage occurs is to sue someone, and that includes any government entities and individuals employed by them who may be involved in trying to fix the problem. For both the emergency responder and the emergency manager, however, knowing where the legal limits lie and complying with the law's requirements permit utilizing their organization's full resources to the edges of the envelope. This approach does not guarantee freedom from lawsuits – it merely makes it most likely that one will prevail in litigation and allows one to sleep at night with the fewest worries possible.

W.C.N.
Durham, N.C.
March 12, 2012

ACKNOWLEDGMENTS

My thanks as always go to my beloved wife, Nancy Schweda Nicholson, for her professional yet loving support in this and all my scholarly endeavors. As I have been writing and revising this text, Nancy's kind assistance has helped me to move forward at a steady pace. I wish to thank the renowned authors who kindly submitted some of the outstanding materials contained in this book, including: wise and innovative emergency manager and educator Lucien Canton who coauthored with me the chapter on "Quantifying Legal Risk;" Michael S. Herman, Esq. for portions of his excellent law review article on "Gubernatorial Executive Orders;" Joseph G. Jarret, Esq. and Michele L. Lieberman, Esq., for their law review article "'When the Wind Blows': The Role of the Local Government Attorney Before, During, and in the Aftermath of a Disaster." The Law Director for Knox County, Kentucky, Joe Jarret is a truly nice guy as well as probably the most knowledgeable local legal counsel whom I have ever encountered – Joe's expertise in matters of emergency response and emergency management law is unmatched on the county level, and Ms. Lieberman is board certified in city, county and local government law, and practices in Inverness, Florida; and Nancy Schweda Nicholson, Ph.D., internationally respected sociolinguist at the University of Delaware, for her groundbreaking chapter on "Limited English Proficient (LEP) Populations and Emergency Management: Legal Requirements and Interpreting/Translating Assistance." Particular thanks to former Director of the Indiana State Emergency Agency (SEMA) and current Transportation Security Agency General Manager of Field Operations Melvin J. Carraway, who hired me to be General Counsel for Indiana SEMA; New York State Commissioner of the Division of Homeland Security and Emergency Services Jerome Hauer, who first suggested that I approach Charles C Thomas, Publisher, Ltd. to publish my books. My appreciation to those who encouraged and fostered my interest in this subject matter, including: former SEMA Deputy Director Phillip K. Roberts; the Honorable Melanie George, Esq.; LeaNora Ruffin, Esq.; Dr. Harvey McMurray; Tammy Little, Esq.; Dave Zocchetti, Esq.; Bill Cumming, Esq.; The Honorable Gerald Zore; Kenny Cragen; Jim Pridgen; Mike Garvey, Chief of Staff, Indiana Department of Homeland Security; Dave Barrabee; Susan Schweda; Leon J. "Long John" Schweda; Mike Robinson; Leith Hansen, Esq.: Griff Gosnell; Chip Devine; the Eat, Drink and Be Literate Book Club; Kay Goss, CEM; Dr. Frannie Edwards; Dr. Rick Sylves; Dr. Bill Waugh; Dr. Kathleen Tierney; Claire B. Rubin; Dr. Dave McEntire; Dr. Carol Cwiak, Esq.; John Becknell; David Harris; and many other current and former emergency response and emergency management practitioner and educator friends around the nation and world. My thanks and respect to LTC Stephen T. Udovich USAR (ret.) for his service and work with disabled service members and their families. So many veterans become great emergency responders and emergency managers. A special word of gratitude to retired guru B. Wayne Blanchard, Ph.D., C.E.M. for his ceaseless devotion to the FEMA Emergency Management Higher Education Project. Wayne's work, with the support of his assistant Barbara Johnson, created the proper atmosphere for incubation and growth of emergency management and related academic pursuits in institutions of higher learning across the United States. Their effort is almost entirely responsible for the ongoing professionalization of emergency management in our nation. Other professional and personal friends beyond number encouraged me to persevere in getting this book out. Mike Thomas, President of Charles C Thomas Publisher, Ltd. has been helpful during the writing process for which I am grateful. My parents, Joan and Jim Nicholson, continue to inspire me. I appreciate the encouragement of my brothers John, Tom, and Jim. As I strive to innovate, I ponder the accomplishments and insights of William B. Barnes.

W.C.N.

CONTENTS

Mitigation focuses on breaking the cycle of disaster damage,
reconstruction, and repeated damage. It includes "prevention" which
DHS considers to be a fifth phase of emergency management.
Gove v. Zoning Board of Appeals, 444 Mass. 754; 831 N.E.2d 865
(Mass. S. J. Ct. 2005).

Government Negligence Liability Exposure in Disaster Management
Ken Lerner, 23 Urban Lawyer 333 (1991)

SECTION III: TOWARD THE FUTURE

TABLE OF CASES

EMERGENCY RESPONSE AND EMERGENCY MANAGEMENT LAW

Section I

EMERGENCY RESPONSE LAW

All emergencies are local.

EMS and law enforcement act together at a major accident scene. Jeff Forster photo.

Chapter 1

DUTY TO ACT

The duty to act arises at different times for different emergency responders. A first responder coming on a scene without having been sent thereto by a supervising entity bears a different burden from his or her colleague who is dispatched in response to an incident. The duty to act may be extinguished under certain limited circumstances.

PART 1. DUTY TO ACT

Whether a duty to act arose and whether it ever ended are central issues in the *American National Bank & Trust Company* case that follows.

The reader should pay particular attention to the points made in the dissenting opinion, and consider why that perspective did not prevail.

American National Bank & Trust Company
v.
City of Chicago

Supreme Court of Illinois 192 Ill. 2d 274 (Ill. 2000)

Heiple, J., filed a dissenting opinion in which Bilandic and Rathje, J. J., joined.

Justice MILLER delivered the opinion of the court:

The plaintiff, American National Bank and Trust Company, as special administrator of the estate of Renee Kazmierowski, brought the present action in the circuit court of Cook County against the defendants, the City of Chicago and two of its paramedics, John Glennon and Kevin T. O'Malley. Raising several theories of liability, the plaintiff sought recovery for the defendants' alleged failure to respond properly to an emergency call by the decedent for medical assistance. The circuit court granted the defendants' motion for dismissal of the complaint. The appellate court affirmed that judgment in an unpublished order. No. 1-97-1212 (unpublished order under Supreme Court Rule 23). We allowed the plaintiff's petition for leave to appeal (177 Ill.2d R. 315(a)), and we now affirm in part and reverse in part the judgment of the appellate court and remand the cause to the circuit court for further proceedings.

The following factual summary is derived from the allegations in the plaintiff's amended complaint and from the information contained in its accompanying exhibits, which include a transcript of the decedent's emergency call and the paramedics' report. At around 7:55 A.M. on April 24, 1995, the decedent, Renee Kazmierowski, suffered an asthma attack while at home at her apartment in Chicago. She called 911 to request help. She provided her address and telephone number and said that she lived on the third floor of the building. The 911 operator replied that paramedics were on the way; the operator did not attempt to keep the decedent on the telephone while the paramedics were responding to the call.

Two paramedics, John Glennon and Kevin T. O'Malley, were directed to respond to what they were told was a "heart attack" victim. They were allowed into the decedent's apartment building by a neighbor in the building and went to the third floor. They asked the neighbor whether he had summoned help, and the neighbor replied that he had not. The paramedics then knocked on the door of the only other apartment located on the third floor, but they received no response. The neighbor escorted the firefighter, who had also responded to the call, through his apartment to the back of the building. The firefighter knocked on the back door, but he received no response and was not able to see into the apartment. While the firefighter was checking the back of the building, the paramedics called the dispatcher, who confirmed that they were at the correct address. In response to the paramedics' questions, the dispatcher also said that the caller had not provided her age, and that an attempt to return the call had reached an answering machine. The neighbor told the paramedics that the apartment was occupied by a young couple, who did not appear to have any medical problems. The paramedics concluded that they were not needed at the address in question, and they left the scene. That afternoon, the same paramedics returned to the apartment, again in response to an emergency call. On this occasion, a man let the paramedics into the apartment, and they found the decedent lying dead on the floor.

The plaintiff's amended complaint comprised a total of eleven counts. These alleged negligence and willful and wanton misconduct, and sought recovery from the City and the two paramedics under the Wrongful Death Act and the Survival Act. An additional count sought to impose liability on the City under a federal civil rights provision. The complaint alleged that the 911 operator acted negligently, willfully, and wantonly in failing to keep decedent on the phone while the paramedics responded. The complaint further alleged that the front door of the decedent's apartment was unlocked when the paramedics responded to her call, and that the paramedics acted negligently, willfully, and wantonly in failing to try the unlocked door and enter the apartment.

The defendants moved to dismiss the amended complaint under section 2-615 of the Code of Civil Procedure (735 ILCS 5/2-615). The defendants argued that they were immune from liability for the decedent's death under the Emergency Medical Services (EMS) Systems Act (EMS Act) (210 ILCS 50/1 through 33). The defendants further contended that the plaintiff had failed to adequately allege that they owed the decedent a special duty of care or engaged in willful and wanton misconduct. The circuit court granted the defendants' motion, ruling that the defendants were immune from liability under the EMS Act and that the plaintiff had failed to adequately allege a special duty or willful and wanton misconduct.

The appellate court affirmed the circuit court judgment in an unpublished order. The appellate court rejected the defendants' contention, raised for the first time on appeal, that section 5-101 of the Local Government and Governmental Employees Tort Immunity Act (Tort Immunity Act) (745 ILCS 10/5-101) granted immunity to the defendants. The appellate court concluded, however, that section 17 of the EMS Act (210 ILCS 50/17) immunized the defendants from liability for the decedent's death. The appellate court also held that the plaintiff had failed to adequately allege that the defendants owed the decedent a special duty or that their conduct was willful and wanton. We allowed the plaintiff's petition for leave to appeal. 177 Ill.2d R. 315.... The final count remaining for our consideration is count X, which alleges willful and wanton misconduct by the two paramedics....

In support of the appellate court's and circuit court's rulings, the defendants contend that the present action is barred by the immunity provision found in section 5-101 of the Tort Immunity Act (745 ILCS 10/5-101). The defendants did not raise this contention in the circuit court, but they argued the point before the appellate court in support of the circuit judge's favorable ruling. The appellate court disagreed with the defendants, who renew the contention before this court, arguing that section 5-101 of the Tort Immunity Act applies to this case and grants them immunity in these circumstances. Section 5-101 provides:

> Neither a local public entity nor a public employee is liable for failure to establish a fire department or otherwise to provide fire protection, rescue, or other emergency service.

> As used in this Article, "rescue services" includes, but is not limited to, the operation of an ambulance as defined in the Emergency Medical Services. (EMS) Systems Act. 745 ILCS 10/5-101

The appellate court ruled that section 5-101 provides immunity "only where a public entity chooses not to provide any fire protection, rescue, or emergency services at all, and not where a public entity offers these services in general but fails to provide them in a particular case." The defendants challenge the appellate court's interpretation, arguing that the statute immunizes a local public entity that establishes a rescue service but fails to competently use that service when summoned by a particular plaintiff. We do not agree.

We believe that section 5-101 immunizes only a local public entity that has not established a fire department or rescue service, or has not instituted a system for otherwise providing fire or rescue services....

We next consider the provisions of the EMS Act and the immunity afforded by it. As an initial matter, the plaintiff argues that the provision is not applicable at all to this case. At the time of the decedent's death, in April 1995, section 17(a) of the EMS Act provided as follows:

> Any person, agency or governmental body licensed or authorized pursuant to this Act or its rules, who in good faith provides life support services during a Department approved training course, in the normal course of conducting their duties, or in an emergency shall not be civilly or criminally liable as a result of their acts or omissions in providing

those services unless the acts or omissions, including the bypassing of nearby hospitals or medical facilities for the purpose of transporting a trauma patient to a designated trauma center in accordance with the protocols developed pursuant to section 27 of this Act, are inconsistent with the person's training or constitute willful or wanton misconduct. 210 ILCS 50/17(a)

Section 17(b) of the EMS Act similarly provided immunity for acts or omissions "in connection with administration, sponsorship, authorization, support, finance, or supervision of emergency medical services personnel, where the act or omission occurs in connection with their training or with services rendered outside a hospital unless the act or omission was the result of gross negligence or willful misconduct." 210 ILCS 50/17(b). We note that both provisions have been amended since the decedent's death. The legislature has now omitted, from section 17(a), the reference to acts or omissions "inconsistent with the person's training," and has reformulated, in section 17(b), the standard of non-immunized conduct to be that involving "willful and wanton misconduct." See 210 ILCS 50/3.150(a), (b). The parties agree that the operative provisions in this case are the ones that were in force at the time of the decedent's death.

The plaintiff contends that section 17(a) applies only when paramedics have actually rendered life support treatment to a patient; the plaintiff maintains that the failure of the responding paramedics in this case to administer any treatment at all to the decedent means that the provision has no application here. In support of this interpretation, the plaintiff cites the provisions appearing in sections 4.02, 4.06, and 4.20 of the EMS Act, which respectively define the terms "Advanced Life Support–Mobile Intensive Care Services," "Basic Life Support Services," and "Intermediate Life Support Services." 210 ILCS 50/4.02, 4.06, 4.20. The plaintiff notes that each of these levels of life support treatment involves the performance of acts or procedures directly involving patient care. The plaintiff thus construes these definitions as signifying that the reference to life support services in the immunity provision of section 17(a) must refer to the actual rendition of medical treatment, and not to the conduct at issue in this case, in which no treatment was ever administered.

We do not believe that the scope of section 17(a) is as narrow as the plaintiff believes it to be. We conclude that the provision applies to this case, even though the acts and omissions alleged here do not relate to the actual rendition of life support treatment. Although the EMS Act does not define the general term "life support services," we do not believe that we are limited, in interpreting section 17(a), by the specialized meanings assigned to the terms "advanced life support–mobile intensive care services," "basic life support services," and "intermediate life support services." Those definitions are designed to distinguish one level or form of care from another, and the legislature could reasonably have decided to omit from the definitions conduct that is common to them all or, though preparatory to the actual rendering of medical care, is no less an integral part of providing life support services. Moreover, section 17(a) also refers to the transportation of patients. If transporting a patient to a hospital is an aspect of life support services, then so too should locating a patient in the first place. Other provisions in the EMS Act also demonstrate that the immunity provisions of section 17 apply in these circumstances. Elsewhere, the Act regulates matters such as communications, response time, and standards for ambulance operation. 210 ILCS 50/7, 7.1, 9. These additional measures are evidence of the Act's broad scope, and of the equally broad meaning we believe must be given to the term "life support services" in the immunity provisions.

The parties also dispute the scope of the immunity provided by section 17(a) of the EMS Act. That provision, by its terms, affords immunity to two distinct types of activity: conduct that is "in consistent with the person's training" as a paramedic, and conduct that is willful and wanton. The parties do not agree on the reach of the former category, involving conduct that is "inconsistent with the person's training." The plaintiff insists that the provision does not afford immunity for conduct that is contrary to or violates training received by a paramedic. The de-

fendants, in contrast, construe this exception more narrowly, contending that it withholds immunity merely for conduct that is beyond the level of a paramedic's training; the defendants maintain that the provision otherwise provides complete immunity for negligence.

We agree with the defendants' interpretation. We believe that this portion of section 17(a) means that conduct that is beyond the level of a paramedic's training is not immunized, while conduct that merely deviates from a paramedic's training and constitutes negligence is subject to immunity.... Under the plaintiff's broad reading, which would withhold immunity for conduct that violates a person's training, the exception would threaten to supplant the immunity rule, for virtually any negligent act could be said to be inconsistent with or in violation of a person's training. The plaintiff has not alleged that the actions of the paramedics were beyond their level of training, and therefore we conclude that section 17(a) operates to immunize them for their conduct, unless it rises to the level of willful and wanton misconduct.

We next consider whether the plaintiff's amended complaint sufficiently alleges willful and wanton misconduct. The circuit court believed that the allegations were merely conclusory, and the appellate court agreed. The defendants argue that the lower courts' determinations were correct.

It is the plaintiff's duty to sufficiently allege conduct that falls within the scope of a recognized cause of action. Moreover, mere conclusory allegations are not sufficient. We believe, however, that the allegations contained in the plaintiff's amended complaint are sufficient to withstand the defendants' motion to dismiss.

This court has previously defined "willful and wanton misconduct" in the following terms:

> A willful or wanton injury must have been intentional or the act must have been committed under circumstances exhibiting a reckless disregard for the safety of others, such as a failure, after knowledge of impending danger, to exercise ordinary care to prevent it or a failure to discover the danger through recklessness or carelessness when it could have been discovered by the exercise of ordinary care. *Ziarko v. Soo Line R.R. Co.*, 161 Ill.2d 267,

273, 204 Ill.Dec. 178, 641 N.E.2d 402 (1994), quoting *Schneiderman v. Interstate Transit Lines, Inc.*, 394 Ill. 569, 583, 69 N.E.2d 293 (1946)

We believe that the allegations in the plaintiff's amended complaint are sufficient, creating a question for the trier of fact to determine whether the defendants' conduct was willful and wanton. According to the complaint, the decedent, Renee Kazmierowski, called 911 on the morning of April 24, 1995, and requested help, explaining that she was having an asthma attack. She provided her address and telephone number and said what floor she lived on. She also told the dispatcher, "I think I'm going to die. Hurry." The dispatcher did not attempt to keep Renee on the line, however, as required by applicable standards.

Moreover, according to the allegations in the plaintiff's amended complaint, paramedics routinely receive instruction on how to respond to calls and, in particular, on how to locate persons in need of emergency medical treatment. The plaintiff alleges that training given to paramedics in Chicago, like training given to paramedics everywhere else, includes the following:

> [I]nstruction, training and enforcement of the standard "Try Before You Pry" which dictates that firefighters, paramedics and other rescue personnel should always attempt to open a shut door by turning the door knob before engaging in destructive methods to gain access, or before exiting the scene altogether, without gaining access in order to ensure delivery of emergency health care services to the critically ill caller.

In the present case, the victim's door was unlocked. If the paramedics had been following these vital and basic precepts of their training, as alleged, they would have found the victim inside the residence, and perhaps then they could have saved her life. Locating a person in need of emergency medical treatment is the first step in providing life support services. Not even that first step was taken here. We believe that the portions of the amended complaint that allege willful and wanton misconduct by the defendants are sufficient to withstand the defendants' motion to dismiss, and we therefore remand the action to the circuit court for further proceedings.

For the reasons stated, the judgment of the appellate court, affirming the judgment of the circuit court of Cook County, is affirmed in part and reversed in part, and the cause is remanded to the circuit court of Cook County for further proceedings.

Affirmed in part and reversed in part; cause remanded.

Justice HEIPLE, dissenting:

On April 24, 1995, Renee Kazmierowski (decedent) suffered an asthma attack at her apartment in Chicago. Decedent called 911 and stated: "I need help. I'm having an asthma attack. I think I'm going to die. Please hurry." The 911 operator told decedent that paramedics were on the way, but failed to keep decedent on the phone while the paramedics were responding. When the defendant paramedics arrived at the reporting apartment, they knocked loudly but received no response. A next-door neighbor escorted one of the responding officers through his apartment to check the back door of the reporting apartment, but the officer received no response there either.

The neighbor told the paramedics that in the apartment lived a young couple who appeared to have no medical problems. The paramedics called the dispatcher, who confirmed that they were at the correct address. Concluding that there was no indication that they were needed at the address, the paramedics reported back in service. Later that day, the defendant paramedics were again called to the same apartment. This time, a man let them into the apartment and showed them on the floor the dead body of his girlfriend, who had died of an asthma attack.

The majority holds that the complaint filed by the administrator of decedent's estate adequately alleged willful and wanton misconduct under section 17(a) of the Emergency Medical Services (EMS) Systems Act (EMS Act) (210 ILCS 50/17(a)). Contrary to the majority's holding, this statutory section does not even apply to this case.

Section 17(a) states that a person who "provides life support services" shall not be liable for the results of their acts or omissions unless those acts or omissions are "inconsistent with the person's training or constitute willful or wanton misconduct." 210 ILCS 50/17(a). The majority errs in holding that the defendants in this case provided life support services within the meaning of the statute. The statute defines life support services as "emergency medical care," and lists as examples of such services "airway management, cardiopulmonary resuscitation, control of shock and bleeding, and splinting of fractures." 210 ILCS 50/4.02, 4.06, 4.20. According to the allegations of the complaint, the defendant paramedics provided no life support services to the decedent. Although the 911 dispatcher promised to provide such services, the services were never provided because the paramedics did not locate the decedent. The majority seems to believe that because the statue applies to the *transportation* of patients, it necessarily applies to the *locating* of patients as well. There is absolutely nothing in the statue to support such a reading. Defendants never even began to provide life support services, because they did not even see the patient. Because the statute immunizes only "acts or omissions in providing [life support] services," the statute does not apply to any acts or omissions of the instant defendants, whether willful and wanton or otherwise. 210 ILCS 50/17(a).

The City argues, however, that even if this immunity statute does not apply, it is still immune from liability in the instant case under the common law "public duty" rule. The public duty rule prevents units of local government from being held liable for their failure to provide adequate governmental services such as police and fire protection.... Under the public duty rule, then, the City is presumptively immune from liability for its failure to promptly locate and treat the decedent.

This court has recognized, however, an exception to the public duty rule known as the "special duty" doctrine. Under this doctrine, a municipality may be held liable for its failure to provide adequate governmental services if the legislature has not granted immunity to the municipality. *Harinek*, 181 Ill.2d at 347, 230 Ill. Dec. 11, 692 N.E.2d 1177. To invoke the special duty doctrine, a plaintiff must prove the following elements:

1. that the municipality was uniquely aware of the particular danger or risk to which the plaintiff was exposed;
2. that the municipality engaged in specific acts or omissions that were affirmative or willful in nature; and
3. that the injury occurred while the plaintiff was under the direct and immediate control of municipal employees or agents....

Assuming *arguendo* that the complaint in the instant case satisfies the first of the requirements for a special duty, I would hold that it fails to satisfy the remaining elements. The complaint fails to establish that the City or its employees engaged in affirmative or willful acts or omissions in connection with decedent's death. The complaint simply alleges that the 911 dispatcher failed to keep decedent on the phone while the paramedics were responding and that the paramedics failed to try the door knob in order to enter decedent's apartment. At most, these allegations show that the city employees neglected to perform certain tasks. There is no indication of a conscious decision by the employees not to perform the tasks. Rather, the employees attempted to locate decedent but neglected to try the doorknob to gain entry to her dwelling. This conduct was neither affirmative nor willful.

Furthermore, the complaint fails to adequately allege that decedent was under the direct and immediate control of the City. Although decedent called 911 for assistance, the paramedics had not located her and were not physically in her presence at the time of her death.... [I]n the instant case the City had no part in initiating the harm that befell decedent. Although the complaint may establish that the City employees performed their duties incompetently, it fails to establish that they owed decedent a special duty. The public duty rule therefore immunizes City from liability for her death.

For these reasons, I respectfully dissent.

Justices BILANDIC and RATHJE join in this dissent.

QUESTIONS

The Illinois tort immunity statute protects local public entities from liability for failure to make available emergency rescue services.

Note the discussion of alleged improper dispatch procedures. Also remember that this matter may be viewed as a "failure to transport" case. These topics are discussed at greater length later in the text.

1. What is the difference between not providing a rescue service at all and failing to use it, once put in place, competently?
2. What arguments exist to support immunity for actions inconsistent with training being defined to prevent immunity for conduct that is beyond a person's training, but not for conduct that deviates from training?
3. Under what circumstances should conduct beyond training be penalized?
4. The Court mentions, regarding the standard for finding paramedics liable, that the criterion changed after the events in the case took place. "The legislature has now omitted, from section 17(a), the reference to acts or omissions 'inconsistent with the person's training,' and has reformulated, in section 17(b), the standard of non-immunized conduct to be that involving `willful and wanton misconduct." What is the effect of the changes? What reasons might the legislature have had for making the changes?
5. The "public duty" and "special duty" concepts mentioned in the dissent are frequently the basis of avoidance of liability by emergency response organizations. As the dissent points out, they are well-established rules of law. This case involved emergency medical services.
   a. What other entities receive protection from these rules?
   b. From whom are they protected?
   c. What are the potential consequences for emergency responses if the special duty doctrine continues to be restricted?

PART 2. WHO MUST ACT?

The following case is an interesting contrast to the *American National Bank* matter.

Dwyer v. Bartlett

Supreme Judicial Court of Maine

472 A.2d 431 (1984)

NICHOLS, Justice.

The Plaintiffs, Gale Dwyer and Paul Dwyer, appeal from an order of the Superior Court (Aroostook County) on July 7, 1983, dismissing their complaint against the Defendants, the Town of Orient and two of its selectmen, Fritz Bartlett and Clayton McKissick, for their alleged failure to seasonably report to appropriate authorities a fire in that town which was then destroying certain of the Plaintiffs' real and personal property. Because the facts alleged do not give rise to any existing common law or statutory duty on the part of these Defendants to protect the Plaintiffs' property by reporting the fire, we conclude that the complaint was properly dismissed.

As private *individuals*, the Defendants Bartlett and McKissick could not be found liable for failing to report the fire because mere nonfeasance is not actionable at common law. See W. Prosser, *Handbook of Law of Torts* (4th ed. 1971) § 56 at 339. Furthermore, there is neither statutory nor common law authority for the proposition that, in their capacity as selectmen, the Defendants owed any duty to protect the private property of other Town residents by reporting a fire. *Compare* 30 **M.R.S.A.** § § 3773 and 3774 (1978) (imposing certain obligations with respect to fire protection on municipal fire chiefs and firefighters).

In the absence of an official duty to act, the Plaintiffs have no cause of action against these Defendants as selectmen. Consequently, the Plaintiffs do not have a cognizable claim against the Town of Orient on a theory of *respondent superior. CG*

Accordingly, the entry must be:

Appeal denied. Judgment affirmed.

All concurring.

_____

QUESTIONS

1. What differences do you see between the entities whose help was desired in the two above cases: paramedics and town selectmen?

2. What duties do individuals have toward others in time of emergency?

3. What duties does an elected official have to his or her constituents in time of emergency? (This topic is more fully considered later in the text.)

4. Contrast the idea of legal duty with that of "moral duty."
   a. Can the latter ever be enforced in Court?
   b. What other avenues might exist to enforce a "moral duty" on behalf of these public officials?

5. The theory of *respondent superior* mentioned in the *Dwyer* case is the method by which an employer may be found to be legally responsible for the acts of employees. Explain why it is or is not a good idea to force the employer, a municipality supported by taxpayer dollars, to pay for errors.

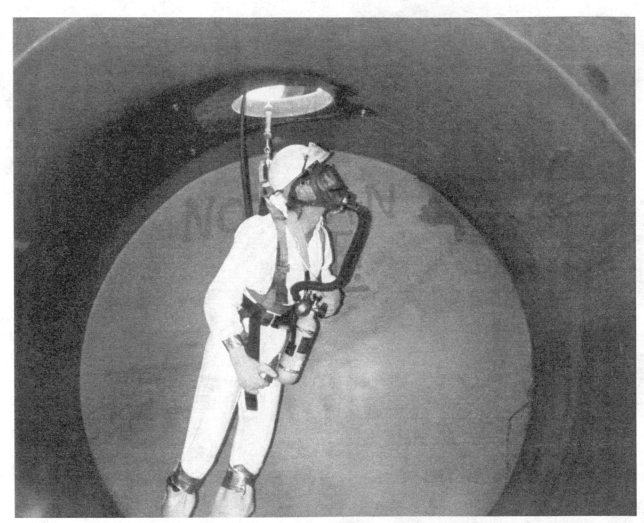

A HAZMAT trainee suited up in a tank. Courtesy Spill Recovery of Indiana.

Chapter 2

TRAINING ACCIDENTS

All emergency response entities must engage in training to keep their skills current, learn new techniques, and understand the proper operation of modern equipment. As discussed later, emergency management and Local Emergency Planning Committees (LEPCs) are also required to sponsor training activities. When death or injury occurs during these events, the legal responsibility for such damages may be the basis for a dispute. The unfortunate result of improper preparation for training, including legal preplanning, may be a nasty surprise for the responder or his/her survivors.

The following case resulted from a poor understanding of the legal relationships between the entities that provided and sponsored the training event. As you read the case, think about what steps could have been taken to assure proper medical care for the injured and protection for the survivors of the deceased. Also consider the innate dangers of training for SWAT operations and what safety precautions are appropriate during such training.

Brassinga v. City of Mountain View

Court of Appeal of California, Sixth Appellate District

66 Cal. App. 4th 195 (1998)

MIHARA, J.

The Mountain View, Palo Alto, and Los Altos Police Departments each had a special weapons and tactics (SWAT) team. In order to ensure an adequate response to "incidents," the three departments cooperated to form the "North County Regional" SWAT team (hereafter the Regional Team) which was composed of all three departments' SWAT teams. The Regional Team trained together and responded to incidents as a single team, although the members of the Regional Team remained employed by their individual departments.

Regional Team training exercises often utilized the services of "role players." During a Regional Team training exercise, Theodore Brassinga, a Palo Alto reserve police officer who was not a SWAT team member but was assigned to serve as a role player, was shot to death by Greg Acton, a Mountain View police officer who was a Regional Team member. Plaintiffs, Brassinga's heirs, filed a wrongful death action against Mountain View and Acton. Mountain View asserted as an affirmative defense that workers' compensation benefits were the exclusive remedy because either Mountain View was Brassinga's "special employer" at the time of his death or the Regional Team was the "special employer" of both officers. Acton's motion for summary adjudication was granted, but Mountain View's summary judgment motion was denied. The case was tried before a jury, but the trial court granted plaintiffs' motion for a directed verdict on liability in *respondeat superior*. The only issue put before the jury was the level of damages. It returned a $3,250,000 verdict for plaintiffs. Mountain View claims on appeal that the superior court erred in denying summary judgment and the trial court erred in directing the verdict because either there was a legitimate factual question that should have been put before the jury or the evidence indisputably established the affirmative defense. Plaintiffs' cross-appeal challenges the summary adjudication in favor of Acton. We conclude that the superior court erred in granting Acton's motion and the trial court erred in directing the verdict. Therefore, we reverse the judgment.

The Pleadings

Plaintiffs' complaint alleged that Acton had been negligent and that Mountain View was liable by way of *respondeat superior* for Acton's negligent acts. It also alleged that Mountain View had been directly negligent in "the hiring, retention, training, and supervision of Acton, the inspection of Acton's weapon, and the failure to establish procedures" to prevent accidental shootings. Mountain View pled as an affirmative defense that Brassinga was Acton's "coworker" and Mountain View's "employee" at the time of his death and therefore his exclusive remedy was provided by the Workers' Compensation Act.

Summary Judgment Motion

Mountain View and Acton moved for summary judgment or summary adjudication. They argued that workers' compensation was the exclusive remedy because Brassinga was a "special employee" of either the Regional Team or Mountain View. The affidavits submitted by Mountain View and Acton in favor of their summary judgment motion established the following facts.

The Regional Team regularly trained together and responded to incidents as a single team although the members of the Regional Team remained employed by and were paid by their individual departments. The Regional Team itself had no employees. The equipment used by the Regional Team was purchased by and stored at the individual departments. The departments attempted to complement each other by not duplicating equipment.

On May 15, 1994, the Regional Team held a training exercise at the Gilroy train yard. Brassinga was employed by Palo Alto as a reserve police officer and paid for his services. Palo Alto paid Brassinga on May 15 for "helping the regional team" during its day-long training exercise. He was assigned to act as a "role player" during the training exercise.[1] Brassinga had served in the same capacity at a previous Regional Team training exercise. Role players were brought to the May 15 training exercise by both Palo Alto and Mountain View. Mountain View brought four Mountain View employees who were not members of the SWAT team to act as role players.

At the beginning of the training session, a weapons inspection was conducted by "range masters" who had been "trained to insure that the weapon is empty." Early in the training session, the Regional Team split into two groups. One group was composed of the Palo Alto members while the other was composed of the Mountain View and Los Altos members. The plan was for these two groups to train separately and then "blend together" and train as a group. The role players, including Brassinga, were initially utilized in the Palo Alto group's training. When the Palo Alto group took a break, Mountain View Police Officer David Worley, who was supervising the Mountain View/Los Altos group's training, asked Brassinga and the other role players "if they minded coming on over, instead of taking a break" and acting as role players in the Mountain View/Los Altos group's training. Brassinga and the other role players agreed to do so.

These role players included the four Mountain View employees, Brassinga, and a second Palo Alto reserve police officer. Worley gave the role players five to ten minutes of instructions about what they should do. He continued to give them instructions as the training progressed. Two or more Amtrak employees requested permission to join the role players. Worley allowed them to participate. Regional Team members from Palo Alto and Mountain View were also acting as role players at times for the Mountain View/Los Altos group. Worley had the power to discipline

---

[1] Palo Alto reserve police officers may refuse an assignment without discipline, but Brassinga did not decline this assignment.

any of the role players by removing them from their role-playing duties. If any of the role players had decided that they no longer wished to participate, they could have discontinued their participation without criticism or discipline.

While he was carrying out his role-playing duties, Brassinga was sometimes supervised by Palo Alto officers and at other times supervised by a Mountain View officer. Brassinga's role playing included the use of a "toy gun" which had been provided to him by Palo Alto. While acting as a role player in the training of the Mountain View/Los Altos group, Brassinga was shot to death by Acton. Plaintiffs received workers' compensation death benefits from Palo Alto.

Ruling on Summary Judgment Motion

Judge Jeremy D. Fogel denied Mountain View's motion for summary judgment. However, he granted Acton's motion for summary adjudication on the ground that "as a matter of law based upon the undisputed evidence" Acton and Brassinga "were general employees of their respective police agencies and special employees of the Regional SWAT Team, a joint enterprise among the cities of Los Altos, Mountain View, and Palo Alto." Fogel concluded that, because Brassinga and Acton "were in effect co-employees of the same joint enterprise, plaintiffs cannot recover against Acton directly unless Acton's conduct comes within one of the exceptions contained in Labor Code *section 3601(a)*, in which event *respondent superior* liability against City would be barred by Labor Code *section 3601(b)*." Nevertheless, Fogel found that "Acton's immunity does not extend to City...." He stated "somebody has got to be responsible if there is tort conduct committed by Officer Acton."

Trial

Judge Richard C. Turrone was the trial judge. He denied Mountain View's request that the court bifurcate the affirmative defense and try it before the rest of the issues in the case. Mountain View conceded that Acton had been negligent. It moved for judgment on the pleadings on the

ground that it was immune from liability under Government Code sections 815 and 815.2. This motion was denied. Judge Turrone ruled that Judge Fogel's finding that Acton and Brassinga were special employees of the Regional Team was not binding on him.

Trial Evidence

The Regional Team was formed because the individual departments lacked adequate resources and personnel to handle emergencies. The Regional Team had about twenty-seven members. It had a single standard for membership and a "unified chain of command." This meant that a Regional Team member from one department could supervise Regional Team members from another department. The Regional Team did not pay the salary of any of its members, owned no property, and had no office. It trained in four week-long sessions each year. Although all of the equipment used by the Regional Team was owned by and stored at individual departments, the departments attempted to complement each other by not duplicating equipment and having interchangeable equipment. The team members "interchanged equipment all the time...." The members of the Regional Team wore identical "team SWAT uniforms." These uniforms were worn only for Regional Team activities.

The Regional Team regularly used "role players" in its training. The role players were either members of the Regional Team or other "law enforcement agency" people. Usually, the role players were police department employees. These role players could be supervised by any member of the Regional Team. The Regional Team commanders were in command of the role players "regardless [of whether the role player] was a Palo Alto employee, a Mountain View employee, or a Los Altos employee."

On May 15, 1994, the Regional Team met for what was planned to be a day-long training exercise at the Gilroy Caltrain yard. The intent was for the Regional Team members from all three departments to work together. In order to simulate actual situations, the training exercise was planned to include role players, and Palo Alto and Mountain View had brought police department employees to act as role players.[3] Brassinga was a Palo Alto reserve police officer. He was not a member of the Regional Team or the Palo Alto SWAT team. Brassinga was assigned to act as one of the role players to assist the Regional Team in this training exercise. He had acted as a role player during a previous Regional Team training exercise. Brassinga was paid $7.40 per hour for his services.

Acton was employed by Mountain View, and he was a member of the SWAT team and the Regional Team. He became a "range master" in 1994. Acton had been asked to become a range master because the range masters on the Regional Team were predominantly from Palo Alto, and Mountain View wanted to "take a little bit of the burden off of them." A range master "is someone who controls the overall supervision of weapons and training." Mountain View did not have a weapons inspection policy which set forth how a weapons inspection was to be done. On May 15, the weapons inspection was done simultaneously for the entire Regional Team. All of the team members lined up at once. The Mountain View officers' weapons were checked by Mountain View range masters while other individuals checked the weapons of the Los Altos and Palo Alto officers. Acton and another Mountain View officer were the range masters who inspected the weapons of the Mountain View officers. Like his fellow officer, Acton intended to have each of the officers whose weapons he inspected inspect his weapon. However, unlike his fellow officer, he neglected to have *anyone* inspect his weapon. Thus, prior to the shooting of Brassinga, no one discovered that Acton had never unloaded his weapon. Acton felt that one of the "factors" contributing to his failure to inspect his weapon or have his weapon inspected was that another Mountain View officer was hurrying the weapons inspection along.

A Regional Team member from Palo Alto explained to Brassinga and the other Palo Alto reserve officer what the two men would be doing as role players. After the weapons inspection, the entire Regional Team went on an extensive tour of the trains on which they would be training. Brassinga and the other Palo Alto reserve police officer accompanied the Regional Team members on this tour. Then, the Regional Team split into two equal groups. One group was made up of the Palo Alto members and the other group was made up of the Mountain View and Los Altos members. The two groups began training in different train cars. Brassinga and the other role players worked with the Palo Alto group for about an hour. The role players were wearing "civilian clothes."

When the Palo Alto group took a break, Mountain View Police Sergeant David Worley asked the role players to assist the Mountain View/Los Altos group with its training.[4] Worley was the Regional Team's assistant tactical commander, and he was in command of the Mountain View/Los Altos group's training exercises. The group of role players included Brassinga, the other Palo Alto reserve police officer, four Mountain View Police Department employees, and several members of the Regional Team command staff. The role players agreed to assist the Mountain View/Los Altos group in its training. These role players were joined by two Amtrak employees who "volunteered" to act as role players in the exercise. Brassinga could have discontinued his role-playing duties at any time without discipline or criticism. Any Regional Team member could have complained about Brassinga's conduct in performing his role-playing duties and caused Brassinga to be disciplined by directing a complaint through the "chain of command" back to Palo Alto. A Regional Team member could also have removed Brassinga from his role-playing duties as discipline for inappropriate conduct.

Throughout his role-playing duties, Brassinga was given instructions on how he should act. While Brassinga and the other role players were role playing with the Mountain View/Los Altos group, they were "subject to the command" of Worley, and Worley gave them instructions on

---

[3] These role players were later supplemented by some Amtrak employees who "apparently volunteered to act as role players.

[4] There was also testimony that it was not Worley who asked them to work with the Mountain View/Los Altos group but a Palo Alto team member who made this request.

how to fulfill their role-playing duties. He would explain the scenario to the role players and tell them how much of a challenge to provide and what type of behavior to engage in, but he would allow the role players to decide among themselves who would play which role and where to conceal themselves. Worley talked to the role players after each scenario and gave them instructions. The role players provided the same services to the Mountain View/Los Altos group that they had provided to the Palo Alto group.

In the course of one of the scenarios conducted by the Mountain View/Los Altos group, about halfway through the day's training exercises, Acton shot Brassinga. At that point, it was discovered that Acton's weapon was fully loaded.

Motions for Nonsuit and Directed Verdict

After both sides had rested, plaintiffs moved for nonsuit on Mountain View's affirmative defense. Mountain View opposed plaintiffs' motion and made its own motion for a directed verdict. Each side argued that it had established its case as a matter of law. However, Mountain View also argued, in opposition to plaintiffs' motion, that the issue of whether Brassinga was a special employee "needs to go before the jury." Judge Turrone found that "the facts in this case, in large part, are not in dispute." "It's uncontradicted [Acton's] an employee of Mountain View. He's working on his job as a Mountain View police officer, paid by Mountain View. [P] At the time of the shooting it's conceded that his chain of command, that is to say his supervisors, were Mountain View officers.... And that he was never under the command of Palo Alto. [P]...[Brassinga's] a reserve officer of Palo Alto. I think it's significant that he's not a member of the regional team.... And I think significantly he was a volunteer to be a role model *[sic]* for Palo Alto as well as later Mountain View when they needed a role player." "The role players, without exception, were, as I understand it, volunteers.... I think that's significant. I mean, is one to argue that the Amtrak volunteers were special employees of the City of Mountain View? I have some trouble with that." "[T]he Palo Alto reserve

officers were doing the same thing they had done for Palo Alto. No one ever countermanded, 'Forget everything that Palo Alto told you; you are now subject to our rulings." "And significantly, everything that transpired that day, except for that latter portion of Mountain View's training with role players, was done separate and distinct." "This regional team had no office, no payroll, no equipment of its own." "Each member of the team and the victim was not a member of the team, but each member of the team was paid by the respective police agencies. The equipment that [Brassinga] was using belonged to Palo Alto...." "The evidence is just not there that [Brassinga]...ever consented to a new employment relationship." "Mountain View's right to control the details of what [Brassinga] did were minimal at best."

Judge Turrone granted plaintiffs' motion for nonsuit on Mountain View's affirmative defense and denied Mountain View's motion for a directed verdict. He directed a verdict "on negligence in *respondent superior*" in favor of plaintiffs. "This case is uncontradicted Acton was negligent, he was working for the city. They are responsible. There's no immunity." The case went to the jury solely on damages, and the jury returned a verdict in the amount of $3,250,000 for plaintiffs. Judgment was entered on the jury's verdict. Mountain View's motions to vacate the judgment for a new trial and for judgment notwithstanding the verdict were denied. Mountain View filed a timely notice of appeal. Plaintiffs filed a timely notice of cross-appeal challenging the summary adjudication of its action against Acton.

Discussion

The only defense litigated by Mountain View was its affirmative defense based on the exclusive remedy provisions of the Workers' Compensation Act. "[T]he right to recover [workers'] compensation is [with certain exceptions not applicable here]...the sole and exclusive remedy of the employee or his or her dependents against the employer..." (Lab. Code, § 3602, subd. (a).) "[Workers' compensation is, with certain exceptions,] the exclusive remedy for injury or death of an employee against any other employee of the employer acting within the scope of his or her employment...." (Lab. Code, § 3601, subd. (a).) "In no event...shall the employer be held liable, directly or indirectly, for damages awarded against, or for a liability incurred by the other employee...." (Lab. Code, § 3601, subd. (b).).

It was undisputed that Palo Alto was Brassinga's general employer, but this fact did not preclude a finding that Mountain View or the Regional Team was Brassinga's special employer. "'Where an employer sends an employee to do work for another person, and both have the right to exercise certain powers of control over the employee, that employee may be held to have two employers – his original or "general" employer and a second, the "special" employer.'... [P] If general

and special employment exist, 'the injured workman can look to both employers for [workers'] compensation benefits. [Citations.] If work[er]'s compensation is available, it constitutes, with an exception not pertinent here, the work[er]'s sole remedy against the employer. [Citation.] Thus where there is dual employment the work[er] is barred from maintaining an action for damages against either employer." (*Kowalski v. Shell Oil Co.* (1979) 23 Cal. 3d 168, 174-175 [151 Cal. Rptr. 671, 588 P.2d 811], fn. omitted.) The basic issues presented in this appeal and the cross-appeal are: (1) did the evidence establish as a matter of law that *either* Mountain View or the Regional Team *was* Brassinga's special employer; and (2) did the evidence establish as a matter of law that *neither* Mountain View *nor* the Regional Team was Brassinga's special employer....

B. REGIONAL TEAM COULD NOT QUALIFY AS AN "EMPLOYER"

Mountain View argues that the Regional Team was the special employer of both Brassinga and Acton. Plaintiffs maintain that the Regional Team was not an entity which could qualify as an "employer" under the Workers' Compensation Act. Plaintiffs' claim is correct.

The Labor Code states that, for workers' compensation purposes, "'employer' means: [P] (a) The State and every State agency. [P] (b) Each county, city, district, and *all public and quasi public corporations and public agencies therein.* [P] (c) Every person including any public service corporation, which has any natural person in service. [P] (d) The legal representative of any deceased employer." (Lab. Code, § 3300, italics added.)

Mountain View concedes that the Regional Team was not a "joint powers agency" with independent existence, but it characterizes the Regional Team as a "joint enterprise' arising out of a 'mutual aid agreement' of the type authorized by Gov. C. § 8616-8617." It describes a "joint enterprise" as an informal undertaking in which the "associates" share control. Government Code sections 8616 and 8617 do not authorize the creation of a public agency. Instead, these statutes permit public agencies to cooperate. In contrast,

Government Code sections 6506 and 6507 do authorize public agencies to create a public agency or entity to administer a joint powers agreement. As it is conceded that there was no joint powers agreement in this case, none of these Government Code sections are applicable....

The Labor Code definition of "employer" is quite specific about the nature of "public" employers. Unquestionably the Regional Team was not a "State agency" or "public [or] quasi public corporation," and the only question remaining is whether it was a "public agenc[y]" within the meaning of Labor Code section 3300. We can find nothing in the record which even suggests that the Regional Team was "public agenc[y]" within what we construe to be the meaning of Labor Code section 3300. As plaintiffs point out, the undisputed evidence established that the Regional Team had no office, property, employees, letterhead, phone number, or any other indicia that it functioned as a "public agenc[y]." Clearly, the Regional Team was not an entity that could have been subjected to any claim for recompense because there would have been no way to contact "it." Under these facts, we agree with plaintiffs that the evidence presented below established as a matter of law that the Regional Team could not have been an "employer" of anyone within the meaning of Labor Code section 3300.

C. WAS MOUNTAIN VIEW BRASSINGA'S SPECIAL EMPLOYER?

Mountain View's alternative theory is that the evidence either established as a matter of law that, or raised a triable issue of fact as to whether, Mountain View was Brassinga's special employer. Plaintiffs assert that the evidence established as a matter of law that (1) Brassinga was acting as a "volunteer," (2) even if Brassinga was a special employee of Mountain View, Mountain View would still be liable under *Marsh*, and (3) Brassinga was not a special employee of Mountain View."[10] The trial court accepted all of plaintiffs' assertions.

1. "Volunteer" Claim

Plaintiffs convinced the trial court that Brassinga had been acting as a "volunteer" at the time of his death.

The term "employee" is defined broadly, and, ordinarily, "[a] person who renders service to another is presumed to be an 'employee." (*County of Los Angeles v. Workers' Comp. Appeals Bd.* (1981) 30 Cal. 3d 391, 396 [179 Cal. Rptr. 214, 637 P.2d 681])

It was undisputed that Brassinga was receiving hourly wages in return for acting as a role player for the May 15 Regional Team training exercises. Although the evidence disclosed that Brassinga could, but did not, decline his assignment by Palo Alto to this duty or Worley's request for his services employment during the training exercises, this did not establish that Brassinga was a "volunteer." The mere fact that an employee is given the freedom to decline a particular assignment does not mean that, when the employee accepts an assignment and engages in compensated work, he or she is trans-

formed into a volunteer. Slavish obedience is not a requisite element of a relationship.

It is immaterial that Worley characterized his request for role players as seeking "volunteers." ... [T]he evidence did not demonstrate that Brassinga, as a compensated employee, was free to make a unilateral decision to donate the time for which he was being compensated by Palo Alto to another entity. Therefore, the evidence did not establish as a matter of law that the role playing services Brassinga was providing at the time of his death were "voluntary services" within the meaning of Labor Code section 3352, subdivision (i)....

3. Factual Dispute Whether Mountain View Was Brassinga's Special Employer

A. The Law

The key to the existence of a special employment relationship is control.

Plaintiffs argue that a special employment relationship cannot be created unless the general employer relinquishes "all" control to the special employer and the special employer becomes solely responsible for the special employee's actions.... Where general and special employers *share control* of an employee's work, a 'dual employment' arises, and the general employer remains concurrently and simultaneously, jointly and severally liable for the employee's torts." (*Marsh, supra*, at pp. 494-495, italics added.)... The applicable legal principle is that a general employer's relinquishment of partial control over its employee to another entity may create a special employment relationship....

---

[10] Plaintiffs also argue that Labor Code section 3362. 5 establishes that Brassinga was solely Palo Alto's employee. That statute provides that a person who is appointed to act as a reserve city police officer and "assigned specific police functions" by a city is an employee of the city "while performing duties as a peace officer." (Lab. Code, § 3362.5.) This statute tells us only that Brassinga was Palo Alto's employee if his duties could be characterized as those of a peace officer on May 15. There was never any dispute as to whether Brassinga was a Palo Alto employee. However, the statute is silent as to whether such an officer might also be a special employee of some other entity. As the statute is nonexclusive, it does not, in and of itself, preclude a conclusion that a reserve officer was, at some point, a special employee of an entity other than the one which appointed him or her.

The special employment relationship and its consequent imposition of liability upon the special employer flows from the *borrower's power to supervise the details of the employee's work.* Mere instruction by the borrower on the result to be achieved will not suffice... [citations omitted]

[S]pecial employment is most often resolved on the basis of '*reasonable inferences* to be drawn from the circumstances shown.' Where the evidence, though not in conflict, permits conflicting inferences,... the existence or nonexistence of the special employment relationship barring the injured employee's action at law is generally a question reserved for the trier of fact.... However, if neither the evidence nor inferences are in conflict, then the question of whether an employment relationship exists becomes a question of law which may be resolved by summary judgment.... [citations omitted]

B. Summary Judgment

The first question is whether the superior court erred in denying Mountain View's motion for summary judgment. We conclude that it did not because there was a triable issue of fact as to whether Mountain View was Brassinga's special employer.

The relevant factors other than control were mixed. Some favored the existence of a special employment relationship while others tended to negate the existence of a special employment relationship between Brassinga and Mountain View.... On the key question of Mountain View's right or power to control Brassinga, the evidence was also mixed....

The conflicting inferences that arose from these facts precluded summary judgment. A jury could have concluded that Worley did not have sufficient control over the details of Brassinga's discharge of his role-playing duties to justify the creation of a special employment relationship between Brassinga and Mountain View. Similarly, the circumstances other than control were decidedly mixed. Although Brassinga was performing unskilled work that was part of Mountain View's regular business, Palo Alto was paying his wages and the period of the alleged special em-

ployment was very brief. Brassinga agreed to serve as a role player under Worley's supervision, but the only tool, a toy gun, was provided by Brassinga's general employer, Palo Alto. Mountain View had no power to discharge Brassinga from his Palo Alto reserve police officer position although it could discharge him as a role player for Mountain View. Because the evidence on the control issue was not definitive and could have supported conflicting inferences and the other factors also supported conflicting inferences, Mountain View had not established its defense on its motion for summary judgment, and the superior court did not err in denying its motion.

C. Directed Verdict

The next question is whether the disputed factual issues remained based on the trial evidence such that the trial court erred in granting plaintiffs a directed verdict on liability based on *respondent superior.* We conclude that the question of whether Mountain View was Brassinga's special employer was not established as a matter of law based on the trial evidence, and the trial court therefore erred in granting a directed verdict on this issue.

...In sum, the factors other than control continued to be mixed based on the trial evidence.

Conflicting inferences could also have been drawn from the trial evidence regarding Mountain View's right or power to control Brassinga....

The same conflicting inferences which precluded summary judgment precluded a directed verdict. The evidence presented at trial detailed above could have supported a jury finding that Worley's control over Brassinga's discharge of his role-playing duties was indicative of the existence of a special employment relationship between Brassinga and Mountain View. The factors other than control could have supported either a finding of a special employment relationship or a finding that there was no special employment relationship. Brassinga's duties were unskilled and part of the regular business of Mountain View, but his wages were not paid by Mountain View and the period of the alleged special employment was brief. Brassinga consented to act as a role player for Mountain View, but the toy gun was provided by Palo Alto. Mountain View could discharge Brassinga from his role-playing duties but not from his Palo Alto reserve police officer position. On this record, the trial court erred in directing a verdict in favor of plaintiffs on liability in *respondent superior*. Consequently, we must reverse the trial court's judgment.

D. Plaintiffs' Cross-appeal

Plaintiffs appeal from the superior court's order granting Acton's motion for summary adjudication and dismissing him from this action. This order was expressly based on the superior court's finding that the undisputed evidence established as a matter of law that the Regional Team was the employer of both Acton and plaintiff. As we have already discussed, the evidence presented below established that the Regional Team did *not* qualify as an "employer" under Labor Code section 3300. Acton's motion was based solely on the argument that he and Brassinga were co-employees of the Regional Team or of Mountain View. The undisputed facts do not support this argument. The Regional Team was not shown to be an entity that could qualify as an employer, and the evidence presented in support of Acton's motion did not establish as a matter of law that Mountain View was Brassinga's special employer. Consequently, the superior court erred in dismissing Acton from this action. The same triable issues of fact that precluded Mountain View from obtaining summary judgment precluded Acton from succeeding on his motion.

Disposition

The judgment is reversed and remanded for further proceedings in accordance with the views expressed herein. The parties shall bear their own costs.

COTTLE, P. J., and BAMATTRE-MANOUKIAN, J., concurred. A petition for a rehearing was denied September 18, 1998.

QUESTIONS

1. SWAT operations frequently expose officers engaged therein to significant danger. Discuss the legal pluses and minuses of realistic training for SWAT teams in particular.
2. Consider the balance that must be maintained between safety and realism in training for any emergency response entity. Where do legal considerations fit in that equation?
3. What dangers exist in "live burn" fire service training, and what steps are taken to assure that those dangers are controlled? What legal steps may be taken to lessen potential liability risk in such training?
4. What are the positive and negative legal aspects of using "toy guns" (nonoperational weapons with the same characteristics as operational weapons) in such simulations?
5. Emergency Medical Services (EMS) members often train using moulange. How effective is this tool? Contrast the legal issues in such reality based EMS training with those in live burn and SWAT training as described in this case.
6. Think about the result in the trial court that led to the appeal resulting in this decision. Did the result there correspond to the expectations of the parties? Why or why not? What about the result of the appeal?
7. Why did the Regional Response Team not qualify as an employer in this matter?
8. Consider the "special employee" discussion. What other approaches to the work of the Regional Team might obviate the need for this discussion?
9. Based on this decision, what might the outcomes in the retrial on remand be, and why?
10. If you were advising a Regional Response Team during its formation, what methods would you suggest to avoid a future case such as this?

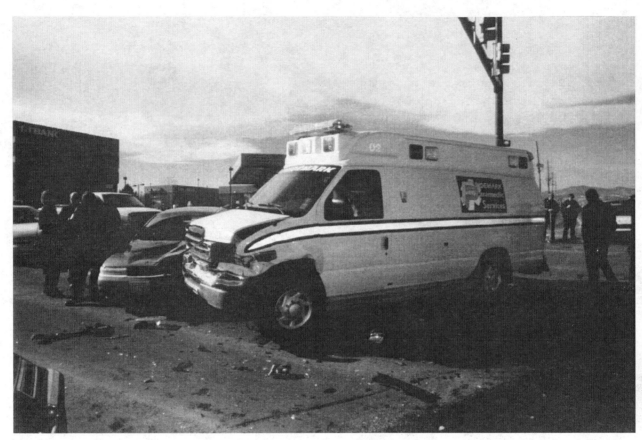

An ambulance accident enroute to an accident scene. Jeff Forster photo.

Chapter 3

VEHICLE ISSUES

PART 1. HOT PURSUIT

Hot pursuit is one of the most controversial topics on law enforcement response. As the following cases illustrate, the legal aftermath of death or injury resulting from hot pursuit varies widely from state to state. The Supreme Court has answered definitively whether the Due Process clause applies to such activities. As you read the cases, pay particular attention to the policy arguments that are made on both sides of the issue.

City of Pinellas Park v. Brown

Supreme Court of Florida

604 So. 2d 1222 (1992)

Rehearing Denied.

KOGAN, J.

After running a red light in Pasadena, Florida, John Deady attempted to elude a sheriff's deputy in a high-speed chase. Before this chase ended on a stretch of U.S. 19, it would pass along a twenty-five-mile course in Pinellas County, through which normal urban traffic also was passing. Thirty-four separate traffic signals – at least some of which were ignored by this ill-fated caravan – were encountered along the way, thereby endangering everyone lawfully passing through those intersections. The route stretched from the suburbs of St. Petersburg, northward through the urban areas surrounding Clearwater, and on beyond the fringes of Dunedin. This is part of the densely populated Tampa-St. Petersburg urban area.

As the chase continued, the sheriff's deputy was joined by at least fourteen and as many as twenty separate police or sheriff's vehicles, each of which was pursuing Deady at speeds that varied between eighty and 120 miles per hour. Although the chase was begun by a Pinellas sheriff's deputy, officers from Kenneth City and the City of Pinellas Park also joined. However, most of the officers involved were from the sheriff's department.

At some point, the Pinellas County Sheriff's Department ordered its officers to discontinue the chase. For unknown or unstated reasons, this order was not obeyed.[2]

By this time, the caravan was approaching the intersection of U.S. 19 and State Road 584 at very high speeds. At this intersection, Sheriff's Corporal Daniel Rusher was waiting in the turn lane, ready to move onto the highway Deady and the caravan were traveling. In the through-lane

---

[2] The Second Amended Complaint alleges that "an order to terminate the pursuit had been given by the supervisor which was disregarded all in contravention of General Order A-9 applicable to the PINELLAS COUNTY SHERIFF'S DEPARTMENT." Second Amended Complaint, at 8.

immediately next to Rusher was a vehicle occupied by two sisters, Susan and Judith Brown. Rusher made no attempt to block the intersection or to prevent the Browns from proceeding into the intersection. Rather, he was preparing to become part of the caravan.

When the light turned green, Rusher moved his vehicle onto U.S. 19 so he could wait for Deady to pass and join the chase. At the same time, the vehicle containing the Brown sisters moved forward into the intersection to pass through it. Deady's vehicle illegally entered the intersection at this precise moment and struck the Browns' vehicle at ninety miles per hour. Deady and Susan Brown died instantly, and Judith Brown died three days later. Id. According to the second amended complaint, the Pinellas Sheriff's Department at the times in question maintained a written policy, contained in General Order A-9, that required the discontinuance of certain "caravan-type" pursuits. This policy applied, says the complaint, whenever the area's citizenry was being endangered by hard pursuit, especially if the pursuit was prompted by a traffic violation. Thus, the complaint alleges that deputies directly violated this policy based on the facts at hand. The complaint alleges that this policy was further violated when the deputies disregarded the order to cease pursuit that had been given them.

In addition, the City of Pinellas Park also is alleged to have maintained a written policy on this question, contained in General Order Number 45, at the times in question. The complaint states that this policy required the termination of pursuit after consideration of a variety of factors. These are: (a) the identity of the fleeing individual has been ascertained, e.g., through a license-plate check; (b) the time of day and the amount of traffic was such that pursuit of the fleeing vehicle was dangerous; (c) the fleeing vehicle was clearly outdistancing the pursuit vehicles; (d) the seriousness of the crime was such that, it would not warrant the risk to innocent bystanders, the officer, or the occupants of the fleeing vehicle; or (e) the number of vehicles involved in the pursuit had become too great. The complaint alleges that, based on this policy, pursuit should have been discontinued; and the officers therefore violated the written policy.

Likewise, the complaint alleges that the Kenneth City Police Department at the relevant times maintained an oral policy prohibiting participation in highspeed chases by its officers. This policy also was violated, says the complaint.

The issues before us today are (a) whether the police owed a legal duty to the Brown sisters, (b) whether the activities of the police officers described above were shielded from all liability by the doctrine of sovereign immunity in spite of any duty owed the Browns, and (c) whether there is a sufficient allegation of proximate causation to create a jury question in this instance.

Duty

In *Kaisner v. Kolb*, 543 So. 2d 732, 735 (Fla. 1989), this Court held that where a defendant's conduct creates a foreseeable zone of risk, the laws generally will recognize a duty placed upon defendant either to lessen the risk or see that sufficient precautions are taken to protect others from the harm that the risk poses.

Petitioners argue that *Kaisner* should be factually distinguished and that the present case is controlled by *City of Miami v. Horne*, 198 So. 2d 10 (Fla. 1967). We cannot agree.

While the facts of *Kaisner* indeed differ from those at hand, it is clear from the plain language of the *Kaisner* opinion that it was describing the general mariner in which a duty of care arises under Florida law. We have so indicated in a recent opinion that directly relied upon *Kaisner* in making the following observation.

As the risk grows greater, so does the duty, because the risk to be perceived defines the duty that must be undertaken.

The statute books and case law, in other words, are not required to catalog and expressly proscribe every conceivable risk in order for it to give rise to a duty of care. Rather, each defendant who creates a risk is required to exercise prudent foresight whenever others may be injured as a result. This requirement of reasonable, general foresight is the core of the duty element.

McCain v. Florida Power Corp., 593 So. 2d 500, 503 (Fla. 1992) (citation omitted). In the present case, we think it manifest that a high-speed chase involving a large number of vehicles on a public thoroughfare is likely to result in injury to a foreseeable victim, and that the discontinuance of this chase by police is likely to diminish the risk. In other words, some substantial portion of the risk is being created by the police themselves, notwithstanding any contributory negligence of the person being chased. Accordingly, we believe the law must recognize a duty in this context even though the accident did not involve a police vehicle.

We find nothing in *Horne* supporting a contrary conclusion. As the district court below correctly noted, *Horne* expressly stands for the proposition that hot pursuit by police officers does not always give rise to liability, but sometimes can. In *Horne* we stated:

> It seems reasonably clear that the complaint [in *Horne*] charged that the pursuit itself constituted reckless and wanton conduct rather than that, although pursuit per se was lawful, the manner of pursuit, the conduct of the officers in otherwise discharging a necessary duty, was reckless and wanton.

Horne, 198 So. 2d at 12. The issue addressed in *Horne*, in other words, was whether a valid complaint is stated if the plaintiff alleges only that hot pursuit is per se negligence. Rejecting this claim, we simply held that a plaintiff must allege that the police engaged in hot pursuit in a negligent or wanton manner. We stated:

> We think the rule is that the officer should take such steps as may be necessary to apprehend the offender but, in doing so, not exceed proper and rational bounds not act in a negligent, careless or wanton manner. Here, "rational bounds" clearly were exceeded under the facts alleged.

We emphasize, however, that even in the absence of the hot-pursuit policies quoted above or the order to cease pursuit, we believe the chase described in the Second Amended Complaint clearly would give rise to a duty under the principles described in *Kaisner*. In no sense should this opinion be read as penalizing to any degree only those law enforcement agencies that have adopted hot-pursuit policies. The acts alleged here describe a situation in which motorists in Pinellas County were placed in deadly peril by as many as twenty police vehicles attempting to chase down a single man who had run a red light. On its face, this allegation alone definitely makes out a case for a duty owed to all persons who might encounter the police caravan that was chasing Deady. Such a duty would have existed whether or not any hot-pursuit policy existed and whether or not the police had been ordered to cease their pursuit.

Sovereign Immunity

The next question is whether the police were immune from liability notwithstanding the duty placed upon them by the law. Again, our most recent pronouncement on this issue is contained in *Kaisner*, where we noted that sovereign immunity does not shield acts that are "operational" in nature but only those that are "discretionary." As to this question, we held that an act is operational if it "is one not necessary to or inherent in policy or planning, that merely reflects a secondary decision as to how these policies or plans will be implemented." Id. at 737 (emphasis added). Governmental acts are "discretionary" and immune, on the other hand, if they involve "an exercise of executive or legislative power such that, for the court to intervene by way of tort law, it inappropriately would entangle itself in fundamental questions of policy and planning." Id.

Based on the plain language of *Kaisner*, we cannot accept petitioners' argument in favor of sovereign immunity in this case. We utterly fail to see how the events alleged in this complaint are anything but "operational." Taking the allegations in the complaint as true, we are faced with a situation in which officers engaged in flagrantly dangerous conduct that went far beyond what was necessary to vindicate the laws of Florida. Moreover, this conduct cannot honestly be characterized either as "policy" or "planning," because it self-evidently was contrary to both. In fact, the plaintiffs have alleged that each of the police agencies had adopted a policy to the contrary. Accordingly, the actions of the police in this instance are not entitled to sovereign immunity.

We agree that the actual execution of a hot-pursuit policy is entitled to a high degree of judicial deference consistent with reason and public safety. *Kaisner* specifically noted that special deference is given to pressing emergencies, and that certain police actions may involve a level of such urgency as to be considered discretionary and not operational. However, this does not mean that state agents can escape liability if they themselves have created or substantially contributed to the emergency through their own negligent acts or failure to adhere to reasonable standards of public safety.

To fall within the *Kaisner* exception, the serious emergency must be one thrust upon the police by lawbreakers or other external forces, that requires them to choose between different risks posed to the public. In other words, no matter what decision police officers make, someone or some group will be put at risk; and officers thus are left no option but to choose between two different evils. It is this choice between risks that is entitled to the protection of sovereign immunity in appropriate cases, because it involves what essentially is a discretionary act of executive decision-making (exercises of executive power are sovereignly immune).

Nevertheless, in the absence of such an emergency, the method chosen for engaging in a hot pursuit will remain an operational function that is not immune from liability if accomplished in a manner contrary to reason and public safety. As we stated in *Kaisner*, when government agents create a zone of risk through operational functions, then the governmental unit will not be shielded by sovereign immunity.

Here, the complaint alleges an enormous overreaction by sheriff's and police officers-one reminiscent of the most violent, daredevil films that Hollywood stunt men have produced. Solely because a man ran a red light, suddenly the innocent citizens of Pinellas County were subjected to a threatening stream of publicly-owned vehicles hurtling pell-mell, at breakneck speed, down a busy roadway in one of Florida's most densely populated urban areas. This caravan stormed through red lights for some twenty-five miles, gathering more and more police vehicles as it sped along. By the time the tragic chase ended, between fourteen and twenty police vehicles were included, only magnifying the risk to Pinellas County's innocent and unsuspecting residents. The reasons for

these actions can only be dubious. Were there no more reasonable means of vindicating Florida's law against running a red light than this?

Surely there is only one answer to this question. The police simply could have taken the violator's license-plate number together with a description of the car and driver, and then stopped the pursuit. Later, the violator could be located in some less dangerous setting, arrested, and brought to justice. And even if he continued to elude police, surely everyone must agree that this result is far better than the deaths of innocent persons. In the balance, the desire to bring Deady to justice for running a red light is far less important than the lives of the Brown sisters.

We do not suggest, however, that the police must allow every lawbreaker to escape merely because a hot pursuit is occurring. Deference will be shown to the reasonable decisions of law officers to maintain pursuit of certain offenders who are reasonably thought to be violent or to pose a danger to the public at large. What is required is for police to use reasonable means in light of the nature of the offense and threats to safety involved. For example, a high-speed chase is likely to be justifiable if its object is a gang of armed and violent felons who probably will harm others. As we have stated elsewhere, deference will be shown to police conduct when officers must choose between two different risks that both will adversely affect public safety....

For the foregoing reasons, the opinion under review is approved, and this cause is remanded to the trial court for further proceedings consistent with this opinion.

It is so ordered.

GRIMES, J., concurring.

DISSENT:

OVERTON, J., dissenting.

I dissent. While a clear, definitive policy in regard to car chases needs to be established by the executive and legislative branches of our government, the majority opinion goes much too far and effectively places that policy decision solely in the judicial branch. The majority effectively makes the taxpayers of the state of the insurer for injuries caused by a lawbreaker's attempt to avoid arrest. That policy also places law enforcement in a straitjacket that, in my view, will result in more harm than good....

The majority relies on hindsight in determining whether there should be liability, fending that, had the law enforcement officers involved broken off their pursuit, the fatal collision would not have resulted. Inherent in this assumption is the majority's belief that the fleeing driver would have ceased his reckless driving once the police stopped their attempt to apprehend him. The decisions made on the street by police officers are not easy ones, and hindsight is never available until it is too late. As a result of this decision, an officer must terminate the pursuit of a vehicle that is observed to be driven recklessly, refuses to stop, and increases its speed in an effort to avoid arrest. Because of concern of liability, the decision will preclude officers from preventing potentially fatal accidents by stopping a reckless drunk driver who initially seeks to avoid arrest. Furthermore, the majority opinion apparently adopts the policy that law enforcement should not try to prevent injuries caused by reckless drivers but, instead, arrest the driver the next day. Waiting until the next day, needless to say, will create the problem of proving the identity of the driver and prevent the driver's arrest for driving while under the influence of alcohol....

Further, I am unable to understand why a governmental entity that has a written policy on high-speed pursuits is penalized for violation of that policy by allowing the violation as a basis of negligence, while a governmental entity that has no such policy is not subject to such a negligence claim. The majority decision in this regard places governmental entities that have adopted these types of policies at a disadvantage over those entities that have not adopted such in-house regulations....

In conclusion, police vehicles must be driven with ordinary care, and government is responsible under the present law for damages caused by the negligent use of police vehicles. The majority's decision takes government responsibility a step further by making the governmental entity pay for the damages caused by a criminal offender trying to avoid apprehension. The decision improperly restricts law enforcement and sends the wrong message to criminal offenders and reckless drivers. The message encourages those individuals to flee from the police rather than accepting apprehension. At the very least, I would affirm the summary judgment as to the City of Pinellas Park and the City of Kenneth City because there are no allegations of any specific conduct of their officers that proximately caused the accident. Merely joining in the chase should not make the officers and the cities liable.

McDONALD, J., concurs.

McDONALD, J. dissenting.

HARDING, J. dissenting.

I dissent. An imaginative author would have a difficult time dreaming up a more bizarre and tragic scenario of events than those revealed by the record in this case. The facts make the call to change what I believe to be an appropriate principle of law most appealing.

Egregious though the fact may be, I would continue to adhere to this Court's holding in *City of Miami v. Horne*, 198 So. 2d 10, 13 (Fla. 1967) (footnote omitted), wherein we stated:

> The rule of governing the conduct of police in pursuit of an escaping offender is that he must operate his car with due care and, in doing so, he is not responsible for the acts of the offender. Although pursuit may contribute to the reckless driving of the pursued, the officer is not obliged to allow him to escape.

The majority opinion places an unwarranted chilling effect on law enforcement. It draws a line too obscure for an officer to clearly know whether to pursue or to cease pursuit. While I grieve that two precious lives have been needlessly taken from their loved ones, I do not agree that the Respondents can seek compensation from the Petitioners in this case because the police are not responsible for the acts of the escaping offender.

McDONALD, J., concurs.

QUESTIONS

1. The opinion's author states that "In no sense should this opinion be read as penalizing to any degree only those law enforcement agencies that have adopted hot-pursuit policies." Do you agree?
2. The Court states that sometimes "... no matter what decision police officers make, someone or some group will be put at risk; and officers thus are left no option but to choose between two different evils." Discuss the challenges facing a court that must evaluate such decisions.
3. The opinion states: "Deference will be shown to the reasonable decisions of law officers to maintain pursuit of certain offenders who are reasonably thought to be violent or to pose a danger to the public at large. What is required is for police to use reasonable means in light of the nature of the offense and threats to safety involved." Explain how this process would work in the real world.
4. The dissent states that the "majority effectively makes the taxpayers of the state the insurer for injuries caused by a lawbreaker's attempt to avoid arrest." Explain why this statement is or is not correct.
5. In this case, the lawbreaker's vehicle was involved in the fatal accident. Should the issue of liability be different if the vehicle causing the deaths were a police vehicle in pursuit? Why or why not?

The following Supreme Court case considers the argument that deliberate or reckless indifference to life in a high-speed chase aimed at apprehending a suspected offender violates substantive due process rights. The case was brought pursuant to 42 U.S.C. § 1983, the federal statute used to pursue violation of civil rights under color of law.

County of Sacramento v. Teri Lewis and Thomas Lewis

Supreme Court of the United States

523 U.S. 833 (1998)

SYLLABUS:

JUSTICE SOUTER delivered the opinion of the Court.

The issue in this case is whether a police officer violates the Fourteenth Amendment's guarantee of substantive due process by causing death through deliberate or reckless indifference to life in a high-speed automobile chase aimed at apprehending a suspected offender. We answer now, and hold that such circumstances only a purpose to cause harm unrelated to the legitimate object of arrest will satisfy the element of arbitrary conduct shocking to the conscience, necessary for a due process violation.

I

On May 22, 1990, at approximately 8:30 P.M., petitioner James Everett Smith, a Sacramento County sheriffs deputy, along with another officer, Murray Stapp, responded to a call to break up a fight. Upon returning to his patrol car, Stapp saw a motorcycle approaching at high speed. It was operated by 18-year-old Brian Willard and carried Philip Lewis, respondents' 16 year-old decedent, as a passenger. Neither boy had anything to do with the fight that prompted the call to the police.

Stapp turned on his overhead rotating lights, yelled to the boys to stop, and pulled his patrol car closer to Smith's, attempting to pen the motorcycle in. Instead of pulling over in response to Stapp's warning lights and commands, Willard slowly maneuvered the cycle between the two police cars and sped off. Smith immediately switched on his own emergency lights and siren, made a quick turn, and began pursuit at high speed. For 75 seconds over a course of 1.3 miles in a residential neighborhood, the motorcycle wove in and out of oncoming traffic, forcing two cars and a bicycle to swerve off of the road. The motorcycle and patrol car reached speeds up to 100 miles an hour, with Smith following at a distance as short as 100 feet; at the speed, his car would have required 650 feet to stop.

The chase ended after the motorcycle tipped over as Willard tried a sharp left turn. By the time Smith slammed on his brakes, Willard was out of the way, but Lewis was not. The patrol car skidded into him at 40 miles an hour, propelling him some 70 feet down the road and inflicting massive injuries. Lewis was pronounced dead at the scene.

Respondents, Philip Lewis's parents and the representatives of his estate, brought this action under Rev. Stat. § 1979, 42 U.S.C. § 1983 against petitioners Sacramento County, the Sacramento County Sheriff's Department, and Deputy Smith, alleging a deprivation of Philip Lewis's Fourteenth Amendment substantive due process right to life.[1] The District Court granted summary judgment for Smith, reasoning that even if he violated the Constitution, he was entitled to qualified immunity, because respondents could point to no "state or federal opinion published before May, 1990, when the alleged misconduct took place, that supports [their] view that [the decedent had] a Fourteenth Amendment substantive due process right in the context of high-speed police pursuits." App. to Pet. for Cert. 52.

The Court of Appeals for the Ninth Circuit reversed, holding that "the appropriate degree of fault to be applied to high-speed police pursuits is deliberate indifference to, or reckless disregard for, a person's right to life and personal security," and concluding that "the law regarding police liability for death or injury caused by an officer during the course of a high-speed chase was clearly established" at the time of Philip Lewis's death. Since Smith apparently disregarded the Sacramento County Sheriff's Department's General Order on police pursuits, the Ninth Circuit found a genuine issue of material fact that might be resolved by a finding that Smith's conduct amounted to deliberate indifference:

> The General Order requires an officer to communicate his intention to pursue a vehicle to the sheriff's department dispatch center. But defendants concede that Smith did not contact the dispatch center. The General Order requires an officer to consider whether the seriousness of the offense warrants a chase at speeds in excess of the posted limit. But here, the only apparent 'offense' was the boys' refusal to stop when another officer told them to do so. The General Order requires an officer to consider whether the need for apprehension justifies the pursuit under existing conditions. Yet Smith apparently only 'needed' to apprehend the boys because they refused to stop. The General Order requires an officer to consider whether the pursuit presents unreasonable hazards to life and property. But taking the facts here in the light most favorable to plaintiffs, there existed an unreasonable hazard to Lewis's and Willard's lives. The General Order also directs an officer to discontinue a pursuit when the hazards of continuing outweigh the benefits of immediate apprehension. But here, there was no apparent danger involved in permitting the boys to escape. There certainly was risk of harm to others in continuing the pursuit.

Accordingly, the Court of Appeals reversed the summary judgment in favor of Smith and remanded for trial.

We granted certiorari to resolve a conflict among the Circuits over the standard of culpability on the part of a law enforcement officer for violating substantive due process in a pursuit case.[3] We now reverse.

---

[1] Respondents also brought claims under state law. The District Court found that Smith was immune from state tort liability by operation of California Vehicle Code § 17004, which provides that "[a] public employee is not liable for civil damages on account of personal injury to or death of any person or damage to property resulting from the operation, in the line of duty, of an authorized emergency vehicle...when the immediate pursuit of an actual or suspected violator of the law." Cal. Veh. Code Ann. § 17004 (West 1971). The court declined to rule on the potential liability of the County under state law, instead dismissing the tort claims against the County without prejudice to refiling in state court.

[3] In *Jones v. Sherrill*, 827 F.2d 1102, 1106 (1987), the Sixth Circuit adopted a "gross negligence" standard for imposing liability for harm caused by police pursuit. Subsequently, in *Foy v. Berea*, 58 F.3d 227, 230 (1995), the Sixth Circuit, without specifically

II

Our prior cases have held the provision that "no State shall...deprive any person of life, liberty, or property, without due process of law," U.S. Const., Amdt. 14, § 1, to "guarantee more than fair process," *Washington v. Glucksberg*, 521 U.S. 702, 138 L. Ed. 2d 772, 117 S. Ct. 2258 (1997) (slip op., at 15), and to cover a substantive sphere as well, "barring certain government actions regardless of the fairness of the procedures used to implement them," *Daniels v. Williams*, 474 U.S. 327, 331, 88 L. Ed. 2d 662, 106 S. Ct. 662 (1986).... The allegation here that Lewis was deprived of his right to life in violation of substantive due process amounts to a such claim, that under the circumstances described earlier, Smith's actions in causing Lewis's death were an abuse of executive power so clearly unjustified by any legitimate objective of law enforcement as to be barred by the Fourteenth Amendment.

Leaving aside the question of qualified immunity, which formed the basis for the District Court's dismissal of their case, respondents face two principal objections to their claim. The first is that its subject is necessarily governed by a more definite provision of the Constitution (to the exclusion of any possible application of substantive due process); the second, that in any event the allegations are insufficient to state a substantive due process violation through executive abuse of power. Respondents can meet the first objection, but not the second....

A

Because we have "always been reluctant to expand the concept of substantive due process," *Collins v. Harker Heights, supra*, at 125, was held in *Graham v. Connor* that "where a particular amendment provides an explicit textual source of constitutional protection against a particular sort of government behavior, that Amendment, not the more generalized notion of substantive due process, must be the guide for analyzing these claims."... Given the rule in *Graham*, we were presented at oral argument with the threshold issue raised in several *amicus* briefs, whether facts involving a police chase aimed at apprehending suspects can ever support a due process claim. The argument runs that in chasing the motorcycle, Smith was attempting to make a seizure within the meaning of the Fourth Amendment, and, perhaps, even that he succeeded when Lewis was stopped by the fatal collision. Hence, any liability must turn on an application of the reasonableness standard governing searches and seizures, not the due process standard of liability for constitutionally arbitrary executive action.... One Court of Appeals has indeed applied the rule of *Graham* to preclude the application of principles of generalized substantive due process to a motor vehicle passenger's claims for injury resulting from reckless police pursuit. See *Mays v. East St. Louis*, 123 F.3d 999, 1002-1003 (CA7 1997).

The argument is unsound. Just last term, we explained that *Graham* "does not hold that all constitutional claims relating to physically abusive government conduct must arise under either the Fourth or Eighth Amendments; rather, *Graham* simply requires that if a constitutional claim is covered by a specific constitutional provision, such as the Fourth or Eighth Amendment, the claim must be analyzed under the standard appropriate to that specific provision, not under the rubric of substantive due process." *United States v. Lanier*, 520 U.S. 259, 272, n. 7, 1997 U.S. LEXIS 2079, *23, 137 L. Ed. 2d 432, 117 S. Ct. 1219, (1997) (slip op., at 13).

---

mentioning *Jones*, disavowed the notion that "gross negligence is sufficient to support a substantive due process claim." Although *Foy* involved police inaction, rather than police pursuit, it seems likely that the Sixth Circuit would now apply the "deliberate indifference" standard utilized in that case, see 58 F.3d at 232-233, rather than the "gross negligence" standard adopted in *Jones*, in a police pursuit situation.

Substantive due process analysis is therefore inappropriate in this case only if respondents' claim is "covered by" the Fourth Amendment. It is not.

The Fourth Amendment covers only "searches and seizures," U.S. Const., Amdt. 4, neither of which took place here. No one suggests that there was a search, and our cases foreclose finding a seizure. . . .

B

Since the time of our early explanations of due process, we have understood the core of the concept to be protection against arbitrary action. . . .

To this end, for half a century now we have spoken of the cognizable level of executive abuse of power as that which shocks the conscience. . . . While the measure of what is conscience-shocking is no calibrated yard stick, it does, as Judge Friendly puts it, "point the way." *Johnson v. Glick*, 481 F.2d 1028, 1033 (CA2), cert. denied, 414 U.S. 1033, 94 S. Ct. 462, 38 L. Ed. 2d 324 (1973).

It should not be surprising that the constitutional concept of conscience-shocking duplicates no traditional category of common-law fault, but rather points clearly away from liability, or clearly toward it, only at the ends of the tort law's spectrum of culpability. Thus, we have made it clear that the due process guarantee does not entail a body of constitutional law imposing liability whenever someone cloaked with state authority causes harm. . . .

Whether the point of the conscience-shocking is reached when injuries are produced with culpability falling within the middle range, following from something more than negligence but "less than intentional conduct, such as recklessness or 'gross negligence,'" 474 U.S. at 334, n. 3, is a matter for closer calls. To be sure, we have expressly recognized the possibility that some official acts in this range may be actionable under the Fourteenth Amendment, *ibid.*, and our cases have compelled recognition that such conduct is egregious enough to state a substantive due process claim in at least one instance. . . .

Rules of due process are not, however, subject to mechanical application in unfamiliar territory. Deliberate indifference that shocks in one environment may not be so patently egregious in another, and our concern with preserving the constitutional proportions of substantive due process demands an exact analysis of circumstances before any abuse of power is condemned as conscience-shocking. What we have said of due process in the procedural sense is just as true here. . . .

Thus, attention to the markedly different circumstances of normal pretrial custody and high-speed law enforcement chases shows why the deliberate indifference that shocks in the one case is less egregious in the other (even assuming that it makes sense to speak of indifference as deliberate in the case of sudden pursuit). As the very term "deliberate indifference" implies, the standard is sensibly employed only when actual deliberation is practical, *see Whitley v. Albers*, 475 U.S. at 320, and in the custodial situation of a prison, forethought about an inmate's welfare is not only feasible but obligatory under a regime that incapacitates a prisoner to exercise ordinary responsibility for his own welfare. . . .

But just as the description of the custodial prison situation shows how deliberate indifference can rise to a constitutionally shocking level, so too does it suggest why indifference may well not be enough for liability in the different circumstances of a case like this one. . . .

Like prison officials facing a riot, the police on an occasion calling for fast action have obligations that tend to tug against each other. Their duty is to restore and maintain lawful order, while not exacerbating disorder more than necessary to do their jobs. They are supposed to act decisively and to show restraint at the same moment, and their decisions have to be made in haste, under pressure, and frequently without the luxury of a second chance. [citations omitted] . . .

[W]hen unforeseen circumstances demand an officer's instant judgment, even precipitate recklessness fails to inch close enough to harmful purpose to spark the shock that implicates "the large concerns of the governors and the governed." *Daniels v. Williams*, 474 U.S. at 332. Just as a purpose to cause harm is needed for Eighth Amendment liability in a riot case, so it ought to be needed for Due Process liability in a pursuit case. Accordingly, we hold that high-speed chases with no intent to harm suspects physically or to worsen their legal plight do not give rise to liability under the Fourteenth Amendment, redressible by an action under § 1983.[13]

The fault claimed on Smith's part in this case accordingly fails to meet the shocks-the-conscience test. In the count charging him with liability under § 1983, respondents' complaint alleges a variety of culpable states of mind: "negligently responsible in some manner," (App. 11, Count one, P8), "reckless and careless" (id., at 12, P15), "recklessness, gross negligence and conscious disregard for [Lewis's] safety" (id., at 13, P18), and "oppression, fraud and malice" (*Ibid.*) The subsequent summary judgment proceedings revealed that the height of the fault actually claimed was "conscious disregard," the malice allegation having been made in aid of a request for punitive damages, but unsupported either in allegations of specific conduct or in any affidavit of fact offered on the motions for summary judgment....

Smith was faced with a course of lawless behavior for which the police were not to blame. They had done nothing to cause Willard's high-speed driving in the first place, nothing to excuse his flouting of the commonly understood law enforcement authority to control traffic, and nothing (beyond a refusal to call off the chase) to encourage him to race through traffic at breakneck speed forcing other drivers out of their travel lanes. Willard's outrageous behavior was practically instantaneous, and so was Smith's instinctive response. While prudence would have repressed the reaction, the officer's instinct was to do his job as a law enforcement officer, not to induce Willard's lawlessness, or to terrorize, cause harm,

or kill. Prudence, that is, was subject to countervailing enforcement considerations, and while Smith exaggerated their demands, there is no reason to believe that they were tainted by an improper or malicious motive on his part.

Regardless whether Smith's behavior offended the reasonableness held up by tort law or the balance struck in law enforcement's own codes of sound practice, it does not shock the conscience, and petitioners are not called upon to answer for it under § 1983. The judgment below is accordingly reversed.

It is so ordered.

CONCUR BY:

REHNQUIST; KENNEDY; BREYER; STEVENS; SCALIA

JUSTICE SCALIA, with whom JUSTICE THOMAS joins, concurring in the judgment.

Respondents provide not textual or historical support for this alleged due process right.... Needless to say, if it is an open question whether recklessness can *ever* trigger due process protections, there is no precedential support for a substantive-due-process right to be free from reckless police conduct during a car chase.

To hold, as respondents urge, that all government conduct deliberately indifferent to life, liberty, or property, violates the Due Process Clause would make "the Fourteenth Amendment a font of tort law to be superimposed upon whatever systems may already be administered by the States." [citations omitted] ... Here, for instance, it is not fair to say that it was the police officer alone who "deprived" Lewis of his life. Though the police car did run Lewis over, it was the driver of the motorcycle, Willard, who dumped Lewis in the car's path by recklessly making a sharp left turn at high speed. (Willard had the option of rolling to a gentle stop and showing the officer his license and registration.) Surely

---

[13] Cf. *Checki v. Webb*, 785 F.2d 534, 538 (CA5 1986) ("Where a citizen suffers physical injury due to a police officer's *negligent use* of his vehicle, no section 1983 claim is stated. It is a different story when a citizen suffers or is seriously threatened with physical injury due to a police officer's *intentional misuse* of his vehicle") (citation omitted).

Willard "deprived" Lewis of his life in every sense that the police officer did. And if Lewis encouraged Willard to make the reckless turn, Lewis himself would be responsible, at least in part, for his own death. Was there contributory fault on the part of Willard or Lewis? Did the police officer have the "last clear chance" to avoid the accident? Did Willard and Lewis, by fleeing from the police, "assume the risk" of the accident? These are interesting questions of tort law, not of constitutional governance.

If the people of the State of California would prefer a system that renders police officers liable for reckless driving during high-speed pursuits, "they may create such a system...by changing the tort law of the State in accordance with the regular lawmaking process."...[citations omitted] It is the prerogative of a self-governing people to make that legislative choice....In allocating such risks, the people of California and their elected representatives may vote their consciences. But for judges to overrule that democratically adopted policy judgment on the ground that it shocks *their* consciences is not judicial review but judicial governance....

_____

QUESTIONS

1. The *Lewis* decision mentions California state law as well as the federal law of substantive due process. The majority opinion states that "To say that due process is not offended by the police conduct described here is not, of course, to imply anything about its appropriate treatment under state law." Why do you think that the Court declined to opine on the state law's treatment the police conduct?

2. Consider the reasoning behind the immunities provided by California law to public employees engaged in hot pursuit. What limits the actions of law enforcement personnel if state law excuses them from tort liability?

3. The Court holds that "high-speed chases with no intent to harm suspects physically or to worsen their legal plight do not give rise to liability under the Fourteenth Amendment." Describe a hot pursuit case in which you think that the intent required by the Court would exist.

4. The Supreme Court held that federal due process does not apply to hot pursuit if pursuing officers' actions do not "shock the conscience." Explain why you agree or disagree. What other standards would you suggest for a federal guideline?

5. The dissent states that to hold all government conduct deliberately indifferent to life, liberty, or property, violates the Due Process Clause would make the Fourteenth Amendment a "font of tort law" to be superimposed upon whatever systems may already be administered by the States. Do you agree? Why or why not?

PART 2. VEHICLE ACCIDENT

Emergency vehicles frequently operate at high speed. The need for prompt response and transport may require weaving in and out of slower traffic and operating contrary to otherwise applicable traffic rules. Sometimes, vehicles are involved in accidents. Studies establish that emergency vehicle accidents constitute fully 50 percent of all pre-hospital claims for Emergency Medical Services (EMS) providers. *See, e.g.*, Soler, et al., "The Ten Year Malpractice Experience of a Large Urban EMS System," *Annals of Emergency Medicine 14*: 982, 985 (1985); Goldberg, et al., "A Review of Prehospital Care Litigation in a Large Metropolitan EMS System," *Annals of Emergency Medicine, 19*:557, 558 (1990). Taking pro-active steps to assure their avoidance, therefore, should be a high priority for all emergency response services.

As you read the case below, consider the pressures on emergency vehicle operators as they strive to safely arrive at a scene or transport their patient as rapidly as possible.

Hoffert v. Luze

Supreme Court of Iowa

578 N.W.2d 681 (1998)

SNELL, Justice.

OPINION:

In this appeal the primary issue argued is what legal standard of care applies to the driver of an ambulance being operated under emergency conditions....

I. BACKGROUND FACTS AND PROCEEDINGS

Defendant Michael D. Luze is a volunteer emergency medical technician for the defendant City of Aplington. On January 19, 1995, Luze was operating an ambulance owned by the City, transporting an individual exhibiting heart attack symptoms from Aplington to a hospital in Waterloo. At the time of the accident, the ambulance was traveling eastbound on Broadway Street in Waterloo, a multi-lane divided highway with a fifty-five mile per hour speed limit. Luze was utilizing the ambulance's flashing lights, but not the siren. Hoffert was traveling northbound on an exit ramp from Iowa Highway 218. The accident occurred at a "T" intersection between the two roadways, which is controlled by traffic lights. When Luze approached the intersection, he saw he had a red light and applied his brakes. Luze saw the car in front of Hoffert's vehicle pass in front of him before he reached the intersection. He observed a gap of approximately five car lengths between the previous car and Hoffert's vehicle and thus released the brakes to proceed through the intersection. Luze saw Hoffert's car enter the intersection, but was unable to brake in time to avoid the collision. The ambulance hit the driver's side of Hoffert's car.

Hoffert filed suit, seeking recovery for personal injuries and property damage. He alleged Luze had been negligent in his actions and that the City was vicariously liable for Luze's actions. Luze and the City raised affirmative defenses premised on Iowa Code chapter 668 (1995) and sections 321.231, 670.4(11) and 670.12. Hoffert filed an amended petition asserting recklessness as an additional theory of recovery and as a basis for an award of punitive damages. The defendants asserted an additional affirmative defense in response to the claim for punitive damages under Iowa Code section 670.4(5).

Defendants filed a motion for summary judgment claiming: (1) defendants were immune under Iowa Code section 670.4(11) because Hoffert's claims arose out of acts or omissions in connection with an emergency response; (2) to the extend the immunity provided by section 670.4(11) may be limited by section 321.231, Luze was not reckless in his operation of the ambulance as a matter of law; and (3) defendants were immune from a claim for punitive damages under Iowa Code sections 670.4(5) and 670.12.

District Court Judge James L. Beeghly found there was insufficient evidence to support a finding of recklessness regarding Luze's operation of the ambulance and thus granted the motion for summary judgment with regard to Hoffert's claims for recklessness and punitive damages. However, the court denied defendants' motion for summary judgment based on Iowa Code sections 670.4(11) and 321.231 and the case proceeded to trial on Hoffert's claim of negligence. This court denied defendants' motion for interlocutory appeal.

At trial, Judge Todd E. Geer refused to reconsider Judge Beeghly's ruling on the motion for summary judgment as requested by both parties, concluding it established the law of the case. Defendants moved for a directed verdict at the end of Hoffert's case and at the close of all the evidence, which the court denied. The jury found both parties were negligent, attributing fifty percent fault to each party, and calculated Hoffert's damages as $127,196.55. The court entered judgment in favor of Hoffert for $63,598.28. Defendants filed a motion for judgment notwithstanding the verdict, which the court denied. Defendants filed notice of appeal of all adverse rulings of the district court. Hoffert does not appeal the district court's adverse rulings as to his recklessness and punitive damages claims.

II. ISSUE

Whether the district court erred in instructing the jury to decide liability based on a standard of negligence rather than recklessness.

III. DISCUSSION

The main issue here involves the interplay of Iowa Code sections 670.4(11) and 321.231 and the resulting standard of care applicable to operators of emergency vehicles. Iowa Code section 670.2 is a general statute subjecting municipalities to liability "for its torts and those of its officers and employees acting within the scope of their employment or duties." Section 670.2 also includes within the definition of employee "a person who performs services for a municipality whether or not the person is compensated for the services." Therefore, Luze was an employee in this instance even though he worked as a volunteer.

Section 670.4 provides several exemptions from the liability imposed in section 670.2. Section 6704 provides in pertinent part:

> The liability imposed by section 670.2 shall have no application to any claim enumerated in this section. As to any such claim, a municipality shall be liable only to the extent liability may be imposed by the express statute dealing with such claims and, in the absence of such express statute, the municipality shall be immune from liability....
>
> 11. A claim based upon or arising out of an act or omission in connection with an emergency response including but not limited to acts or omissions in connection with emergency response communications services.

From the introductory paragraph to section 670.4, it appears, and the plaintiff argues, that section 321.231 is an express statute dealing with claims regarding emergency response vehicles. Therefore, Luze and the city can be held liable "only to the extent liability may be imposed by" section 321.231. Section 321.231 provides in pertinent part:

1. The driver of an authorized emergency vehicle, when responding to an emergency call or when in the pursuit of an actual or suspected perpetrator of a felony or in response to an incident dangerous to the public or when responding to but not upon returning from a fire alarm, may exercise the privileges set forth in this section....
2. The driver of a fire department vehicle, police vehicle or ambulance may:
   a. Proceed past a red or stop signal or stop sign, but only after slowing down as may be necessary for safe operation.
   b. Exceed the maximum speed limits so long as the driver does not endanger life or property.
3. The exemptions granted to an authorized emergency vehicle under subsection 2 and for a fire department vehicle, police vehicle or ambulance as provided in subsection 3 shall apply only when such vehicle is making use of an audible signaling device meeting the requirements of section 321.433, or a visual signaling device approved by the department....
4. The foregoing provisions shall not relieve the driver of an authorized emergency vehicle from the duty to drive with due regard for the safety of all persons, nor shall such provisions protect the driver from the consequences of the driver's reckless disregard for the safety of others.

Plaintiff argues the proper standard of care to be derived from section 321.231(5) is negligence, based on the wording "duty to drive *with due regard* for the safety of all persons." (Emphasis added.) Defendants argue that the latter portion of section 321.231(5) provides the pertinent standard of care: "nor shall such provision protect the driver from the consequences of the driver's *reckless disregard* for the safety of others." (Emphasis added.) In its instructions to the jury, the court used a standard of negligence, to which the defendants objected. It is on this ground that they appeal.

Defendants cite *Morris v. Leaf* 534 N.W.2d 388 (Iowa 1995), in support of their position that recklessness is the appropriate standard of care. In *Morris*, the plaintiff was injured when her car was struck by a vehicle driven by a suspect attempting to flee from the police. The suspect and the police were engaged in a high-speed chase. *Morris*, 534 N.W.2d at 389. The district court granted summary judgment for the city and police officer on the ground that the officer had no duty to protect the injured party from the fleeing suspect and that the officer's acts were not a proximate cause of the collision. Id. On appeal, our court discussed the applicable standard of care, noting that section 321.231(5) mentions both "due regard," or negligence, and "reckless disregard." Id. We found that "section 321.231 requires a level of culpability beyond mere negligence to support liability." Id. We cited a New York Court of Appeals decision

interpreting a similar statute in support of this conclusion. That court found that "the only way to apply the statute is to read its general admonition to exercise 'due care' in light of its more specific reference to 'recklessness." *Saarinen v. Kerr*, 84 N.Y.2d 494, 644 N.E.2d 988, 991, 620 N.Y.S.2d 297 (N.Y. 1994). We concluded "that a police officer should not be civilly liable to an injured third person unless the officer acted with 'reckless disregard' for the safety of others." *Morris*, 534 N.W.2d at 390 (citing *Saarinen*, 644 N.E.2d at 991). This conclusion appears reasonable based on the plain language of the statute. *See* Iowa Code § 321.231(5) ("Nor shall such provisions protect the driver from the *consequences* of the driver's reckless disregard for the safety of others.") (Emphasis added).

Plaintiff argues that *Morris* should be limited to its facts, specifically to the scenario of a high-speed chase in which the plaintiff is injured by the fleeing suspect. Defendants argue that such a reading of *Morris* is too narrow and that our interpretation of section 321.231(5) providing a recklessness standard should apply in the case of a driver being struck by an emergency response vehicle as well.

IV. RESOLUTION

The meaning of statutes is a legal question and our review is therefore at law.

In the case at bar, it is the manner of driving that is the issue. As in *Morris*, the level of culpability to retain the statutorily granted immunity is negligence. Id. Regarding the standard of recklessness we said:

> In order to prove recklessness as the basis for a duty under section 321.231(3)(b), we hold that a plaintiff must show that the actor has intentionally done an act of unreasonable character in disregard of a risk known to or so obvious that he must be taken to have been aware of it, and so great as to make it highly probable that harm would follow. [citations omitted]

We also addressed the relevance of the statute on municipality liability, stating:

The plaintiffs also rely on Iowa Code section 670.2 to establish liability. That statute is general in its scope, making municipalities liable for the torts of their employees. However, that statute does not create liability for the acts of police officers that involve mere negligence and therefore do not provide a basis for liability of the employee personally. That is the case here; we find no duty owing from the employee to the public and therefore conclude that section 670.2 is inapplicable.

Id.

We find that our interpretation of these statutes relating to authorized emergency vehicles is applicable to an ambulance as well as to a police vehicle. Iowa Code section 321.1(6), covering definition of words and phrases, includes in the definition of "authorized emergency vehicle," vehicles of the fire department, police vehicles, ambulances, and other emergency vehicles.

We hold that the legal standard of care applicable to the conduct of an ambulance driver as a driver of an authorized emergency vehicle under Iowa Code section 321.231 is to drive with due regard for the safety of all persons, but the threshold for recovery for violation of that duty is recklessness, not negligence. The district court erred in its jury instruction on this issue which requires a reversal. Because of our resolution of this issue, we do not consider other issues raised on the appeal.

The district court granted defendants' summary judgment motion with regard to plaintiff's claims of recklessness and for punitive damages. No appeal was taken from the ruling so we do not order a new trial.

The case is reversed and remanded for entry of judgment for defendants.

REVERSED AND REMANDED.

QUESTIONS

1. The Court stated: "We hold that the legal standard of care applicable to the conduct of an ambulance driver as a driver of an authorized emergency vehicle under Iowa Code section 321.231 is to drive with due regard for the safety of all persons, but the threshold for recovery for violation of that duty is recklessness, not negligence." Does this approach make sense? Why or why not?

2. The Iowa statute provides that "The driver of a fire department vehicle, police vehicle or ambulance" may operate the authorized emergency vehicle in ways that otherwise would be unlawful. Why are all three entities permitted the same latitude in vehicle operation?

3. Some states permit privately owned vehicles to display colored lights that operate while on the way to an emergency situation. For example, in Indiana, volunteer firefighters are permitted to display blue lights and emergency medical services personnel may display green lights. Both must obtain proper permissions to do so, and their vehicle operations must not be in violation of otherwise applicable law. What advantages and disadvantages do you see in allowing such private emergency vehicle lights? What standards should apply to the drivers of such vehicles?

PART 3. CRIMINAL LIABILITY FOR EMERGENCY VEHICLE ACCIDENT

Sometimes, a vehicle accident can result in criminal charges for the driver. Think about what the standards should be for emergency vehicle drivers involved in accidents resulting in death or serious injury.

Ohio v. Montecalvo

Court of Appeals of Ohio, Ninth Appellate District, Lorain County

1990 Ohio App. LEXIS 3942

DECISION AND JOURNAL ENTRY:

...Michael Montecalvo, a certified paramedic and ambulance driver, appeals his convictions for failing to proceed cautiously past a red or stop signal in violation of R.C. 4511.03 and for involuntary manslaughter, R.C. 2903.04(B). We affirm.

Montecalvo while driving an ambulance in response to an emergency call, drove through a red light and struck an automobile at the intersection of Broadway Avenue and Cooper Foster Park Road. Neither Montecalvo nor his fellow paramedic assistant were injured, however, the driver of the automobile, Angela Robinson, who was pregnant at the time, died as a result of the accident.

...Montecalvo was found guilty of involuntary manslaughter and failing to proceed with due regard in an emergency vehicle past a red or stop signal as found in counts two and four of the indictment....

R.C. 2903.04(B) provides that no person shall cause the death of another as a proximate result of committing or attempting to commit a misdemeanor. Appellant contends that misdemeanor as used in this statute does not include minor misdemeanor. However, there is no support from case law or the legislative history to suggest the exclusion of vehicular minor misdemeanors from the operation of the statute.... Most traffic offenses are minor misdemeanors....

A claim of insufficient evidence requires a determination that any rational trier of fact could find all the essential elements of the offense beyond a reasonable doubt after viewing, but not weighing, the probative evidence and construing reasonable inferences most favorable to the prosecution. *State v. Martin* (1983), 20 Ohio App. 3d 172, 175. In this case, there was sufficient probative evidence to support the defendant's conviction....

Counsel for defendant requested an instruction on the word emergency. In reviewing the defendant's proposed instruction on an emergency, the trial court noted:

THE COURT: Then we will-I think that in this case, after under emergency call, we add the first two lines of your charge, which says, "an emergency is a serious and urgent situation which demands immediate use of a public safety vehicle," period.

"Then we go to the second or we go to the Charge that says, 'if a driver proceeds with no reasonable grounds for believing that an emergency exists, even though he is using a siren and flashing, rotating lights when he approaches the intersection he forfeits or gives up his rights of an emergency vehicle."

MR. STERNBERG: You are making it better, Your Honor. No question about it. I don't know if I could start with a negative, the answer is better and more point to the law.

At defendant's urging, and without further objection the trial court modified the proposed instruction. The claim of error is not well taken.

In reviewing defendant's proposed instruction on accident, the trial court reasoned.

THE COURT: In reviewing this particular Charge, I feel that this goes to argument rather than to what is appropriately charged and I don't think that it is appropriate to the Charge.

"Certainly, it goes to argument and certainly Counsel can argue that there are accidents that are beyond the control and so on and so forth."

The defendant suffered no material prejudice from the exclusion of the proposed instruction.

In our review of the court's instruction to the jury as a whole, we find that the charge adequately covered the law and issues presented by the evidence. Thus, the fourth assignment of error is overruled and the judgment of the trial court is affirmed....

CACIOPPO, J., DISSENTS SAYING:

I must respectfully dissent from the decision announced today regarding appellant's second and third assignments of error and would reverse Montecalvo's conviction for involuntary manslaughter.

Violations of municipal ordinances and state laws where the punishment is by fine alone have been recognized as quasi-criminal offenses and not strictly criminal cases....

I agree with the majority of this panel that most traffic offenses are minor misdemeanors. However, numerical majority alone does not indicate that the General Assembly consider all traffic offenses equally culpable....

The imprisonment of a paramedic ambulance driver who committed an act of simple negligence that resulted in a fatal accident will not avenge a young woman's tragic death. The decision reached today represents neither the law nor justice. Therefore, I must respectfully dissent.

The reader offended by the Court's conclusion in the *Montecalvo* matter will be relieved to learn that the case was reversed on appeal. *Ohio v. Montecalvo*, 671 N.E.2d 236 (Ohio 1996). The fact that drivers may be found criminally liable for accidents in their emergency vehicles should be borne in mind, however, by all emergency responders.

QUESTIONS

1. The dissent points out that criminal negligence and civil negligence are substantially different from one another. What differences do you see between them?
2. The dissent implies that the verdict in this matter was the result of a desire for vengeance. What do you see to support or oppose this position in the majority opinion?
3. Under what circumstances, if ever, would criminal liability for an emergency vehicle accident be appropriate?

Emergency Medical Dispatch in action. Photo courtesy Priority Dispatch.

Chapter 4

DISPATCH ISSUES

The first step in answering the call to an emergency scene is dispatch. Common areas of concern include delayed dispatch, pre-arrival instructions and standards for dispatch.

PART 1. DELAYED DISPATCH

The following case came about when a delayed ambulance dispatch resulted in significantly worse injury to the patient. As you read the case, pay particular attention to the words of the dispatcher and the reactions of the patient and his family.

Koher v. Dial

Court of Appeals of Indiana, Fourth District

653 N.E.2d 524 (1995)

RILEY, Judge

FACTS:

A little before 2:00 P.M. on August 24, 1991, Neil Koher experienced chest pains while mowing the lawn. Neil had a history of heart problems beginning in 1989, and thus, upon feeling pain, took two nitroglycerin tablets which his physician had previously prescribed. The medicine did not relieve the pain.

Sandra Koher sent for her sister, Cynthia White, a nurse who lived across the street from the Kohers. Cynthia and her husband, Roger White, arrived at about 2:00 P.M. Twice, Roger attempted without success to telephone the Noble County Sheriff's Department. Sandra then dialed 911, but received no answer. Sandra redialed 911 and was connected with the 911 dispatcher. Sandra told the dispatcher that her husband was having a heart attack, that he was in and out of consciousness, and that he needed an ambulance

immediately. She gave directions to her house and told the dispatcher that someone would be in front of the house to direct the ambulance. The 911 dispatcher promised Sandra that an ambulance would be dispatched immediately. Sandra's call was received by the 911 dispatcher at 2:10 P.M.

Sandra posted individuals at the roadside to direct the ambulance when it arrived. Sandra, Cynthia and Neil waited. Relying on the promise of the dispatcher, they did not transport Neil to the hospital themselves, nor did they make additional attempts to seek ambulance services directly.

As a member of the Noble County Volunteer Fire Department, Roger carries a pager with which he can monitor the dispatch of the Noble County Emergency Medical Squad (EMS). After Sandra called 911, Roger did not hear a dispatch of an ambulance to the Koher home on his pager.

At 2:17 P.M., Cynthia telephoned the Noble Township Fire Department at Wolf Lake where the Wolf Lake Branch of the EMS was holding a

43

meeting at the Fire Department. An ambulance was immediately dispatched and arrived at the Koher homes at 2:18 P.M.

At all times relevant to this action, the County was in charge of and operated the 911 Emergency telephone number and was responsible for dispatching emergency medical service in Noble County.[2] Three weeks later, Sandra spoke with Chief Deputy Sheriff Harlan Miller of the Noble County Sheriff's Officer who told her that the reason his department failed to respond to the telephone call and failed to dispatch an ambulance promptly as requested and promised was that on August 24, the sheriff's department was having its annual picnic, and an officer, unfamiliar with the "equipment was brought in to operate the dispatching of the emergency services and that said operator was unable to properly and efficiently operate the equipment." (R. at 88).

The County filed a motion for summary judgment on the grounds that it did not have any special relationship with Koher such as to give rise to a private duty. (R. at 61). After a hearing, the trial court granted the motion and entered judgment for the County, citing to *Lewis v. City of Indianapolis* (1990), Ind.App., 554 N.E.2d 13, trans. denied (summary judgment granted to municipality because there was no evidence showing that plaintiff had a special relationship with municipality or its services giving rise to a special, individualized duty on municipality's part toward plaintiff).

Koher appeals.

DISCUSSION

In this appeal we address the narrow issue of whether Koher had a special relationship with the County giving rise to a special, individualized duty....

The existence of a duty is a question of law which is properly decided by the trial court. The County asserts that Koher was required to establish that the County owed him a private duty in this action. Standing alone, a governmental entity's dispatch of emergency services does not create a private duty. Thus, in order to prevail, Koher had to establish some special relationship between himself and the County which created a private duty....

In *Mullin*, 639 N.E.2d 278, our supreme court held that when a "governmental entity is aware of the plight of a particular individual and leads that person to believe that governmental rescue services will be used, and the individual detrimentally relies on that promise, it would be unfair to leave that individual worse off than if the individual had not sought assistance from the government at all." Id. at 285. The court delineated the following elements for imposition of a private duty on governmental defendants:

1. an explicit assurance by the municipality, through promises or actions, that it would on behalf of the injury party;
2. knowledge on the part of the municipality that inaction could lead to harm; and
3. justifiable and detrimental reliance by the injured party on the municipality's affirmative undertaking.

Id. at 284.

A review of the evidence most favorable to Koher reveals that Sandra spoke to the 911 dispatcher. She told the dispatcher that Neil was having a heart attack, that he was in and out of consciousness, and that he needed an ambulance immediately. This is sufficient to establish knowledge on the part of the municipality that inaction could lead to harm. The dispatcher assured Sandra that an ambulance would be dispatched immediately. This is sufficient to establish an explicit assurance by the municipality that it would act on behalf of the injured party. There is also evidence of detrimental reliance. After Sandra spoke to the dispatcher, she posted individuals at roadside to direct the ambulance to her house. She waited, and relying on the promise of the dispatcher, did not transport Neil to the hospital herself, nor did she seek ambulance services directly. During the delay,

---

[2] IND. CODE 34-4-16.5-3 (1993) provides that a governmental unit is not liable for its acts in connection with the operation and use of "an enhanced emergency communication system." During the relevant period of time, Noble County did not have an enhanced emergency communication system.

Neil was deprived of oxygen and medication and suffered permanent damage to his heart and his quality of life has been greatly diminished. Under these circumstances, Koher established the existence of a private duty owed to him by the County.

The County argues that Koher was not lulled into inaction by the promise of an ambulance because he was aware, at all times, that no ambulance had been dispatched. This argument is without merit. The mere fact that Roger could monitor dispatches of the Noble County EMS on his pager does not show that Koher did not rely on the promise of the dispatcher. In fact, the relatively short amount of time between when Sandra contacted the dispatcher and when Cynthia contacted the Fire Department directly is probably due in large part to the family's ability to monitor ambulance dispatches. The short time span between phone calls does not diminish Koher's reliance on the dispatcher's assurance that he would send an ambulance out immediately.

The trial court erred in finding that the County did not owe a private duty to Koher. We reverse and remand this action to the trial court for proceedings in accordance with this decision.

QUESTIONS

1. Develop a time line for the actions described in this case.
2. Compare the words of the dispatcher, the actions of the dispatcher, and the actions of response units. How, if at all, did they differ?
3. To what extent do patients rely on prompt dispatch of emergency services from a 911 center?
4. The issue of detrimental reliance by the patient was important to the court. How might a dispatching entity avoid detrimental reliance?
5. How might the events described have turned out had the dispatcher had given more accurate information? What possible legal effects might such information have caused?
6. Recall the allegations of improper dispatch activity in the *American National Bank and Trust Company* case in Chapter 1. Compare the actions of the dispatchers in that case and those in the current matter.
7. Compare the private duty discussion in this case with that in the *American National Bank* case dissent in Chapter 1.

PART 2. DISPATCH STANDARDS, INFLICTION OF MENTAL DISTRESS, AND CREATION OF NEW LAW

The following case examines the failure of dispatchers to perform their duties. Note the discussion of whether adoption of a more liberal standard for intentional infliction of emotional distress is appropriate, and consider its implications for emergency response.

Hammond v. Central Lane Communications Center

Supreme Court of Oregon

816 P.2d 593 (1991)

VAN HOOMISSEN, J.

OPINION:

This is a tort action for damages based on theories of negligent and reckless infliction of severe emotional distress. Plaintiff alleges that she sustained psychic and emotional injuries as a result of the manner in which defendants responded to a 9-1-1 call that she made concerning her husband and as a result of the manner in which defendants designed the Lane County 9-1-1 emergency telephone system. The trial court granted defendants' motions for summary judgment. ORCP 47. The Court of Appeals affirmed.... Because plaintiff suffered no physical injury from defendants' alleged negligence and because she has not shown that defendants' conduct was anything more then negligent, we also affirm.

We view the record in the light most favorable to plaintiff, the party opposing defendants' summary judgment motions.

Plaintiff awoke to find her 67-year-old husband, who suffered from congestive heart failure, lying on the kitchen floor. He was not breathing, had no apparent pulse, was cold to the touch, and was bluish in color.

Plaintiff dialed 9-1-1. She spoke to a 9-1-1 operator, who asked her questions about the nature of the emergency. At the operator's request, plaintiff checked to see if her husband was breathing. He was not. The operator asked plaintiff to try to find a pulse. She found none. On the basis of the information provided by plaintiff, the operator concluded that plaintiff's husband was dead from natural causes. The operator told plaintiff that someone would be at her house "in just a couple of minutes." During the time plaintiff waited for help to arrived, her husband made "rasping breathing sounds."

An electronic message, reporting information from plaintiff's call and the 9-1-1 operator's assessment of the nature of the call, was immediately dispatched via computer to the Eugene Fire Department's Dispatch Center and to the Oregon State Police (OSP).[2] Because plaintiff's call originated from an unincorporated area of the county and indicated that plaintiff's husband was dead from natural causes, OSP was the agency responsible for responding to the call. Central Lane Communications Center (CLCC) personnel who made follow-up calls to OSP were told

that no state trooper was available to respond immediately to a call of a non-emergency nature, that a deputy sheriff would soon be available and, therefore, the deputy would respond to plaintiff's call. The deputy, who also was advised that this was a "deceased person" call, arrived at plaintiff's house about 45 minutes after plaintiff called 9-1-1 and found plaintiff's husband dead. Between plaintiff's 9-1-1 call and the arrival of the deputy at her house, plaintiff's son-in-law called 9-1-1 to inquire when an ambulance would be arriving. He indicated to the 9-1-1 operator that plaintiff's husband was, in fact, dead.

Plaintiff's complaint first alleges that defendants were negligent in the following particulars: in treating her call as a "deceased person" call; in advertising that, if she called 9-1-1, emergency medical services would be delivered to her home, when, in fact, defendants knew that, because of a defect in the original design of the 9-1-1 system, such services would never have been sent to the unincorporated area where she lived; in failing to provide those services as advertised; in misleading her to believe that those services would be provided "in just a couple of minutes," when defendants knew that those services would not be provided to the unincorporated area where she lived; and in failing to ensure that those services were provided as advertised. Plaintiff's complaint next alleges that defendants' conduct was "extreme and outrageous" and reckless.[3]

In granting defendants' motions for summary judgment, the trial court held that plaintiff had no claim against defendants for negligence, because plaintiff was not a "direct victim" of defendants' alleged negligence, and that plaintiff has no claim against defendants for reckless infliction of severe emotional distress, because defendants' handling of plaintiff's call entailed no misconduct that a

---

[2] Defendant Central Lane Communications Center (CLCC) operates the 9-1-1 Telephone Answering and Call Routing Service. CLCC itself does not physically respond to calls. Rather, it routes calls via computer to responding agencies according to the nature and location of the call. Those agencies include defendants county sheriff and various city fire and police departments and the state police. When calls of a medical nature are received, the CLCC operator utilizes an emergency medical dispatch priority card system to ensure that appropriate questions are asked and to identify pre-arrival care information.

[3] This court has disapproved of the label "outrageous conduct," preferring the phrase "intentional infliction of severe emotional distress." *Patton v. J. C. Penney Co.*, 301 Or 117, 119 n 1, 719 P2d 854 (1986); *Humphers v. First Interstate Bank*, 298 Or 706, 709 n 1, 696 P2d 527 (1985). In this case, the appropriate labels are, respectively, "negligent infliction of severe emotional distress" and "reckless infliction of severe emotional distress."

trier of fact could find was more than negligent. The Court of Appeals affirmed.

Negligent Infliction of Severe Emotional Distress

Plaintiff first contends that the trial court erred in sustaining defendants' motions for summary judgment on her claim for negligent infliction of severe emotional distress.

This court has recognized common law liability for psychic injury alone in three situations. First, where the defendant intended to inflict severe emotional distress.... Second, where the defendant intended to do the painful act with knowledge that it will cause grave distress, when the defendant's position in relation to the plaintiff involves some responsibility aside from the tort itself.... Third, where the defendant's conduct infringed on some legally protected interest apart from causing the claimed distress, even when that conduct was only negligent.... However, in *Norwest v. Presbyterian Intercommunity Hosp.*, 293 Or 543, 558-59, 652 P2d 318 (1982), this court explained:

> This court has recognized common law liability for psychic injury alone when defendant's conduct was either intentional or equivalently reckless of another's feelings in a responsible relationship, or when it infringed some legally protected interest apart from causing the claimed distress, even when only negligently. * * * But we have not yet extended liability for ordinary negligence to solely psychic or emotional injury not accompanying any actual or threatened physical harm or any injury to another legally protected interest. (Footnotes and citations omitted.)

Further, the court stated that:

> ordinarily negligence as a legal source of liability gives rise only to an obligation to compensate the person immediately injured, not anyone who predictably suffers loss in consequence of that injury, unless liability for that person's consequential loss has a legal source besides its foreseeability.

Plaintiff asserts, however, that she,

> not her husband – chose to use this service. In a contractual sense, [she] accepted the offer of emergency medical services made by the 9-1-1 service. The failure of the 9-1-1 system directly affected her since it was she who contracted for the service, and it was 9-1-1 that promised [her] they would 'have someone there in just a couple of minutes * * *.'

Thus, she argues, she was the direct victim of defendants' negligence....

Plaintiff also argues that...she can recover for emotional distress as a direct victim if her distress was a foreseeable consequence of a negligent injury to her husband. Although foreseeability is a prerequisite to liability, plaintiff must first point to some "legally protected interest" of hers that defendants violated. She has not done so.

...[A]lthough emotional injury to a bystander may be foreseeable, the bystander's emotional well-being is not a legally protected interest unless liability for the injury "has a legal source besides its foreseeability." Here, plaintiff points to no legal source of liability for her emotional jury other than its foreseeability.... Viewing the record in the light most favorable to plaintiff, we conclude that plaintiff may not recover for defendants' alleged negligence, because she sustained no physical injury.

In the alternative, plaintiff asks us to abandon the impact rule...we decline plaintiff's invitation.

This court will not lightly overturn precedent, especially when the precedent has been followed for a long time....

Plaintiff has failed to demonstrate affirmatively that any of the...premises...justifies this court's reconsideration of the impact rule in this case. Under such circumstances, and in the absence of any other special considerations, we decline to undertake a reconsideration of that rule.

Reckless Infliction of Severe Emotional Distress

Plaintiff next contends that the trial court erred in sustaining defendants' motions for summary judgment on her claim for reckless infliction of severe emotional distress.

A plaintiff may recover for severe emotional distress without accompanying physical injury by showing recklessness or a reduced level of intent.... Such a responsibility generally constitutes a "special relationship" between the parties or the violation of a legally protected interest or known legal right.

Plaintiff argues that she and defendants were in a "special relationship," because defendants held themselves out as providers or insurers of the prompt delivery of emergency medical services and because of her old age and her emotional strain on finding her husband lying on the kitchen floor, and that defendants owed her an independent legal right or duty arising out of ORS 401.710 et seq. (providing for 9-1-1 emergency telephone systems). Therefore, she argues, she need only show that defendants acted "recklessly" to support her claim.

We need not determine whether plaintiff had a "special relationship" with defendants, or whether defendants owed plaintiff an independent legal right or duty arising out of ORS 401.710 et seq., because we conclude, as did the Court of Appeals, that plaintiff has not shown that defendants' conduct was anything more than negligent. . . .

In summary, plaintiff may not recover under a theory of negligent infliction of emotional distress, because she suffered no physical injury from defendants' alleged negligence. Plaintiff may not recover under a theory of reckless infliction of severe emotional distress, because she has not shown any misconduct on defendants' part that a trier of fact could find was anything more than negligent. Therefore, the trial court properly granted defendants' motions for summary judgment.

The decision of the Court of Appeals and the judgment of the circuit court are affirmed.

DISSENT:

UNIS, J., concurring in part and dissenting in part.

I agree with the majority's conclusion that the trial court properly granted defendant's motion for summary judgment on plaintiff's claim for reckless infliction of severe emotional distress. I also agree with the majority that *present* Oregon common law does not recognize negligent infliction of severe emotional distress as an independent cause of action. I disagree, however, with the majority's disposition of plaintiff's claim for negligent infliction of severe emotional distress. This court exercised its discretionary power to grant review in this case to determine whether to extend common law tort liability to protect Oregon citizens from negligently inflicted severe emotional distress, without any accompanying actual or threatened physical harm or any injury to another legally protected interest. . . .

A growing number of jurisdictions in the United States recognize and protect a person's right to be free from negligently inflicted severe emotional distress without requiring either physical injury or an independent underlying tort. . . . I believe that this court should join those jurisdictions and recognize negligent infliction of severe emotional distress as an independent cause of action. In my view, psychic well-being is entitled to as much legal protection as physical well-being. The right to be free from severe emotional distress should not depend on whether the distress was intentionally, recklessly, or negligently inflicted. Under present Oregon law, we allow a finder of fact to judge the validity of a claim for intentional infliction of severe emotional distress. The finder of fact does so by evaluating the quality and genuineness of proof and by relying on modern medical technology for proof of the cause, severity of the motional distress, and the ability of the court and jury to weed out dishonest claims. Yet, we do not allow a factfinder to make such a resolution if the severe emotional distress is negligently inflicted, unless the emotional injury is accompanied by an actual or threatened physical harm or an injury to another legally protected interest.

Fear of imaginary or false claims is often cited as a primary reason for the "physical injury" requirement. . . . Apart from some quite untenable notions of causal connection, the theory appears to be that the "bodily injury" affords the desired guarantee that the mental disturbance is genuine. . . . The distinction between physical and psychological injury, however, is artificial and arbitrary. One can imagine readily the harsh and arbitrary effects of the "physical injury" requirement. If an automobile careens off the road as a result of the negligence of its driver and by barely missing a party causes that party to suffer a severe psychological trauma, there would be no recovery under present Oregon law. If the side mirror of that same automobile brushed the party's arm, slightly scratching the skin, the party could then recover damages. Another stated reason for the "physical injury" requirement is the concern that allowing recovery for solely emotional injury would result in an avalanche of new litigation. There is a two-fold response to this argument. First, there is insufficient proof that such a result has occurred in jurisdictions that have abandoned the "physical injury" rule. Second, it is a fundamental concept of our judicial system that a judicial forum be available for vindication of citizens' rights....

The requirement that a plaintiff must prove that he or she suffered *severe* emotional distress provides a level of protection from imaginary or false claims. Another alternative to protect against unfounded claims would be to require the plaintiff to prove the element of severe emotional distress by *clear and convincing* evidence.

For these reasons, I respectfully dissent.

———————————————

QUESTIONS

1. Discuss whether the use of emergency medical dispatch in this case was correctly performed.
2. The Court "disapproved of the label 'outrageous conduct,' preferring the phrase 'intentional infliction of severe emotional distress.' In this case, the appropriate labels are, respectively, 'negligent infliction of severe emotional distress' and 'reckless infliction of severe emotional distress.'" Using these labels, how and why do you evaluate the dispatch agency's conduct?
3. The Court concludes that the actions of the dispatch agency in this matter were negligent, that is, they fell below the standard of what a reasonable person would have done in the same situation. Explain why this failure of performance should or should not create the basis for a lawsuit.
4. Discuss the implications of dispatch functions being performed by a private entity rather than a unit of government.
5. Evaluate the pros and cons of the "physical injury" rule discussed in the case and the dissent.
6. This case involves dispatch of EMS and law enforcement. Discuss how dispatch of other emergency response entities might be delayed.
7. People involved in emergency situations frequently feel emotional distress. Consider the following fact pattern:

Two police officers are first to respond to a call of a victim in distress following an armed robbery. Arriving on the scene, they find the victim, a 50-year-old 300-pound man, lying on the sidewalk, blue in the face, gasping for breath, and clutching his heart. His distraught wife is at his side, sobbing. The officers both pat her shoulders to reassure her. The victim's wife tells the officers, "The attacker showed us a knife and took my purse. My husband tried to run after him, then fell down gasping. He says it's his heart." "Don't worry ma'am. EMS is on the way," the woman officer reassures her. Her partner pokes the victim's distended belly, and says under his breath to the woman officer, "These fat pigs. What do they expect when they try to get physical? That heart attack is his own fault."

The wife overhears the comment, and is shaken by the idea that her husband, and not the wrongdoer, brought on his own heart attack.

Later, the victim and his wife sue the officers and their department, alleging both negligent infliction of severe emotional distress and reckless infliction of severe emotional distress, based on the male officer's remarks regarding the responsibility for the victim's heart attack.

Analyze whether the wife and the husband, individually, would succeed in their claims against each of the officers and department, and explain your conclusions.

Would your conclusions be different if the emergency responders involved had been EMTs engaged in providing medical treatment at the time one of them indicated the patient's responsibility for his heart attack? Explain.

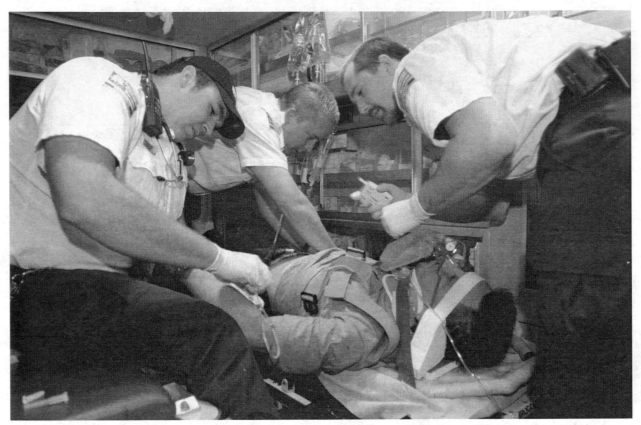
Working on a patient in the back of an ambulance. Jeff Forster photo.

Chapter 5

EMERGENCY MEDICAL SERVICES

A number of legal topics are of particular interest to EMS, including documentation, patient consent, "good Samaritan" acts, treatment of minor patients, and delayed response. The law on these subjects is in a state of evolution, as the following materials and cases demonstrate.

PART 1. DOCUMENTATION AND LIABILITY PROTECTION

William C. Nicholson

All emergency personnel, their supervisors, and attorneys should note that, while the following discussion is directed at EMS personnel, it applies equally to all other emergency services.

When members of the EMS community are asked why they entered the EMS field, the first reason that comes to mind is not typically a love of writing. Indeed, when polled on their least favorite aspect of EMS practice, the paperwork required of them is frequently the top objection. Generally, EMS practitioners shrug their shoulders and grumble at the "paper loving bureaucrats" at the management or state regulatory level. Then, unless trained otherwise, they do a minimal job on their run reports and other required documentation, eager to "get back to their REAL work."

In fact, however, far from being their enemy, documentation can prove to be EMS personnel's finest friend. The reasons for this statement spring from the laws of evidence that must be complied with in the trial of a lawsuit that might arise from allegations of wrongdoing by EMS.

In essence, a lawsuit may be viewed as a contest between two differing views of reality. The plaintiff alleges that EMS committed some wrong, which generally involves one of three categories:

1. failing to perform an indicated procedure;
2. improperly performing an indicated procedure; or
3. performing a procedure for which training was not received (or, in other words, beyond the EMS member's scope of practice).

The way to avoid the case going against one is to present sufficient evidence so that the other party fails to carry his or her burden of proof. In a civil case such as might result from allegations of failing in one of the three categories enumerated above, the burden of proof will be a preponderance of the evidence. This has been characterized as the "more likely than not" standard. In order to prevail under this standard, the plaintiff must show more weight on his/her side of the balance. The evidence that constitutes that weight may include many types of items: oral statements made before the Court under oath; physical things, like pieces of a broken vehicle; photographs of the incident scene; and written records.

The following discussion uses the Federal Rules of Evidence (FRE) as its referent point. State evidence codes parallel the Federal Rules, for the most part so closely that they use the same numbering system. The non-attorney or law student need not be concerned. Understanding of this subject matter by all affected persons is of such importance that it is presented in the plainest possible language.

The Hearsay Rule

Under FRE Rule 801, written or oral statements, other than ones made by a declarant while testifying at a trial or hearing, made to prove the truth of what the matter asserted are defined as "hearsay." Such statements are generally excluded from evidence under Rule 802. The reasoning behind the hearsay exclusion is the desire to prevent admission of evidence that might be made up by one side or another to support its view in a case. Some types of out of court statements, however, are regarded by the rules as more credible, and hence are admitted in evidence as exceptions to the hearsay rule.

Three exceptions to the hearsay rule are of particular import to EMS personnel involved in a trial:

The first, under Rule 803(2), involves statements of physical condition.

The second, under Rule 803(4), includes statements for purposes of medical diagnoses or treatment, and includes statements of medical history, past or present symptoms, pain or sensations, or the inception or general character of the cause or external source thereof insofar as reasonably pertinent to diagnoses or treatment.

The third, under Rule 803(6), entails records of regularly conducted activity. Included are any "memorandum, report, record, or data compilation, in any form, of acts, events, conditions, opinions, or diagnoses, made at or near the time by, or from information transmitted by, a person with knowledge, if kept in the course of a regularly conducted business activity, and if it was the regular practice of that business activity to make the memorandum, report, record, or data compilation.... "

The Rule 803(6) exception is the route by which EMS records such as run reports are admitted into evidence. The reason that such records are an exception to the hearsay rule is that the law regards them as likely to be true. Records kept in the regular course of business, as the reasoning goes, are probably accurate, since they are made for reasons other than in anticipation of trial. Hence, the run report may be viewed by the jury as the most accurate record of events.

The Jury's Memory

Studies of memory indicate that the principals of primacy and recency closely affect a person's memory. That is, one is most likely to remember the thing first seen and the thing last seen. The fact that the run report may be taken into the jury room may mean that it is the last version of the events at issue that the jury sees. Its version of events may therefore be the one most accessible to their minds.

Admission of these records into evidence may profoundly affect the outcome of a trial. One reason that these documents are so vital is that they are exhibits, which the jury may ask to examine in the sanctity of the jury room. The jury may, therefore, be examining the version of the events contained in the run report without the interference of opposing statements, which may fade in the jurors' memories.

Preparing the Run Report

As demonstrated above, the run report may well be the most important piece of evidence at a trial. The pro-active EMS manager and his/her attorney should therefore consider how to draft it in a way that will make the most impact in the event of potential litigation. One important thing to bear in mind is that the record will be more accurate when it is prepared as closely as possible to the events it describes. Ideally, a run report should be finalized immediately after a run, before the next dispatch. In no event should a run report be written any later than the end of the shift during which the run occurred.

The first thing to bear in mind is that the physical document says a great deal about the nature of the person who drafted it. Consider the following suggestions for the form of the run report:

1. The writing must be legible. This shows a concern for later readers.
2. The writing must be neat. This demonstrates that the writer is precise and careful.

3. All sections of the form must be filled in, using "N.A." or the equivalent for fields not utilized. This demonstrates that the writer is thorough.

The content of the run report is, of course, the meat of the matter. When drafting the report, think about the following suggestions for content:

1. Do not use abbreviations. They may result in a document that the lay person on the jury cannot understand.
2. Include everything that occurred during the run with regard to patient care. As instructors are wont to say, "If you didn't write it down, it didn't happen."
3. When getting instructions from medical control, document carefully what was said to medical control, precisely what instructions were received, and exactly how those instructions were carried out. In the event of later dispute, a plaintiff would generally prefer to sue a doctor rather than the less wealthy medic or EMT who was following the doctor's possibly erroneous instructions.
4. Take all vitals at the prescribed intervals and record them properly. Doing so shows that the patient was diligently observed to assure that his or her condition did not deteriorate.

5. Whenever there is any deviation whatever from standard operating procedures, protocols, or standing orders, explain in detail why the normal way of doing things was not followed. Such variations are presumptively the basis for liability. If not thoroughly explained in the run report, any later attempt at justification during a trial sounds like a lame excuse, regardless of whether that explanation is the truth.

While Rule 803(6) allows run reports to be included as evidence at a trial, such inclusion may be a two-edged sword for the sloppy or unwary. Rule 803(7) states that absence of an entry in a record kept pursuant to Rule 803(6) may be evidence to prove the "nonoccurence or nonexistence of the matter, if the matter was of a kind of which a...record...was regularly made." In other words, all those instructors were correct when they informed aspiring members of EMS that "If you didn't write it down, it didn't happen."

Taking steps to ensure proper documentation is the single best litigation preparedness step that an agency can take over and above acting correctly in every situation encountered by emergency responders.

PART 2. PATIENT CONSENT

As you read the following case, consider the policy decisions that underlie the rules for patient consent outlined therein. While the following case involves treatment by a physician in an emergency room, the rules of law are identical or virtually so for members of EMS in the field.

Shine v. Vega

Supreme Judicial Court of Massachusetts

709 N.E.2d 58 (1999)

OPINION:

MARSHALL, J.

In this wrongful death case, we must resolve the conflict between the right of a competent adult to refuse medical treatment and the interest of a physician in preserving life without fear of liability. In 1990, an invasive procedure, intubation,[3] was forcibly performed on Catherine Shine (Catherine), a life-long asthmatic in the

---

[3] Intubation is a procedure by which a tube is inserted through either the nose or the mouth into the windpipe. The tube enables oxygen to be delivered directly into the lungs, typically by means of a ventilator.

midst of a severe asthma attack. Dr. Jose Vega, an emergency physician at Massachusetts General Hospital (MGH), initiated the intubation without Catherine's consent and over her repeated and vigorous objections. In 1993, Dr. Ian Shine, Catherine's father and the administrator of her estate, brought a multi-count complaint against Dr. Vega and MGH seeking damages for tortious conduct and the wrongful death of his daughter.[5] He alleged that Catherine was traumatized by this painful experience, and that it led to her death two years later. On that occasion, Catherine again suffered a severe asthma attack but refused to go to a hospital because, it was claimed, she had developed an intense fear of hospitals. Her father alleged that Catherine's delay in seeking medical help was a substantial factor in causing her death.

At trial the defendants took the position that, confronted with a life-threatening emergency, Dr. Vega was not required to obtain consent for treatment from either Catherine or her family. A judge in the Superior Court agreed, and charged the jury that no patient has a right to refuse medical treatment in a life-threatening situation. She also instructed that in an emergency the physician need not obtain the consent of the patient or her family to proceed with invasive treatment. A jury returned verdicts for the defendants on all counts. Dr. Shine appeals from the judgment entered on the jury verdicts, and from the denial of his motion for judgment notwithstanding the verdict or a new trial. He contends that the trial judge incorrectly instructed the jury that (1) a patient's right to refuse medical treatment does not apply in an "emergency" medical situation; (2) it is not battery for a physician to treat a patient without obtaining consent if the treatment is necessary to prevent death or serious bodily harm; and (3) it is not false imprisonment forcibly to restrain a patient in a life-threatening situation. He also challenges the judge's ruling excluding certain notes Catherine made concerning her treatment at MGH. We transferred the case here on our own motion. We conclude that the instructions were erroneous, and that the errors were prejudicial. We vacate the judgment and remand the case to the Superior Court for a new trial.

I.

At approximately 7 A.M. on Sunday, March 18, 1990, twenty-nine-year-old Catherine Shine arrived at the MGH emergency room seeking medical help for an asthma attack. Catherine had been asthmatic throughout most of her life, a condition she controlled through prescription medication. The daughter of a physician, Catherine had educated herself about her condition and was well informed about her illness. Her asthmatic attacks were characterized by rapid onset, followed by a rapid remission. She had never required intubation in the past.

Earlier that morning, Catherine had suffered a severe asthma attack at her sister Anna's apartment. Despite believing that her condition was improving after using her prescription inhaler, Catherine agreed with Anna's suggestion to go to MGH, but on the condition that she be administered only oxygen. After Anna received assurances from an MGH representative that Catherine would be treated with just oxygen, Catherine entered the MGH emergency department, accompanied by Anna.

Catherine initially was given a nebulizer, a mask placed over her mouth which delivered oxygen and medication. She complained to Anna that the medication was giving her a headache, removed the mask and indicated that she wished to leave the hospital. Catherine's behavior alarmed the nurse who was treating her. An arterial blood

---

[5] Dr. Shine alleged negligence, assault and battery, false imprisonment, intentional infliction of emotional distress, wrongful death, violation of Catherine's civil rights, and violation of the Massachusetts Patient Bill of Rights Act and cognate Federal rights. The complaint also sought damages for tortious conduct against Catherine's sister, Anna Shine. Dr. Shine later voluntarily dismissed all counts of the complaint relating to Anna.

gas test, measuring the levels of oxygen and carbon dioxide in her blood, was drawn at approximately 7:15 A.M. The results, obtained at approximately 7:30 A.M., showed that Catherine was "very sick." Dr. Vega, the only emergency room attending physician on staff at MGH that morning, examined Catherine and concluded that she required intubation. Catherine resisted, and Dr. Vega initially agreed to try more conservative treatment with the oxygen mask. Catherine continued to disagree with the medical staff concerning her treatment.

Anna, frustrated by what she felt was a medical staff unwilling to listen to her sister, telephoned her father, Dr. Shine, who was in England. Dr. Shine had treated Catherine when she was a child and was familiar with Catherine's condition. Dr. Shine spoke to an MGH physician and told him that Catherine was intelligent and "very well-informed" about her illness, and he urged the physician to listen to Catherine and to try to obtain her consent for any treatment. Dr. Vega testified that he told Dr. Shine that Catherine was in "the midst of an extremely severe asthma attack," and that he unsuccessfully had tried to avoid intubation. Dr. Vega testified that Dr. Shine asked him to wait until he flew to Boston before intubating Catherine. He also testified that he had made a "conscious decision" not to tell Catherine that her father had opposed intubation.

Anna returned to Catherine's room to find her in "heated" argument with the MGH staff. Catherine's condition had improved somewhat, and she was able to talk and to breathe more easily. At approximately 7:40 A.M., during a moment when the doctors left Catherine and Anna alone together, Catherine told Anna to "run." They ran down the corridor to the emergency room exit doors, where they were forcibly apprehended by a physician and a security guard. Catherine was "walked back" to her room where Dr. Vega immediately ordered that she be placed in four-point restraints, in part because she had refused treatment and attempted to leave the emergency room.[7] Catherine and Anna were forcibly separated. Dr. Vega initiated the process of having Catherine intubated. At approximately 8:00 A.M. the results of a second blood gas test became available, showing that Catherine's condition had improved somewhat. Dr. Vega testified that the results, even if he had read them (he had not), would not have changed his decision to intubate Catherine. At approximately 8:25 A.M., the intubation procedure commenced, approximately forty-five minutes after Catherine had been strapped in four-point restraints. Catherine never consented to this treatment. Dr. Vega testified that he never discussed with Catherine the risks and benefits of intubation. Neither Anna, who was still in the hospital, nor Dr. Shine was asked to consent to the intubation. Catherine was released from MGH the following day.

Catherine's family testified that she was traumatized by these events. She had nightmares, cried constantly, and was unable to return to work for several months. For the first time in her life, they said, she was obsessed about her medication and what she ate. Catherine became suspicious of physicians, and repeatedly "swore" she would never go to a hospital again. In July, 1992, Catherine suffered another severe asthma attack while at home with her fiance and her brother. She did not want to go to a hospital. After she became unconscious, her brother called an ambulance. Despite two days of medical treatment at South Shore Hospital, she died.

II

Dr. Shine's central claim both at trial and on appeal is that Dr. Vega and MGH wrongfully restrained and intubated Catherine without her consent.[8] He sought to show that Catherine's

---

[7] Dr. Vega testified that Catherine's patient chart contained the reason for her forcible restraint: "Patient became more confused and combative, refusing treatment and suddenly ran down the hallway and nearly out of the [emergency ward] and brought back."
[8] Dr. Vega explained that in his judgment Catherine was too "confused" to give her assent and did not appreciate "the severity of her illness." He testified that he considered Catherine's "combative" behavior, refusal of treatment and attempt to flee as indications of

mental abilities at the relevant times were not impaired, and that she was not facing a life-threatening emergency when she was restrained and intubated.[9] The defense took the position that Dr. Vega was confronted with a life-threatening emergency, and Catherine's consent was not necessary.'[10] On appeal they argue that a medical emergency operates as a limitation on the "abstract right" of a patient to refuse treatment, and that in this situation a doctor may override a patient's right to refuse treatment."[11]

The judge instructed the jury that "under Massachusetts law a patient has the right to refuse medical treatment except in an emergency, life-threatening situation" (emphasis added). It was therefore up to the jury, she said, to determine whether or not such a situation existed. She repeated that instruction, emphasizing "once again" that "the law in Massachusetts is that a patient has the right to refuse medical treatment except in an emergency, life-threatening situation." She told the jury that this was not a case of "informed consent" because Catherine's right to refuse to be intubated was "not an absolute right." It is a right constrained by "the right of the state or the obligation of the state to preserve the lives of its citizens . . . a right that exists in an emergency room setting to perform treatment without the consent of the patient," she charged.

The judge repeated this several times. On the element of negligence, she instructed that "if there is a life-threatening circumstance, then the hospital, its employees, and Doctor Vega have the right to treat Catherine Shine without getting her consent or anybody else's consent, whatever her condition. . . . In other words," she said, "a physician who . . . has reason to believe that the failure to conduct a procedure such as intubation would create a likelihood of serious harm to the patient by reason of a life-threatening situation may perform that procedure without the consent of the patient."

As to the assault and battery count, she instructed "that a doctor and/or a hospital does not commit an assault and battery when they treat a patient without her consent if the treatment is necessary to save her life or to prevent serious bodily harm." On the charge of false imprisonment, the judge instructed that medical personnel may confine a patient without her consent, "if there is reason to believe that a person in an emergency room is suffering from a life-threatening situation." She repeated that "it is lawful for the hospital and for Doctor Vega to have confined Catherine Shine if she is experiencing a life-threatening emergency. That is justified."[12]

---

her increasingly confused mental state. The plaintiff offered compelling evidence that Catherine was not incapable of giving her consent and, even if she had been, her family was readily available for consultation.

[9] Expert witnesses for Dr. Shine testified that the intubation procedure was not an appropriate treatment for Catherine, that MGH medical staff failed properly to evaluate Catherine's competency to consent to treatment, and that failure to comply with unwanted treatment does not necessarily indicate lack of competence. The plaintiff's experts also testified that the situation was not an emergency, that Catherine was able to make rational decisions, and that intubation should be used only if absolutely necessary because the patient may develop fear of future intubation. There was also expert testimony that Catherine's treatment at MGH was below the appropriate standard of care because no determination of her competence was made, and that, if she was incompetent, the treating physician should have but did not seek consent from her family.

[10] Several experts testified on behalf of the defendants that the actions of the MGH staff were appropriate, that if Catherine had been given only oxygen, as she requested, she likely would have died, and that Catherine's treatment at MGH was not the type of experience that could produce posttraumatic stress disorder.

[11] In a nonemergency setting, the right of an incompetent patient to consent to or to refuse medical treatment is protected by a judicial, "substituted-judgment" proceeding. *Rogers v. Commissioner of the Dept of Mental Health*, 390 Mass. 489, 504, 458 N.E.2d 308 (1983). The medical best interest of the patient is not the touchstone of a substituted judgment decision. Rather, the determination is "'that which would be made by the incompetent person, if that person were competent' . . . and giving 'the fullest possible expression to the character and circumstances of that individual." Id. at 500, quoting *Superintendent of Belchertown State Sch. v. Saikewicz*, 373 Mass. 728, 747, 752-753, 370 N.E.2d 417 (1977).

[12] The judge further instructed: "[A] doctor, a hospital and its employees are permitted a wide range in the exercise of their professional judgment concerning the treatment to be given a patient as long as the exercise of that professional judgment is in accordance with the duty of care as I have described it to you. Once again, ordinarily a physician must obtain the consent of a patient before treatment. However, in an emergency room situation, a physician may undertake treatment provided in that what he does is within the customary practice of physicians practicing his specialty in similar circumstances; and I have defined an emergency for you as a life-threatening situation or something akin to that."

A

The defendants first argue that Dr. Shine did not adequately preserve his challenge to the jury instructions on the emergency exception to tort liability because he failed to convey to the judge the definition of an "emergency" he espouses on appeal — that there must be a life-threatening situation and the patient must be unconscious or otherwise incapable of giving her consent. We conclude that the issue was not waived: the question whether Catherine's consent was required before intubation was a live issue throughout trial, and was properly preserved. The plaintiff objected to the judge's refusal to instruct the jury on Catherine's competence, and he objected to the defendants' instructions on the emergency exception espoused by the defendants and adopted by the judge. Moreover, the defendants had the burden of proving that an exception relieved them of tort liability....

B

In *Norwood Hosp. v. Munoz,* 409 Mass. 116, 121, 564 N.E.2d 1017 (1991), we considered in what circumstances a "competent individual may refuse medical treatment which is necessary to save that individual's life." We described in that case both the common law and constitutional bases for our recognition of the "right of a competent individual to refuse medical treatment." Id. at 122.... We recognized the "every competent adult has right 'to [forgo] treatment, or even cure, if it entails what for [her] are intolerable consequences or risks however unwise [her] sense of values may be in the eyes of the medical profession." *Harnish v. Children's Hosp. Medical Ctr.,* 387 Mass. at 154, quoting *Wilkinson v. Vesey,* 110 R.I. 606, 624, 295 A.2d 676 (1972).

In *Norwood Hosp. v. Munoz,* 409 Mass at 122-123, we also described how the "right to bodily integrity" had developed through the doctrine of informed consent .... Under that doctrine, "a physician has the duty to disclose to a competent adult 'sufficient information to enable the patient to make an informed judgment whether to give or withhold consent to a medical or surgical procedure." *Norwood Hosp. v. Munoz,* 409 Mass. at 123....

Dr. Vega and MGH concede that Catherine exercised her right to refuse medical treatment and never consented to intubation. But, they argue, Dr. Vega could override Catherine's wishes as long as he acted "appropriately and consistent with the standard of accepted medical practice" and "to save and preserve her life in an emergency situation." It was not necessary, they argue, to instruct the jury on a competent patient's right to refuse medical treatment because it was, in their words, "largely irrelevant" to the critical liability question — whether Catherine faced a life-threatening situation.

The emergency exception to the informed consent doctrine has been widely recognized and its component elements broadly described.... In *Canterbury v. Spence,* 150 U.S. App. D.C. 263, 464 F.2d 772 (D.C. Cir. 1972), a seminal case, the court explained that the emergency exception.

> Comes into play when the patient is unconscious or otherwise incapable of consenting, and harm from a failure to treat is imminent and outweighs any harm threatened by the proposed treatment. When a genuine emergency of that sort arises, it is settled that the impracticality of conferring with the patient dispenses with need for it. Even in situations of that character the physician should, as current law requires, attempt to secure a relative's consent if possible. But if time is too short to accommodate discussion, obviously the physician should proceed with the treatment. 464 F.2d at 788-789.

See Restatement (Second) of Torts § 892D (a) (1979) ("emergency makes it necessary or apparently necessary, in order to prevent harm to the other, to act before there is opportunity to obtain consent from the other or one empowered to consent for him"). Consistent with other courts that have considered the issue, we recognize that the emergency-treatment exception cannot entirely subsume a patient's fundamental right to refuse medical treatment. The privilege does not and cannot override the refusal of treatment by a patient who is capable of providing consent. If the patient is competent, an emergency physician must obtain her consent before providing treatment, even if the physician is persuaded that, without the treatment, the patient's life is threatened. If the patient's consent cannot be obtained because the patient is unconscious or otherwise incapable of consenting, the emergency physician should seek the consent of a family member if time and circumstances permit.... We are aware of no other court that has sanctioned the sweeping emergency privilege the defendants advocated here.

In the often chaotic setting of an emergency room, physicians and medical staff frequently must make split-second, life-saving decisions. Emergency medical personnel may not have the time necessary to obtain the consent of a family member when a patient is incapable of consenting without jeopardizing the well-being of the patient.

But a competent patient's refusal to consent to medical treatment cannot be overridden because the patient faces a life-threatening situation.[19]

To determine whether an "emergency" existed sufficient to insulate Dr. Vega and MGH from all tort liability, the jury should have been required to decide whether Catherine was capable of consenting to treatment,[20] and, if not, whether the consent of a family member could have been obtained. It is up to the jury to determine whether the treating physician took sufficient steps, given all of the circumstances, to obtain either the patient's informed consent, or the consent of a family member.... In this case the judge's charge foreclosed the jury from making those necessary determinations.... On this record, there is no basis on which to conclude that the error was not prejudicial. A new trial is required....

The judgment is vacated and the case is remanded to the Superior Court for a new trial.. So ordered.

QUESTIONS

1. Do you think that Dr. Vega's actions in this case were reasonable? Why or why not?
2. This case considers the tension between the need for informed consent and the emergency exception to that requirement. When do the requirements of an emergency situation overcome a patient's refusal to treatment?
3. Under what circumstances might the patient's wishes regarding treatment reasonably be ignored or disobeyed?
4. If you were advising an EMS provider, what standards would you suggest for consent in the field?

PART 3. TREATMENT OF MINOR PATIENTS*

Patient consent is perhaps the thorniest legal issue to confront EMS on a daily basis. Like many of the issues which caregivers face, consent affects all EMS services, whether they serve rural or urban environments. In most states, a minor under eighteen (18) years of age may generally

---

[19] In *Norwood Hosp. v. Munoz, supra* at 127, we said that the State's interest in maintaining the ethical integrity of the profession does not outweigh the patient's right to refuse unwanted medical treatment: "The ethical integrity of the profession is not threatened by allowing competent patients to decide for themselves whether a particular medical treatment is in their best interests." Id., citing *Matter of Conroy, supra* at 352.

[20] A physician, and a jury, may reasonably take into account a patient's refusal to consent to lifesaving medical treatment in determining whether the patient is competent to consent to or refuse treatment, but this factor is not dispositive. See *Lane v. Candura,* 6 Mass. App. Ct. 377, 383, 376 N.E.2d 1232 (1978) (patient's refusal to consent to life-saving amputation in itself was not sufficient to render her legally incompetent for purposes of appointing guardian).

 * This material is adapted from: William C. Nicholson, *To Transport or not to Transport: Treatment of Seemingly Healthy Minor-Age Patients.* OUR WATCH p. 7 (May–June 1996).

not legally give or withhold consent. The proper course of action is to obtain consent or release from the minor's parent or guardian.

Extensive studies establish that pre-hospital claims fall into three categories:

1. vehicle accidents: 50% of claims;
2. failure to transport/refusal to transport: 25% of claims;
3. all other claims (including malpractice): 25% of claims.

See, e.g., Soler, et al., "The Ten Year Malpractice Experience of a Large Urban EMS System," *Annals of Emergency Medicine 14*: 982, 985 (1985); Goldberg, et al., "A Review of Prehospital Care Litigation in a Large Metropolitan EMS System," *Annals of Emergency Medicine, 79*, 557, 558 (1990). The failure to transport/refusal to transport claims may all be fairly categorized as arising from consent issues.

Lawsuits arising from minor consent issues arise frequently. For example, a car full of 16- and 17-year-olds is in a fender-bender, and there is no detectable harm to any of the patients. How can the EMS caregiver end its responsibility for the minors without incurring possible liability for failure to treat? This subject is covered by state law, such as Indiana Code 16-36-1-3 through 5, which mandates consent by the minor's parent or guardian.

What if the minor's parent or guardian can not be found? Can the crew just do a "check out" and release the patient? This situation brings to the fore EMS scope of practice, which has three (3) elements: assessment, stabilization/treatment, and transportation. Not included is diagnosis – and a diagnosis is what the crew is doing if one of its members "checks out" a patient. Diagnosis is generally reserved for licensed physicians. If the EMS crew member looks at the patient and says, "He or she is OK," the EMS person has just performed a diagnosis and violated the law in most states by performing beyond the scope of training. Such activity also exposes the medical director, under whose license EMS personnel practice when performing treatment, to possible revocation of his or her license.

The caregiver may try to persuade law enforcement to act *in loco parentis*, and take the responsibility for refusal of treatment. If this approach doesn't work, the EMS caregiver should try to keep the minors on the scene or transport them until the appropriate release is obtained. Sometimes minors may refuse to stay at the scene or to be transported. This is a difficult situation for the EMS caregiver, an entity without authority to hold or transport persons against their will. The assistance of law enforcement should be sought in this situation. Always remember to fully document the run, particularly including any patient departures against the advice of the EMS caregiver. Transporting apparently uninjured minors may cause bad feelings on the part of the minors and their parents. Fees are incurred that would not be incurred if there were a proper release. But legally, the caregiver has no other option, because the EMS caregiver's scope of practice does not include diagnosis, which is required to find that the minors are uninjured.

If an injury is life-threatening, implied consent may apply. The law assumes that in an emergency, a person would consent to having his/her life saved. Remember this when confronted with an intoxicated person, whether minor or adult, who refuses to consent to assistance. State law, such as Indiana Code 16-36-1, addresses consent issues.

Some EMS caregivers obtain refusals of treatment from parents or guardians over the telephone. This is a very dangerous practice. A well-documented written release provides the best protection against a later lawsuit.

The refusal must be express and informed and must be made by the proper person, just as consent to treatment must be. All witnesses to the refusal should sign the refusal form. Planning response to issues like minor transport is one of the EMS provider's heaviest responsibilities. The EMS provider, in consultation with legal counsel and medical control, must anticipate potential legal problems, establish strong protocols, and insist on thorough documentation. Taking these steps will help insure against a lawsuit due to a botched minor refusal of treatment.

QUESTIONS

1. If you were advising an EMS provider, what standards would you suggest for minor consent in the field?
2. How should standards for minor consent differ, if at all, from those for adult consent?
3. Describe an ideal minor consent regulation for an EMS regulatory entity. How would such a standard differ from that of an individual provider?

PART 4. GOOD SAMARITAN ACTS

Various states have enacted statutes or regulations to protect persons not otherwise obligated who come to the aid of their fellow citizens. Before beginning this case, consider your current understanding of how "Good Samaritan" acts work. As you read, think about the differences between interpreting a statute like that at issue here and relying solely on case law as precedent. You may wish to look ahead in the materials on Using Volunteer Resources in Chapter 8 to consider how they relate to "Good Samaritan" acts.

Evelyn M. Burks v. Lars T. Granholm

C.A. No. 99C-06-012, LEXIS 364
(Del. Super. 2000)

JUDGES: T. HENLEY GRAVES, JUDGE

OPINION:

This Letter Opinion constitutes the Court's decision on the Plaintiff's Motion in Limine. Based on the affirmative defenses the Defendant raises in his Answer, the Plaintiff asks the Court to exclude any argument or evidence of the Defendant's status as a paramedic and any protections he may gain from 16 Del. C. § 9813. She also seeks to exclude any argument or evidence of an unavoidable accident or sudden emergency. For the reasons set forth below, the Court GRANTS the Motion in part and DENIES the Motion in part.

STATEMENT OF FACTS

On June 21, 1997, Evelyn M. Burks ("Plaintiff") was driving on County Road 382 in Sussex County, Delaware. Lars T. Granholm ("Defendant") was traveling on the same roadway and was in front of the Plaintiff. The Plaintiff and Defendant were separated by two vehicles. The group of four cars approached a stop sign. In turn, each stopped and continued on Road 382. Just beyond the stop sign, the Defendant noticed another vehicle stopped along the opposite shoulder with the driver slumped over the steering wheel.

The Defendant, a paramedic, suspected that the driver of that vehicle could be in distress and decided to offer assistance. To do so, he pulled off onto the right shoulder so that he could make a U-turn and return to the other vehicle. It appears from the record before the Court that the Defendant activated his left turn signal to indicate his intentions. However, while he waited for the other two vehicles to pass him safely, the Defendant failed to see the Plaintiff's vehicle and began his U-turn. The Defendant struck the side of the Plaintiff's vehicle.

After impact, the Defendant pulled to the side. He asked the passenger riding with him to check on the Plaintiff while he went to see if the driver on the opposite side of the road did, indeed, need assistance. That car, however, had driven off by this time and apparently needed no assistance.

DISCUSSION

Applicability of 16 Del. C. § 9813. The Plaintiff asks this Court to exclude any argument or evidence that the Defendant, as a paramedic, is protected from liability for mere negligence by 16 Del. C. § 9813. This Statute provides that:

No paramedic who in good faith attempts to render or facilitate emergency medical care authorized by this chapter shall be liable for any civil damages which occur as a result of any act or omission of the paramedic in the rendering of such care; unless such paramedic is guilty of wilful and wanton, reckless or grossly negligent conduct.

16 Del. C. § 9813(b).

The issue here is whether an off-duty paramedic is protected from liability for negligent acts that may occur prior to, or in preparation for, rendering emergency medical care. More specifically, does the act of turning a vehicle around so that he may offer aid constitute an "attempt to render or facilitate emergency medical care" as contemplated by the statute? For the reasons stated below, this Court finds that the statute is not so broad as to cover acts merely preparatory to the physical act of administering medical care to the injured.

This is a case of statutory construction and an issue of first impression as no other Court has issued an opinion interpreting the coverage of this statute.

The goal of statutory construction is to determine and give effect to legislative intent. If a statute is unambiguous, there is no need for judicial interpretation, and the plain meaning of the statutory language controls. If a statute is ambiguous, it should be construed in a way that will promote its apparent purpose and harmonize with other statutes.

Eliason v. Englehart, Del. Supr., 733 A.2d 944, 946 (1999) (citations omitted).

The Court believes that the "plain meaning" of the statutory language requires a finding that the General Assembly intended to protect only those acts involved in the actual rendering of emergency medical care and that only a strained interpretation of the language would broaden the scope to encompass acts of preparation. Nevertheless, even if the Court were to find the statute was ambiguous on its face as to its scope, the same result would be reached by identifying its apparent purpose and harmonizing it with other statutes.

Here, the statute's primary purpose is to protect a paramedic from liability for mere negligence in rendering emergency medical care. The only statement of the General Assembly's intent in enacting this statute is found in the Synopsis to House Amendment Two to Senate Substitute One to Senate Bill Number One, 135th General Assembly.

This amendment, as revised, does the following:

....7. Changes the measure of liability from "reasonable care and ordinary diligence" to "willful and wanton, reckless or grossly negligent conduct" in order to make this Act consistent with provisions governing immunity for rendering emergency care (16 Del. C. § 6801) and the Tort Claim Act (10 Del. C., Ch. 40) which utilize the higher standard.

This synopsis states the legislature's intent that persons, now specifically including paramedics, rendering emergency medical care will be liable for injuries they cause only if their acts were wilful or wanton, reckless, or grossly negligent. This synopsis does not, however, speak to the scope of activities falling within the umbrella of "rendering emergency medical care."

It is in harmonizing this statute with others that the Court is convinced that the statute, as written, does not cover the act of making a U-turn to go to the scene of a perceived emergency. First, the statute covers "emergency medical care authorized by this chapter." 16 Del. C. § 9813(b) (emphasis added). By statute, a paramedic may provide those services as set forth in the paramedic's certificate if they are provided under the supervision of a physician, direct or by radio or telephone. Where radio or telephone contact with a physician is not possible, the paramedic may provide certain care when the life of the patient is in immediate danger. See 16 Del. C. § 9807. This gist of this statute is to require that paramedics render emergency medical care under the supervision of a physician unless that is impossible and the life of the patient is in immediate danger. The supervision of a physician would not be needed in traveling to the scene of a medical emergency and thus indicates that the General Assembly intended "emergency medical care" to mean the actual act of providing care and not merely preparatory acts.

Delaware Code, Title 21, Motor Vehicles, also provides some guidance in resolving this issue. Under 21 Del. C. § 4106, authorized emergency vehicles are allowed to violate certain "rules of the road" in responding to emergencies. Moreover, the driver is not liable for damage or injury unless he or she acts in a wilful or wanton or grossly negligent manner. These protections only arise, however, where there is an "authorized emergency vehicle" and such vehicle is making use of audible or visual signals. See 21 Del. C. § 4106(c). This statute allows emergency personnel, in responding to emergencies, to violate certain rules of the road in order to meet the exigencies of the circumstances. Moreover, the statute sets forth the circumstances under which the motor vehicle laws may be ignored. Finally, even if the emergency vehicle is privileged to avoid the traffic laws, it must do so in a safe manner. See 21 Del. C. § 4106(b).

In the present case, the Defendant was in his personal vehicle and, thus, not driving an "authorized emergency vehicle." He also was not operating "audible or visual signals meeting the requirements of this title." For these reasons, he would not fall within the protections provided by 21 Del. C. § 4106. Because even ambulances responding to emergencies must drive in a safe manner while violating the normal rules of the road, this Court finds that interpreting 16 Del. C. § 9813 in a manner so that it covers the Defendant's use of a motor vehicle in this case would be in derogation of the motor vehicle laws.

This Court cannot adopt the strained interpretation of this statute that the Defendant proposes. Under 16 Del. C. § 9813, a paramedic will be protected from liability for simple negligence in providing the emergency medical care authorized under the chapter (i.e. under the supervision of a physician unless exempted). A paramedic will not, however, be able to use the statute as a shield against liability for injuries caused in the operation of a private motor vehicle while responding to a perceived emergency. For these reasons, the Court finds, as a matter of law, that the statute would not apply to this Defendant and GRANTS the Plaintiff's Motion to exclude any argument or evidence of protection under 16 Del. C. § 9813.

The "Emergency Doctrine." The Plaintiff asks this Court to exclude any argument for, or evidence of, the applicability of the "emergency doctrine." In Delaware, when a person, without negligence on her part, without time for reflection, is confronted with an emergency situation, the law does not hold her to as a high a standard of care as it would if she had more time to consider what to do to avoid the emergency. If in the exigency of the moment the defendant chose the course of action which, from hindsight, proved not to be the best means of avoiding the danger, defendant cannot be charged with negligence if defendant acted as a reasonably prudent person would under the same circumstances.

Basher-Lee v. Bradley, Del. Supr., 577 A.2d 751 (1990) ORDER.

The Court is skeptical, based on the facts before it, that the emergency doctrine would apply to the Defendant's actions. First, this does not appear to be the type of "emergency" the doctrine addresses. Moreover, there is some evidence that the Defendant had ample opportunity to reflect on his choices because he let other cars pass safely before attempting the U-turn. The Court, however, is not prepared to rule at this point that the doctrine will not apply. To that end, the Plaintiff's motion on this issue is DENIED and the Court reserves judgment until the close of evidence on whether an instruction on this issue is appropriate.

CONCLUSION

While the Defendant's sense of duty and compassion is certainly commendable. this Court finds that the Defendant's act of making a U-turn in this case is not covered by 16 Del. C. § 9813. The Court also finds that, despite skepticism about its applicability, it is not appropriate at this stage to rule as a matter or law that the "emergency doctrine" will not apply to this case. Considering the foregoing, the Plaintiff's Motion in Limine is GRANTED in part and DENIED in part.

IT IS SO ORDERED.

T. Henley Graves

_____

QUESTIONS

1. Why are Paramedics protected from negligence liability when performing their duties?
2. Was Paramedic Granholm, the Defendant in this case, in good faith attempting to render or facilitate emergency medical care when he u-turned in his private vehicle to render aid? Why did the Court reject this view?
3. What is the emergency doctrine, and why did it not apply in this case?
4. In what way is a Paramedic who happens upon an accident on a highway like an untrained Good Samaritan in the eyes of the law? In what ways is he or she different from an average person coming upon an accident scene? Of what consequence are the differences?

Clarken v. United States of America

United States District Court for the District of New Jersey

791 F. Supp. 1029 (1992)

BISSELL, District Judge

This matter arises before the Court pursuant to the defendant's motion in limine to determine whether the plaintiff must prove gross negligence.

FACTS AND BACKGROUND

Plaintiff Nancy Clarken seeks damages for the death of her husband allegedly as a result of the negligence of emergency medical technicians ("EMTs") employed by a government-owned hospital. It is presently scheduled for trial on January 14, 1992.

On March 1, 1987, Matthew Clarken suffered a cardiac arrest while a guest at the Thayer Hotel, located on the campus of the United States Military Academy at West Point, New York. The hotel staff notified Keller Army Community Hospital, and an ambulance was dispatched. The ambulance was staffed by two U.S. Army medics, Privates Carlos Smith and John Stratiff. Defendant alleges that these two medics have training equivalent to that of a New York State basic emergency medical technician.

Mr. Clarken was initially conscious when the medics arrived, but thereafter lost consciousness. He was eventually resuscitated at the Keller Army Hospital, but suffered severe brain damage causing him to remain in a vegetative state until his death on January 6, 1988. The plaintiff has offered six theories to support her claim of negligence:

1. Neither Stratiff nor Smith attempted to take a blood pressure reading after decedent became unconscious.
2. No cardiopulmonary resuscitation ["CPR"] was begun until decedent was placed in the ambulance and only after it was ordered by a physician via radio communication between the ambulance and the hospital – approximately six minutes after the need for it clearly was apparent.
3. Decedent vomited but there was no suctioning of the oral cavity nor any placement of any oral airway.
4. The medical personnel deviated from the acceptable standard of care with tragic consequences. These deviations included:
   a. Failure to take blood pressure;
   b. A failure to secure the airway by inserting an oral airway;
   c. A failure to provide supplemental oxygen in a timely manner;
   d. A failure to provide supplemental oxygen by the appropriate method, i.e., bag-valve-mask;
   e. A failure to initiate [CPR] in a timely fashion; and
   f. A failure to administer proper CPR.
5. Neither decedent's medical history nor the illness which he suffered on March 1, 1987,

would have precluded Mr. Clarken from continuing his employment until retirement, had he been properly cared for on March 1, 1987.
6. Neither decedent nor plaintiff contributed to the injuries and losses herein by any improper action or failure to act.

(Pretrial Order at 5-6 (Plaintiff's Contested Allegations))

The defendant presently moves for an order determining whether plaintiff has to prove gross negligence, rather than ordinary negligence, in order to recover. The basis of this request is defendant's position that under the Federal Tort Claims Act, courts apply the law of the place where the allegedly negligent act occurred, which is New York. Defendant further argues that under New York law, specifically, the Good Samaritan Law found at Section 3013 of the New York Public Health Law, plaintiff cannot recover against voluntary ambulance personnel unless she proves gross negligence.

The plaintiff concedes that New York law applies to this action. However, plaintiff argues that the defendants are not "voluntary ambulance personnel" because they were under a preexisting duty to render emergency service and are therefore not within New York's Good Samaritan Law.

II. DISCUSSION

A. The Parties' Arguments

Section 3013 of the New York Public Health Law provides, in relevant part:

1. Notwithstanding any inconsistent provision of any general, special or local law, a voluntary ambulance service described in subdivision three of section three thousand one of this article and any member thereof who is an emergency medical technician or an advanced emergency medical technician and who voluntarily and without the expectation of monetary compensation renders medical assistance in an emergency to a person who is unconscious, ill or

injured shall not be liable for damages for the death of such person alleged to have occurred by reason of an act or omission in the rendering of such medical assistance in an emergency unless it is established that such injuries were or such death was caused by gross negligence on the part of such emergency medical technicians or advanced emergency medical technician.
4. A certified emergency medical technician or advanced emergency medical technician, whether or not he is acting on behalf of an ambulance service, who voluntarily and without the expectation of monetary compensation renders medical assistance

in an emergency to a person who is unconscious, ill or injured shall not be liable for damages alleged to have been sustained by such person or for damages for the death of such person alleged to have occurred by reason of an act or omission in the rendering of such medical assistance in an emergency unless it is established that such injuries were or such death was caused by gross negligence on the part of such emergency medical technician or advanced emergency medical technician.

(Public Health Law, § 3013(1)). The relevant definitions are as follows:

3. "Voluntary ambulance service" means an ambulance service (i) operating not for pecuniary profit or financial gain, and (ii) no part of the assets or income of which is distributable to, or enures to the benefit of, its members, directors or officers except to the extent permitted under this article.
  * * *
6. "Emergency medical technician" means an individual who meets the minimum requirements established by regulations pursuant to section three thousand two and who is responsible for administration or supervision of initial emergency medical assistance and handling and transportation of sick, disabled or injured persons.

(Public Health Law, § 3001(3), (6)).

The United States argues that these provisions shield the medics from liability except to the extent that plaintiff proves gross, rather than ordinary, negligence. The medics were on duty for the Army, and received Army pay. (Defendant's Br. at 4). The medics were actually hospital orderlies, who manned ambulances whenever use of the ambulance was necessary. (Id. at 1). The U.S. Army did not charge the plaintiff for emergency medical services, does not earn a profit from such services, and does not offer such services to the public in general except on an emergency basis. (Id. at 4). "None of the assets or revenue of Army medical facilities enures to the benefit of any of the U.S. Army's personnel." (Id.)

In addition, the United States argues that the fact that the medics receive ordinary Army salary does not take this ambulance service out of the Good Samaritan Law. (Id. at 7-8). In particular, "they are paid for fulfilling various medical functions within the military medical service, principally those of hospital orderlies, and only incidentally do they provide emergency ambulance service to civilians." (Id. at 8). Finally, the United States argues that the policy underlying the Good Samaritan Law would be served by protecting these defendants from liability for ordinary negligence:

The Army's medical services are provided for the military community at West Point, and are not offered to the general public. Defendant is protected from liability when providing medical services to military personnel.... The military has no duty to care for civilians who become ill while visiting military facilities.... The purpose of New York's Good Samaritan Law is to encourage volunteers to provide emergency medical services for the public's benefit. If defendant is to be encouraged to open its facilities to civilians defendant should be protected in the same manner as any other community's voluntary service is protected.

(Id. at 8-9). Thus, the United States argues that the plaintiff must prove gross negligence, rather than ordinary negligence, in order to recover.

In opposition, the plaintiff argues that the Good Samaritan Law does not protect these medics because the statute does not protect those who have a pre-existing duty to act. (Plaintiff's Br. at 6). Plaintiff argues that "it was in the normal course of the duties of Specialists Stratiff and Smith to render emergency assistance to individuals injured or ill within the jurisdiction of Keller Army Hospital." (Id. at 8).

Plaintiff relies on several cases in support of her position that an EMT with a pre-existing duty to act is not within the Good Samaritan statute. First, plaintiff contends that the present matter is not within the intent of the legislature in enacting the Good Samaritan statute, relying on *Rodriguez v. New York City Health and Hospital Corporation*, 505 N.Y.S.2d 345 (N.Y. Sup. Ct. 1986). In *Rodriguez*, plaintiff brought suit against the treating hospital and a doctor for the death of her husband. The doctor moved for summary judgment on the basis that he never treated the deceased and that there was never any doctor-patient relationship between them. (Id. at 346). The court granted the motion on the basis of the Good Samaritan statute, since the record revealed that the doctor lived in the same building in which the deceased was the superintendent.... Since there was no evidence of gross negligence, the court granted the motion. (Id.)

Plaintiff herein distinguishes *Rodriguez* on the basis that Stratiff and Smith had a pre-existing duty to render aid, and did not act voluntarily but were under an obligation to perform that duty. (Plaintiff's Br. at 5-6). In further support of her position, plaintiff relies on various out-of-state cases which recognize that the Good Samaritan statute does not protect those who are under a pre-existing duty to act. One such case is *Henry v. Barfield*, 367 S.E.2d 289 (Ga. App. 1988), in which the Georgia court described Georgia's Good Samaritan statute:

> The basic premise of the statute is "to induce voluntary rescue by removing fear of potential liability which acts as an impediment to such rescue." Thus, they are directed at persons who are not under some pre-existing duty to rescue. If the doctor had a particular employment duty to aid the patient at the hospital or had a pre-existing doctor-patient relationship to the patient he aided, then he had a duty to the patient to begin with; and in such case he does not need a special inducement to offer aid, the aid he offers is not 'voluntary' in the sense of a Good Samaritan, and public policy would be ill-served if he were relived of the usual physician's duty of care and given immunity in such a case. Good Samaritan statutes are directed at persons, including physicians, who by chance and on an irregular basis come upon or are called upon to render emergency care....

This Court finds that these cases such as *Henry* are persuasive, even though the Good Samaritan statute may not be identical, on the question of preexisting duty. This Court agrees that the policy to be served by enacting a Good Samaritan statute is to encourage the voluntary provision of services to those in need....

This does not render the defense of the Good Samaritan statute invalid in the present matter,

however. The question remains as to what sort of "preexisting duties" remove the defendants from the protection of the statute. The plaintiff argues that so long as the medics owed a duty to someone, in this case, the Army, then their conduct is outside the protection of the statute. In this Court's view, however, the duty exempting the actors from statutory protection must flow to the injured. In other words, Stratiff and Smith cannot claim Good Samaritan protection if they owed a duty to Mr. Clarken to come to his aid.

In *Bunting v. United States*, 884 F.2d 1143 (9th Cir. 1989), the Ninth Circuit was faced with the question of whether a Coast Guard helicopter crew and doctor were within the protections of Alaska's Good Samaritan statute. The court noted that under 28 U.S.C. § 2674, the United States is liable for Federal Tort Claims Act claims to the extent that private individuals are liable under the same circumstances. (Id. at 1145)....

[T]he Good Samaritan statute had been amended in Alaska, changing the qualifying actor from "a person who, without expecting compensation, renders care" to "a person at a hospital or any other location, who renders emergency care." (Id. at 1146). Thus, the Alaskan statute provided protection to individuals who receive compensation for emergency care rendered at hospitals. (Id.) The policy behind broadening the protection of the statute was to encourage the creation and maintenance of emergency medical facilities. (Id.). Thus, the Ninth Circuit held that a private physician in Duckworth's position would be within the protection of the Good Samaritan statute. (Id.)

The next question for the *Bunting* court was whether the Coast Guard had a pre-existing duty to rescue, placing it beyond the protection of the Good Samaritan statute. (Id.) Under 14 U.S.C. § 1 et seq., the Coast Guard is under a duty to operate rescue facilities. (Id., citing 14 U.S.C. § 2). The Coast Guard is also authorized to perform any and all acts necessary to those needing aid on the seas. (Id. at 1146-47 (citing 14 U.S.C. § 8)). However, the statute did not pose a duty on the Coast Guard to rescue, and therefore it is protected by the Good Samaritan statute. (Id.)

The *Bunting* court's analysis is particularly persuasive in the present matter. The question for the

Ninth Court was not whether the Coast Guard physician was under some duty to act (he was, as he was required to act by the Coast Guard on behalf of the Coast Guard), but whether he was under a pre-existing duty to render aid to the plaintiff. Otherwise, the statute's purposes are not served.

Similarly, in *Henry*, quoted above, the court referred to "a particular employment duty to aid the patient at the hospital" and a doctor-patient relationship obligation. 367 S.E.2d at 290. This language suggests that only where the employer is obligated to the injured, and the actor is obligated to the employer, does a duty to the injured arise between the actor and the injured. Indeed, both of these duties go to the actor's obligation to the injured party, not the actor's obligation to his employer or some other person....

As the quoted provisions reveal, New York's statute is specifically designed to address the negligence of voluntary ambulance services and EMTs employed thereby. The ordinary factual scenario addressed by the statute is that of a community which is served by an all-volunteer ambulance service, which is run not-for-profit. Where such services exist, the legislature sought to limit their liability in order that existing services, manned by volunteers, might survive in the face of increasing amounts of litigation. In addition, the Good Samaritan statute also encourages the creation of voluntary services by eliminating a major cost of such services.

In this Court's view, the present situation is within the Good Samaritan statute of New York. The military is under no duty to provide emergency relief to civilians present on its premises, but does so voluntarily. The fact that the medics owe a duty to the military to perform the tasks assigned to them, for which they are compensated by the military, (but not the recipient of their EMT services) does not take their conduct here outside of the protection of the Good Samaritan statute. The EMTs and ambulance operating out of the Keller Army Community Hospital are the virtual equivalent of the volunteer ambulance unit and its trained members, particularly when answering this call to help a civilian occupant of a hotel on post.

The plaintiff has been unable to provide the Court with anything suggesting that the military owed such a duty to her or her decedent as would take this case outside of either the letter of the New York Good Samaritan Law or the underlying policies upon which it is based.

Furthermore, the fact that plaintiff received a bill of $882.00 from the hospital does not remove this matter from the statute's protection. Plaintiff does not assert that any portion of that bill was attributed to the emergency services performed prior to arrival at the hospital. Even if there were such an amount so allocated, there is no assertion from plaintiff that such an amount would be "profit," "financial gain," or "monetary compensation" rather than reimbursed expenses.

II. CONCLUSION

For the reasons set forth above, defendant's motion in limine for an order declaring that the plaintiff must prove gross negligence in order to recover is granted.

QUESTIONS

1. Find your state's Good Samaritan Act and compare it to the statutes discussed in this case.
2. The Court states that "Stratiff and Smith cannot claim Good Samaritan protection if they owed a duty to Mr. Clarken to come to his aid." What kinds of duties might operate to void the act in this case?
3. Why are Good Samaritan Acts limited to volunteers.
4. Compare this case with the *American Nat'l Bank & Trust Co.* matter in Chapter 1. Why

did these responders have a duty to act when dispatched, while these did not?

5. What does the term volunteer mean in the context of the Good Samaritan Act? Does it include volunteer fire companies? If so, under what circumstances?

6. Consider law enforcement officers who carry defibrillators in the vehicles. In what ways might their actions when using the defibrillator or otherwise rendering first aid assistance be considered volunteer acts?

PART 5. DELAYED RESPONSE

The following case points up the liabilities that may arise from delayed emergency response. Note the actions of the responders who were under contract to the unit of government compared with those of the private service.

Regester v. Longwood Ambulance Company, Inc.

Commonwealth Court of Pennsylvania

751 A.2d 694 (2000)

OPINION BY JUDGE FLAHERTY

Alice K. Regester, in her own right and as Administratix of the Estate of George E. Regester, III, et al. (collectively Regester), appeal from an order of the Court of Common Pleas of Chester County (trial court) which granted summary judgement in favor of Longwood Fire Company (Longwood) Inc., and Southern Chester County Medical Center (Medical Center). We affirm in part and reverse in part.

Longwood is a Pennsylvania corporation which is the volunteer fire company for Kennet Township. Longwood also provides ambulance service and responds to the Chester County Emergency Services 911 System. Longwood's geographical area of responsibility includes the address of George Regester. The Medical Center is a Pennsylvania corporation. The Medical Center is a hospital and provides, among other services, mobile critical care medical services, which includes Paramedic Unit 94-3 which responds to Chester County Emergency Service 911 transmissions.

George Regester suffered a heart attack on September 8, 1996 at 7:07 P.M. His family called Chester County 911 and began administering cardiopulmonary resuscitation and restored George Regester's pulse. The 911 dispatcher called Longwood and the Medical Center at 7:15 P.M.

In response, both Longwood and the Medical Center dispatched their paramedics. Although the County 911 Dispatcher repeated directions to the Regester home twice, neither the paramedics from Longwood nor from the Medical Center followed the directions to the Regester home given to them. Instead, they inexplicably traveled to the southern end of the county and as such did not reach the Regester home until approximately 7:30 P.M. Normal travel time for the paramedics to reach the Regester home from their bases would have been 3–4 minutes. Although George Regester survived the cardiac arrest, at 7:27 P.M., he vomited and choked to death on his vomit.

Regester filed a complaint against, among others, Longwood and the Medical Center alleging that had the paramedics followed the directions given to them by the 911 dispatcher, George Regester's airway would have been secured by intubation and his ability to survive would not have been compromised. Regester also noted that on December 4, 1995, only nine months prior to September 8, 1996, the date of the fatal heart attack, George had suffered a prior heart attack and the Medical Center's paramedics had responded to the Regester home.

Longwood filed a motion for summary judgment asserting that it was immune from suit based on its status as a local agency and the immunity provided to local agencies under the popularly called Political Subdivision Torts Claim Act, 42 Pa. C.S. § § 8541-8542. The trial court granted this motion after reconsideration by order dated September 18, 1998. The Medical Center moved for summary judgment based upon the defense of the Emergency Medical Services Act, Act of July 3, 1985, P.L. 164, 35 P.S. § § 6921-6938 (EMSA). The Medical Center alleged that it was immune pursuant to Section 11 of EMSA, 35 P.S. § 6931 (j)(2), absent an allegation of gross or willful negligence. *See* Medical Center's Motion for Summary Judgment, Reproduced Record at 153a. Section 11 of EMSA, 35 P.S. § 6931(j)(2) provides that no emergency medical technician or EMT-paramedic or health professional who in good faith attempts to render or facilitate emergency medical care authorized by this act shall be liable for damages as a result of any acts or omissions, unless guilty of gross or willful negligence. This provision shall apply to students enrolled in approved courses of instruction and supervised pursuant to rules and regulations.

In response to the Medical Center's summary judgment motion, Regester sought leave to amend the complaint to include allegations of gross or willful negligence. On January 21, 1999, in its order granting the Medical Center's motion for summary judgment, the trial court denied Regester's motion to amend the complaint to include allegations of gross or willful negligence.

From the trial court's grant of summary judgment in favor of Longwood and the Medical Center, Regester appeals to this court. Appellate review over the grant or denial of summary judgment is limited to determining whether the trial court committed an error of law or abused its discretion. *Pickett Construction Inc. v. Luzerne County Convention Center Authority*, 738 A.2d 20 (Pa. Cmwlth. 1999).

The first issue raised by Regester is whether the trial court erred when it concluded that the Medical Center was entitled to immunity under Section 11 of EMSA, 35 P.S. § 6931(j)(2), where the Medical Center is not a licensed entity pursuant to that provision.

Essentially, Regester argues that 35 P.S. § 6931(j)(2) by its plain language covers only those individuals enumerated therein, namely, emergency medical technicians, EMT-paramedics, and health professionals. Because the Medical Center is not one of these, EMSA does not protect it from liability according to Regester.

The trial court found that the immunity provided by EMSA for the enumerated individuals applies to the employer of the enumerated individuals as well as to the specifically enumerated individuals. In doing so, the trial court relied upon *D'Amico v. VFW Post 797 Volunteer Ambulance Association*, 8 Pa. D. & C.4th 113 (C.P. 1990), aff'd, 413 Pa. Super. 660, 596 A.2d 256 (Pa. Super. 1991) (without published opinion). In *D'Amico*, the estate of a deceased individual sued a volunteer ambulance service. The ambulance service argued that it fell within the protections of EMSA. The estate argued that EMSA contained no limitation of liability in favor of an ambulance service.

The trial court in *D'Amico* held the ambulance service to be within the protections of EMSA.

The Medical Center argues that the legislative intent to enacting EMSA was to establish and maintain an effective and efficient emergency medical services system and that the object of court construction of statutes is to effectuate the legislative intent. The Medical Center urges that it would further this legislative intent if this Court construed EMS to include the Medical Center within its coverage. Specifically, the Medical Center argues that it appears self evident that the Emergency Medical Services Act provides immunity from civil liability to emergency medical personnel due, in large part, to the nature of the work as well as to this Commonwealth's ongoing interest in establishing, and maintaining, an effective and efficient emergency medical services system. Hence, as in *D'Amico*, to provide immunity to employees, but not to hospitals, medical centers and other healthcare entities, would be nonsensical. Therefore, as in *D'Amico*, SCCMC [the Medical Center] is entitled to the protections afforded emergency healthcare providers under the Act. Only by protecting those entities which employ emergency services workers can the purposes of the Act be effectuated.

Medical Center's brief at pp. 20-21.

Given the Medical Center's Motion for Summary Judgment below, we are confronted with the question of whether 35 P.S. § 6931(j)(2) covers the Medical Center. In order to determine this, our first inquiry is whether the language of 35 P.S. § 6931(j)(2) provides such protection.... The operative terms of § 6931(j)(2) are all defined by EMSA. An "Emergency medical technician" is defined as "an individual who is trained to provide emergency medical services and is certified as such by the department [i.e., the Department of Health of the Commonwealth] in accordance with the current national standard curriculum for basic emergency medical technicians...." 35 P.S. § 6923. An "Emergency medical technician-paramedic" referred to as an "EMT-paramedic" is defined as "an emergency medical technician specifically trained to provide advanced life support services who is certified as such by the department in accordance with the current national standard curriculum for emergency medical technician-paramedics...." Section 3 of EMSA, Act of July 3, 1985, P.L. 164, as amended, 35 P.S. § 6923 "Health professional" is defined as a "licensed physician or professional registered nurse who has education and continuing education in advanced life support and prehospital care." None of these definitions includes a hospital. Moreover we note that the legislature was not unmindful of hospitals in enacting EMSA. The legislature specifically provided for a definition of hospital as being "an institution having an organized medical staff which is primarily engaged in providing to inpatients, by or under the supervision of physicians, diagnostic and therapeutic services or rehabilitation services. Our review of the plain language leads to the conclusion that the Medical Center as a "hospital" is not included within the language of 35 P.S. § 6931(j)(2). The Medical Center does not argue otherwise, but instead urges this court to consider the legislative intent in enacting EMSA so as to engage in the task of statutory construction leading to the conclusion that 35 P.S. § 6931(j)(2) includes hospitals within its protections. Because the plain language of the statute is unambiguous, the law prohibits this court from doing so. *See, e.g., Primiano v. City of Philadelphia,* 739 A.2d 1172 (Pa. Cmwlth. 1999).

The Medical Center's invitation to consider the legislature's intent in enacting EMSA so as to include hospitals within the ambit of § 6931(j)(2) when the plain language does not include hospitals runs afoul of the rule that courts are not to utilize rules of statutory construction, including trying to ascertain the intent of the legislature, when the words of the statute are not ambiguous....

Because the trial court held that the Medical Center was within the protections of the language of 35 P.S. § 6931(j)(2) and the clear language of that section does not so provide, the trial court committed an error of law and the grant of summary judgment to the Medical Center must be reversed.

Next, we address Regester's appeal of the trial court's grant of summary judgment to Longwood. The trial court found that Longwood was entitled to immunity pursuant to the commonly called Political Subdivision Tort Claims Act, 42 Pa. C.S. §§ 8541-8542 because Longwood as the fire company for the Kennet Township is a "local agency" pursuant to *Wilson v. Dravosburg Volunteer Fire Department No. 1,* 101 Pa. Commw. 284, 516 A.2d 100 (Pa. Cmwlth. 1986).

Regester argues that Longwood cannot be a "local agency" within the meaning of 42 Pa.C.S. § 8541 because in Longwood's contract with Kennet Township whereby Longwood agrees to "furnish fire and rescue protection to the residents of [Kennet] Township" (R.R. at 100a), there is a paragraph which provides that "notwithstanding anything to the contrary herein, [the Longwood] Fire Company is an independent contractor and shall not be deemed an agency, servant or employee of Township." R.R. at pp. 101a. Regester asserts that because the contract provides that Longwood is an "independent contractor" there is no agency relationship between Longwood and Kennet Township and therefore Longwood cannot be a "local agency." Regester is engaging in the fallacy of equivocation which is just because the same word or form of the same word is used in two different contexts, it must mean the same thing in both contexts. As Longwood points out, simply because under the contract with Kennet Township, Longwood cannot be deemed Kennet's "agent" it does not follow that Longwood cannot be a "local agency" within the meaning of 42 Pa.C.S. § 8541. Local agency status is accorded to volunteer fire companies not because they are otherwise deemed agents of the local government unit under traditional concepts of principal-agency law but rather are traditionally "accorded local agency status because the duties performed by fire fighters are of a public character." *Eger v. Lynch*, 714 A.2d 1149, 1152 (Pa. Cmwlth. 1998), allocatur denied, Pa. _, _ A.2d_, 1999 Pa. LEXIS 630 (March 12, 1999) ("Eger II") (citing cases). Hence, merely because the contract may preclude a finding of a principal-agent relationship between Kennet Township and Longwood, it does not follow that Longwood is not a "local agency" within the meaning of 42 Pa.C.S. § 8541. Thus, the existence of the contractual language does not, as a matter of law or of fact, preclude a conclusion that Longwood is a "local agency" contrary to Regester's arguments.

However, Regester next argues that there was a question of material fact as to whether Longwood was the official volunteer fire company because merely introducing the resolution of Kennet Township designating Longwood as such is insufficient to merit the grant of summary judgment. Regester relies upon *Eger II*.

The analysis for determining whether Longwood is a local agency entitled to immunity was set forth in the case of *Guinn* wherein the Supreme Court stated that "a volunteer fire company created pursuant to relevant law and legally recognized as the fire company for a political subdivision is a local agency" within the intendment of 42 Pa.C.S. § 8541. In *Kniaz v. Benton Borough*, 164 Pa. Commw. 109, 642 A.2d 551 (Pa. Cmwlth. 1994) we recognized that the Supreme Court's statement in *Guinn*, set up a two part test asking whether the volunteer fire company was created pursuant to relevant law (which is not presently disputed) and whether the volunteer fire company is legally recognized as the official company for the local political subdivision. It is not necessarily sufficient merely

to present into evidence a resolution by the local political subdivision which designates the volunteer company as the company to provide fire services to that political subdivision. *Eger II*, 714 A.2d at 1150. In *Eger v. Lynch* (No. 1560 C.D. 1994, filed March 14, 1996) (published memorandum opinion, disposition of which is noted at 677 A.2d 1335 (Pa. Cmwlth. 1996) (Eger I)), the volunteer fire company introduced a resolution of the local political subdivision dated from 1939 which designated that volunteer fire company as the provider of fire services for the political subdivision. *See Eger II*, 714 A.2d at 1150 n.2. However, the incident giving rise to the suit in *Eger I* occurred in 1992, some 53 years after the resolution designating the volunteer fire company as the provider of services for the political subdivision. *Eger II*, 714 A.2d at 1159 (quoting *Eger I*, slip op. at pp. 7–8)....

Contrary to Regester's contention, we do not find that anything in *Eger II* requires a finding here that Longwood was not entitled to immunity as a local agency. In the case at hand, Longwood introduced the resolution of Kennet Township designating Longwood as its provider of fire protection which was dated January 2, 1996. R.R. at 103a. The incident giving rise to this case occurred on September 8, 1996 which is a significant difference from the facts of the *Eger* case. Moreover, the record here includes the agreement between Longwood and Kennet whereby Longwood agreed to provide fire protection and ambulance services to Kennet. That agreement was dated January 1, 1996 and provided that the agreement was in force from January 1, 1996 until December 1, 1996. Accordingly, unlike in the *Eger* case, there is no factual question as to whether at the time of the incident giving rise to this litigation that Longwood was the official fire company of Kennet. Thus, pursuant to *Guinn*, the two-pronged test was met herein and Longwood was entitled to immunity as a local agency.

Regester argues in the alternative that even if Longwood is entitled to immunity as a local agency, an exception applies herein, namely the exception to immunity found at 42 Pa.C.S. § 8542(b)(1), the so-called vehicle exception. This exception provides in relevant part that the following acts by a local agency or any of its employees may result in the imposition of liability on a local agency:

(1) Vehicle liability. The operation of any motor vehicle in the possession or control of local agency....

Longwood argues that the exception to immunity does not apply because Regester's allegations of negligence relate not to the "operation" of the ambulances involved but rather to other acts of negligence, e.g., negligently failing to follow the correct directions given by the 911 dispatcher to Regester's house. Longwood argues that nothing about the operation of the emergency vehicle by Longwood Fire Company contributed to Regester's death. Plaintiffs have not claimed that the vehicle was driven negligently such that it was involved in an accident, was driven off the roadway, or in any other fashion some careless act in the movement of the vehicle caused it to be delayed in arrival at the Regester home. Nor is it claimed that the vehicle broke down or was disabled because of some prior negligence in maintenance of the vehicle.

Longwood's brief at p. 10.

Regester responds that Longwood's interpretation of the exception ignores the recent Supreme Court's construction of the exception in *Mickle v. City of Philadelphia*, 550 Pa. 539, 707 A.2d 1123 (198) (*Mickle II*). In *Mickle*, the plaintiff, while being transported by a Philadelphia fire vehicle to the hospital, was injured when the rear wheels of the vehicle fell off. The parties stipulated that there was "nothing negligent about the manner in which the firefighter actually drove/operated the fire rescue vehicle" *Mickle II*, 550 Pa. at 541, 707 A.2d at 1125. They also stipulated that the cause of the rear wheels falling off was negligent maintenance and repair of the vehicle by the defendant, City of Philadelphia. Upon the stipulated facts, both the plaintiff and the City of Philadelphia moved for summary judgment.... The trial court granted the plaintiff summary judgment. This court affirmed. *Mickle v. City of Philadelphia*, 669 A.2d 520 (Pa. Cmwlth. 1996) (*Mickle I*), aff'd, 550 Pa. 539, 707 A.2d 1124

(1998). In affirming this court, the Supreme Court reasoned that since there is not dispute that the van was in operation at the time of the accident as required by *Love v. City of Philadelphia*, 518 Pa. 370, 543 A.2d 531 (1988)] we must determine whether Mickle's injury was caused by the City's **negligent acts with respect to the van's operation**. The City argues that there was no such negligent act because it is undisputed that the firefighter did not drive the van in a negligent manner.

Negligence related to the operation of a vehicle encompasses not only how a person drives but also whether he should be driving a particular vehicle in the first place. The motor vehicle exception does not say that liability may be imposed only where the operator's manner of driving is negligent. **Rather, it requires that the injury is caused by a negligent act with respect to the operation of a motor vehicle**. 42 Pa. Cons.Stat. § 8542(a),(b)(1)....

We must ask whether the failure of the driver of Longwood's ambulance to follow directions to Regester's house was a negligent act with respect to the operation of a motor vehicle. Because Longwood does not dispute that failing to follow directions was negligent, the only issue is whether such failure was an act with respect to the operation of a vehicle.

We conclude that the Longwood paramedics' failure to follow the directions of the 911 dispatcher was not an act with respect to the operation of a vehicle as a matter of law.... the giving of instructions was not the operation of the vehicle....

"[W]e find that the failure of Longwood's paramedics to follow directions was not an act with respect to the actual operation of a vehicle, but rather...was an act with respect to the giving and receiving of directions which does not constitute "operation" of a vehicle. Likewise, the decision of the paramedics to not follow the explicit directions given by the 911 dispatcher...[was not an act] with respect to the "operation" of a vehicle.

Because the trial court did not commit an error of law or abuse its discretion in granting the summary judgment motions of Longwood, the trial court's order doing so is affirmed.

QUESTIONS

1. The Court reasoned that paramedics were named as protected from liability for delayed response by the statute, but that neither their employers nor hospitals were named on the list of protected entities. Discuss public policy reasons that might underlie the legislature's decision to limit immunity to individual emergency responders.

2. Compare the actions of the private provider with those of the volunteer entity contracted with by the local unit of government.

3. Evaluate the following statement: "In the *Regester* case, the status of the entities involved rather than their actions determines their liability."

4. Recall the allegations of improper dispatch activity in the *American National Bank and Trust Company* case in Chapter 1. Consider the facts in the *Koher* delayed dispatch case from Chapter 4. Compare the actions of the dispatchers in those cases and the current matter. What other sorts of events might cause a delayed response?

5. What length of delay should result in liability?

6. What factors, if any, should courts consider in particular cases that might result in excusable delay?

Shawnee, Okla., January 18, 2006 – Oklahoma Emergency Managers Bob Worrell, Steve Palladino, and Melvin Potter coordinate firefighting efforts from the Incident Command Post set up in the Shawnee Expo Center. Bob McMillan/FEMA Photo.

Chapter 6

MANAGING A RESPONSE: STANDARD OPERATING PROCEDURES, MUTUAL AID, AND THE INCIDENT MANAGEMENT SYSTEM

I. MANAGING A RESPONSE*

Three invaluable management tools operate together at any large emergency scene to preserve the safety of responders and the public. Emergency response organizations utilize standard operating procedures ("SOPs") to guide their members during daily operations. Good SOPs are the sturdy foundation of safety. When a response requires resources beyond those available to an individual organization, well-written mutual aid agreements ("MAAs") tie together the good SOPs of multiple response organizations on a sizable emergency scene. The incident management system ("IMS") roofs over the structure of safety, assuring that on-scene organization reinforces both good SOPs and well-written MAAs. The symbiotic functioning of these three elements results in efficient use of resources and maximum safety for responders and the public. These tools are designed so that the failure to properly utilize any single element will not result in unsafe conditions. Rather, they provide checks and balances for one another to assure scene safety.

II. BACKGROUND OF THE INCIDENT MANAGEMENT SYSTEM: HAZWOPER

After a very bad wildfire season during the 1970s, California fire managers decided that a better way was needed to respond to emergencies. In many incidents, lack of interagency cooperation resulted in unsafe conditions and improper allocation of resources. The managers noted several specific problems: (1) lack of communication due to differing radio code; (2) no command system existed. Every agency depended on the personality of the leader in charge at any given moment; (3) lack of common terminology – even when communication was possible, misunderstandings arose; (4) no way to effectively assign resources – logistics was a product of luck;[1] (5) no clear definition of functions, and how functions related to one another. The incident command system developed in a response to these challenges. ICS evolved into the universally accepted way of integrating response to emergencies.[2]

---

*This material is adapted from: William C. Nicholson, "Legal Issues in Emergency Response to Terrorism Incidents Involving Hazardous Materials: The Hazardous Waste Operations and Emergency Response (HAZWOPER) Standard, Standard Operating Procedures, Mutual Aid and the Incident Management System," *Widener Symposium I, J.*, Vol. 9, No. 2, 2003.

[1] PAUL M. MANISCALO AND HANK T. CHRISTEN, UNDERSTANDING TERRORISM AND MANAGING THE CONSEQUENCES 24-5 (2001).

[2] William C. Nicholson, The Incident Command System: Legal and Practical Reasons for Incident Management, OUR WATCH, July–Sept. 1999 at 3, 5., See generally ALAN V. BRUNACINI, FIRE COMMAND (1985). Brunicini has been termed "the Godfather of

Incidents continued to grow in size and complexity after ICS was first created. The involvement of multiple response agencies and leaders on these scenes provided the impetus for a further refinement in scene management. The incident management system ("IMS") utilizes a management model rather than a command model. IMS emphasizes consensus among leaders, and operates with representatives of involved agencies working together to provide group leadership. This approach allows the specialized knowledge of all commanders regarding their groups' members to be part of the decision matrix. A strong indication that IMS will continue to be the standard is the Homeland Security Act of 2002's requirements to build a comprehensive national incident management system with Federal, State, and local government personnel, agencies, and authorities, to respond to terrorist attacks and other disasters.[3]

A further development, utilized to date only in California, is the Standardized Emergency Management System ("SEMS"). SEMS incorporates ICS, multi- and inter-agency coordination, mutual aid, and an operational area concept for flexible response to extremely large incidents. SEMS has been predicted to be the standard of the future.[4] Currently, IMS is the standard,[5] and the text will refer to IMS and ICS interchangeably hereafter.

The HAZWOPER standard requires all HAZMAT responses to utilize ICS.[6] The following characteristics are part of a good ICS: modular organization, integrated communications, common terminology, a unified command structure, consolidated action plans, a manageable span of control, designated incident facilities, and comprehensive resource management.[7]

HAZWOPER includes important additional and very specific requirements for ICS.[8] It specifies that the senior emergency response official responding to an emergency shall become the individual in charge of site-specific ICS (henceforth "incident commander" or "IC").[9] The standard recognizes that incidents evolve and that the actual individual in command may change as additional resources arrive on-scene.[10] One tragic aspect of the September 11, 2001 attack on new York was the death of the people comprising the New York Fire Department's Incident Command structure when the towers collapsed. An important lesson learned from that tragedy is the need to set up back-up command structures at terrorism responses.[11] A defined command

---

incident command." His "well-respected" work places him as the "pre-eminent expert in incident command." Telephone Interview with Tracy Boatwright, Indiana State Fire Marshal (April 24, 2002). Marshal Boatwright served on the Executive Board of the National Association of State Fire Marshals from 1995–2000, and was Secretary/Treasurer from 1999–2000. A long time paid and volunteer firefighter, Boatwright has been State Fire Marshal since 1993.

[3] HR 5005 Sec. 501(5).

[4] ROBERT A. JENSEN, MASS FATALITY AND CASUALTY INCIDENTS, A FIELD GUIDE 4-7 (2000).

[5] NATIONAL FIRE PROTECTION ASSOCIATION 472: STANDARD FOR PROFESSIONAL COMPETENCE OF RESPONDERS TO HAZARDOUS MATERIALS INCIDENTS requires the incident commander to implement IMS as the first step of implementing the pre-planned response to a HAZMAT incident. NFPA 472 at 472-24. Further, NFPA 1600 STANDARD ON DISASTER/EMERGENCY MANAGEMENT AND BUSINESS CONTINUITY PROGRAMS 1600-6 (2000) requires that an incident management system shall be utilized.

[6] 29 CFR § 1910.120 (q)(3)(i) requires that during an emergency response the most senior emergency response official becomes the individual in charge of a site-specific Incident Command System (ICS). All emergency responders and their communications shall be coordinated and controlled through the individual in charge of the ICS assisted by the senior official present for each employer.

[7] See e.g., JENSEN at 3, William C. Nicholson, Beating the System to Death: A Case Study in Incident Command and Mutual Aid, 152 FIRE ENGINEERING, 128 at 129-130 (1999).

[8] 29 CFR § 1910.120 (q)(3) requires these characteristics at all HAZMAT response sites.

[9] 29 CFR § 1910.120 (q)(3)(i). NFPA 472 requires use of IMS and contains detailed competencies for the IC at 472-22 - 472-25.

[10] Note to (q) (3) (i) specifies that the "senior official" at an emergency response is the most senior official on the site who has the responsibility for controlling the operations at the site. That person is the senior officer on the first-due piece of responding emergency apparatus to arrive on the incident scene. More senior arriving officers (i.e., battalion chief, fire chief, state law enforcement official, site coordinator, etc.) assume the position, which is passed up the previously established line of authority.

[11] Telephone Interview with Rick D. Schlegel, EAI Corporation, Deputy Program Manager, Incident Command Responder Course, Anniston, AL (April 24, 2002).

transfer process must be put in place well before an incident to avoid potential chaos and danger to responders and the public.[12]

Whether leadership is exercised by a single IC or falls on a group under IMS, the characteristics required of the person(s) in charge are the same. The IC must apply command and control efforts to achieve results, rather than for the ego gratification of being in charge.[13] Further, the IC must be suited by disposition to the task. Important personality characteristics for the IC include:

1. respect for the task;
2. ability to stay cool under pressure;
3. knowledge of command;
4. an inclination to command, not act;
5. the ability to provide a positive example;
6. psychological stability;
7. physical fitness;
8. treating all involved in a fair manner;
9. ability to communicate clearly;
10. willingness to take reasonable risks;
11. concern for all personnel;
12. knowing limitations of self, others, equipment, and approaches to the situation;
13. respect for command;
14. organizational ability; and
15. ability to act in a consistent, disciplined manner.[14]

The IC must identify, to the extent possible, all hazardous substances or conditions present and address site analysis, use of engineering controls, maximum exposure limits, hazardous substance handling procedures, and use of any new technologies.[15] The IC's duties at this point include both identifying the substance and controlling the hazard, related but not duplicative tasks.[16]

The IC must implement appropriate emergency operations, and assure that the personal protective equipment ("PPE") worn is appropriate for the hazards present.[17] There are special requirements for breathing equipment.[18] Attempting to hold one's breath and trying not to take too many breaths is not alternate means of compliance.[19]

The number of emergency response personnel at the emergency site, in those areas of potential or actual exposure to incident or site hazards, must be limited to those who are actively performing emergency operations.[20] The IC, of necessity, has complete control over who is on the HAZMAT scene and what they do. This authority gives the IC a tool to control the emergent volunteers who show up in the aftermath of any large incident. Such volunteers may range in background from civilians with chain saws to trained responders, including basic firefighters or even HAZMAT technicians. When trained responders come as the organized result of a mutual aid agreement (discussed in detail below), the result can be helpful resources to deal with the incident. Responders will also arrive at the site individually or in mass without being requested, as happened both in New York and at the Pentagon after the September 11, 2001 attacks.[21] For the IC to maintain control of the scene in such a situation requires both organization and tact. One of the first tasks for an IC is establishment of a perimeter, which should be controlled by law enforcement. Persons attempting to enter the perimeter without proper authority

---

[12] BRUNACINI at 121-125.

[13] Id. at 7.

[14] Id. at 10.

[15] 29 CFR § 1910.120 (q)(3)(ii).

[16] Secretary of Labor v. Victor Microwave, Inc., No. 94-3024, 1996 OSAHRC Lexis 57, at *39-40 (O.S.H.R.C.A.I.J. June 17, 1996). In the aftermath of a release of hazardous gases, the IC failed to identify hazardous gases present or to monitor or utilize engineering or other controls such as ventilating the building.

[17] 29 CFR § 1910.120 (q)(3)(iii) requires personal protective equipment to meet, at a minimum, the criteria contained in 29 CFR 1910.156(e) when worn while performing fire fighting operations beyond the incipient stage for the incident or site.

[18] 29 CFR § 1910.120 (q)(3)(iv).

[19] See Victor Microwave, Inc., 1996 OSAHRC Lexis at *40-41. Such an approach found to be a serious violation.

[20] 29 CFR § 1910.120 (q)(3)(v).

[21] See e.g., Dan Barry, After the Attacks: The Search; A Few Moments Of Hope In A Mountain Of Rubble," N.Y. TIMES A-1 (September 13, 2001). "There were volunteers everywhere, arguably more than were needed."

must be stopped and sent to a remote staging area.[22] At that location, their training and abilities can be evaluated, they can be rostered and they can either be incorporated into the response or politely turned away. Failure to set up this entry mechanism early in a response can result in an unsafe scene.

The buddy system in groups of two or more must be utilized.[23] Working within the buddy system requires that one is available to observe and, if necessary, rescue the other.[24] Back-up personnel must be ready to provide assistance or rescue. They may be within hearing range, including presumably radio range, although they need not be in visual contact with the person within the hazardous area.[25] Advance first aid support personnel must also stand by with medical equipment and transportation capability.[26]

Perhaps the most important requirement is designation of a safety officer who is knowledgeable in the operations being implemented at the emergency response site. He or she possesses specific responsibility to identify and evaluate hazards and to provide direction with respect to the safety of operations for the emergency at hand.[27] The safety official has the authority to alter, suspend, or terminate those activities. The safety official must immediately inform the IC of any actions needed to be taken to correct these hazards at an emergency scene.[28] Case law makes clear that the safety officer must be an individual distinct from the IC him or herself.[29]

After emergency operations have terminated, the IC is responsible for implementation of appropriate decontamination procedures.[30]

The timing of when emergency response ceases and moves to post-emergency cleanup is an issue that may provoke controversy. During emergency response, scene stabilization and containment of the release are the central concerns. When actions change in character to routine cleanup activities, the law requires additional protective planning, in the nature of a site control program.[31] Failure to recognize the change of circumstances may lead to penalty for the violator.[32]

III. EARLY FEDERAL GUIDELINES FOR INCIDENT MANAGEMENT AT A TERRORISM RESPONSE

In the aftermath of the 1993 attack on the World Trade Center and the bombing of the Murrah Building in Oklahoma City, numerous programs arose to address the need to better prepare for terrorism events.[33]

In the 1998 Appropriations Act[34] and accompanying report, Congress expressed its concern over the potentially catastrophic effects of chemical or biological acts of terrorism. The legislature recognized the fact that the federal government's

---

[22] See BRUNACINI at 23-24 for example of standard operating procedures for staging at a fire scene.

[23] 29 CFR § 1910.120 (q)(3)(v).

[24] See Victor Microwave, Inc., 1996 OSAHRC at *41-43. One member of a pair entering into the hazardous area of release while the other waits out of sight found to be insufficient. Such an approach found to be a serious violation.

[25] Id. at *43-44.

[26] 29 CFR § 1910.120 (q)(3)(vi).

[27] 29 CFR § 1910.120 (q)(3)(vii).

[28] 20 CFR § 1910.120 (q)(3)viii).

[29] See Victor Microwave, Inc., 1996 OSAHRC at *44-47 Failure to designate a separate safety officer found to be a serious violation.

[30] 29 CFR § 1910.120 (q)(3)(ix).

[31] 29 CFR § 1910.120 (d)(1) and (2). A site control program for preventing contamination of employees shall be developed during the planning stages of a hazardous waste operation clean-up.

[32] Westinghouse Haztech, Inc., No. 88-2458, 1989 OSHARC Lexis 205, at *8-12 (O.S.H.R.C.A.IJ. JUNE 7, 1989). Failure to prepare a site map and mark off contaminated areas resulted in a less than serious violation and a fine of $100.

[33] See e.g., David Lore, Federal Bucks Flow to Fight Terrorism, COLUMBUS (OH) DISPATCH, August 7, 2000 at B5.

[34] Department of Commerce, Justice, and State, The Judiciary, and Related Agencies Appropriations Act of 1998, Pub. L. No. 105-119, 111 Stat. 2440 (1997).

role revolves around preventing and providing supportive response to such threats. In reality, state and local public safety personnel respond first to the scene of these incidents. Congress therefore directed the Attorney General to aid state and local public safety personnel in acquiring the advanced training and equipment needed to safely respond to and manage weapons of mass destruction ("WMD") terrorist events. On April 30, 1998, the Attorney General delegated authority to the Office of Justice Programs ("OJP") for development and administration of programs for training and help with equipment for state and local emergency response agencies. In response to the Attorney General's instruction, the Office of Justice Programs created the Office for Domestic Preparedness ("ODP") to develop and administer the Domestic Preparedness Program.[35]

As part of its efforts to improve upon the abilities of emergency responders to deal with terrorism events, the ODP established the Center for Domestic Preparedness ("CDP"), at the previous home of the U.S. Army Chemical School, Fort McClellan in Anniston, Alabama. The CDP trains state and local emergency responders to control and lessen the effect of WMD incidents. The CDP curriculum includes two courses of instruction: WMD HAZMAT Technician and WMD Incident Command. The Incident Command course is accompanied by a Guide (henceforth the "IC Guide").[36] The IC Guide contains federally suggested guidelines for managing the response to terrorism events?[37]

Many of these guidelines echo those found in NFPA 472 and HAZWOPER.[38] These include scene safety, command, control and communications, patient care, decontamination, and resource management.[39] The IC must be aware that terrorists may plant secondary devices, which may be explosives, intended to kill or injure emergency responders?[40] Terrorists use secondary devices to hinder the response, with the goal of frightening the public into believing that the government cannot protect them. Since the terrorism incident is a federal crime scene, the IC Guide suggests preplanning with the local FBI field office to assure preservation of evidence.[41]

IV. LEGALLY SOUND MUTUAL AID AGREEMENTS

Emergency responses, particularly to large and/or complex HAZMAT incidents, frequently require resources beyond those available to the entity tasked with the first response to an incident. HAZWOPER requires that the emergency response plan include pre-emergency planning and coordination with outside parties.[42] The means for procuring such coordinated assistance is typically a mutual aid agreement, although sometimes the emergency planning and coordination happens at the site before committing responders to enter the hazardous area.

The fire, emergency medical services, and emergency management communities are encouraged,

---

[35] This program was transferred to the Department of Homeland Security upon Congress enacting President Bush's initiative. HR 5005 Sec. 502 (2002) provides "In accordance with title VIII, there shall be transferred to the Secretary [of Homeland Security] the functions, personnel, assets, and liabilities of the following entities –
... (2) the Office for Domestic Preparedness for the Office of Justice Programs, including the functions of the Attorney General related thereto;

[36] Center for Domestic Preparedness, CHEMICAL, ORDINANCE, BIOLOGICAL AND RADIOLOGICAL ("COBRA") INCIDENT COMMAND COURSE RESPONDER GUIDE ("IC GUIDE"), (2002).

[37] See also Presidential Decision Directive ("PDD") 39, Policy on Counterterrorism, (June 21, 1995) Specifies the use of an on-scene coordinator during acts and responses to terrorist activities.

[38] IC GUIDE at RG-7 - 08. "The principles for site safety and control are the same as for a HAZMAT incident."

[39] IC GUIDE at RG-7 - 07.

[40] IC GUIDE at RG-7 - 08.

[41] IC GUIDE at RG-7 - 25.

[42] 29 CFR § 1910.120 (q)(2)(i) requires that the emergency response plan include planning and coordination with outside parties.

both by law[43] and by common sense, to enter into mutual aid agreements. The reasoning behind mutual aid is sound: in this era when emergency service providers are being told to do more with less, combining forces to battle a common foe preserves resources and allows more efficient response to crisis situations. There is a definite national trend towards greater reliance on mutual aid agreements, as well as better planning for such agreements.[44] Many smaller municipalities have reached MAAs with neighboring larger cities, although such agreements may be forced on one entity or the other by circumstances.[45] A variety of players, including state and local governments, the private sector, and federal agencies and departments, enter into MAAs with one another.[46] Mutual aid, joint powers, and intergovernmental assistance agreements have been characterized as "expected and the norm" among municipalities.[47] In the aftermath of a catastrophic terrorism event, the role of the military will be very significant.[48] A preexisting MAA will greatly facilitate use of that assistance.

MAAs may be either verbal "handshake" understandings or written documents. Verbal MAAs may prove to be dangerous invitations to potential liability.[49] Written mutual aid agreements are preferable for several reasons. First, a written agreement provides a guideline for response during a crisis. Second, a written agreement clearly defines who is responsible for various expenses during and after the response. Third and most important, written mutual aid agreements avoid conflict between agencies that need to be mutually supportive in future responses.

During a response, many important issues must be addressed immediately. who is in overall command? Who is in command of responding units from the assisting jurisdiction? Who is responsible for the actions of responding agencies' employees? After the response, other issues arise. Who pays for the costs of response? Who is accountable for equipment damaged in the response? Who pays for medical expenses of injured responders? Who is responsible for workman's compensation for injured responders? At a HAZMAT scene, as demonstrated above, HAZWOPER answers many of these questions. State law may also address some of these matters. Some state laws, however, allow the parties to an agreement to vary the duties otherwise imposed by state law. For example, in Indiana, unless otherwise provided for by agreement, a requesting unit of government is responsible for the costs of a party responding to a request for mutual aid.[50]

A written mutual aid agreement allows these and all other issues that might arise to be addressed

---

[43] See, e.g., Indiana Comprehensive Emergency Management Plan III.0 (2002), which suggests mutual aid be requested prior to state assistance being approved.

[44] Granting the Consent of Congress to the Emergency Management Assistance Compact, Pub. L. No. 104-312, 110 Stat. 3877 (1996) Congress consents to the Emergency Management Assistance Compact to which 43 states and territories are currently signatures, further information at http://www.nemaweb.org/emac/index.htm.

[45] Jared Eigerman, "California Counties: Second-Rate Localities or Ready-Made Regional Governments?" 26 *Hastings Const. L.Q* 621, 651 (Spring, 1999) citing Frank P. Sherwood, Some Major Problems of Metropolitan Areas, in Governor's Commission on Metropolitan Area Problems, Metropolitan California 19, 95 (Ernest A. Engelbert ed., 1961). As an example, Sherwood explained that the City of Los Angeles, out of self-interest, had to help the City of Vernon protect its factories from fire, but that the Vernon Fire Department was all but useless to the City of Los Angeles.

[46] See e.g., Francis A. Delzompo, Warriors the Fire Line: The Deployment of Service Members to Fight Fire in The United States, Army Law. 51, 55 (April, 1995). Mutual aid agreements for use of firefighting assets between military installations and local units of government are common.

[47] Howard D. Swanson, THE DELICATE ART OF PRACTICING MUNICIPAL LAW UNDER CONDITIONS OF HELL AND HIGH WATER, 76 N.D. L. Rev. 487, 497 (2000).

[48] Barry Kellman, Catastrophic Terrorism – Thinking Fearfully, Acting Legally, 20 MICH. J. INT'L L. 537, 546-547 (1999). For information on military assistance after the Oklahoma City bombing, see Jim Winthrop, The Oklahoma City Bombing: Immediate Response Authority and Other Military Assistance to Civil Authority (MACA), 1997 Army Law. 3 (July, 1997).

[49] William C. Nicholson, Legally Sound Mutual Aid Agreements, 2 EMS BEST PRAC., 41, 46 (1999).

[50] Indiana Code § 36-1-7-7(b) (1999) Requires, in the absence of a written agreement, fire service and law enforcement entities that request mutual aid to pay for the travel expenses of the responding units. The responding units are also under the supervision of the requesting unit. This statute may frequently not be observed in the field, where "handshake" agreements and "paying one's own way" are the norm among many rural volunteer fire departments.

in a calm atmosphere, with all entities involved able to present their points of view. The time to consider these matters is not during an emergency response, when saving lives and preserving property are very correctly the top priority. Unfortunately, handshake agreements and discussions in the heat of emergency response have all too often been the cause of subsequent disagreement over who is responsible for expenses and legal liabilities. Indeed, long histories of friendship and co-operation have ended due to different understandings of handshake agreements. Some states have statutory guidelines for mutual aid agreement contents.[51] The written MAA is typically memorialized as a Memorandum of Understanding or Memorandum of Agreement ("MOU" or "MOA").

As mentioned above, the lack of clear definition of response functions, and how functions related to one another, numbered among the emergency response problem that IMS was devised to address. Traditionally, MAAs involve entities addressing the same functions, such as fire suppression. The more advanced view is to look at MAAs in light of how they will support IMS at a response scene. From this perspective, it is advisable to address other functions reasonably anticipated to be needed at the scene. Such MAAs might involve multiple response entities, including the fire service for the fires suppression function, Emergency Medical Services ("EMS") for the mass care function, the Red Cross for the sheltering function, amateur radio for the support of the communications function, law enforcement or volunteer emergency management for the traffic control function, and so on.[52] Such pro-active MAAs facilitate thought and action in the IMS mode.

V. STANDARD OPERATING PROCEDURES

IMS and written mutual aid agreements are designed to work closely together to achieve the goals of safety for emergency responders and the public and efficient use of resources. One key purpose of the MAA is to assure that responding entities, whether public or private, adhere to standard operating procedures ("SOPs") or standard operating guidelines ("SOGs") during mutual aid responses. When drafting SOPs, MAAs should be considered – similarly, when drafting MAAs, SOPs must be evaluated.[53] The symbiotic nature of these documents must be understood for them to properly work together and provide maximum safety. Many other documents, plans, and agreements need to be considered as well when developing SOPs, including the requirements of HAZWOPER.[54] Like MAAs, SOPs must be written to be effective.[55] They must also be enforced to work properly.[56]

The National Fire Protection Association ("NFPA") defines an SOP as "an organizational directive that establishes a standard course of action.[57] A complete set of SOPs sets out in a detailed manner how an emergency response organization will function during an event, functioning as a "game plan" before the event.[58]

---

[51] See, e.g., Indiana Code § 10-4-1-9 authorizing the director of the local emergency management organization to develop intrastate mutual aid agreements and to assist in negotiation of intrastate agreements, which must be signed by the Governor. The guidelines for other local agencies are contained at Indiana Code § 36-1-7. Firefighting and law enforcement agencies have particular limitations on agreement content, and agreements involving state agencies or units of government of other states must have particular approvals.

[52] Telephone Interview with Rick D. Schlegel, EAI Corporation, Deputy Program Manager, Incident Command Responder Course, Anniston, AL (April 24, 2002).

[53] FEDERAL EMERGENCY MANAGEMENT AGENCY – UNITED STATES FIRE ADMINISTRATION, DEVELOPING EFFECTIVE STANDARD OPERATING PROCEDURES FOR FIRE & EMS DEPARTMENTS 9 (1999). (Subsequently "DEVELOPING SOPs")

[54] Id. at 67-75, 92-93.

[55] BRUNACINI at 16-17.

[56] Id. at 17.

[57] DEVELOPING SOPs at 1.

[58] BRUNACINI at 16.

Properly utilized, SOPs are a key element in assuring personal safety and protection from liability. SOPs must be written with intelligent management of risks as their primary goal to ensure that safety becomes the standard expected by all involved.[59] Safety-specific SOPs are absolute mandates that must be followed, no matter what other circumstances may obtain.[60]

For individual responders, SOPs provide understandable statements of employer requirements and give detailed explanation of expectations. Manager use SOPs for a number of purposes: examining their operations from a strategic perspective, noting needed changes, documenting regulatory compliance, establishing intentions, improving training, and measuring performance. They provide a way to communicate legal and administrative requirements to members of emergency response organizations.[61] At a large scene with many responding agencies involved, SOPs become even more important.[62]

Fire departments were the first emergency response organizations to develop and use SOPs during emergency responses. As departments grew beyond their informal roots, they began to address safety considerations through internal controls. These tenets, originally termed "rules of engagement," protected firefighters during daily fire operations. The more modern terminology for these guidelines is standard operating procedures. One example of an early SOP that persists to this day is the requirement that not all firefighters at a scene enter a burning structure at the same time. As time went on, different fuel combinations and more elaborate equipment led to the need for more elaborate SOPs. As fires have become more complex, SOPs have evolved from procedures that are "chiseled in stone" to SOGs. SOGs allow increased flexibility in responding to complex fire scenes, encouraging full utilization of firefighters' knowledge, skills, and abilities. Other emergency response organizations have learned from the experience of the fire service, similarly developing ever more sophisticated SOPs and SOGs.[63] For HAZMAT responses, employers must incorporate SOPs in their written safety and health program.[64]

VI. THE NATIONAL RESPONSE FRAMEWORK (NRF) AND THE NATIONAL INCIDENT MANAGEMENT SYSTEM (NIMS)[65]

The success of the newly created incident management system ("IMS") led to it being adopted by a variety of players, including law enforcement, public health, public works, and the private sector.[66] Emergency management groups also embraced the system,[67] and recommended in 2000 that all levels of government utilize it for response to weapons of mass destruction

---

[59] Id. at 222.

[60] Id. at 228.

[61] DEVELOPING SOPs at 2.

[62] William C. Nicholson, Standard Operating Procedures: the Anchor of On-Scene Safety, OUR WATCH 2 (Winter 1999).

[63] Id. at 5.

[64] 29 CFR 1910.120 (b)(1)(ii)(F)

[65] This section incorporates some materials from William C. Nicholson, "Seeking Consensus on Homeland Security Standards: Adopting th National Response Plan and the National Incident Management System" *Widener Law Review*, Vol. 12, No. 2, 491-559 (2006). Reproduced by permission.

[66] Pat West, NIMS: *The Last Word on Incident Command?*, FIRE CHIEF, Mar. 5, 2004, *available at* http://firechief.com/ar/firefighting_nics_last_word/index.html. ("What we've said now with the NIMS document is that it's not just a fire service issue. We're expanding (incident management) to include all the agencies involved in response to emergencies – beyond police, EMS and fire – to include all the government agencies that will respond to a disaster as well as some private organizations.").

[67] "[T]he National Emergency Management Association adopted a position in September 1996 adopting the National Interagency Incident Management System and its Incident Command System (ICS) as the model for all risk/hazard response activities by state and local governments..." NAT'L EMERGENCY MGMNT. ASS'N, TERRORISM COMM., A RESOLUTION ADVOCATING THE INCIDENT COMMAND/MANAGEMENT SYSTEM FOR ALL WMD OPERATIONS BY ALL LEVELS OF GOVERNMENT (2000), http://knxas1.hsdl.nps.navy.mil/ homesec/docs/legis/nps08-112603-18.pdf.

("WMD") events.[68] Credentialing bodies have universally accepted IMS as the pattern for integrating emergency response.[69] Despite IMS' demonstrated value, however, some emergency response groups were slow to adopt it.[70]

A. Why the NRP/NRF and NIMS?

The NRP and NIMS are the product of a process that began with the passage of the HS Act. Signed into law by President Bush on November 25, 2002, the HS Act considerably alters the national approach to terrorism and all other emergency events. The HS Act's most visible effect was the creation of DHS by uniting 180,000 federal workers from twenty-two agencies into a single organization.[71] As written, the mission of DHS in the HS Act revolves around terrorism.[72] A further challenge posed by the law, therefore, was how to combine the efforts of DHS with those of state and local governments as well as emergency responders into a truly national system reaching beyond terrorism

to include "all hazards" emergency preparedness and response.[73]

On February 28, 2003 President Bush put out marching orders for the implementation of the HS Act. On that date, he issued Homeland Security Presidential Directive 5 ("HSPD 5").[74] HSPD 5 instructed all federal agencies to take specific steps for planning and incident management.[75] The Directive also mandated setting emergency responder performance standards and established sanctions for responders who fail to conform to those standards.[76]

HSPD 5 directed all federal agencies to collaborate with DHS to establish a NRP and a NIMS.[77] HSPD 5 also set out a timetable for those actions.[78] NIMS is the operational part of the NRP.[79] In this fashion, authority for creation of the NRP/NRF and the NIMS flows from the HS Act through HSPD 5 to DHS.[80]

Some observers initially worried that creation of DHS might be a bureaucratic nightmare.[81] The new organization, some feared, might be rife with infighting, inflexible, and sluggish in

---

[68] Nat'l Emergency Mgmt. Ass'n, Terrorist Comm. note 4.

[69] William C. Nicholson, Legal Issues in Emergency Response to Terrorism Incidents Involving Hazardous Materials: The Hazardous Waste Operations and Emergency Response ("HAZWOPER") Standard, Standard Operating Procedures, Mutual Aid and the Incident Command System, 9 WIDENER L. SYMP. J. 295, 308-09 & n.116 (2003).

[70] "Clearly, ICS is gaining momentum, though there's still a long road before it's a truly universal structure and language for managing incidents." Scott Baltic, ICS For Everyone, 3 HOMELAND PROTECTION PROF., Jan./Feb. 2004, at 1, 22.

[71] Press Release, White House, Fact Sheet: President Highlights a More Secure America on First Anniversary of Department of Homeland Security (Mar. 2, 2004), http://www.whitehouse.gov/news/releases/2004/03/20040302-4.html. ("On March 1, 2003, approximately 180,000 personnel from 22 different organizations around the government became part of the Department of Homeland Security – completing the largest government reorganization since the beginning of the Cold War.").

[72] "The primary mission of the Department is to – (A) prevent terrorist attacks within the United States; (B) reduce the vulnerability of the United States to terrorism; [and] (C) minimize the damage, and assist in the recovery, from terrorist attacks that do occur within the United States; . . ." 6 U.S.C. § 111(b) (1) (2004).

[73] Homeland Security Presidential Directive 5 mandates the address of all hazards. Press Release, White House, Homeland Security Directive/HSPD 5, ¶ 16 (Feb. 28, 2003), http://www.whitehouse.gov/news/releases/2003/02/20030228-9.html [hereinafter, HSPD 5] ("[The NRP] shall integrate Federal Government domestic prevention, preparedness, response, and recovery plans into one all-discipline, all-hazards plan.); *Id.* at ¶ 6("The Secretary will also provide assistance to State and local governments to develop all-hazards plans and capabilities, including those of greatest importance to the security of the United States, and will ensure that State, local, and Federal plans are compatible.").

[74] Id.

[75] Id. at ¶ 3.

[76] *Id.* at ¶ 17(b).

[77] *Id.* at ¶¶ 15-16.

[78] *Id.* at ¶¶ 17-20.

[79] HSPD 5 at ¶¶ 15-16.

[80] William C. Nicholson, *The New (?) Federal Approach to Emergencies*, HOMELAND PROTECTION PROF., Aug. 2003, at 8, 15.

[81] Jeffrey Manns, *Legislation Comment: Reorganization as a Substitute for Reform: The Abolition of the INS*, 112 YALE L.J. 145, 151(2002) ("The creation of this superagency may result in little more, however, than forcing a host of agencies to order new letterhead and change their seals. Worse still, the Department of Homeland Security may become a bureaucratic juggernaut, whose unmanageability may magnify the shortcomings of each component agency.").

responding to events.[82] Emergency responders were concerned that DHS would focus on law enforcement needs. Such anxiety seemed reasonable, since acts of terrorism, which are crimes,[83] were the impetus for the creation of DHS. NRP 1 and NIMS 1 were viewed as logical outgrowths of an agency insensitive and unresponsive to the needs of the non-law enforcement responders it was tasked with serving.[84]

Concern that the nation needed a universal approach to management of incidents led Congress, in the Homeland Security Act of 2002 ("HS Act"), to require adoption of IMS. In response to the Congressional mandate, during the summer of 2003 the Department of Homeland Security ("DHS") released a draft of the National Response Plan and National Incident Management System ("NRP 1" and "NIMS 1") to emergency response groups as well as to state and local government representatives.

Concerned stakeholders voiced several criticisms of the NRP 1 and NIMS 1. The language in the documents confused many observers. The text appeared to create new structures that could have the effect of complicating emergency response and coordination of resources. Some experienced emergency management professionals worried that the focus rested too heavily on terrorism, with a consequent lessening of the "all hazards" approach to risk management. In particular, they were bothered that natural hazards did not appear to receive sufficient emphasis. The NRP 1 and NIMS 1 incorporated problematic approaches to emergency responder issues. Emergency response organizations as well as representatives of state and local governments believed that their input should have been part of the process of creating the documents from their inception. Instead, DHS only asked for their feedback after several drafts had been created

and circulated internally at the agency. Some affected constituencies worried that this history boded ill for any attempts to try to rectify what many viewed as badly flawed blueprints for national preparedness and response.

DHS solicited input from affected groups, sending out preliminary versions of the National Response Plan (NRP or adopted NRP) and National Incident Management System (NIMS or adopted NIMS). August 1, 2003, was the date set by the agency for receipt of feedback from interested parties. So emphatic and critical was the response that DHS rapidly put together an NRP/NIMS State and Local Working Group in an attempt to obtain state and local endorsements without which the effort could not succeed. The Group first met to discuss the NRP line by line during the week of August 11, 2003. The Initial National Response Plan ("INRP") issued September 30, 2003, resulted from their efforts. In September, the group met again, this time to go over NIMS 1. The result of the second meeting was a significantly revised National Incident Management System – Coordination Draft ("NIMS 2"), which was issued for comment on December 3, 2003. On March 1, 2004, the first anniversary of the Department of Homeland Security, DHS Secretary Tom Ridge announced publication of an adopted version of NIMS.

On February 25, 2004, DHS released to state homeland security advisors and other homeland security partners a reworked National Response Plan Draft #1 ("NRP Draft #1"). Subsequently, DHS put out a number of additional draft versions of the NRP. These include NRP Draft #2, issued April 28, 2004, and the NRP "Final Draft" issued on June 30, 2004. The document has steadily grown in size as different stakeholders have provided feedback. Despite its nomenclature, the NRP "Final Draft" is not the same as the adopted document.

---

[82] Elishia L. Krauss, *Building a Bigger Bureaucracy: What the Department of Homeland Security Won't Do; Views You Can Use*, 32 PUBLIC MANAGER 57 (2003). ("Unfortunately, with bureaucracy we only create more layers of inefficiency and bureaucratic red tape, instead of streamlining the processes and focusing resources directly on the mission. Specifically, homeland security requires an organization that is collaborative, quick acting, and efficient. These are not qualities inherent in federal bureaucracies.").

[83] Terrorism is a federal crime. *See generally* Note, *Responding To Terrorism: Crime, Punishment, and War*, 115 HARV. L. REV. 1217, 1224 (2002) ("[T]he United States has traditionally treated terrorism as a crime. The U.S. Code contains criminal statutes that define and establish punishments for terrorism.").

[84] William C. Nicholson, *The New (?) Federal Approach to Emergencies*, HOMELAND PROTECTION PROF., Aug. 2003, at 18-21.

DHS finally adopted and issued the NRP on November 16, 2004. The NRP was reviewed and reissued. Eventually, it was replaced with the National Response Framework.

B. From NRP to NRF

The various versions of the NRP continued to be criticized as less than useful on the basis that it was not a plan. After Hurricane Katrina, DHS was required to establish the National Response Framework (NRF) by the Post-Katrina Emergency Management Reform Act, which laid out its contents specifically. Its first draft was greeted by a "wholesale outcry against the NRF from local, state and tribal emergency managers, and first responders who said that the plan ignored the important role of on-the-ground responders, who must implement the plan, and said that they were cut out of the mandated process of consultation." Under pressure from House Subcommittee on Economic Development, Public Buildings and Emergency Management, FEMA and DHS extended the comment period and heavily revised the NRF, which the International Association of Emergency Managers supported when it came out in early 2008.

The tension did not diminish. Rather, after FEMA's failure following Hurricane Katrina came calls for FEMA to be separated from DHS (see Chapter 12).

C. The National Response Framework

[To access the entire NRF, go to http://www.fema.gov/pdf/emergency/nrf/nrf-core.pdf]

Overview of the NRF[85]

General Approach. The NRF is part of a national strategy for homeland security. It provides the doctrine and guiding principles for a unified response from all levels of government, and all sectors of communities, to all types of hazards regardless of their origin.[86] Although the primary focus of the NRF is on response and short-term recovery, the document also defines the roles and responsibilities of the various actors involved in all phases of emergency management.[87]

The NRF is not an operational plan that dictates a step-by-step process for responding to hazards. Rather, the NRF appears to be an attempt to build flexibility into response efforts by setting up a framework that DHS believes is necessary for responding to hazards. Within this framework, the NRF gives users a degree of discretion as to how they choose to respond to the incident.

Components of the NRF Document. The NRF is organized into five parts. The introductory chapter presents an overview of the entire document and explains the evolution of the NRF, and identifies the various actors involved in emergency and disaster response. The chapter also discusses the concepts undergirding emergency preparedness and response by providing a list of what DHS describes as the "five key principles" of response doctrine.[88]

The first chapter of the NRF, entitled "Roles and Responsibilities," provides an overview of the roles and responsibilities of federal, state, and local governments, the nonprofit and private sectors, and individuals and households. The first chapter also discusses the roles and responsibilities

---

[85] Excerpted from Congressional Research Service "The National Response Framework: Overview and Possible Issues for Congress" Order Code RL34758 (November 20, 2008). Found on line at: http://assets.opencrs.com/rpts/RL34758_20081120.pdf

[86] U.S. Department of Homeland Security, "National Response Framework Released," press release, January 22, 2008, [http://www.dhs.gov/xnews/releases/pr_1201030569827.shtm].

[87] These phases are mitigation, prevention, response, and recovery.

[88] These are: (1) Engaged Partnership: the NRF advocates for open lines of communication among various emergency management entities and for support partnerships during preparedness activities so that when incidents take place, these various entities are able to work together.

(2) Tiered Response: responses to incidents begin at the local level. When local capacity is overwhelmed, state authorities assist the locality. Likewise, should the state be overwhelmed, assistance from the federal government is requested.

(3) Scalable, Flexible, and Adaptable Operational Capabilities: as incidents change in size, scope, and complexity, there needs to be a corresponding change in the response apparatus.

of those who hold various positions within these entities.

The second chapter, entitled "Response Actions," describes and outlines key tasks as they pertain to what DHS calls the "three phases of effective response." These phases include "prepare," "respond," and "recover." Preparing includes planning, organizing, equipping, training, exercising, and conducting evaluations. Activities related to responding include gaining and maintaining situational awareness,[89] activating and deploying resources and capabilities, coordinating response actions, and demobilizing. "Recover" activities are broken down into two broad categories. These are short-term and long-term recovery.

The third chapter of the NRF, entitled "Response Organization," discusses the organizational structure and staffing used to implement response actions, all of which are based on NIMS and ICS.[90] The NRF describes the organization and staffing structure of every entity responsible for preparedness and response in detail. The fourth chapter, entitled "Planning," describes the process of planning as it pertains to national preparedness and summarizes planning structures relative to the NRF. The chapter describes the criteria for successful planning and offers example scenarios for planning.

The fifth and final chapter of the NRF, entitled "Additional Resources," describes the 15 Emergency Support Function (ESF) Annexes to the NRF, eight Support Annexes, and seven Incident Annexes.[91] These annexes are listed in the table below. The final chapter also explains that the NRF and its annexes are posted online through the *NRF Resource Center*, which allows for ongoing revisions to the document.[92]

Annexes to the National Response Framework

Table 1. List of ESF Annexes

Emergency Support Function (ESF) Annexes
ESF #1 Transportation
ESF #2 Communications
ESF #3 Public Works and Engineering
ESF #4 Firefighting
ESF #5 Emergency Management
ESF #6 Mass Care, Emergency Assistance, Housing, and Human Services
ESF #7 Logistics Management and Resource Support
ESF #8 Public Health and Medical Services
ESF #9 Search and Rescue
ESF #10 Oil and Hazardous Materials Response
ESF #11 Agriculture and Natural Resources
ESF #12 Energy
ESF #13 Public Safety and Security
ESF #14 Long-Term Community Recovery
ESF #15 External Affairs

Table 2. Lists of Incident and Support Annexes

Incident Annexes	Support Annexes
Biological Incident	Critical Infrastructure and Key Resources
Catastrophic Incident	
Cyber Incident	Financial Management
Food and Agriculture Incident	International Coordination
	Private-Sector Coordination
Mass Evacuation Incident	Public Affairs
Nuclear/Radiological Incident	Tribal Relations
Terrorism Incident Law Enforcement and Investigation	Volunteer and Donations Management
	Worker Safety and Health

D. The National Incident Management System (NIMS)

What NIMS Is:

- A comprehensive, nationwide, systematic approach to incident management, including

---

(4) Unity of Effort Through Unified Command: a clear understanding of the roles and responsibilities of each entity is necessary for effective response. Moreover, effective response requires a unit of effort within the emergency management chain of command.

(5) Readiness to Act: all emergency management agencies, to the extent possible should anticipate incidents and make preparations to respond swiftly to them.

[89] "Situational awareness" refers to monitoring information regarding actual and developing incidents.

[90] ICS establishes a management system that helps agencies responding to an incident work together in a coordinated and systematic approach.

[91] ESFs provide the structure for coordinating federal interagency support for responses involving multiple federal agencies. Support Annexes describe how federal, state, tribal, and local entities, as well as nongovernmental organizations (NGOs) and the private sector, coordinate and execute the common functional processes and administrative requirements for incident management. Incident Annexes are specific hazard scenarios that require specialized and specific response efforts.

[92] NRF Resource Center, [http://www.fema.gov/emergency/nrf/#].

the Incident Command System, Multiagency Coordination Systems, and Public Information
- A set of preparedness concepts and principles for all hazards
- Essential principles for a common operating picture and interoperability of communications and information management
- Standardized resource management procedures that enable coordination among different jurisdictions or organizations
- Scalable, so it may be used for all incidents (from day-to-day to large-scale)
- A dynamic system that promotes ongoing management and maintenance

What NIMS Is NOT:

- A response plan
- Only used during large-scale incidents
- A communications plan
- Only applicable to certain emergency management/incident response personnel
- Only the Incident Command System or an organization chart
- A static system[93]

This text is not intended to be a detailed primer on the contents or workings of the NRF and NIMS. Perusal of the contents of FEMA's NIMS Resource Center http://www.fema.gov/emergency/nims/ will reveal multiple resources for those interested in further study of the system.

For further insight regarding the NRF and NIMS within this text, see the case in Part 8. How Far Do NIMS' and the NRF's Writ Run? and APPENDIX A: Fiscal Year 2009 NIMS Implementation Objectives.

VII. PUTTING IT ALL TOGETHER*

As you read the following case, consider and compare the reality of a wildfire response with the approaches outlined in the materials above.

IN THE UNITED STATES DISTRICT COURT FOR THE DISTRICT OF IDAHO

Deanna C. Buttram, et al. Plaintiffs

v.

United States of America

CIVIL CASE No. 96-0324-S-BLW
CIVIL CASE No. 96-0452-S-BLA
CIVIL CASE No. 97-0129-S-BLW

FINDINGS OF FACT AND CONCLUSIONS OF LAW

INTRODUCTION

William Buttram and Joshua Oliver lost their lives fighting a wildfire known as the Point Fire. Their families, the plaintiffs in this consolidated action, claim that the agency supervising the firefighters, the Bureau of Land Management (BLM), is responsible for the deaths. The plaintiffs brought suit under the Federal Tort Claims Act, challenging a broad range of decisions made by the BLM. The Court held a court trial beginning on December 7, 1998, and ending December 15,

---

[93] FEMA, National Incident Management System at 6. (December 18, 2008). Found on line at: http://www.fema.gov/pdf/emergency/nims/NIMS_core.pdf

*The author wishes to thank Jan Amen of the Texas Forestry Service for providing this valuable case.

1998. The Court received the final post-trial briefs on January 26, 1999.

On the basis of the evidence and legal arguments, the Court finds that the BLM and the Kuna Rural Fire District (Kuna RFD) both committed negligence that was the proximate cause of the deaths of Buttram and Oliver. The Court finds that the Kuna RFD bears 65% of the responsibility and that the BLM bears 35 percent of the responsibility. The Court's damage award is set forth in detail at the end of this decision. The findings of fact and conclusions of law supporting this decision are set below.

FINDINGS OF FACT

1. Late in the afternoon July 28, 1995, lightning sparked a fire in the dry grasses and sagebrush desert land about 16 miles southwest of Boise, Idaho.
2. The fire was burning on BLM land, and the BLM fire crews responded.
3. The first BLM crew to reach the fire was headed by David Kerby who, by virtue of being the first crew chief to reach the fire, was designated as the Incident Commander, the person with overall responsibility for fighting the fire. Other crews soon followed. The fire became known as the Point Fire.
4. Kerby's decision as Incident Commander (IC) would be guided by BLM fire suppressor policies that depended in part on the fire's location. The Point Fire was burning near the Snake River Birds of Prey National Conservation Area, a 500,000-acre sanctuary for the largest concentration of raptors in the world. To protect the birds, the BLM had made a policy decision years earlier to "aggressively attack and suppress all wildfires" in this area. That policy is contained the BLM's Boise District Fire Management Activity Plan (FMAP).[94]
5. The FMAP also played an important role in determining what resources Kerby had to work with in fighting the fire. The FMAP includes a Lightning Operations Plan that is triggered when lightning fires are occurring, and was in effect during the Point Fire.
6. The Lightning Operations Plan describes the resources that should be made available for different types of lightning-caused fires, and provides a guide for rating the severity of the fire. The rating system – known as the "Burning Index" – rates fires in a numerical scale depicting the amount of effort needed to contain the fire given the time of year, fuel conditions, and other factors. The Burning Index increases as the days become hotter and the fuel conditions become more incendiary. The higher the Burning Index number, the greater the resources that should be made available to fight the fire. For example, a fire with a Burning Index between 0 and 34, burning early or late in the fire season, would be rated as a Response Level I fire, and the FMAP dictates that the "typical response would be single unit or crew and a detection aircraft, if available." Another section of the FMAP also recommends that two fire engines be dispatched along with the detection aircraft.

---

[94] The FMAP is the result of a bottom-up process of developing fire management plans with the BLM. These plans were developed first at the District level and then consolidated into statewide plans what were in turn consolidated into a BLM National Plan. The Bolse District FMAP was complete in July of 1994. It was consolidated into BLM's Idaho FMAP, which was completed in January of 1994. It was consolidated into the BLM's Idado FMAP, which was completed in January, 1995, and approved by the BLM in March, 1996. Two other plans relevant to the Point Fire were the BLM's Lower Snake River Ecosystems Fire Preparedness Plan, issued in June, 1995, and the South Canyon Fire Abatement Plan issued in May, 1995. Further guidance that was available at the time of the Point Fire was contained in two publications of the National Wildfire Coordinating Group (NWCG): (1) The Fireline Handbook, a "nuts and bolts" pocket field guide for a firefighting techniques; and (2) The Wildland Fire Qualifications Subsystem, a set of standards for training. The NWCG is a collection of federal agencies, including the BLM, and the National Association of State Foresters.

7. The Point Fire was burning during the summer fire season, with temperatures near the high 90s. Because of higher-than-normal spring moisture, cheat grass growth was especially dense and mature sagebrush added to the fuel load.

8. The fire was caused by dry lightning, and the National Weather Service in Boise had forecast – both the day before and the morning of July 28, 1995 – that "gusty erratic winds of 55 MPH" could be "near any thunderstorms."

9. At noon on July 28, 1995, before the Point Fire began, the National Weather Service in Boise issued a "Fire Weather Watch" for dry lightning, and forecast that thunderstorms would be moving through western Idaho "late this afternoon and tonight."

10. Given these facts, the Fire Management Specialist with BLM, William Casey, Jr., gave the Point Fire a Burning Index of 31, requiring the implementation of a Response Level I.

11. In fact, the BLM deployed even more resources to the Point Fire that would be called for in a Response Level I Fire with a Burning Index of 31. The BLM dispatched five engines, a bulldozer, a tender, and a detection helicopter.

12. IC David Kerby arrived at the fire at 7:00 P.M.

13. Upon arriving, IC Kerby examined the fire from the ground, and then went up in a BLM helicopter to get an aerial view.

14. Shortly after IC Kerby returned from his helicopter tour, the Kuna RFD's Fire Chief, Richard W. Cromwell, contacted Kerby and asked him if he needed any assistance. Kerby responded that he could use a brush truck and a water tender. Cromwell responded that he would comply with that request.

15. A brush truck is a four-wheel-drive vehicle designed to work directly on the fire line of a wildland fire. It carries 1,500 gallons of water, and has nozzles attached to the front bumper that the driver can operate from inside the vehicle to spray water on the fire. There is also a hose that can be manually operated at the rear of the vehicle.

16. A water tender, on the other hand, is not designed to attack fires, but is rather a mobile water reservoir. It contains a larger water tank, about 2,500 gallons, and is typically stationed away from the fire line to permit the brush trucks to refill their tanks.

17. The Kuna RFD identified their vehicles by number. The brush trucks were 620 and 622, and the tender was 625.

18. In response to IC Kerby's request, Kuna RFD Chief Cromwell radioed Kuna RFD Captain Doyle McPherson and told him to (1) respond to the Point Fire with two brush trucks and a water tender; (2) to assume command of the Kuna firefighters who responded; and (3) to instruct the occupants of the two brush trucks to keep their trucks together.

19. McPherson said he would do so, but questioned Cromwell whether Cromwell also wanted to send Kuna's squad cars. The squad cars were smaller than the brush trucks, but also were equipped with water tanks. In an earlier conversation Cromwell told McPherson that the squad cars should be deployed with the brush trucks to provide a safety backup. But on this occasion, Cromwell told McPherson not to deploy the squad cars.

20. As of July 28, 1995, the chain of command at Kuna RFD, in order of authority, was as follows: (1) Chief Cromwell; (2) Assistant Chief Darwin Taylor; (3) Captain McPherson; and (4) Captain Joseph Steer.

21. Under Kuna RFD policies, a firefighter was qualified to drive a vehicle as soon as he learned to operate the vehicle.

22. The typical assignment was two firefighters per vehicle. The Kuna RFD leadership did not assign certain firefighters to certain trucks. The pairings were random; the first two firefighters who could get in a vehicle were a team, assuring one was qualified to drive.

23. Bill Buttram and Josh Oliver, both volunteer firefighters, took truck 620.

24. Buttram was qualified to drive and he took the driver's seat while Oliver took the passenger seat. McPherson was also a passenger.

25. Buttram started as a volunteer fireman with the Kuna RFD in October of 1994, nine months prior to the Point Fire. Buttram did not, however, have nine months of experience fighting wildfires by the time of the Point Fire. The fire season is May through September, and thus Buttram largely missed the 1994 fire season; the 1995 fire season was his first wildfire season.

26. At the time of the Point Fire, Buttram was 31 years old, and was employed by the Idaho Department of Corrections as a prison guard earning $28,273 (on an annualized basis).

27. Oliver started as a volunteer fireman with the Kuna RFD in October of 1994, and was also experiencing his first wildfire season when he responded to the Point Fire. He was 18 years old at the time of the Point Fire.

28. Before Buttram, Oliver, and McPherson left the station in truck 620, they heard Chief Cromwell tell them over the radio that (1) truck 620 should stay together with truck 622, and (2) that when they arrived at the fire site, they should "switch to channel 16 [the BLM radio channel], and talk to BLM and find out where they want you."

29. The BLM was communicating by radio over Channel 16, also known as the BLM channel.

30. Responding in truck 622 were volunteer firefighter Michael Law and Robert Black. Law had worked as a volunteer fireman since 1976; Black since 1970.

31. Responding in the tender truck 625 were Captain Joseph Stear and Jenny Taylor. Captain Stear had more than a decade of experience while Taylor had very little experience.

32. Captain McPherson testified that he would have liked more experience in truck 620, but he felt that Buttram and Oliver were qualified and so did not make any truck assignment changes.

33. The Kuna vehicles reached the Point Fire at about 7:30 P.M. They met together on Swan Falls Road about 1,000 feet from the northeast perimeter of the fire, where Captain McPherson gave them a briefing.

34. In the briefing, Captain McPherson repeated Chief Cromwell's admonitions for 620 and 622 to stay together, and for all Kuna personnel to take orders from the BLM IC. Captain McPherson told them to maintain radio contact with BLM, to stay in safe zones, to get into the black burned-out areas if there was any trouble, and to be aware of escape zones.

35. Captain McPherson then told 620 and 622 to proceed north on Swan Falls Road and find out what the BLM IC wanted them to do.

36. Captain McPherson decided to stay with the water tender, 625.

37. Immediately after 620 and 622 left the initial staging area, Captain McPherson discovered that tender 625's radio was not capable of reaching BLM channel 16.

38. Chief Cromwell was also not monitoring BLM channel 16.[95]

39. Thus, the only channel that Captain McPherson could monitor and transmit messages over was the Kuna channel, and the only channel that Chief Cromwell was monitoring was the Kuna channel.

---

[95] While Captain Cromwell's squad car could tune into BLM channel 16, Chief Cromwell did not use that radio to monitor BLM channel 16 once he arrived back at the Kuna station house, as this exchange in his deposition shows:

Question: [by Government counsel]: After you drove away from the fire scene, did you continue to-did you yourself monitor[BLM] channel 16 on your radio so you could sort of attempt to keep track of what was going on out there?

Answer [by Chief Cromwell]: No. When we got back to the fire station at that time, we didn't have [BLM channel] 16 in our bay station, and we had gotten out of the truck and went into the station and we were sitting outside in one of the bay doors watching and listening to conversation between our trucks. At that time we had no radio on that had BLM.

See Deposition of Cromwell at 104, 11. 11-21.

40. The radio on 620 and 622 had scan capability. When set in a scan mode, the radios automatically switched between BLM's channel 16 and the Kuna frequency, broadcasting whatever messages came over those channels, and allowing the occupants of the trucks to monitor both channels without having to manually operate the radio. If both channels were broadcasting at the same time, Kuna frequency would "step on" BLM's channel 16, so that only the Kuna broadcast could be heard.

41. If the radios were not set on scan mode, the occupants could only hear the single frequency that they had manually chosen.

42. As 620 and 622 pulled away from the initial staging area, Captain Stear noticed that 620 had some loose gear on the back of the truck. He tried to reach 620 on Kuna channel but 620 did not respond. This is some evidence that 620's radio was either not working properly or that 620 had manually selected the BLM channel, and not put its radio on scan mode.

43. 620 and 622 proceeded south on Swan Falls Road and reached the northeast perimeter of the fire at about 7:30 P.M.

44. 622 then contacted IC Kerby on BLM channel 16 and asked Kerby what they should do.

45. IC Kerby directed the Kuna RFD engines to "bump in behind" a BLM vehicle and work the northern perimeter of the fire. Another BLM vehicle would bring up the rear, sandwiching the two Kuna trucks.

46. The procession of these four vehicles began working the northern perimeter of the fire.

47. 622 and 620 applied water through the front bumper nozzles to the flames, and had an immediate effect – the flames diminished in the area where the trucks were working, and the smoke changed color, indicating that the fire was being suppressed.

48. After moving a little more than a quarter mile in a westerly direction along the fire's northern perimeter, IC Kerby radioed 622 and told them to get closer to the perimeter, and that they should continue to proceed around the entire perimeter of the fire. Both trucks complied with those orders.

49. The trucks moved around the west side of the fire, and then proceeded east along the fire's southern perimeter.

50. At about 8:15 P.M., the procession had nearly circumnavigated the fire's entire perimeter, and had reached the fence at the southeast corner of the fire.

51. There, someone from one of the Kuna trucks asked BLM's Bryan Barney for further directions. Barney radioed IC Kerby who told them to turn around and return around the fire's perimeter, soaking it down as they went. The procession complied, turned around, and started reworking the southern perimeter.

52. Just after 8:15 P.M., the National Weather Service called the BLM dispatch office and issued a red flag warning for the Snake River Valley area, including the area where the Point Fire was burning, warning of an approaching thunderstorm with accompanying winds of over 50 M.P.H.

53. At 8:22 P.M., the BLM dispatch office called IC Kerby over the BLM channel 16 and told him that a red flag warning had been issued by the National Weather Service and to expect high winds of up to 50 M.P.H.

54. IC Kerby felt that since the red flag warning had come over the BLM channel 16, all the firefighters had heard it and there was no need for him to ensure that everyone had heard the warning.

55. At the time of the red flag warning, 8:22 P.M., 620 and 622 were on the fire's southern perimeter traveling westerly, about a quarter mile from the fence where they had turned around.

56. It is unclear if 620 heard BLM dispatch's red flag warning message. The Kuna channel transcript shows that at 8:24 P.M., 620 radioed Captain McPherson on the Kuna channel and said "[W]e're basically

just doing mop-up. Is it okay for Josh to get some drive time just doing mop-up?" Captain McPherson said no, and 620 responded that they were "tucked in behind 622." If the times on the transcript and BLM dispatch log book are precise, then just two minutes after the red flag warning was given, Bill Buttram was obviously unconcerned and asking Captain McPherson to allow an unqualified beginning fireman to drive. This shows either that (1) Buttram and Oliver did not hear the red flag warning, or (2) that they did not appreciate the seriousness of the red flag warning. If, however, the times on the logs are off by a minute or so, it is possible that Buttram was calling on the Kuna channel at the same time that the red flag warning was given, in which case Buttram's transmission would "step on" the red flag warning, then 622 should not have heard the red flag warning either. Although Robert Black in 622 does not remember getting the red flag warning over their radio, Mike Law does think it came over the radio, although it is unclear whether he heard the BLM dispatch warning or a warning at a later time during a conversation with other employees. *See* Law Deposition at 73. The evidence is inconclusive, and the Court can therefore not reach any definitive conclusion as to whether 620 heard the red flag warning.

57. After the red flag warning given at 8:22 P.M., 620 and 622 continued along the western perimeter.

58. When they reached the western perimeter, 622 ran out of water. Black radioed IC Kerby on the BLM channel and asked for direction. IC Kerby told him to refill and stand by because a windstorm was to pass through.

59. Law testified that immediately after IC Kerby gave these orders, 620 came on the radio and stated that they still had water and would stay on the line. However, this testimony was contradicted by Law's partner, Black, who testified that it was his impression that 620 was empty and would be following 622 in to refill. Captain Stear testified he saw 622 at the refill site; Black told him that 620 was right behind them coming in to refill. The Court finds Black's testimony most credible because he was the person operating the radio in 622, and because his account is corroborated. In addition, Law did not recall any such "we-will-stay-on-the-line" transmission from 620 in his deposition taken on March 3, 1998. It was not until he testified at trial, that Law gave the account. For these reasons, the Court finds that 620 never told 622 that they still had water and would stay on the fire line.

60. 622 then proceeded east through the black burned-out area until it turned north just before the fence in order to get to the fence break that was located at the northeast corner of the fire perimeter.

61. Just before 622 reached the fence break, 620 radioed 622 and stated that they were overheating and requested assistance. At this time, 620 did not appear to be in any distress. 622 responded that 620 should remove a screen on the front of the vehicle. 620 responded to the effect that they heard the message and would check the screen. That is the last radio contact 622 had with 620.

62. 622 proceeded through the fence and onto Swan Falls Road, stopping by the tender 625 to refill. Tender 625 was located at this time on Swan Falls Road at the southeast corner of the fire perimeter.

63. Black told Captain McPherson and Captain Stear that 620 was right behind them and coming to refill.

64. When 620 did not arrive at the refilling site, Captain McPherson started walking north on Swan Falls Road to look for the truck. When Captain McPherson had reached the middle of the fire's eastern perimeter, he saw 620 in the black burned-out area heading towards him, retracing the route taken by 622. When McPherson first saw 620, it was about 600

fcct away from McPherson, heading for the northeast perimeter of the fire with the apparent intent to follow 622's path through the break in the fence.

65. In a wildfire, any black burned-out area might be a safety zone depending on how much unburned fuel remains there; a lack of unburned fuel ensures that area will not be burned over a second time. But the black's functions as a safety zone is also dependent on two other variables; the wind, and the movement of the firefighters. A high wind will kick up dust and ash in the black area, obscuring visibility. This poses little danger if the black contains no unburned fuel and the firefighters are stationary. However, when the firefighters are moving in the black towards the perimeter of the fire, beyond which lies unburned fuel in the path of the high winds, the black is much less of a safety zone. The complete lack of visibility that accompanies high winds in the black makes the firefighters prey to disorientation, not a problem while they are sitting still, but a major concern when they are moving toward unburned fuels in the path of winds.

66. McPherson felt that 620 was just minutes from reaching the fence break, and the truck was close enough that McPherson could see passenger Oliver "real well."

67. McPherson started walking toward the fence break with the idea that he would meet up with 620 there.

68. At this point the winds had increased dramatically. Visibility in the black burnt-out area was reduced to almost nothing as the winds kicked up the dust and ash.

69. 620, now close to the fence, turned north attempting to find the fence break. McPherson could hear 620 bouncing and moving north at a high rate of speed.

70. Due to a combination of obscured visibility, disorientation, and panic 620 overshot the fence break and drove into unburned cheat grass and sagebrush due north of the fire's northern perimeter.

71. The winds revived the fire, pushing it northward at a furious rate.

72. 620 was still moving at a fast rate, but it was now trying to outrun the fire that was close behind it. After driving 1,750 feet from the fire's northern perimeter, 620 stalled in the middle of unburned cheat grass and sagebrush.

73. Bill Buttram got on the radio and relayed a frantic message to Captain McPherson over the Kuna channel: "We're on the north line, Doyle, we got fire coming hard, this thing has died."

74. Captain McPherson responded inaudibly, and Buttram said "It's not going to let us out of here."

75. Chief Cromwell then came on the radio and asked 620 to identify the problem.

76. Buttram responded that "we're surrounded by fire." When asked to repeat his message, Buttram stated "The truck has been overtaken by fire."

77. That was the last radio communication anyone received from 620.

78. Shortly thereafter, 620 was overtaken by fire.

79. Bill Buttram and Josh Oliver were found dead in the front seats of 620.

80. At that time, 620 was located 713 feet west of Swan Falls Road and 1,750 north of the northern fire perimeter as it existed just prior to the fire's blow-up.

81. About 25 minutes elapsed between the red flag warning given over the BLM channel at 8:22 P.M. and the high winds that swept through at 8:46 P.M.

82. Joshua's mother, Darla Reber, was forty-one at the time of his death and she had a life expectancy at that time of thirty-two years.

83. Plaintiff Michael Oliver was Joshua's father. Michael Oliver and Darla Reber divorced when Joshua was eleven years old. At that time, Michael moved to Illinois and Darla moved to New Mexico. Michael testified that Darla did not make it difficult for him to see or talk to Joshua, but nevertheless Michael rarely even spoke

to Joshua during the next seven years. Joshua did come to live with Michael for a month in Illinois, and there were two other extended stays. In addition, Michael was always current on his support payments.

84. Deanna C. Buttram and William Buttram were married on April 27, 1985, and thereafter resided as husband and wife. When William Buttram died, Deanna was 31 years old and their son Jeremiah R. Buttram was one year old.

LEGAL STANDARDS

85. The plaintiffs, Joshua Oliver's father and mother and William Buttram's wife and son brought this suit against the BLM under the Federal Tort Claims Act (FTCA).

86. Under the FICA, the government is liable "in the same manner and the same extent as a private individual under like circumstances." 28 U.S.C. § 2674.

87. Whether a private person would be liable under like circumstances is determined by application of "the law of the place where the act or omission occurred." 28 U.S.C.§ 1346(b).

88. Because this case arose in Idaho, Idaho tort law will govern the results.

89. Under Idaho law, a plaintiff must show the following by a preponderance of the evidence to prove negligence: (1) a duty, recognized by law, requiring a defendant to conform to a certain standard; (2) a breach of that duty; (3) a causal connection between the defendant's conduct and the resulting injuries; and (4) actual loss or damage. See *West v. Sonke*, 968 P.2d 228 (1998).

90. Under Idaho law, an employer "has a duty to exercise reasonable care commensurate with the nature of its business in order to protect employees from hazards to the employment and to provide him with safe tools, appliances, machinery, and working places," although the employer has no duty to warn of a danger that was not reasonably foreseeable. *West*, 968 P.2d at 237.

91. A danger is reasonably foreseeable if it "is apparent, or should be apparent, to one in the position of the actor. The actor's conduct must be judged in the light of the possibilities apparent to him at the time, and not by looking backward 'with the wisdom of the event." W. Page Keeton et al., Prosser and Keeton on the Law Torts §31, at 170 (5th ed. 1984). "In light of the recognizable risk, the conduct, to be negligent, must be unreasonable. No person can be expected to guard against harm from events which are not reasonably to be anticipated at all, or are so unlikely to occur that the risk, although recognizable, would commonly be disregarded." Id. "On the other hand, if the risk is an appreciable one, and the possible consequences are serious, the question is not one of mathematical probability alone.... It may be highly improbable that lightning will strike at any given place or time; but the possibility is there, and it may require precautions for the protection of inflammables. As the gravity of the possible harm increases, the apparent likelihood of its occurrence need be correspondingly less to generate a duty of precaution." Id.

92. The term "proximate cause" means "a cause which, in natural probable sequence, produced the damage complained of. It need not be the only cause. It is sufficient if it is a substantial factor concurring with some other cause acting at the same time, which in combination with it, causes the damage." *Fussel v. St. Clair*, 120 Idaho 591, 595, 818 P.2d 295, 298 (1991).

93. "There may be one or more proximate causes of the injury. When the negligent conduct of two or more persons contribute concurrently as substantial factors

in bringing about an injury, the conduct of each may be a proximate cause of the injury regardless of the extent to which each contributes to the injury." Idaho Jury Instructions #230.

94. The comparative fault of the person or his legal representative which is as great as the comparative fault of the defendant bars recovery of the damages for his death or injury. *See* Idaho Code §6-801. Comparative fault which is not as great as the comparative fault of the defendant diminishes the damages allowed in proportion to the amount of comparative fault attributable to the person recovering. Id.

95. When the Court is apportioning negligence under the comparative fault provisions of §6-801, it may include parties to the transaction which resulted in the injury, whether or not those parties to the lawsuit. See *Pocattello Industrial Park Co. v. Steel West, Inc.*, 101 Idaho 783, 621 P.2d 399 (1980).

96. In an heir's action for wrongful death, the negligence of the decedent is imputed to the plaintiff. See *Adams v. Krueger*, 124 Idaho 74, 856 P.2d 864 (1993).

97. The elements of damage in a death case are as follows: (1) the reasonable expenses incurred for the decedent's funeral; (2) the reasonable value of the hospital/ambulance care received prior to decedent's death; (3) the reasonable value of the loss of decedent's services, comfort, society, and conjugal relationship, and the present cash value of such loss that is reasonably certain to occur in the future, taking into consideration the decedent's life expectancy and circumstances; and (4) the amounts decedent would have contributed to support plaintiff, and the present cash value of the amount of money decedent would have been reasonably certain in the future to have contributed to the support

of plaintiff, taking into consideration decedent's life expectancy and circumstances. *See* Idaho Jury Instruction 911-1.

98. Damages are not awarded for any grief or sorrow plaintiff may have suffered by reason of the death of decedent or for any pain or suffering of the decedent before he died. *See* Idaho Jury Instructions 911-2.

99. In a parent's action for wrongful death of a child, the trier of fact may consider the parent's degree of intimacy with his or her child in setting damages. See *Gardner v. Hobbs*, 69 Idaho 288, 206 P.2d 539 (1949).

100. Noneconomic damages are limited to $400,000, plus the percentage amount by which the Idaho Industrial Commission adjusts the average annual wage. *See* Idaho Code § 6-1603.

101. As of this date, the noneconomic damages cap under Idaho Code §61603 is $590,291.26.

102. Under Idaho Code §6-1603, the jury is not to be instructed on the statutory cap. The statute thus contemplates that a jury should award more than the cap. If the jury also attributes fault to multiple defendants, or to the plaintiff and defendants, an issue arises as to whether the verdict should be reduced to the cap amount first before the comparative fault reductions are made, or whether the comparative fault reductions should be made first, and if the verdict remains above the cap, only at that point reduce the verdict to the cap amount. The statute and the Idaho courts offer no guidance on this issue.[3] The plaintiffs urge the Court to apply comparative fault before reducing any award to cap amount, and they cite Colorado Supreme Court decision in support of their argument. In

---

[3] The lack of Idaho case law interpreting this Idaho statute raise a question whether the issue should be certified to the Idaho Supreme Court. It would not make sense, however, to delay this decision to await an answer on certification. First, the Idaho Supreme Court will need this decision to place the issue in its proper context. Second, the parties are free to file motion to reconsider after receiving this decision, and the Court can determine at that point whether the issue should be certified. The Court expressed no opinion at this point whether certification is appropriate.

General Electric Company v. Niemet, 866 P.2d 1361 (1994), five members of the Colorado Supreme Court interpreted a similar Colorado statute to require that the court reduce the verdict before applying the statutory cap. Two justices dissented, finding that such a result would allow plaintiffs to get around the statutory cap when they were injured by multiple defendants. The Court finds that the dissent has the better arguments. The majority opinion relies heavily on legislative history rather than the language of the statute. There is, however, no similar legislative history available concerning the Idaho statute. Instead, the Court is left with nothing but the language of Idaho Code §6-1603. That language is intended to limit a plaintiff's recovery regardless of the number of defendants involved; there is no exception for cases where the plaintiff is injured by multiple defendants. Plaintiffs' interpretation, however, would essentially read such an exception into the statute. For example, if a victim is injured by one defendant and has $2 million in noneconomic damages, the victim would collect the statutory cap, $590,291.26. If another victim with noneconomic damages of $2 million is injured by two defendants, each 50 percent at fault, the Court would –

under plaintiffs' proposal – apply the comparative fault first, reducing the award against each defendant to $1 million, and would then apply the statutory cap, reducing each award to $590,291.26, permitting the plaintiff a total recovery of over $1 million. Thus, the plaintiff injured by two defendants would collect twice the cap amount. That result essentially places an exception in the statute that the legislature did not intend. The statute is a cap on the plaintiff's total recovery with no exception for multiple defendant cases. Thus, the Court rejects plaintiffs' interpretation of Idaho Code § 6-16-2. *See General Electric*, 866 P.2d at 1369 (Cj. Rovira dissenting) (finding that the Colorado legislature could not have intended, in drafting a similar statute, to allow, "by the sheer fortuity of being injured by multiple defendants, one plaintiff [to] collect more damages than a second plaintiff with identical injuries who had the misfortune of being injured by only one person.").

103. Payments that Ms. Buttram received under Idaho's Workers' Compensation law, the federal Social Security Administration benefit programs, and private life insurance, are not deductible under Idaho Code § 61606 as compensation received from collateral sources for personal injury.

ANALYSIS

Definition of BLM's Duty

104. The BLM, through IC Kerby, had a duty to exercise reasonable care to protect Buttram and Oliver from foreseeable hazards incident to fighting the Point Fire.

105. More specifically, the BLM had the following duties:

106. First, the BLM IC had a duty to ensure that Buttram and Oliver were assigned duties commensurate with their ability

and with the Kuna RFD's qualifications. This duty is imposed on the BLM in part due to its own regulation found in §9215.115 of the BLM *Manual on Fire Training and Qualification*. It is also imposed on the basis that the BLM IC had supervisory authority over Buttram and Oliver, and was in the best position to know of the fire's extent and dangers. When the BLM is fighting a fire on BLM property, the BLM IC has access to much more information about the fire than the volunteer firemen supplied by rural fire districts with limited budgets. On the other hand, if the BLM IC assigns tasks to volunteers with no consideration of whether the volunteers' qualifications are sufficient to safely perform the tasks, then the BLM IC has not complied with the general duty to provide for the safety of those volunteers. Thus, the BLM IC must make some determination about the volunteers' qualifications. At the same time, it must be recognized that the BLM IC is forced by the emergency nature of firefighting to make quick decisions. The BLM IC cannot be expected to conduct a sit-down interview with each volunteer, but must be allowed to rely on certain indicators of qualifications. In other words, the BLM IC complies with his or her duty when he or she makes reasonable assumptions about the qualifications of volunteers based on the appearance and conduct of the volunteers, or other factors that bear on qualification. However, when the BLM IC makes reasonable assumptions about volunteers' qualifications, rather than depending on actual knowledge, the BLM IC is thereafter operating under a relatively higher duty to provide for the safety of those volunteers. The level of duty varies on a sliding scale depending on the BLM IC's knowledge. The duty is at its highest when the BLM is making reasonable assumptions without any actual knowledge, and lessens as the BLM IC makes appropriate assignments based on actual knowledge of the qualifications. In the present case, the BLM IC was proceeding on reasonable assumptions about their qualifications without having actual knowledge.[4]

107. Second, the BLM IC has a duty to fully instruct rural fire district volunteers, before fire suppression efforts begin, about the nature of the fire, fuel conditions, weather information, safety reminders, command structure, and radio use. This information is crucial to provide for the volunteers' safety, especially when the IC is making assumptions about their qualifications without any actual knowledge. The BLM IC is in the best position to provide such information. In this case, IC Kerby had seen the fire from a helicopter, and could describe the nature and extent of the fire. He had information from the National Weather Service that included a fire weather watch. He knew the common safety practices that should be reviewed even by highly trained firefighters. Specifically, the existence of a fire weather watch made it imperative to review with the volunteers the significance of a red flag warning, and how the high winds that accompany such heavy weather would intensify the fire and

---

[4] With the benefit of the trial evidence, the Court reaffirms its earlier summary judgment decision that the BLM IC's assumptions that Buttram and Oliver were qualified to work the fire line in a brush truck was not a policy decision protected by the discretionary function exception. Based on the trial evidence, as set out in paragraphs 106 and 100 above, the Court will clarify its reasoning. In the summary judgment proceedings, the Government argued that BLM IC was entitled to assume qualifications of volunteers from the mere fact that they were dispatched by the rural fire district. The Court rejected that contention and found that the BLM IC was not entitled to make such assumptions under §9215.11E. After hearing the trial evidence, the Court is now convinced that §9215.11E does not permit the BLM IC to make certain reasonable assumptions based on outward indicators of the volunteers; qualifications. However, the Court continues to reject the BLM's contention that the mere dispatch of volunteers by a rural fire district to a BLM fire entitles the BLM IC to assume that the volunteers are qualified to do anything. Section 9215.11E clearly puts the responsibility on the BLM to either determine the volunteers' qualifications or make reasonable assumptions from outward indicators. The mere fact that volunteers were dispatched, standing by itself, is not a sufficient outward indicator on which to make any reasonable assumption. In this case, however, there was more than a mere dispatch of volunteers. As the Court will discuss above, Buttram and Oliver arrived in a brush truck, a vehicle designed to work directly on the fire line. The BLM IC therefore made a reasonable assumption that Buttram and Oliver were qualified to work on the fireline. That decision was not a "legislative [or] administrative decision grounded in social, economic, [or] political policy" and hence was not protected by the discretionary function exception. *United States v. Gaubert*, 499 U.S. 315, 323 (1991).

obscure visibility. Briefing about the command structure was important, because the BLM IC knew he was working with volunteers from a rural fire district, and so needed to reduce any chance of confusion as to who was in charge. Finally, he knew the importance of everyone monitoring the same radio channel during wildland fires where the firefighters are spread out over a large geographic area. IC Kerby had a duty to impart his information to the Kuna RFD firefighters before they started their suppression efforts.

108. Third, the BLM IC had a duty to (1) ensure that all firefighters heard the red flag warning that was issued at 8:22 P.M.; (2) ensure that Kuna RFD volunteers understood the significance of that warning, i.e., that the high winds could dangerously intensify the fire and substantially decrease the visibility; and (3) ensure the RFD volunteers knew that the high winds would be coming immediately and from the south. The Government's own expert, Ronald Johnson, agreed that the BLM IC had a duty to ensure that the firefighters got the red flag warning. The two additional aspects of this duty (items (2) and (3) listed above) arise in this case largely because the BLM IC failed to hold a safety briefing before work began, and had no idea if the Kuna RFD volunteers knew what a red flag warning was.

109. Fourth, the BLM IC had a duty after the red flag warning was issued to take all reasonable steps to ensure that 620 stayed away from the Point Fire's northern perimeter. Because he failed to hold an initial safety briefing, and knew nothing of Buttram's and Oliver's experience, the BLM IC had a special duty to Buttram and Oliver to warn them to stay away from the northern perimeter of the Point Fire because approaching high winds would drive the fire in that direction.

BLM's Breach of Duty

110. BLM did not breach its duty to ensure that Buttram and Oliver were assigned to duties commensurate with their qualifications and the qualifications of the Kuna RFD. Buttram and Oliver responded to the fire in a brush truck. The brush truck is specifically designed to work the fire line. IC Kerby reasonably assumed that two Kuna volunteers driving a brush truck were qualified by Kuna RFD standards to work the fire line in that brush truck. It was also reasonable for IC Kerby to assume that Kuna RFD had maintained 620 in a safe condition. However, that was all IC Kerby could reasonably assume without further inquiry. In other words, IC Kerby could not assume from Buttram's and Oliver's appearance in the brush truck that they knew about (1) the Point Fire's nature and extent; (2) the most recent weather report; (3) fuel conditions; (4) safety guidelines; (5) who was in command; (6) what radio channel was being used; and (7) that a red flag warning would mean the imminent approach of winds that would increase the intensity of the fire and obscure visibility.

111. The BLM IC breached his duty to hold a briefing on the issues listed in paragraph 107. IC Kerby held no briefing at all before assigning Buttram and Oliver to the fire line. Because IC Kerby was assuming that Buttram and Oliver were qualified without inquiring into their training or asking any questions about their equipment, it was especially important for IC Kerby to review all issues listed in paragraph 107 before suppression efforts began.

112. The BLM IC breached his duty to (1) ensure that all firefighters heard the red flag warning that was issued at 8:22 P.M.; (2) ensure that RFD volunteers understood the significance of that warning, i.e., that high winds would dangerously intensify the fire and substantially obscure

visibility; and (3) ensure that the RFD volunteers knew that the high winds would be coming immediately and from the south. Given the lack of any initial safety briefing, and the BLM IC's ignorance of the Kuna firefighters' actual qualifications, it was especially important to ensure not only that the Kuna RFD firefighters heard the red flag warning but that they understood the significance of the warning. The BLM IC failed to comply with his duty.

113. The BLM IC breached his duty to warn Buttram and Oliver after the red flag warning to stay away from the fire's northern perimeter because high winds would drive the fire in that direction. Instead of so warning 620, the BLM IC instructed them to refill. He knew that to refill, 620 would most likely drive toward the fence break at the northeast corner of the fire. In other words, the BLM IC directed 620 toward the northern perimeter of the fire at a time when high winds were forecast to drive the fire in that very direction. As the Court discussed previously in paragraph 65, the black's function as a safety zone is partly dependent on whether the firefighters are moving through the black towards the unburned fuels that are in the path of the upcoming winds. The BLM contends the IC Kerby gave 620 "the safe assignment of going to the road and staying there." *See* BLM's Trial Brief at 11. The BLM asserts that "it does not make sense that [Buttram] would drive wildly around if he could see at all." Id. From this the BLM concludes there is no proof that the "BLM proximately caused Buttram to drive out of the safe zone and into harm's way." Id. The Court disagrees. At the time IC Kerby gave his refill order, it was foreseeable that the high winds could kick up the dust and ash in the black and completely obscure Buttram's and Oliver's visibility. It was also foreseeable that Buttram and Oliver would be very near the fire's northern perimeter — and moving toward that perimeter — at the time the high winds were due to come through the area. Finally, it was foreseeable that the lack of visibility could cause panic and disorientation that would expose the firefighters to great risk because of their close proximity to the dangerous northern perimeter of the fire. By instructing 620 to refill, the BLM IC placed Buttram and Oliver in a foreseeably dangerous position, and thereby breached his duty to provide for their safety.

Definition of Kuna RFD's Duty

114. The Kuna RFD had a duty to exercise reasonable care to protect Buttram and Oliver from foreseeable hazards incident to fighting the Point Fire. This duty continued even after IC Kerby assigned duties to Buttram and Oliver and the two volunteers began working the fire line. While IC Kerby was the leader with primary responsibility, the Kuna RFD had a continuing responsibility to monitor its firefighters, warn them of dangers, and make recommendations to IC Kerby based on Kuna RFD's superior knowledge of Buttram's and Oliver's lack of experience. This duty arises because Captain McPherson sent 620 manned by two men experiencing their first fire season, to the BLM for duty assignment (1) without notifying the BLM IC that 620 was operated by rookies, (2) knowing that the BLM IC would have no time himself to inquire into Buttram's and Oliver's experience and training, and (3) knowing virtually nothing about the extent of the fire and the weather forecast. Under these circumstances, it was especially important that Chief Cromwell and Captain McPherson monitor the suppression efforts and the weather, so that they could suggest courses of action to IC Kerby that would ensure the safety of Buttram and Oliver.

115. More significantly, the Kuna RFD had the following duties:

116. First, the Kuna RFD had a duty to provide within reasonable limits the equipment necessary to ensure the firefighters' safety. Specifically, the Kuna RFD had a duty to provide adequate radios to its firefighters. Chief Cromwell knew that the Point Fore was a grass/wildland fire, and that suppression efforts could spread his firefighters out to the point where they would be out of sight of leadership. Adequate radios were the only way that Chief Cromwell and Captain McPherson could comply with their duty to monitor their firefighters during the suppression efforts, as discussed above.

117. Second, the Kuna RFD had a duty to make personnel assignments that were designed to ensure the firefighters' safety. Specifically, the Kuna RFD had a duty to (1) send only qualified volunteers to fight any fire, and (2) pair firefighters in two-person trucks in a manner that reasonably provided for the firefighters' safety.

118. Third, the Kuna RFD had a duty to obtain a weather report from the National Weather Service before proceeding to fight the Point Fire. Kuna RFD Assistant Chief Darwin Taylor testified that he believed that the National Weather Service Reports were only available to federal agencies, but he was mistaken – the unrebutted testimony of Richard Ochoa, a staff meteorologist for the National Weather Service, established that the National Weather Service Reports would have been easily and immediately available to the Kuna RFD on July 28, 1995. These reports contained the most detailed weather information, and the testimony at trial established that such information is absolutely crucial to the safety of firefighters during wildland fires.

119. Fourth, the Kuna RFD had a duty to ensure that its firefighters received a briefing either by Kuna RFD personnel or BLM personnel before suppression efforts began about the nature of the fire, fuel conditions, weather information, safety reminders, command structure, and radio use. This duty arises in part because Chief Cromwell and Captain McPherson knew that (1) the Point Fire was a wildland fire, (2) the Kuna RFD firefighters had received very little formal training on wildland fires, and (3) 620 was manned by two firefighters experiencing their first fire season.

120. Fifth, the Kuna RFD had a duty to train its volunteer firefighters to fight wildland fires in a safe effective manner. This duty arises because the main type of fire fought by the Kuna RFD in the time period of the Point Fire was the grassland/wildland type fire. This duty becomes particularly significant when the Kuna RFD elects to assign their volunteer firefighters to work with other agencies who will have no knowledge of the training that such firefighters possess.

Breach of Kuna RFD's Duty

121. The Kuna RFD breached its continuing duty to endure the safety of Buttram and Oliver during their suppression efforts. Chief Cromwell and Captain McPherson should have know that IC Kerby was unaware that Buttram and Oliver were rookies. Yet Chief Cromwell and Captain McPherson took no steps to inform IC Kerby of this fact. That failure would not necessarily have been negligent if Chief Cromwell and Captain McPherson had taken extra care to monitor Buttram and Oliver, and monitor any dangerous fire or weather conditions. Both Chief Cromwell and Captain McPherson recognized the importance of this monitoring when they testified that if they had known that high winds were approaching, they would have pulled 620 off the fire line. Yet despite recognizing the importance or monitoring, neither monitored the BLM channel for warning about dangerous

conditions. Thus, when the high wind warning was given over the BLM channel, neither Chief Cromwell nor Captain McPherson heard the warning. Captain McPherson had no capability to tune into the BLM channel; Chief Cromwell had the capability but failed to do so. The end result was that both men missed a crucial warning, and lost the opportunity to (1) recommend to IC Kerby that he put 620 in a safe zone, and (2) talk to Buttram and Oliver about the significance of the warning. These circumstances constitute a breach of Kuna RFD's continuing duty to ensure the safety of Buttram and Oliver during their fire suppression efforts.

122. The Kuna RFD breached its duty to pair firefighters in two-person trucks in a manner that reasonably provided for the firefighters' safety. Allowing two rookies to pair up is not necessarily a negligent decision if, for example, the rookies were operating a water tender far from the fire line. But here the rookies were operating a brush truck that was designed to work directly on the fire line. Captain McPherson had an opportunity at the initial staging to switch Oliver and Black. That would have put a seasoned firefighter with a rookie in each truck. In addition, Assistant Chief Darwin Taylor, an experienced firefighter, was available to ride in 620, but was not used in that capacity. Thus, the Kuna RFD was not faced with a lack of resources that often, and understandably, plague small fire protection units. The Kuna RFD already had the resources – that is, experienced firefighters – but negligently failed to deploy them in a way that would ensure the safety of Buttram and Oliver. This was also not a situation where it was reasonable to

assume that two rookies could work together because the fire presented little danger. He also knew or should have known that Captain McPherson would be unable to monitor the BLM channel.[5] Under these circumstances, it was unreasonable for the Kuna RFD to allow two rookies to operate 620. Thus, the Kuna RFD breached its duty to pair firefighters in two-person trucks in a manner that reasonably provided for the firefighters' safety.

123. The Kuna RFD breached its duty to obtain a weather report from the National Weather Service before proceeding to fight the Point Fire. The National Weather Service report was available but was not obtained by anyone from Kuna RFD.

124. The Kuna RFD breached its duty to train its volunteer firefighters to fight wildland fires in a safe and effective manner. Despite the fact the main type of fire fought by the Kuna RFD was the grass/wildland fire, the Kuna RFD had only two training sessions in fighting these types of fires.

Definition of Buttram's and Oliver's Duty

125. Buttram and Oliver had a duty to exercise reasonable care to provide their own safety while fighting the Point Fire.

Buttram's and Oliver's Compliance With Their Duty

126. At the time the Point Fire blew up at about 8:46 P.M., Buttram and Oliver were following directions. They were returning to Swan Falls Road to refill as directed by IC Kerby. Their conduct after the high winds started was a product of

[5] Chief Cromwell testified that he knew-before sending his firefighters to the Point Fire-that 620 and no BLM channel capability. He testified that he was not concerned, however, because 620 and 622 had BLM channel capability. But he should have known that Captain McPherson – his choice to be IC for the Kuna firefighters – would have no capability to monitor the BLM channel. While Captain McPherson rode to the fire in 620, nobody comtemplated that he would operate as Kuna RFD IC in the cramped confines of 620, a truck designed for only two men. It was much more foreseeable that Captain McPherson would operate near tender 625, far from 620 and 622 working on the fire line.

disorientation and panic caused by the obscured visibility and fast-moving fire. At no time did Buttram and Oliver negligently fail to exercise reasonable care to provide for their own safety.

Causation

127. The BLM's negligence was a proximate cause of the deaths of Buttram and Oliver.
128. The Kuna RFD's negligence was a proximate cause of the death of Buttram and Oliver.
129. The negligence of the BLM and the Kuna RFD caused Buttram and Oliver to be near the northern perimeter of the fire at the time of the fire's blow-up. The fact that Buttram and Oliver were in this location at the time of the high winds was a substantial factor in their deaths.
130. The BLM's failure to hold an initial safety briefing, failure to explain the significance of the red flag warning, and failure to warn 620 to stay away from the northern perimeter in combination with its refill instruction that directed 620 toward the fire's northern perimeter, were substantial factors in Buttram and Oliver being exposed to an unreasonable risk of danger that led to their deaths.
131. The Kuna RFD's failure to (1) ensure the safety of Buttram and Oliver, (2) provide adequate equipment, (3) make safety personnel assignments, (4) obtain a weather report, (5) properly train Buttram and Oliver in wildland fire suppression safety, and (6) advise the BLM IC of Buttram's and Oliver's limited training and experience, were substantial factors in Buttram and Oliver being exposed to an unreasonable risk of danger that led to their deaths.

Comparative Responsibility

132. The BLM and the Kuna RFD were both responsible for the death. Each had knowledge the other lacked.

133. The BLM knew a great deal about the fire but nothing about the experience and knowledge of Buttram and Oliver.
134. The Kuna RFD knew a great deal about Buttram and Oliver but nothing about the fire.
135. If either had shared its knowledge, this tragedy could have been avoided.
136. Ultimately, the Kuna RFD was in the best position to use its knowledge to ensure the safety of Buttram and Oliver. The Kuna RFD's failure to do so was therefore the greatest contributing factor to their deaths.
137. The Kuna RFD's decision to allow two rookies to operate a brush truck was the key threshold decision that set in motion the chain of events that would end in tragedy. By presenting the rookies to the BLM in a brush truck, the Kuna RFD was essentially vouching for their qualifications to work directly on the fire line – that is, after all, the purpose of the brush truck. Thereafter, knowing that the rookies would be in harm's way, the Kuna RFD never informed the BLM of their rookies status or took the special care necessary to monitor their firefighting and the fire's condition. Of course, the BLM was also negligent because it had a special duty to monitor as well because it knew it had little knowledge of the Kuna RFD's qualifications, and no actual knowledge of the training, qualifications, and experience of the Kuna RFD firefighters assigned to the Point Fire. The percentages of responsibility might be allocated equally if the failure to monitor was the only negligence. But the Kuna RFD's pairing decision and its failure to advise the BLM IC of Buttram's and Oliver's lack of training and experience, places more of the responsibility on the Kuna RFD.
138. The Kuna RFD is therefore allocated 65 percent of the responsibility of the deaths and the BLM is allocated 35 percent.

Damages[6]

139. Plaintiffs Deanna Buttram and Jeremiah Buttram have sustained economic damages in the amount of $961,092. They have sustained damages for out-of-pocket funeral expenses in the amount of $6,074.85. Jeremiah Buttram has sustained damages for loss of society and companionship in the amount of $900,000 that must be reduced to the statutory cap of $590,291.26. Deanna Buttram has sustained damages for the loss of consortium, society, and companionship in the amount of $900,000 that must be reduced to the statutory cap of $590,291.26. The total damages sustained by Deanna Buttram and Jeremiah Buttram are $2,147,749.37.

140. Plaintiff Darla Reber has sustained damages for loss of society and companionship of her son in the amount of $300,000.

141. Plaintiff Michael Oliver has sustained damages for the loss of the society and companionship of his son in the amount of $50,000.[7]

142. With regard to plaintiffs Deanna Buttram and Jeremiah Buttram, the BLM is liable in damages for 35 percent of their total damages of $2,147,749.37, or $751,712,28.

143. With regard to plaintiff Darla Reber, the BLM is liable in damages for 35 percent of her total damages of $300,00, or $105,000.

144. With regard to plaintiff Michael Oliver, the BLM is liable in damages for 35 percent of his total damages of $50,000 or $17,500.

145. The Court will issue a separate Judgment in accord with Federal Rules of Civil Procedure 58.

Dated this 19th day of February, 1999.

/s/_____

B. Lynn Winmill
United States District Court

_____

QUESTIONS

1. Discuss the legal effects of the use of IMS, Mutual Aid, and SOPs in this case as they relate to the safety of the responders.

2. *Buttram* is an unpublished slip opinion. How does this affect its value as a legal precedent?

3. The Court analyzed the *Buttram* matter by legal standards defined by state OSHA law. Discuss the implications of this approach for liability in other cases involving death or injury to firefighters in the performance of their duties.

4. By what route did the Court find that the requirements of IMS applied in this case? Where did the Court obtain the IMS standard that it used?

_____

[6] The worth of a life cannot be set with mathematical precision. If it is our own life we are discussing, we would not give it up at any price. William Bradford, in writing of those who died during the crossing of the Mayflower, said that "the loss of honest and industrious men's live cannot be valued at any price." Yet the law requires in this case that the Court, in determining the loss of society and companionship suffered by the plaintiffs, set a price on two men's lives, a nearly impossible task. With regard to the Buttrams, the Court did not attempt to make any distinction between the loss suffered by the wife and the loss suffered by the son. The Court started from the premise that their loss was clearly above the statutory cap and thus would eventually be reduced to the cap level, no matter where the figure was initially. The Court then proceeded to approximate the loss of society and companionship by equating it roughly to the economic loss suffered by the family. With regard to Darla Reber, the Court balanced the fact that Joshua was on his own and no longer a minor child. The award to Michael Oliver is explained in footnote 7.

[7] The issue of Michael Oliver's is very difficult for the Court. On the one hand there is evidence that Michael was estranged from his son. As discussed in the findings of fact, Michael went for many years without contacting Joshua. Michael's lack of contact appears to be the result of his own choice since he testified that Darla did not make it difficult for him to contact Joshua. On the other hand, Joshua did come to live with Michael for a month in Illinois, and there were other extended stays. In addition, Michael was always current on his support payments. In the whole, however, an award of damages to Michael that comes anywhere near the award to Darla would ignore the plain fact that Darla raised Joshua for half of his life without any help from Michael, other than his financial assistance. In light of this, the Court decided to award Michael a sum that was substantially lower than that awarded to Darla Reber.

5. Describe the ways in which you think use of IMS could have been improved in this case.
6. *Buttram* came out before the 9-11 attacks and the NRF and NIMS, which were conceived in response to the terrorist events. What effect does this have on *Buttram* as a legal precedent?
7. Compare the MAA in this case with the standards discussed in the introductory material.
8. How, if at all, and why would you change the procedure by which the BLM and Kuna RFD agreed that aid would be provided for a particular incident?
9. Discuss the legal requirements for an MAA. Where do they originate? How do the legal requirements relate to standards for responder safety?
10. How did the Court learn of the Kuna RFD's SOPs?
11. What changes would you make in the Kuna RFD's SOPs to prevent a recurrence of the events in this case?
12. How can SOPs be made truly "standard" beyond a single department?
13. Explain why this is or is not a valuable approach.

VIII. HOW FAR DO NIMS' AND THE NRF's WRIT RUN?

Herbert Freeman, Jr., et al. vs. United States Department of Homeland Security, et al.

2007 U.S. Dist. LEXIS 31827 (E.D. La. 2007)

OPINION

I. BACKGROUND

Plaintiffs in these consolidated cases are the relatives and/or representatives of Ethel Freeman, John J. DeLuca, and Clementine Eleby ("the Decedents"). Each of the Decedents passed away in the days immediately following Hurricane Katrina. Plaintiffs contend that the deaths were caused by Defendants' failure to follow their lawful duties in the aftermath of Hurricane Katrina. Defendants move for dismissal pursuant to Rules 12(b)(1) and 12(b)(6) of the Federal Rules of Civil Procedure.

Civil Action 06-4846

This action was filed by Herbert Freeman, Jr., individually and in his capacity as representative of the Estate of Ethel Freeman. Prior to Katrina, Mrs. Freeman was elderly and infirm and required a wheelchair for mobility. The Freemans rode out the storm in their New Orleans home but the home flooded in the aftermath of Katrina when the levees gave way. Mr. Freeman secured a boat from a friend and placed Mrs. Freeman, and her wheelchair in the boat. Mr. Freeman proceeded to higher ground and New Orleans

Police officers eventually directed the Freemans to the New Orleans Convention Center.

Mr. Freeman alleges that the Convention Center was not manned or equipped to handle evacuees. He complains that the conditions were squalid and that no triage, food, water, or medical assistance was provided. Mrs. Freeman died on Wednesday, September 1, 2005.

Freeman alleges that Defendants failed to follow their non-discretionary duties under the National Response Plan ("NRP") to take appropriate steps to marshal resources and to aid the city, the state, and evacuees like Mrs. Freeman. Freeman claims that Defendants' negligence in failing to provide adequate shelter, medical services, triage, evacuation, and transport to facilities where appropriate medical care could be provided resulted in Mrs. Freeman's wrongful death.

Freeman also claims that by failing to follow "the public law, the Stafford Act and public policy, the National Response Plan" Defendants denied Mrs. Freeman equal protection under the law because she was discriminated against because of her age and infirmity. Freeman alleges that the violation of Mrs. Freeman's equal protections rights under the Fourteenth Amendment proximately caused her death.

Freeman submitted an administrative tort claim to FEMA, DHHS, and DOD. DHHS denied the claim but FEMA and DOD have yet to issue denials. Freeman seeks wrongful death and survivor action damages for the loss of his mother.

Civil Action 06-5689

This action was filed by Frances Lodriguss, the sister of the decedent John J. DeLuca. Prior to Katrina, Mr. DeLuca was a seventy-seven-year-old mobility impaired, chronically ill man who resided at the Nazareth Inn, an independent and assisted living facility located in New Orleans East. Mr. DeLuca rode out Katrina at the Nazareth Inn but was trapped by flood waters in the storm's aftermath. Mr. DeLuca was taken by helicopter to a location on Interstate 10 at its interchange with Causeway Boulevard.

According to the complaint, there was no shelter, food, water, security, medical services or transportation at this location. Buses began to arrive on August 31 but no one was in charge to ensure the evacuation of the elderly and infirm. On September 2, 2005, Mr. DeLuca collapsed and was taken by helicopter to the airport. Mr. DeLuca died on September 3, 2005.

Lodriguss alleges that Defendants owed Mr. DeLuca a duty of care to provide transportation and medical care in the aftermath of the storm. Lodriguss alleges that Defendants' failure to deliver a rapid and appropriate federal response was the proximate cause of Mr. DeLuca's death and was in violation of his rights to equal protection under the Fourteenth Amendment to the U.S. Constitution, federal law, and public policy. Lodriguss alleges that Defendants' negligence and/or their adoption of an unconstitutional policy which deprived the elderly and infirm of

an equal chance of survival violated her brother's constitutional rights and caused his death. Lodriguss seeks wrongful death and survivor action damages for the loss of her brother.

Civil Action 06-5696

This action was filed by Barbara Eleby Lee, Brenda Bissant, Glenda Eleby, Griffin Eleby, Jr., Rosalie Brooks, Dorothy Beal, Earline Coleman Ethel Jackson, and Nancy Eleby, the children of the decedent Clementine Eleby. Prior to Katrina, Ms. Eleby was a seventy-nine-year-old bedridden paralyzed woman who was residing with her daughter Barbara Eleby Lee in New Orleans East. The Elebys rode out the storm in their New Orleans East home but were trapped by flood waters in the storm's aftermath. The Elebys were eventually transported to the Convention Center where they arrived on August 31, 2005.

Plaintiffs allege that the Convention Center was not manned or equipped to handle evacuees. They complain that the conditions were squalid and that no triage, food, water, or medical assistance was provided. Mrs. Eleby died at the Convention Center on September 1, 2005.

Plaintiffs allege Defendants' failure to deliver a rapid and appropriate federal response was the proximate cause of Mrs. Eleby's death and was in violation of her rights to equal protection under the Fourteenth Amendment to the U.S. Constitution, federal law, and public policy. Plaintiffs allege that Defendants' negligence and/or their adoption of an unconstitutional policy which deprived the elderly and infirm of an equal chance of survival violated their mother's constitutional rights and caused her death. Plaintiffs seek wrongful death and survivor action damages for the loss of their mother.

II. THE PARTIES' CONTENTIONS

Defendants' motion pertains to the claims against the United States and the individual defendants *in their official capacities only.* Defendants contend that dismissal is proper under Rule 12(b)(1) (lack of subject matter jurisdiction) because the Court lacks jurisdiction over these

claims because the United States has not waived its sovereign immunity with respect to the claims at issue. Specifically, Defendants contend that neither the Stafford Act, the Administrative Procedure Act, nor the Federal Tort Claims Act ("FTCA") provide a waiver of sovereign immunity.

Further, Defendants contend that the waiver of immunity contained in the Stafford Act and FTCA does not apply because the provision of disaster assistance in the aftermath of Katrina is a discretionary act excepted from the waiver of sovereign immunity contained in those statutory provisions.

Next, Defendants contend that dismissal of the Fourteenth Amendment claims is proper under Rule 12(b)(6) (failure to state a claim upon which relief can be granted) because the Fourteenth Amendment applies only to States' actions and not to actions by the United States. Finally, Defendants contend that Plaintiffs' claims (with one exception as to Freeman) are not actionable under the FTCA because they have failed to satisfy the jurisdictional prerequisites to sue the United States in tort.

In opposition, Plaintiffs clarify that they do not oppose dismissal of the Fourteenth Amendment claims against Defendants in their official capacities because the guarantee of equal protection binds only the States.

Plaintiffs argue that Congress did not confer total immunity on the United States when it passed the Stafford Act. Rather, the Act includes discretionary function exception language much like that contained in the FTCA.

Plaintiffs concede that Freeman is the only Plaintiff with a FTCA claim currently pending in this litigation and that this claim is directed at FEMA's conduct in the aftermath of Katrina. Freeman contends that his complaint alleges the breach of non-discretionary duties and that the Court cannot resolve any underlying fact disputes under the guise of ruling on a 12(b)(1) or 12(b)(6) motion. Freeman contends that additional discovery is needed to determine more precisely the facts regarding the failures of FEMA in the preparation for and response to the storm. Freeman contends that the federal government has no discretion when it comes to its duty to comply with mandatory policies and regulations dictating the measures to be taken to protect the citizenry from the aftermath of disasters.

In reply, Defendants point out that concessions in Plaintiffs' opposition narrow the focus of the pending motion to whether the discretionary function exceptions within the Stafford Act and the FTCA apply.

III. DISCUSSION

1. General Legal Principles

In general where subject matter jurisdiction is being challenged under Rule 12(b)(1), the trial court is free to weigh the evidence and resolve factual disputes in order to satisfy itself that it has the power to hear the case. *Montez v. Dep't of Navy*, 392 F.3d 147, 149 (5th Cir. 2004) (citing Land v. Dollar, 330 U.S. 731, 67 S. Ct. 1009, 91 L. Ed. 1209 (1947)). No presumptive truthfulness attaches to the plaintiff's allegations and the court can decide disputed issues of material fact in order to determine whether or not it has jurisdiction to hear the case. Id. However, where issues of fact are central both to subject matter jurisdiction and the claim on the merits, the trial court must assume jurisdiction and proceed to the merits of plaintiff's case under either Rule 12(b)(6) or Rule 56. Id. (citing *Williamson v. Tucker*,

645 F.2d 404 (5th Cir. 1981)). Under well-settled standards governing Rule 12(b)(6) motions to dismiss, a claim may not be dismissed unless it appears certain that the plaintiff cannot prove any set of facts that would entitle him to legal relief. In re Supreme Beef Processors, Inc., 468 F.3d 248, 251 (5th Cir. 2006) (citing *Benton v. United States*, 960 F.2d 19 (5th Cir. 1992)).

The doctrine of sovereign immunity renders the United States, its departments, and its employees in their official capacities immune from suit except as the United States has consented to be sued. *Williamson v. United States Dep't of Agriculture*, 815 F.2d 368, 373 (5th Cir. 1987). Absent an express waiver of federal immunity by Congress, the United States, its departments, and its employees in their official capacities are immune from suit. Supreme Beef Processors, 468 F.2d at 252. A waiver of the United States' sovereign immunity must be unequivocally expressed in statutory text. *Lane v. Pena*, 518 U.S. 187, 192, 116 S. Ct. 2092, 135 L. Ed. 2d 486 (1996). Waivers of sovereign immunity are narrowly construed in favor of the United States. Id.

Congress partially waived the United States' sovereign immunity from tort claims by enacting the Federal Tort Claims Act ("FTCA"), 28 U.S.C. § 1346(b). *Brown v. United States*, 653 F.2d 196, 198 (5th Cir. 1981). The FTCA allows a plaintiff to pursue tort actions against the federal government and it holds the government liable as if it were a defendant in state court. Supreme Beef Processors, 468 F.3d at 252. The liability of the United States under the Act arises only when the law of the state would impose it. Brown, 653 F.2d at 201. The FTCA is the sole basis of recovery for tort claims against the United States. Supreme Beef Processors, 468 F.3d at 252 n.4. The FTCA is subject to strict limitations one of which is the discretionary function exception codified at 28 U.S.C. § 2680(a). When applicable the discretionary function exception deprives the court of subject matter jurisdiction and it cannot be waived. Id. (citing *Hayes v. United States*, 899 F.2d 438 (5th Cir. 1990)).

The Stafford Act, also known as the Disaster Relief Act of 1974, was enacted to provide federal assistance to states in times of disasters. *Diversified Carting, Inc. v. City of N.Y.*, 423 F. Supp. 2d 85, 90 (S.D.N.Y 2005) (citing 42 U.S.C. § 5121, et seq.). To the extent that the Stafford Act allows for a private right of action against the federal government,[3] the United States retains its sovereign immunity with respect to any activities that are "based on the exercise or performance of or the failure to exercise or perform a discretionary function or duty on the part of a Federal agency or an employee of the Federal Government in carrying out the provisions of [the Act]." 42 U.S.C.A. § 5148 (West 2003). Courts generally use the same discretionary function exception analysis for claims brought under the Stafford

Act as well as the FTCA. See, e.g., McWaters, 408 F. Supp. 2d at 229.

2. Analysis

All claims under attack at this juncture are claims against the United States directly, and claims against federal employees in their *official* capacities, which are in effect claims against the United States. The United States cannot be sued absent an express and unequivocal waiver of sovereign immunity so the first hurdle that Plaintiffs must clear is the sovereign immunity bar. If the United States has not expressly waived its sovereign immunity for the claims at issue, then the Court lacks subject matter jurisdiction over the claims and they are subject to dismissal pursuant to Rule 12(b)(1). The Court does not believe that issues of fact play a key role in the sovereign immunity analysis but to the extent that they do the Court will follow the standards applicable to Rule 12(b)(6) motions to dismiss and construe all disputed facts in Plaintiffs' favor.

The claims against the United States are brought under the FTCA and the Stafford Act. The FTCA includes express statutory text that waives immunity for certain tort claims so long as the claim is not based on a discretionary function or duty. The Stafford Act, on the other hand, contains no express and unequivocal language purporting to waive the United States' immunity from suit. However, § 5148 of the Act is a discretionary function exception, and Plaintiffs reason that Congress must have intended to allow some claims under the Act because § 5148 would not be necessary if all claims under the Act were already barred by sovereign immunity. The Court questions whether such reverse reasoning suffices under the strict requirements imposed by the Supreme Court's jurisprudence in order to find a waiver of sovereign immunity. See, e.g., Lane, 518 U.S. at 192.

---

[3] See *Graham v. Federal Emergency Mgmt. Agency*, 149 F.3d 997, 1001 (9th Cir. 1998) ("[T]he Stafford Act does not provide for a private cause of action upon which the plaintiffs may rely.... "). But see *Diversified Carting, Inc. v. City of N.Y.*, 423 F. Supp. 2d 85, 90-91 (S.D.N.Y. 2005) (implicitly recognizing a private right of action under the Stafford Act); *McWaters v. FEMA*, 408 F. Supp. 2d 221, 228-29 (E.D. La. 2006) (implicitly recognizing a private right of action under the Stafford Act).

Clearly, however, to the extent that a waiver of immunity does exist under the Stafford Act, actions based on discretionary functions are not included in that waiver. See 42 U.S.C.A. § 5148. At this juncture, the Court will *assume* that the Stafford Act does contain a limited waiver of immunity for nondiscretionary functions. Because both the FTCA and the Stafford Act include discretionary function exceptions, the sole issue before the Court today is whether Plaintiffs' claims are based upon the performance (or non-performance) of a discretionary function or duty on the part of Defendants. If the discretionary function exception applies, then sovereign immunity will bar Plaintiffs' claims.

3. Discretionary Function Analysis

Courts use the same Berkovitz discretionary function analysis for claims brought under the FTCA and the Stafford Act. See, e.g., McWaters, 408 F. Supp. 2d at 229. In *Berkovitz v. United States*, the Supreme Court explained and expounded upon the established principles that guide the determination of whether the discretionary function exception bars a suit against the United States. 486 U.S. 531, 536, 108 S. Ct. 1954, 100 L. Ed. 2d 531 (1988). "[I]t is the nature of the conduct, rather than the status of the actor, that governs whether the discretionary function exception applies in a given case." Id. (quoting *United States v. Varig Airlines*, 467 U.S. 797, 813, 104 S. Ct. 2755, 81 L. Ed. 2d 660 (1984)). In examining the nature of the challenged conduct, a court must first consider whether the conduct involves an element of judgment or choice for the acting employee. Id. (citing *Dalehite v. United States*, 346 U.S. 15, 73 S. Ct. 956, 97 L. Ed. 1427 (1953)). The discretionary function exception will not apply when a federal statute, regulation, or policy specifically prescribes a course of action for an employee to follow. Id. In this event the employee has no rightful option but to adhere to the directive. (Id.)

Assuming the challenged conduct involves an element of judgment, the court must then determine whether that judgment is the kind that the discretionary function exception was designed to shield. Berkovitz, 486 U.S. at 536. The discretionary

function exception serves to "prevent judicial 'second guessing' of legislative and administrative decisions ground in social, economic, and political policy through the medium of an action in tort." Id. (quoting Varig Airlines, 467 U.S. at 814). The exception properly construed protects only governmental actions and decisions based on public policy. Berkovitz, 486 U.S. at 537. In sum, the discretionary function exception insulates the United States from liability if the action challenged in the case involves the permissible exercise of policy judgment. (Id.)

Pursuant to Berkovitz, this Court must analyze Plaintiffs' specific allegations of governmental wrongdoing in light of the applicable statutory and regulatory scheme governing the provision of disaster relief following a natural disaster such as Hurricane Katrina. In particular, Plaintiffs must identify a specific *mandatory* directive that Defendants had a clear duty under federal law to perform. And if the specific mandatory directive nevertheless involves an element of judgment, then application of the discretionary function exception will turn on whether the officials making that judgment permissibly exercised policy choice. Berkovitz, 486 U.S. at 545.

The Stafford Act is codified at 42 U.S.C. §§ 5121, et seq. Freeman points to no specific, *mandatory* directive found within the statutory scheme of the Stafford Act or its accompanying federal regulations that Defendants are alleged to have ignored in the aftermath of Katrina. Defendants correctly point out that the Act is replete with discretionary language particularly in those sections that address federal emergency assistance. See, e.g., 42 U.S.C.A. § 5192 (West 2003) ("In any emergency, the President *may* direct any Federal agency . . . to utilize its authorities and the resources granted to it under Federal law . . . in support of State and local emergency assistance efforts to save lives, protect property and public health and safety, and lessen or avert the threat of a catastrophe.") (emphasis added). Freeman cannot meet the threshold requirement for a waiver of immunity under the Stafford Act, i.e., identification of a *mandatory* duty under the law. Thus, Freeman's claims under the Stafford Act are barred by sovereign immunity.

Freeman also relies upon the National Response Plan as the source of a mandatory directive or duty imposed on Defendants under federal law. The National Response Plan ("NRP") was developed by the Secretary of the Department of Homeland ("DHS") at the direction of the President. Homeland Security Presidential Directive/HSPD-5, 2003 WL 604606 (Feb. 28, 2003). The purpose of the HSPD-5 is to "enhance the ability of the United States to manage domestic incidents by establishing a single, comprehensive national incident management system." Id. The main goal of HSPD-5 was to come up with a plan so that all levels of government would have the capability to work together to prevent and respond to terrorist attacks, major disasters, and other emergencies. Id. HSPD-5 expressly notes that "[i]nitial responsibility for managing domestic incidents generally falls on State and local authorities. The Federal Government will assist State and local authorities when their resources are overwhelmed, or when Federal interests are involved." (Id.)

As part of HSPD-5 the President directed the Secretary of Homeland Security to develop the NRP which "shall integrate Federal Government domestic prevention, preparedness, response, and recovery plans into one all-discipline, all hazards plan." Id. The NRP, with respect to domestic incidents, was to provide the structure and mechanisms for national level policy and operational direction for Federal support to State and local incident managers...." Id. The NRP, as envisioned by the President, would include protocols for operating under different threats or threat levels, incorporate existing federal emergency and incident plans. Id. The NRP was to include a "consistent approach to reporting incidents, providing assessments, and making recommendations to the President...." Id.

The version of the NRP relied upon by Freeman is dated December 2004 and a six page excerpt is filed in the record. This document, which was authored by a single agency head, is not an executive order, Congressional statute, or properly promulgated federal regulation so as to constitute federal "law." It is a questionable legal proposition to say the least that this document could be used as a basis for imposing a mandatory, non-discretionary duty under federal law, the likes of which could be used to overcome the United States' sovereign immunity.

Freeman's reliance on the NRP is also problematic because the NRP does not prescribe a specific course of conduct for federal employees to follow. Freeman quotes several passages from the NRP and alleges that the NRP called for DHS and FEMA to deploy vital resources including medical teams and equipment to evacuate patients who needed medical care and to comply with an urgent duty to save lives. Freeman alleges that Defendants had a non-discretionary duty under the NRP to take appropriate steps to marshal resources and to aid the city, the state and evacuees, including Freeman.

The NRP and its Annexes are worded for the most part in precatory language and even where stronger "must" language is used the document is still a very high level overview of the anticipated rollout plan for federal resources during a domestic incident. Likewise, the annex sections quoted by Freeman in his complaint recognize from a practical perspective the things that need to occur following a domestic incident. However, implicit in the structure of this generalized document is the discretion and flexibility that the affected agencies retain to allocate their limited resources in response to a disaster such as Hurricane Katrina. The NRP simply does not provide a source of non-discretionary *duties* imposed on Defendants under federal law. Further, the timing and scope of FEMA's allocation of resources in the aftermath of Katrina not only includes the element of judgment or choice for the acting agency employees but that element of choice is clearly one grounded in social, economic, and public policy.

Plaintiffs assert that Defendants' motion must be denied because Plaintiffs have alleged in their complaint that Defendants breached non-discretionary duties, and Plaintiffs repeatedly point out that they must be afforded an opportunity to conduct discovery prior to dismissal. Plaintiffs' contention is contrary to Berkovitz which makes clear that a mere "assertion" that a federal actor has breached a non-discretionary duty is insufficient to avoid dismissal. Rather, the

Berkovitz analysis demonstrates the need to identify, as a threshold matter, a *specific* mandatory, non-discretionary directive that the federal actor has violated.

Further, the fact-based discovery that Plaintiffs are anxious to obtain only comes into play once they identify the specific directive that Defendants have ignored. Such a directive will be in the public realm and therefore fact discovery will not assist them in meeting this crucial threshold requirement.

In sum, the actions being challenged in this lawsuit are the types of actions that Congress intended to shield from liability under both the Stafford Act and the FTCA. One might contend that the federal decisions made in conjunction with Hurricane Katrina demonstrated nonchalance and/or incompetence on the part of those involved. The Government has publicly admitted that it made many mistakes in the aftermath of Hurricane Katrina. One can only speculate at this point whether these mistakes caused the tragic deaths of the Decedents.

Even assuming that the Government's mistakes caused these deaths, Congress has not exempted such claims from the discretionary function exception. This Court is very sympathetic to the Plaintiffs for the loss of their loved ones, however, this Court is prohibited from changing the laws that Congress has enacted. As such, the Court lacks the authority to award money damages for claims in which the plaintiffs are not legally entitled. Because Plaintiffs have not identified a non-discretionary directive that Defendants failed to follow, no waiver of sovereign immunity applies. Plaintiffs' Stafford Act and FTCA claims against the United States and the remaining defendants in their official capacities must be dismissed for lack of subject matter jurisdiction.

Accordingly, and for the foregoing reasons;

IT IS ORDERED that the Motion to Dismiss (Rec. Doc. 13) filed by the United States of America, Michael Chertoff, in his official capacity as Secretary of Homeland Security, the United States Department of Homeland Security, Michael Leavitt, in his official capacity as Secretary of Health and Human Services, the Department of Health and Human Services, Donald Rumsfeld, in his official capacity as Secretary of Defense, the Federal Emergency Management Agency, and Michael Brown, in his official capacity as former director of the Federal Emergency Management Agency should be and is hereby **GRANTED**.
April 30, 2007

JAY C. ZAINEY

UNITED STATES DISTRICT JUDGE

QUESTIONS

1. What is sovereign immunity?
2. As the court states, "the discretionary function exception insulates the United States from liability if the action challenged in the case involves the permissible exercise of policy judgment." How does one decide if an action is an exercise of policy judgment?
3. Why did the court find that the Stafford Act did not waive sovereign immunity?
4. Regarding the National Response Plan (NRP), the court states that "This document, which was authored by a single agency head, is not an executive order, Congressional statute, or properly promulgated federal regulation so as to constitute federal "law." It is a questionable legal proposition to say the least that this document could be used as a basis for imposing a mandatory, non-discretionary duty under federal law, the likes of which could be used to overcome the United States' sovereign immunity." As discussed previously, the NRP was issued pursuant to Homeland Security Presidential Directive (HSPD) 5. Should that document be considered as a federal "law?" What are the implications of the court's statement that the NRP does not impose mandatory duties?

5. Is the outcome of this case "fair?" What does "fairness" mean in this context?
6. Would this case change the outcome of *Buttram* if that case were brought now? Why or why not?
7. Divide the class into two groups. Have one group advocate in favor of the NRF/NIMS imposing mandatory duties, and one group advocate against it. Use the following scenario:

The State of Texas has elected a majority of legislators and Governor from the Tea Party. They refuse to comply with the National Response Framework and NIMS as well as requirements that their state and local emergency management agencies perform specific tasks in order to receive Federal grant funds. They sue in US District Court for the grant funds, citing *Freeman v. DHS*, stating that withholding them using the NRF and NIMS as a basis for requiring performance of tasks in order to receive them is illegal.

Quoting *Freeman v. DHS*, the State of Texas argues that:

> Th[ese] document[s], . . . [are] not an executive order, Congressional statute, or properly promulgated federal regulation so as to constitute federal "law." It is a questionable legal proposition to say the least that th[ese] document[s] could be used as a basis for imposing a mandatory, non-discretionary duty under federal law[.]

Both sides must include in their discussion the implications for the nation if their approach is adopted by the courts.

Appendix A

 FEMA

FY 2009 NIMS Implementation Objectives

The chart below depicts the 28 NIMS Implementation Objectives prescribed by National Integration Center's Incident Management Systems Integration (IMSI) Division for Federal Fiscal Year (FY) 2009. **State, territorial, tribal, and local jurisdictions must ensure *all* NIMS objectives have been initiated and/or are in progress toward completion.**

NIMS Component	NIMS Implementation Objective	Federal FY Prescribed to:		
		State/ Territory	Tribal	Local
ADOPTION 1.	Adopt NIMS for all Departments/Agencies; as well as promote and encourage NIMS adoption by associations, utilities, nongovernmental organizations (NGOs) and private sector emergency management and incident response organizations.	2005		
2.	Establish and maintain a planning process to communicate, monitor, and implement all NIMS compliance objectives across the State/Territory/Tribal Nation (including Departments/Agencies), to include local governments. This process must provide a means for measuring progress and facilitate reporting.	2006	2008	N/A
3.	Designate and maintain a single point of contact within government to serve as principal coordinator for NIMS implementation jurisdiction-wide (to include a principal coordinator for NIMS implementation within each Department/Agency).	2006	2007	
4.	Ensure that Federal Preparedness Awards [to include, but not limited to, DHS Homeland Security Grant Program and Urban Area Security Initiative Funds] to State/Territorial/Tribal Departments/Agencies, as well as local governments, support all required NIMS Compliance Objectives (requirements).	2005	2008	
5.	Audit agencies and review organizations should routinely include NIMS Compliance Objectives (requirements) in all audits associated with Federal Preparedness Awards.	2006	2008	
6.	Assist Tribal Nations with formal adoption and implementation of NIMS.	2007	N/A	
PREPAREDNESS — Planning 7.	Revise and update emergency operations plans (EOPs), standard operating procedures (SOPs), and standard operating guidelines (SOGs) to incorporate NIMS and National Response Framework (NRF) components, principles and policies, to include planning, training, response, exercises, equipment, evaluation, and corrective actions.	2005		
8.	Promote and/or develop intrastate and interagency mutual aid agreements and assistance agreements (to include agreements with the private sector and NGOs).	2005		
Training 9.	Use existing resources such as programs, personnel and training facilities to coordinate and deliver NIMS training requirements.	2006	2008	
10.	Implement IS-700 *NIMS: An Introduction* training to include appropriate personnel (as identified in the *Five-Year NIMS Training Plan*, February 2008).	2006		
11.	Implement IS-800 *National Response Framework (NRF): An Introduction* training to include appropriate personnel (as identified in the *Five-Year NIMS Training Plan*, February 2008).	2006		
12.	Implement ICS-100 *Introduction to ICS* training to include appropriate personnel (as identified in the *Five-Year NIMS Training Plan*, February 2008).	2006		
13.	Implement ICS-200 *ICS for Single Resources and Initial Action Incidents* training to include appropriate personnel (as identified in the *Five-Year NIMS Training Plan*, February 2008).	2006		
14.	Implement ICS-300 *Intermediate ICS* training to include appropriate personnel (as identified in the *Five-Year NIMS Training Plan*, February 2008).	2007		
15.	Implement ICS-400 *Advanced ICS* training to include appropriate personnel (as identified in the *Five-Year NIMS Training Plan*, February 2008).	2009		
16.	Incorporate NIMS concepts and principles into all appropriate State/Territorial/Tribal training and exercises.	2005		
Exercises 17.	Plan for and/or participate in an all-hazards exercise program [for example, Homeland Security Exercise and Evaluation Program] that involves emergency management/response personnel from multiple disciplines and/or multiple jurisdictions.	2006		
18.	Incorporate corrective actions into preparedness and response plans and procedures.	2006		
COMMUNICATION AND INFORMATION MANAGEMENT 19.	Apply common and consistent terminology as used in NIMS, including the establishment of plain language (clear text) communications standards.	2006		
20.	Utilize systems, tools, and processes to present consistent and accurate information (e.g., common operating picture) during an incident/planned event.	2007		
RESOURCE MANAGEMENT 21.	Inventory response assets to conform to NIMS National Resource Typing Definitions, as defined by FEMA Incident Management Systems Integration Division.	2006		
22.	Ensure that equipment, communications and data systems acquired through State/Territorial and local acquisition programs are interoperable.	2006		
23.	Utilize response asset inventory for intrastate/interstate mutual aid requests [such as Emergency Management Assistance Compact (EMAC)], training, exercises, and incidents/planned events.	2007		
24.	Initiate development of a State/Territory/Tribal-wide system (that incorporates local jurisdictions) to credential emergency management/response personnel to ensure proper authorization and access to an incident including those involving mutual aid agreements and/or assistance agreements.	2008		
COMMAND AND MANAGEMENT — Incident Command System 25.	Manage all incidents/ planned events in accordance with ICS organizational structures, doctrine and procedures. ICS implementation must include the consistent application of Incident Action Planning (IAP), common communications plans, implementation of Area Command to oversee multiple incidents that are handled by separate ICS organizations or to oversee the management of a very large or evolving incident that has multiple incident management teams engaged, and implementation of unified command (UC) in multi-jurisdictional or multi-agency incident management, as appropriate.	2006		
MultiAgency Coordination Systems 26.	Coordinate and support emergency management and incident response objectives through the development and use of integrated MACS, [i.e. develop/maintain connectivity capability between local Incident Command Posts (ICPs), local 911 Centers, local/regional/State/territorial/tribal/Federal Emergency Operations Centers (EOCs), as well as NRF organizational elements.]	2006		
Public Information 27.	Institutionalize, within the framework of ICS, Public Information, [e.g., Joint Information System (JIS) and a Joint Information Center (JIC)] during an incident/planned event.	2006		
28.	Ensure that Public Information procedures and processes can gather, verify, coordinate, and disseminate information during an incident/planned event.	2007		

NIMS IMPLEMENTATION DISCUSSION SUGGESTIONS

Read Appendix A FEMA's "FY 2009 NIMS Implementation Objectives." Note that all of these were to have been completed by the end of 2009. Perform research to determine the extent to which your local emergency response organizations have complied with the following requirements:

1. Adopt and encourage adoption of NIMS
7. Revise and update EOPs, SOPs, SOGs, to incorporate NIMS and NRF components, principles, and policies
9. Use existing resources to deliver NIMS training requirements
10. Implement IS-700 NIMS Introduction training
25. Manage all incidents in accordance with ICS

Aftermath of a HAZMAT truck accident. Photo courtesy Spill Recovery of Indiana.

Chapter 7

HAZARDOUS MATERIALS INCIDENTS

PART 1. HAZARDOUS MATERIALS INCIDENTS AND TERRORISM*

In the aftermath of September 11, 2001 terrorist attacks on the World Trade Center Towers, the scene at "ground zero" – the sixteen-acre site of destruction-was described as the worst environmental disaster ever inside a major city.[1] The scene had "the same scope as a Superfund[2] site," according to New York University Hospital environmental-medicine specialist Max Costa.[3] Although Environmental Protection Agency monitoring determined the aftermath to be safe from an environmental viewpoint, concern over potential effects of hazardous materials exposure to responders was widespread.[4] Given this high profile example, it is not surprising that emergency responders frequently refer to terrorist attacks as "hazardous materials ("HAZMAT") incidents with an attitude." The reasoning behind this nomenclature is sound: a terrorist attack will almost always result in the release of hazardous substances,[5] and the terrorist always has a purposeful attitude, or intent to the criminal law[6] sense of the word. The potential liabilities for

---

* This material is adapted from: William C. Nicholson, Legal Issues in Emergency Response to Terrorism Incidents Involving Hazardous Materials: The Hazardous Waste Operations and Emergency Response (HAZWOPER) Standard, Standard Operating Procedures, Mutual Aid and the Incident Management System, Widener Symposium LJ., Vol. 9, No. 2, 2003.

[1] David France, The Cleanup, NEWSWEEK, October 1, 2001 at 6.

[2] The Comprehensive Environmental Response, Compensation, and Liability Act (CERCLA) section 104 (i), as amended by SARA, requires ATSDR and the EPA to prepare a list, in order of priority, of substances that are most commonly found at facilities on the National Priorities List (NPL) established by the National Contingency Plan (NCP). The NCP provides guidelines and procedures needed to respond to releases and threatened released of hazardous substances, pollutants, or contaminates. Superfund Amendment and Reauthorization Act of 1986 (SARA or Superfund), 42 USC §§ 11001 et seq. This act provided for broad Federal authority to respond directly to releases or threatened releases of hazardous substances that may endanger public health or the environment. In addition, the act provides for a tax on the chemical and petroleum industries, which is then deposited into a trust fund for cleaning up abandoned or uncontrolled hazardous waste sites. *Also see* Note 41 and accompanying material.

[3] France, *supra* note 1, at 6.

[4] Robert Lee Hotz, Gary Polakovic, America Attacked; Environmental Nightmare; Experts Differ On Peril From Smoke; Health: EPA Says The Cloud Rising From the Ruins Is Not Toxic, but Others Aren't So Sure. Rescuers Are Most At Risk For Possible Ill Effects, L.A. TIMES A-5 (September 14, 2001). "Those construction workers, firefighters and cops are being very heavily exposed to dust and asbestos. That [exposure] isn't going to end tomorrow; they'll be heavily exposed for weeks and months."

[5] http://www.epa.gov/superfund/programs/er/hazsubs/cercsubs.htm A CERCLA hazardous substances are defined in terms of those substances either specifically designated as hazardous under the Comprehensive Environmental Response Compensation, and known as the Superfund law, or those substances identified under other laws. In all, the Superfund law includes references to four other laws to designate more than 800 substances as hazardous, and identify many more as potentially hazardous due to their characteristics and the circumstances of their release. A complete 2001 CERCLA Priority List of Hazardous Substances can be found at the following web site http://www.atsdr.cdc.gov/clist.html

[6] Terrorism is a federal crime. See generally, Note: Responding To Terrorism: Crime, Punishment, And War, 115 HARV. L. REV. 1217, 1224 (2002) "the United States has traditionally treated terrorism as a crime. The U.S. Code contains criminal statutes that define and establish punishments for terrorism." (footnotes omitted).

emergency responders to terrorist incidents therefore include the same subject matter as any response to a hazardous materials spill, with the added concern that the incident is also a crime scene.

As with any hazardous materials response, the actions of all responders to terrorism events are closely controlled by extensive regulations of both Occupational Safety and Health Administration ("OSHA")[7] and the Environmental Protection Agency ("EPA").[8] Both public and private entities may be charged with first response to a HAZMAT occurrence. Typically, in-plant response teams will be first at incidents that occur in an industrial setting. Such HAZMAT teams are mandated by OSHA's rule for Process Safety Management of Highly Hazardous Chemicals, whose purpose is preventing or minimizing the consequences of catastrophic releases of toxic, reactive, flammable, or explosive chemicals.[9] On the public side, for spills or airborne releases occurring on public property such as highways or traveling beyond the boundaries of an industrial facility, the first response organization is typically the fire service. The highly dangerous nature of hazardous materials requires sophisticated technical expertise of responders.[10]

PART 2. CFR 1910.120 HAZWOPER AND NFPA STANDARDS FOR HAZMAT RESPONDERS

In 1986, under Congressional mandate,[11] the United States Secretary of Labor promulgated minimum training requirements for hazardous waste workers.[12] Congress also provided that the states may develop their own occupational safety and health regulations absent OSHA standard or, in the alternative, preempt OSHA by submitting a plan to the Secretary of Labor.[13]

OSHA's HAZWOPER standard requires all employers to "develop and implement a written safety and health program for their employees involved in hazardous waste operations. The program shall be designed to identify, evaluate, and

---

[7] In 1970, Congress enacted the federal Occupational Safety and Health Act of 1970 ("OSH Act"). 84 Stat. 1590 (codified at 29 U.S.C. 553, 651-678 (2002)). The OSH Act specifically authorized the Secretary of Labor to promulgate national health and safety standards. 29 U.S.C. 655(a). Occupational Safety and Health Standards 29 CFR § 1910.120 (q) (1998) covers employees who are engaged in emergency response to hazardous substance releases no matter where it occurs except that it does not cover employees engaged in operations specified in paragraphs (a)(1)(i) through (a)(1)(iv) of this section. Nor does it cover those emergency response organizations that have developed and implemented programs equivalent to this paragraph for handling releases of hazardous substances pursuant to section 303 of the Superfund Amendments and Reauthorization Act of 1986 shall be deemed to have met the requirements of this paragraph.

[8] Environmental Protection Agency (EPA) 40 CFR § 372.18 (1995) deals with the enforcement and compliance guidelines for toxic chemical release reporting and community right-to-know.

[9] 29 CFR § 1910.119 deals with preventing or minimizing the consequence of catastrophic release of hazardous materials in the industrial setting.

[10] See generally FEDERAL EMERGENCY MANAGEMENT AGENCY – UNITED STATES FIRE ADMINISTRATION, HAZMAT incidents, including training required for their utilization.

[11] In 1986, Congress ordered the Secretary to "promulgate standards for the health and safety protection of employees engaged in hazardous waste operations" pursuant to authorization granted in 655(a) of the OSH Act. Superfund Amendments and Reauthorization Act of 1986 ("SARA"), Pub. L. 99-499, Title I, 126(a)-(f), 100 Stat. 1613, as amended Pub. L. 100-202, 10(f) Title II, 101 Stat. 1329-198 (codified as amended at historical note following 29 U.S.C. 655 (1988)).

[12] Hazardous Waste Operations and Emergency Response, 29 C.F.R. 1910.120 (1992). The regulations provided that employees in close proximity to hazardous wastes must receive 40 hours of off-site training and have three days of on-site field experience. 29 C.F.R. 1910.120(e)(3)(i). Employees occasionally on-site must receive 24 hours of off-site training and have one day of on-site field experience. 29 C.F.R. 1910.120(e)(3)(ii) & (iii). Supervisors must complete an additional eight hours of training on subjects such as employee safety and spill containment. 29 C.F.R. 1910.120 (e) (4).

[13] 18, 84 Stat. at 1590. OSH Act section 18(a) provided that "nothing in this chapter shall prevent any state agency or court from asserting jurisdiction over any occupational safety or health issue with respect to which no Federal OSHA standard is in effect." 18(a), 84 Stat. at 1590. OSH Act section 18(b) provided that "any state which at any time, desires to assume responsibility for development and enforcement therein of occupational safety and health standards relating to any occupational safety or health issue with respect to which a Federal standard has been promulgated ... shall submit a State plan ... to the Secretary of Labor." 18(b), 84 Stat. at 1590. But see Note 12 and accompanying material.

control safety and health hazards, and provide for emergency response for hazardous waste operations."[14] Even for non-OSHA states, EPA incorporates OSHA's HAZWOPER standard.[15]

The HAZWOPER standard creates duties for individual responders as well as for the organizations that employ them.[16] Indeed, individual responders are charged with knowledge of their duties whether or not they have actual knowledge thereof.[17] The requirement to use the HAZWOPER standard applies to volunteers as well as to paid responders.[18] Federal employees are also bound by HAZWOPER requirements.[19]

HAZWOPER requires that an emergency response plan be developed and implemented to handle anticipated emergencies prior to the commencement of emergency response operations.[20] The HAZWOPER requirement is, however, only one of a number of standards mandating plans for emergency response entities.[21] The HAZWOPER plan may be merged with other necessary emergency plans following the National Response Team Integrated Contingency Plan Guidance to avert duplicative efforts.[22]

Employers who will evacuate their employees from the danger area when an emergency occurs, and who do not permit any of their employees to assist in handling the emergency, are exempt from these requirements if they provide an emergency action plan in accord with the rules.[23] In order to achieve an exemption from the full planning requirements, an employer's plan must call for evacuation only.[24] Ambiguity in the plan as to whether employees are required to respond to an uncontrolled HAZMAT release may be interpreted in favor of exemption, but this is a highly fact-specific inquiry.[25] Further, to obtain exemption, an employer must comply with the separate planning requirement of 29 CFR § 1910.38(a).[26] Any involvement by employees in emergency rescue activities will almost always subject the employer to the pre-planning requirements.[27] A narrow exception is made for voluntary employee rescue.[28] Although planning requirements

---

[14] 29 CFR § 1910.120 (b) Any safety and health programs developed and implemented to meet other Federal, state, or local regulations are considered acceptable in meeting this requirement if they cover or are modified to cover the topics required in this paragraph. An additional or separate safety and health program is not required by this paragraph.

[15] 40 CFR § 311.1 The substantive provisions found at 29 CFR 1910.120 on and after March 6, 1990, and before March 6, 1990, found at 54 FR 9317 (March 6, 1989), apply to State and local government employees engaged in hazardous waste operations, as defined in 29 CFR 1910.120(a), in States that do not have a State plan approved under section 18 of the Occupational Safety and Health Act of 1970.

[16] *Wiley Organics, Inc. v. OSCHRC*, No. 96-3575, 1997 U.S. App. WL 476530, at *20-21 (6th Cir. Aug. 19, 1997).

[17] *Ed Taylor Constr. Co. v. OSHRC*, 938 F.2d 1265, 1272 (11th Cir. 1991).

[18] 40 CFR § 311.2 (1998) An "employee" within Section 311.1 is as a compensated or non-compensated worker who is controlled directly by a State or local government, as contrasted to an independent contractor.

[19] Executive Order 12196 of February 26, 1980 subjected all federal employees to the requirements of OSHA. Executive Order 13225 of September 28, 2001 continues the effect of EO 12196. 66 FR 50291 (2001).

[20] 29 CFR § 1910.120 (q)(1) deals with the requirements for developing and implementing an emergency plan. Section 1910.120(q)(1) requires the plan to be in writing and available for inspection and copying by employees, their representatives and OSHA personnel. It is clear that mere reliance upon the basic human instinct to flee danger is not a plan. *See* Victor Microwave, Inc., 1996 OSAHRC Lexis 57, 37-38; 17 OSHC (BNA) 2141 (OSHRC Docket No. 94-3024 1996).

[21] *See*, e.g., Arnold W. Reitze, Jr. and Randy Lowell, Control of Hazardous Air Pollution, 28 B.C. Envtl. Aff. L. Rev. 229, 331-332 (2001).

[22] See The National Response Team's Integrated Contingency Plan Guidance, 61 Fed. Reg. 28,642 (June 5, 1996).

[23] 29 CFR § 1910.120 (q)(1).

[24] IBP, Inc. v. Iowa Employment Appeals Board, 604 NW 2d 807, 314 (Iowa, 1999).

[25] Akzo Nobel Chemicals, Inc., 1998 OSAHRC Lexis 98, 18-23, 18 OSHC (BNA) 1643, 1998 OSHD (CCH)¶31,695 (OSHRC Docket No. 96-0062 1996).

[26] Id. Pleadings amended to correspond to the evidence and citation affirmed.

[27] 604 NW 2d at 312-313.

[28] OSHA's interpretive rule regarding "voluntary employee rescue," § 1903.14(f), became effective on December 27, 1994. The statement of policy clarifies that: It is not OSHA's policy... to regulate every decision by a worker to place himself at risk to save Another individual. Nor is it OSHA's policy to issue citations to employers whose employees voluntarily undertake acts of heroism to save another individual from Imminent harm, [except in specifically stated circumstances]. FR Doc 94-31625.

are comprehensive, a court may excuse an incomplete plan under limited circumstances.[29]

National Fire Protection Association ("NFPA")[30] 472 "Professional Competence for Responders to Hazardous Materials Incidents" 2002 Edition[31] provides additional valuable detailed standards for HAZMAT response. NFPA 472 requires all occurrences requiring response, including suspected terrorism incidents, to be evaluated by first responders (the lowest level of training) as potential HAZMAT events as part of general situational awareness.[32] NFPA 472 sets out detailed skill sets for all levels of responder training set forth in the HAZWOPER standard.

PART 3. OSHA VIOLATION LIABILITY

As indicated above, OHSA law closely regulates the actions of emergency responders to a HAZMAT incident. In the event that a violation is found to have taken place, significant penalties may ensue. Penalties and fines range as follows:

De Minimis Notice	$0[33]
Nonserious	$0-$7,000[34]
Serious	$1-$7,000[35]
Repeated	$0-$70,000[36]
Willful	$5,000-$70,000[37]
Failure to Abate Notice	$0-$7,000 per day[38]

Even multiple serious and willful violations may not result in an aggregate monetary penalty that approaches the maximum for a single willful violation.[39] Violations of serious or greater gravity will, nonetheless, have significant consequences for the offending employer. Above and beyond the penalties listed above, employers may find themselves losing insurance coverage that may be required to conduct business. Further, a violation of law may be used as proof in a civil trial for damages for personal injury or wrongful death. When the elements of the violation are congruent

---

[29] *Jordan v. Lehigh Construction Group, Inc.*, 269 A.D. 2d 743, 744 (S.Ct.N.Y. App. Div. 4th Dept. 2000). Since decedent was aware of the only means of egress from this office, its omission from the plan required by 29 CFR § 1910.120 was not a proximate cause of his death.

[30] See http://www.nfpa.org/Home/AboutNFPA/index.asp. The mission of the international nonprofit NFPA is to reduce the worldwide burden of fire and other hazards on the quality of life by providing and advocating scientifically based consensus codes and standards, research, training and education. NFPA membership totals more than 75,000 individuals from around the world and more than 80 national trade and professional organizations.

[31] NATIONAL FIRE PROTECTION ASSOCIATION 472: STANDARD FOR PROFESSIONAL COMPETENCE OF RESPONDERS TO HAZARDOUS MATERIALS INCIDENTS (2002). Like many NFPA benchmarks, this standard has been incorporated by reference into law at the state level. See e.g., Wis. State. § 166.215 (2001) (1)(B) A member of a regional emergency response team shall meet the standards for a hazardous materials specialist in 29 CFR 1910.120 (q) (6) (iv) and national fire protection association standards NFPA 471 and 472. Further, many fire departments have adopted it as part of their standard operating procedures. Telephone interview with Jerry Laughlin, Deputy Director, Alabama Fire College, former staff liaison for NFPA to the Hazardous Materials Emergency Response Technical Committee, which is responsible for NFPA 472. (June 20, 2002)

[32] Id. at 472-9 - 472-11 describes requirements for first responders at the awareness level, which includes all emergency responders. Their duties are summed up at 472-57: "First responders at the awareness level are expected to recognize the presence of hazardous materials, protect themselves, call for trained personnel, and secure the area." 472-11 - 472-15 describes competencies for the first responder at the more intensively trained operational level.

[33] Occupational Safety & Health Act of 1970 (OSH Act) § 9(a). Discusses the procedure of issuing a citation to an employer who is in violation of a requirement of Section 5 of OSH Act or Section 6 of the Act. (to be codified at 29 U.S.C. 658).

[34] OSH Act § 17(c) A civil penalty accessed on an employer for violation of either an employer's requirements and duties under Section 5 of the OSH Act or any violations of the Occupational Safety and Health Standards under Section 6 of the Act.

[35] OSH Act § 17(k) Describes a serious violation as existing in a place of employment if there is a substantial probability that death or serious physical harm could result from an existing condition, or from one or more practices, means, methods, operations, or processes which have been adopted or are in use.

[36] Id. at § 17(a) (to be codified at 29 U.S.C. §666).

[37] Id.

[38] Id. at § 17(c).

[39] *Secretary of Labor v. Victor Microwave, Inc.*, No. 94-3024, 1996 OSAHRC Lexis 57, at *97-98 (O.S.H.R.C.A.Lj. June 17, 1996). Total penalties for 30 violations equal $34,000.

to the elements required for civil liability and the burden of proof is the same for both, the only issue in a civil trial may be the measure of damages.

An Incident Commander (and his or her employer) faces potential liability for improperly supervising or directing loaned emergency responders during a common undertaking. An examination of the Second and Seventh Circuit decisions involving main contractor liability in construction cases involving subcontractor employee injuries illustrates a standard that could be construed to apply in a case involving emergency response. The Second Circuit decision found that it was only necessary to show that "a hazard had been committed and that the area was accessible to the employees of the cited employer or those of other employers engaged in a common undertaking."[40] The Second and Seventh Circuit expanded on this rule by holding that "an employer who does control an area containing safety hazards will be held in violation of OSHA regulations, regardless of whether his own employees were exposed to the potential danger."[41] Furthermore, when an employer does not have control of the area, the employer will not be held in violation of the OSHA regulation, even if his own employees are exposed, unless the exposure is to a hazard presenting a likelihood of death or serious harm.[42]

Clearly, like subcontractor employees, when emergency responders to a HAZMAT event are engaged in a common undertaking there exists a specific duty to them on the part of all entities with supervisory authority, regardless of who employs them. During a common undertaking, this specific duty of adequate training, instruction, and supervision is over and above the general duty owed to an IC's own emergency responders.

Consider the above information as you evaluate the following case. Concentrate on its description of OHSA requirements for HAZMAT response and the activities that were found not to comply.

Victor Microwave, Inc.

OSHRC Docket No.: 94-3024

Occupational Safety and Health Review Commission

Administrative Law Judge

1996 OSAHRC LEXIS 57; 17 OSHC (BNA) 2141 (1996)

Before: Administrative Law Judge NANCY J. SPIES

DECISION AND ORDER

Victor Microwave, Inc. (Victor), is a small Massachusetts corporation which assembles and manufactures military components for the United States Army. On March 23, 1994, as a result of an industrial accident at Victor, an "acid cloud" engulfed employees, causing them to evacuate the premises. Two of Victor's workers were hospitalized (Tr. 32, 117). Industrial Hygienists Francis Pagliuca and Mary Hoye of the Occupational Safety and Health Administration (OSHA) investigated the incident from March 23 through June 16, 1994. Pagliuca responded to the original incident, beginning his investigation the afternoon of the accident (Tr. 403). Hoye continued the inspection, returning to Victor on several subsequent dates. On September 12, 1994, the Secretary issued Victor a serious (32 items), a willful (1 item) and an "other" (5 items) citation.

---

[40] *Brennan v. Occupational Safety and Health Review Commission.* 513 F.2d 1032, 1038 (2d Cir. 1975).

[41] Note, Administrative Law-OSHA-On Multiemployer Jobsite, When Employees of any Employer are Affected by Noncompliance with a Safety Standard, Employer in Control of Work Area Violates Act; Employer not in Control of the Area Does Not Violate Act, Even if His Own Employees Are Affected, Provided that the Hazard is "Nonserious." 89 Harv.L.Rev. 793 (1976)

[42] *See* Id. at 796. See generally *Underhill Construction Corporation v. Secretary of Labor and OSHRC,* 526 F.2d 53 (2nd Cir. 1975). *See* generally *Anning-Johnson Co. v. United States Occupational Safety & Health Review Commission,* 516 F2d 1081 (7th Cir. 1975).

Victor contested all asserted violations. Of particular dispute between the parties was whether the "emergency response" standard of § 1910.120 applied to Victor's operation and whether an alleged violation of the hazard communication standard as willful.

Victor is owned and operated by Stephen and Robert Parks, father and son. It has been in business since 1969 (Exh. R-4). Victor partially occupies a building it owns in Wakefield, Massachusetts. It rents out a portion of the building to an unrelated company, Raytheon. Because of financial reversals the company began downsizing in November, 1993. Victor decreased its shared space. Over the months of January and February, 1994, it moved a part of its business operation into the building's basement, where the incident occurred (Tr. 466). Employees performed varied duties at Victor, including machining, cleaning, and soldering parts to military specifications (Tr. 137148). Long-term employees, James Amaral, his brother John Jack) Amaral, and Michael Metheny testified at the hearing, as did both of the Parks....

EMERGENCY RESPONSE STANDARD ALLEGATIONS

Background

Victor's employees worked with solutions of "pickle" (or Turco) and "bright dip." Pickle was used to clean the copper and brass component rods. To perform this operation the rods were submerged into the heated pickle solution. When the components appeared sufficiently clean, usually after an hour, they were removed from the pickle and rinsed. To give the components a bright or shiny finish, they were then placed into a plating solution called "bright dip" for between 5 to 30 seconds (depending on the strength of the mixture) (Tr. 16). When copper or brass (which contains copper) was introduced into the bright dip solution, a yellow-tinged smoke would often be released. This was especially true when the solution was newly mixed (Exh. R-2; Tr. 62-64).

On March 23, 1994, employee Ralph De-Monte mixed up a new batch of bright dip and took it down to James Amaral in the basement. Amaral had just finished cleaning a number of brass rods with pickle. After rinsing, he placed one or more rods into the new batch of bright dip solution. A dense cloud of orange smoke immediately rose from the solution, presumably caused by a chemical reaction between the bright dip and the rods (Tr. 25, 62-63). The emission came "like a blanket [and] within seconds, it had consumed the whole downstairs" (Tr. 134).

The release contained large amounts of nitrogen dioxide gas[7] (Tr. 255). The emission engulfed employees in the basement and soon rose to the upper floor and to the adjoining business. James Amaral described what occurred (Tr. 25-27):

It [the fumes] blew in my face and it was burning and I couldn't see.... And when I opened my eyes, there was this big cloud of orange smoke everywhere. So I had just thought of how Steve had yelled at Mark about Ralph getting burnt with the bright dip, that I proceeded to try and grab the rods, feel for them, because there was a cloud and you couldn't see nothing.

And I started grabbing the rods and pulling them up. Some of them had gone into the tank of bright dip. And it was getting all over me and burning me. And my mind told me, "It's time to leave."... And my brother was also working downstairs and Ralph was at the cutoff [saw]. So I asked Ralph to get me a fan to help push the smoke out the door. And he took off. I never seen him again. And my brother took off and he said, "Just get out of there." And when my brother left, than I left behind him. But rather than go to my left, where the bright dip was, I went to my right to go around the station and then up the stairs....

...And I slipped in the oil. And I knew I was going to fall, so I put my hands out to stop me. And when I hit the floor, I broke this wrist.... And I caught my chin on the cement floor. And it kind of spun me out. I saw stars, I guess you could say.... I just could not [get up].

---

[7] Nitrogen dioxide can be liberated from nitric acid and copper.

James' brother John Amaral warned employees on the first floor of the fast-rising emission. He asked them to "tell the boss," leave the building, and call the fire department" (Tr. 120) John then realized James was not outside with the other employees. As John described the events (Tr. 122-123):

> I didn't know for sure if he was there, but I didn't want to take the chance of wasting any more time because I knew the cloud was so bad, it was so thick, it was going to the floor and then covering the ceiling. . . . I started looking around, but I really couldn't see him. I was feeling my way around trying to feel for him. And I kind of ran down towards the back of the machines and tripped. I tripped on his body. He was on the floor unconscious. . . .

> I tried not to breath [sic] when I went into the cloud, when I first entered it. But when I jerked my brother up off the floor, I had to breath [sic] up the energy and strength to drag him out of there. And I got some of this cloud in me.

> As soon as I took a breath of the stuff in, my lungs basically stopped, froze right up. I couldn't breath [sic]. I felt like I was going to pass out and possibly wouldn't make it out the door with my brother. I thought we were both going to be right there. I thought I was going to die.

Both men were in obvious distress after John finally brought James out of the building. Robert and Stephen Parks advised their employees that they "were handling the situation" (Tr. 153). Raytheon employees called the fire department.

At some time after employees evacuated, but before the fire department arrived, Robert and Stephen Parks donned plastic "garbage" bags to protect their clothing and re-entered the building. They individually made several short efforts to get to the basement, holding their breaths to avoid breathing in the nitrogen dioxide. One of them finally arrived at the basement, pulled the rods out of the five-gallon bucket of bright dip, and carried the bucket outside of the building. Although the bright dip was still emitting vapor, the emission had lessened and was at the point that it could have been handled by the basement sink exhaust system (Tr. 222, 468).

Approximately 20 minutes after the incident, Michael Metheny also reentered the building looking for the Parks. Metheny observed the Amarals gagging and unable to catch their breaths and believed they needed immediate medical attention. Metheny had been told that the Parks did not want anyone to go to the hospital. He sought the Parks to gain permission to take them. Metheny was about halfway down the basement stairs when he saw Stephen Parks. He told Parks that the Amarals needed to go to the hospital. Parks directed his son to have them taken (Tr. 154, 173, 174). The cloud had cleared, but Metheny was not comfortable about being inside the building (173). Later, the fire department arrived and ventilated the building with large exhaust fans (Tr. 467).

Other employees took the Amarals, DeMonte, and Mark Heath to a hospital where they were initially treated as if the exposure was to nitrous oxide (laughing gas) rather than nitrogen dioxide. Robert Parks provided the MSDS for nitric acid (Tr. 32, 136). Two workers from Raytheon were lying on stretchers while the Amarals were at the hospital (Tr. 32-33). James Amaral was also treated for a broken wrist. As of the date of the hearing, neither James or John Amaral had returned to work. They described continued ill effects from the nitrogen dioxide exposure (Tr. 36-39, 129-138).

Coverage Under the Standard

Subpart 1910.120 applies to "hazardous waste operations and emergency response." Victor may arguably be covered only through § 1910.120(a)(1)(iv) of this standard. If it is, it must comply with § (q) (*see* § 1910.120(2)(iv)).

Section 1910.120(a)(1)(iv) covers the following operations: (v) Emergency response operations for releases of, or substantial threats of releases of, hazardous substances without regard to the location of the hazard. Section 1910.120(3) defines "emergency response corresponding to emergencies" as: [A] response effort by employees from outside the immediate release area or by other designated responders (i.e., mutual-aid groups, local fire departments, etc.) to an occurrence which results, or is likely to result, in an uncontrolled release of a hazardous substance.

Both definitions apply to a release of a "hazardous substance(s)." By weight, pickle contains

55 percent phosphoric acid (Exh. C-1). Bright dip is a solution of nitric acid, sulfuric acid, and water. Official notice is taken that Victor's employees regularly used chemicals which are defined as "hazardous substances" by § 1910.120(3), i.e., they are included as hazardous materials under 49 CFR 172.101. Nitrogen dioxide, which was released in large quantities on March 23, is also a "hazardous substance" under § 1910.120(3).

Not every employer which works with hazardous substances has reason to anticipate an emergency release. Victor, however, worked with large quantities of nitric acid solution, which if combined with other substances used in Victor's processes, such as copper, could result in emergency release (Tr. 213, 308-309, 324).

In addition to proving that there was the release of a hazardous substance, there is no coverage unless there is also an "emergency response" (or "response effort") by employees to the release. The Secretary asserts that there were three separate responses to the release: (1) the employees, especially James Amaral, initially attempted

to contain the release rather than evacuating; (2) John Amaral rescued his brother; and (3) both Stephen and Robert Parks returned to the release area to bring out the source of the emission. The first allegation, James Amaral's attempt to contain the emission, was not "from outside the immediate release area," as the standard specifies, but took place at the actual release site. Based on OSHA's subsequent clarification, John Amaral's valiant rescue of James likewise does not constitute an emergency response.[8] Only the Parks' re-entry into the building to remove the emission source arguably falls within the definition of a "response." Without question, the Parks are "employees" of Victor, regardless of their ownership interest. The Parks, who were outside the release area, returned to it to remove and to contain the source of the emission. When they re-entered the building, the hazardous emission was ongoing, although lessening. The Secretary has made a prima facie showing for coverage under the "emergency response" standard for this third alleged instance....

DEFENSES TO EMERGENCY RESPONSE ALLEGATIONS

(1) *Validity of the Standard.* Victor argues that specific language in the standard is misleading and, generally, that the standard is "torturous," confusing, and unenforceably vague.

Victor specifically challenges the validity of the standard arguing that the phrase "or *other designated* responders (i.e., mutual-aid groups, local fire departments, etc.)" (emphasis added) in the definition of "emergency response" misleads an employer into believing that it has no responsibility *unless* it designates an employee to respond to emergencies. It is a well-established principle of statutory construction that the words of a standard are viewed "in context, not in isolation." *Georgia*

Pacific, 16 BNA OSHC 1171, 1174 (No 89-2806, 1993). "Other designated responders" is followed by specific examples, thus removing potential ambiguity. Contrary to Victor's argument, a reasonable reading of the standard provides sufficient notice to the employer of what is required of it.

The more general challenge is also rejected. By necessity, some standards must be broadly worded. A standard is not impermissibly vague simply because it is broad in nature. "External, objective criteria, including the knowledge and preceptions of a reasonable person, may be used to give [the standard] meaning" *J.A. Jones,* 15 BNA OSHC 2200, 2205-2206 (No. 872059, 1993).

---

[8] OSHA's interpretive rule regarding "voluntary employee rescue," § 1903.14(f), became effective on December 27, 1994. Although post-dating the citation, the statement of policy clarifies that: It is not OSHA's policy...to regulate every decision by a worker to place himself at risk to save another individual. Nor is it OSHA's policy to issue citations to employers whose employees voluntarily undertake acts of heroism to save another individual from imminent harm, [except in specifically stated circumstances.] FR Doc 94-31625.

Amaral's rescue does not fall within one of these exceptions, even that of § 1903.14(f)(3), because John Amaral's assigned duties were not "directly related" to the workplace operation where a life-threatening accident was foreseeable.

(2) *Victor's Knowledge.* Victor used large quantities of nitric acid solution. The solution was understood to "react vigorously" with specified substances (Exh. C-5). Stephen Parks had a background in chemical engineering and long experience in the nitric acid process (Tr. 441). He considered the solution to be "very hazardous" (Tr. 450). On a smaller scale, the chemical reaction between bright dip and the copper or brass rods was a common knowledge. Employees who observed the bright dip process often saw a "puff of smoke," a "haze," or yellow vapor (Tr. 64-65, 165). Employees on the first floor smelled the vapors and experienced a burning sensation from breathing fumes when the bright dip and pickle operations were ongoing. Michael Metheny, the machinist, complained to Robert Parks about the fumes. Parks stated that a friend who was in heating and air conditioning would see to it, but nothing had been done (Tr. 165). Victor had no real controls for the procedure. Untrained individuals were permitted to mix the solutions, and untrained employees used the solutions (Exh. C-8, Tr. 17, 47). Conditions existed for a hazardous substance release and for the occurrence of March 23, 1994. A reasonably prudent employer would have recognized this fact. Victor's knowledge informed the terms of the general standard. These facts also bear upon whether Victor knew or should have known of the conditions which constitute the emergency response allegations (items 10-16) for which the Secretary has the burden of proof.

Victor disputes knowledge relying on its theory that James Amaral caused the release. It speculates that Amaral added too many rods to the solution. Thus, Victor contends that the emission was unforeseeable. Amaral testified that he did not place more rods than usual into the solution. He thought additional rods may have fallen into the bucket as he attempted to control the emission (Tr. 70-71, 108). In any event, both parties theorized about the cause of the incident. For example, the Secretary suggested the emission may have resulted from having less water in the solution or a greater quantity in the pail (Tr. 309, 376). Neither party actually attempted to prove the theory. The anticipated hazard is the primary focus, not the immediate cause of the incident.

If placement of additional rods in the solution yielded such an extreme reaction, the volatility of the chemicals is underscored.

(3) *Employee Misconduct.* Victor also characterizes its foreseeability argument as an "employee misconduct" defense. Establishing an employee misconduct defense requires specific proof of, among other things, the existence of a work rule designed to prevent the violative conduct. *Falcon Steel Co.*, 16 NA OSHC 1179, 1193 (No. 89-3444, 1993). Victor argues that it had an oral work rule which would have prevented Amaral from placing too many rods into the solution. James Amaral was not aware Victor had restrictions on the number of rods to be placed in the bright dip solution. He himself made that determination based upon such factors as the job to be done, or the age, strength, or heat of the solution. He had substantial discretion in performing the job (Tr. 50-51, 63, 70-71, 108). Victor has not proven the existence of even an oral work rule covering the subject. Further, the conduct which precipitated OSHA's citation was that the Parks re-entered the site to control the release. Victor had no work rule directed at this conduct.

(4) *The Chemical Composition.* Finally, Victor argues in its brief that the bright dip solution was not purely nitric acid but a "much weaker solution." Although the Secretary sought the MSDS for bright dip during the investigation and discovery, Victor did not provide it. The Secretary did not know that an MSDS for bright dip existed. Victor provided only the MSDS for nitric acid. At the hearing, Stephen Parks testified to the exact composition of the bright dip, referring to its MSDS. Bright dip is purchased from the manufacturer as 54 percent sulfuric acid, 22 percent nitric acid and 24 percent water (Tr. 445-446). Victor diluted the mixture with water by "approximately 20 percent to 30 percent" by volume (Tr. 446). As illustrated, the Secretary was unaware of the existence of a MSDS for the exact composition of the solution (Tr. 452):

Mr. Metzler: Would you go through the procedure that you wrote?

S. Parks: The first advice is to proper protective equipment, gloves, face shield, goggles, shop coat, plastic apron and to read the bright dip MSDS, then to clean the stainless steel container and cover it thoroughly.

Mr. Baskins: Excuse me. I'm sorry to interrupt you. You said the "bright dip MSDS." Is that the nitric acid?

S. Parks: No, that's a different one. It just says "bright dip MSDS."

Mr. Baskins: Bright dip MSDS or bright dip preparation sheet?

S. Parks: No. This is a bright dip MSDS and this is a bright dip preparation or standard operating and mixing procedure.

Mr. Baskins: Excuse me, Mr. Parks. (Reviewing documents) I've never gotten a copy of the document I have in my hand, Your Honor, unless it's exactly identical to . . . – Apparently, no. It certainly isn't. It's absolutely not identical to C-1 [sic]. So this is not a document that was given to me at the deposition.

Mr. Metzler: Your Honor, I don't want to introduce that document.

Victor did not seek to introduce the MSDS for bright dip and did not object to introduction of the MSDS for nitric acid. Victor now claims that the Secretary must show that nitric acid produces the same effects when mixed with sulfuric acid and water. While it is accepted that bright dip contains the two acids and water, it is unfair to penalize a party because it failed to present evidence of a fact the other party prevented it from knowing. Moreover, as Stephen Parks testified, acids react more readily in solution than would straight nitric acid (Tr. 450). Regardless of the exact composition of the solution, employees unquestionably were exposed to significant and dangerous amounts of nitrogen dioxide when gas was liberated on March 23. The argument has no effect on the decision.

Victor's defenses are rejected. It is subject to requirements of § 1910.120(q).

Item 10: § 1910.120(q)(1)

The Secretary asserts that Victor did not have an emergency response plan on how to proceed in case of an emergency. Victor primarily argues that under the terms of the standard it had no obligation to have the plan. The standard provides:

(1) Emergency response plan. An emergency response plan shall be developed and implemented to handle anticipated emergencies prior to the commencement of emergency response operations. The plan shall be in writing and available for inspection and copying by employees, their representatives, and OSHA personnel. Employers who will evacuate their employees from the danger area when an emergency occurs, and who do not permit any of their employees to assist in handling the emergency, are exempt from the requirements of this paragraph if they provide an emergency action plan complying with section 1910.38(a) of this part.

Employees Stephen and Robert Parks engaged in an emergency response. The standard permits an employee to have either an emergency response plan (when there is an emergency response) or a written emergency plan which complies with § 1910.38(a) (if employees will not assist with the emergency). Victor had neither (Tr. 219). Contrary to Victor's suggestion, reliance on the basic human instinct to flee danger is not a "plan." *See Pressure Concrete Constr. Co.*, 15 BNA OSHC 2011, (No. 90-2668, 1992). Failure to have the plan could result in serious injury or death. The violation is affirmed.

Items 11a & 11b: § 1910.120(q)(3)(ii)

Both items allege a violation of the same standard. The Secretary maintains Victor violated § 1910.120(q)(3)(ii) because it failed to perform necessary site monitoring to determine which hazardous substances were present (item 11 a) and did not implement the appropriate responses based on substances which would have been found (item 11b). The standard provides:

(ii) The individual in charge of the ICS [incident command system] shall identify, to the extent possible, all hazardous substances or conditions present and shall address as appropriate site analysis, use of engineering controls, maximum exposure limits, hazardous substance handling procedures, and use of any new technologies.

On March 23, 1994, Victor made no attempt to determine what conditions existed before

Stephen and Robert Parks conducted an emergency response. Specifically, they did not seek to identify the hazardous gases which were expected to be present. They made no effort to utilize engineering or other controls, such as ventilating the building. Neither did they use personal protective equipment to limit their exposure to the nitrogen dioxide (Tr. 222). The standard requires separate conduct: the first requirement is to identify the substances, and the second is to appropriately control the hazard. The requirements are not duplicative, although they are interrelated.

The conduct could result in serious injury or death. Items 11a and 11b are affirmed as violations of the standard. The items were grouped, and the one penalty was recommended for both.

Item 12: § 1910.120(q)(3)(iv)

The Secretary charges that employees did not use self-contained breathing apparatus (SCBA) in violation of § 1910.120(q)(3)(iv). The standard provides:

> (iv) Employees engaged in emergency response and exposed to hazardous substances presenting an inhalation hazard or potential inhalation hazard shall wear positive pressure self-contained breathing apparatus [SCBA] while engaged in emergency response, until such time that the individual in charge of the ICS determines through the use of air monitoring that a decreased level of respiratory protection will not result in hazardous exposures to employees.

The permissible ceiling level for nitrogen dioxide is 5 parts per million. Employees were exposed to no less than 50 parts per million of nitrogen dioxide, ten times the permissible ceiling level (Tr. 225). The specific duration of this exposure is unknown. At the time that the Parks re-entered the building to engage in the emergency response, there was, at the very least, a "*potential* inhalation hazard." No one monitored

the air before the re-entry (Tr. 221). Use of self-contained breathing apparatus (SCBA) would have protected the Parks' respiratory systems. Attempting to hold one's breath, as the Parks did when they made their short entries into the building and trying "not to take too many breaths" obviously does not constitute an alternate means of compliance (Tr. 221-222). Severe respiratory injury is the probable result of exposure. The violation is affirmed as serious.

Items 13 & 14: §§ 1910.120(C3)(v) and 1910.120(q)(3)(vi)

The Secretary asserts that when the Parks re-entered the building, they should have complied with § 1910.120(q)(3)(v)[10] by using a "buddy system." In addition to the buddy system, the Secretary alleges that the Parks should also have had back-up personnel as required by § 1910.120(q)(3)(vi).[11] The "buddy system" is defined in § 1910.120(a)(3) as:

> a system of organizing employees into work groups in such a manner that each employee of the work group is designated to be *observed* by at least one other employee in the work group (emphasis added).

The only credited proof offered to support items 13 and 14 is admission Hoye attributes to Robert Parks. Parks explained to Hoye that he and his father "took turns going down into the basement to secure the area" (Tr. 225). Although both the Parks testified, neither was asked to describe the method by which they re-entered the building. Working within the buddy system would require that someone be available to observe, and if necessary, to rescue the other. Minimally sufficient evidence may establish a prima facie case. *Cf. Falcon Steel Co., supra,* 16 BNA OSHC 1190-91 ("meager" testimony of compliance officer concerning practicality of fall protection sufficient to establish violation). When one of the Parks proceeded alone into the basement

---

[10] Section 1910.120(q)(3)(v) provides:
(v) The individual in charge of the ICS shall limit the number of emergency response personnel at the emergency site, in those areas of potential or actual exposure to incident or site hazards, to those who are actively performing emergency operations. However, operations to hazardous areas shall be performed using the buddy system in groups of two or more.
[11] Section 1910.120(q)(3)(vi) requires:
(vi) Back-up personnel shall be standing by with equipment ready to provide assistance or rescue. Qualified basic life support personnel, as a minimum, shall also be standing by with medical equipment and transportation capability.

to contain the emission, he could not be visually observed by the other from outside the building. The basement area was "hazardous," at least until the emission source was removed. Just a short time earlier, the Amarals illustrated the serious consequences of failing to have an individual available for rescue. Item 13 is affirmed as serious.

The same evidence is not sufficient to support item 14. The Secretary contends that the Parks should have had additional personnel ready for their rescue. Stephen and Robert Parks each served in this capacity for the other. Unlike item 13, it is not necessary that back-up personnel remain in visual contact. Robert Parks was within hearing range of Stephen, as illustrated by Metheny's description (Tr. 155):

> At that point, [Stephen Parks] didn't say anything [to Metheny]. He looked up. He thought he was walking past the door to the outside. You can look to the outside from where I was standing also. And he yelled up to Bob to get someone to take them to the hospital.

The facet that the two "took turns" entering the building does not create a presumption of a violation of item 14. The Secretary did not address whether rescue equipment was available, and its absence will not be presumed. The Secretary failed in his burden of proof. Item 14 was vacated.

Item 15: § 1910.120(q)(3)(vii)

The Secretary alleges that the individual in charge of the incident command system (ICS) did not designate a safety officer as required by § 1910.120(q) (3) (vii). The standard provides:

> (vii) The individual in charge of the ICS shall designate a safety officer, who is knowledgeable in the operations being implemented at the emergency response site, with specific responsibility to identify and evaluate hazards and to provide direction with respect to the safety of operations for the emergency at hand.

The Secretary cited a violation of this standard because employees were allegedly allowed to return to the building before the fire department gave the "all clear." His theory appears to be that a designated safety officer would have prohibited employees from entering into the building until there was proof that the hazardous emission was below 5 parts per million of nitrogen dioxide. Although no employees did, Stephen Parks advised them that they could return to work before the fire department arrive (Tr. 228-230). Two signed employee statements were introduced into the record under on Rule 801(d)(2)(D), Fed.R.Civ.P, to support this conclusion. Excerpted questions and answers are on point (Exhs. C-8 & 9):

> *Q:* On the day of the accident, March 23, 1994, did either Steve Parks or Robert Parks tell you it was okay to go back into the building before the fire department arrived.
> *A1:* Yeah, he did. Steve did. (Ralph DeMonte)
> *A2:* Yes, they said go back to work we'll handle this outside. That's when I jumped in my truck and left. (Robert Onorato)
> *Q:* Did you go back into the building? If yes, was there still an odor?
> *A1:* No. I went to the emergency room. Then I went home.
> *A2:* No, I did not. Even the next morning it was bad. We were still gagging.

Stephen Parks, as the individual in charge, should have designated an appropriate safety officer to monitor the emergency response. Metheny and other employees sought to act on their safety concerns. None had authority. Metheny sought out Stephen Parks in a potentially hazardous area to get authorization to take employees to the hospital. A safety officer should have been available to determine, for example, when employees could safely reenter the building or when exposed employees needed to go to the hospital. The evidence sufficiently establishes the violation. Death or serious respiratory injury is the probable result of failing to designate an individual to emphasize safety during a chemical release response. The violation is affirmed as serious.

Item 16a: § 1910.120(q)(6)(i)(A)

Items 16 a-d relate to training.

In item 16a, the Secretary asserts that employees who were "first responders at the awareness level" were not adequately trained or experienced, in violation of § 1910.120(q)(6)(i)(A).[13] The standard defines responders at this level as individuals:

> who are likely to witness or discover a hazardous substance release and who have been trained to initiate an emergency response sequence by notifying the proper authorities of the release.

The Secretary argues that James and John Amaral and Ralph DeMonte acted as first responders who should have been provided with information and training to allow them to recognize the potential for a release and to take appropriate action. James Amaral and DeMonte worked with substances which reasonably may have been expected to cause an uncontrolled release. Victor, however, never designated either to initiate an emergency response sequence. Thus, the definition of "first responder at the awareness level" is not met. The Secretary has failed to establish that the standard applies to the condition cited. Item 16a is vacated.

Item 16: § 1910.120(q)(6)(iii)

A related charge is that the Parkses were not trained when they functioned in the role of hazardous materials technicians, as required by § 1910.120(q)(6)(iii). The standard requires:

> (iii) Hazardous materials technicians are individuals who respond to releases or potential releases for the purpose of stopping the release . . . [and they] shall have at least 24 hours of training equal to the first responder operations level and in addition have competency in the following areas. . . .

The Parks responded to the release "for the purpose of stopping [it]." The Secretary submits that the Parks' actions during the release provide sufficient proof that they were not trained (Tr. 233). For example, the Parks failed to monitor to determine their expected chemical exposure. Their only protective equipment was garbage bags. The Parks met the definition of hazardous materials technicians. There is a prima facie showing that the Parks were not trained. Victor did not rebut the showing. The standard has been violated. Serious injury or death is the expected result of failing to train those who seek to stop the emission. The standard is affirmed as serious.

Alternative Items 16c and 16d: § § 1910.120 (q)(6)(v)(A) or 1910.120(q)(3)(i)

The Secretary asserts alternative violations of § § 1910.120 (q)(6)(v)(A)[15] and 1910.120(q)(3)(i).[16] Because they are alternative violations, they are more properly designated as alternative violations of 16c. Following the Secretary's designation, he asserts that either Stephen Parks acted as an incident commander without being appropriately trained (item 16c); or alternatively, he did not function as an incident commander and "there should have been an incident plan implemented" (item 16d) (Tr. 234). Although Stephen Parks was the senior member of management present, the Secretary did not prove that he acted as either the "on scene official" (referred to in .120 (q)(6)(v)(A)) or as a "response official" (described in .120(q)(3)). To establish a violation, as opposed to merely asserting one, the Secretary must prove that the definitional terms of the standard are met. As Hoye acknowledged, what the Secretary really contends here is that Victor should have had and followed "the incident plan." The Secretary has cited for this failure in item 10. Since the cited standard does not apply, item 16c/16d is vacated.

---

[13] Section 1910.120(q)(6)(i)(A) provides:
An understanding of what hazardous substances are, and the risks associated with them in an incident.
[15] Section 1910.120(q)(6)(v)(A) provides:
Know and be able to implement the employer's incident command system.
[16] Section 1910.120(q)(3) provides:
Procedures for handling emergency response. (i) The senior emergency response official responding to an emergency shall become the individual in charge of a site-specific Incident Command System (ICS). All emergency responders and their communications shall be coordinated and controlled through the individual in charge of the ICS assisted by the senior official present for each employer.
Note to (q)(3)(I).-The "senior official" at an emergency response is the most senior official on the site who has the responsibility for controlling the operations at the site. Initially it is the senior officer on the first-due piece of responding emergency apparatus to arrive on the incident scene. As more senior officers arrive (i.e., battalion chief, fire chief, state law enforcement official, site coordinator, etc.) the position is passed up the line of authority which has been previously established.

PENALTY FOR EMERGENCY RESPONSE VIOLATIONS

The statutory considerations of size, good faith, and past history have been discussed at item 2. Of the emergency response violations alleged, items 10, 11, 12, 13, 15, and 16b were affirmed. Only one of the asserted instances (the Parks' re-entry) was covered under the standard. James Amaral's attempted containment and John Amaral's rescue, which were penalized by the Secretary as part of his penalty calculations, are excluded from the final penalty assessments.

Considerations of the gravity include the number of persons exposed and the degree of exposure. Exposure to excessive amounts of nitrogen dioxide is considered to be of the highest gravity, capable of causing delayed and severe adverse health effects. If inhaled in sufficient quantities, nitrogen dioxide can cause pulmonary edema or death (Tr. 190). The amount of the hazardous substances Victor worked with, together with the fact that its processes were dependent on a chemical reaction between the copper or brass rods and the bright dip, increased the probability of a release. When the release occurred, the Parks' primary concern appeared to be a desire to minimize the effects of the release on its business operation (Exh. C-8 & C-9). Nevertheless, in spite of this and the severity of the hazard, the amount of penalty is moderated. Failure to have an emergency response plan or a designated safety officer exposed all employees, although the Parks were the most directly affected. Other affirmed violations exposed only the Parks. As owners of Victor, they will ultimately be responsible for the assessed penalties.

Also, the violations of this standard are interrelated. For example, items 11a and 11b required an assessment of conditions, followed by proper use of personal protective equipment; item 12 involved failure to use personal protective equipment; and 15 again required that an assessment be made, this time by a safety officer. The Secretary tacitly recognized this point when in his post-hearing brief he treated all emergency response allegations as if they were one violation. Based upon these considerations, the following penalties are assessed: item 10, $3,000; item 11a & lib, $1,700; item 12, $1,100; item 13, $1,000; item 15, $1,000; item 16b, $700; or a total penalty of $8,500 for items 10-13 and 15-16b....

QUESTIONS

1. Describe your understanding of the requirements for responding to a HAZMAT incident.
2. How do the requirements for response to a HAZMAT incident compare with those for response to other types of incident, such as a large structure fire?
3. Compare the standards for HAZMAT response that apply to a private entity like Victor Microwave with the standards that apply to a public emergency response entity, such as a fire department.
4. Recall the discussion of the Incident Management System with the duties of an Incident Commander found in Chapter 6. Compare the actions in response to the release at Victor Microwave with those standards.
5. The case faults Victor for failure to appoint a safety officer. What implications does this violation have for future HAZMAT responses?
6. Discuss the Administrative Law Judge's rationale for the amounts of fines levied on Victor Microwave.
7. Explain why the fines levied are or are not appropriate.

Fenton, Mo., March 21, 2008 – Local residents and area volunteers band together to fill sand bags and stack them next to businesses and property on to the Meramec River. Jocclyn Augustino/FEMA.

Chapter 8

USING VOLUNTEER RESOURCES

Volunteers can be a significant resource in responding to emergencies. Different types of volunteer, however, provide different levels of support. Trained volunteers who fulfill emergency support function roles in compliance with an emergency response plan increase response capability significantly. Emergent volunteers, who come from "out of the woodwork" in the aftermath of an incident, on the other hand, can be a source of headaches and a danger to both themselves and trained responders. As you read the materials below, recall the earlier discussion of "Good Samaritan" Acts and consider their similarities to these materials.

PART 1. TRADITIONAL LEGAL STANDARDS FOR VOLUNTEERS

Warren v. District of Columbia

District of Columbia Court of Appeals

1980 D.C. App. LEXIS 423

DISSENT:

NEBEKER, Associate Judge, concurring in part and dissenting in part:

…In the classic case, *H.R. Moch Co., Inc. v. Renesselaer Water Co.*, 159 N.E. 896 (1928), then Judge Cardozo delineated the liability of a volunteer:

> It is ancient learning that one who assumes to act, even though gratuitously, may thereby become subject to the duty of acting carefully, if he acts at all. … The hand once set to a task may not always be withdrawn with impunity though liability would fail if it had never been applied at all. … If conduct has gone forward to such a stage that inaction would commonly result, not negatively merely in withholding a benefit, but positively or actively in working an injury, there exists a relation out of which arises a duty to go forward. [159 N.E. at 898.]

The *Moch* case involved a suit against a water company for failure to supply adequate water to fight a city fire. Judge Cardozo found that the failure to provide adequate water to fight the fire constituted, at most, a nonactionable withholding of a benefit. Whatever the omissions and failures of the defendant police officers in this action, those alleged omissions and failures too, constituted no more than a similar withholding of a benefit.

Moreover, volunteer liability is premised in large part upon the assumption that the volunteer is free to assess each rescue situation, weigh the risks involved, and determine whether to shoulder the obligation or leave it to someone else. Police officers clearly are not in a position to make such choices on a case-by-case basis and it would be absurd to presume that an individual assumes a permanent "volunteer" status when he becomes a police officer. Again, in the words of Judge Cardozo:

> An intention to assume an obligation of indefinite extension to every member of the public is seen to be

the more improbable when burden that the obligation promisor will not be deemed the assumption of a risk to trivial reward. [159 N.E. at 89

QUESTIONS

1. How does your understanding of volunteer liability compare with that espoused by Justice Cardozo?

2. Analyze the duty to respond firefighter in light of the ca

PART 2. VOLUNTEER PROTECTION ACT OF 1997

William C. Nicholson

In order to encourage volunteers to participate in organized groups that benefit emergency response and recovery, the Congress enacted the Volunteer Protection Act of 1997 ("VPA")[1] to provide statutory immunity for persons desiring to assist in good works. The Congress found that citizens' willingness to volunteer was deterred by possibility of litigation arising from their volunteer

activities.[2] This law pre-empts state higher levels of liability for volunte negligence,[4] although states may op dition to protection from neglige punitive damages may not be awar volunteer acting within the scope of sponsibilities to non-profit organiz when that volunteer is negligent or g gent.[6] The immunity does not attach nization with which the volunteer is a

---

[1] Pub. L. 105-19, 111 Stat. 218 (codified at 42 U.S.C.A. 14501-14505 (West Supp. III 2002)). As is the case with a reform, the VPA has come in for significant criticism. See e.g., Andrew F. Popper, *A One-Term Tort Reform Tale: Victim erable*, 35 Harv. J. on Legix. 123, 130-137 (Winter, 1998). An underlying principle of tort law is that the threat of pe creates individual accountability and thereby enhances the quality of goods and services. Accordingly, the common minimum level of due care on people who choose to volunteer. The Volunteer Protection Act changes that stand doing, reduces the incentive to provide quality services. Id. at 134-5 (footnotes omitted).

[2] 42 USCS § 14501 Findings and purpose
(a) Findings. The Congress finds and declares that
(1) the willingness of volunteers to offer their services is deterred by the potential for liability actions against them;
(2) as a result, many nonprofit public and private organizations and governmental entities, including voluntary assoc service agencies, educational institutions, and other civic programs, have been adversely affected by the withdrawal from boards of directors and service in other capacities;
(3) the contribution of these programs to their communities is thereby diminished, resulting in fewer and higher c than would be obtainable if volunteers were participating;

[3] "Volunteer" is defined as an individual (including a director or officer) performing services for a "nonprofit org a governmental entity who does not receive compensation (other than reasonable expenses) in excess of $500 14505 (6).

[4] Id. 14503(a)(3).

[5] The opt-out authorization would only apply where all of the parties in a case are residents of the state in question.

[6] The VPA's limits on punitive damages liability and joint and several liability for non-economic damages are not li ters where the volunteer acted with a required license or was not caused by a motor vehicle. The prohibition on or gence actions and limits on punitive damages against volunteers do not apply to civil cases brought by a governmental entity against affiliated volunteers. The limitation is not contained in the provisions limiting non-eco ages in joint and several liability cases. See 42 U.S.C. 14503 (a), (c), (e) & 14504.

[7] The Volunteer Protection Act does not provide any direct liability protections to the nonprofit organizations or governm Id. 14503(c).

Significantly, the Act does not exempt volunteers from liability for any harm caused while driving a motor vehicle.[8] This exclusion is significant, since, by some counts, half the claims involving emergency response organizations involve vehicle accidents.[9] While the Volunteer Protection Act changes the basis for a lawsuit, it probably does not affect administrative actions taken on a negligence basis. Therefore, laws specifying negligent conduct endangering persons as a basis for administrative penalties,'" continue to be valid.

QUESTIONS

1. What will the effect of the VPA be on the willingness of people to volunteer?
2. Note that the definition of "volunteer" is limited by a $500 per year cap on remuneration. List the types and amounts of money and things that a volunteer firefighter or EMS person might receive during a year, and add them up. What is the effect of the $500 yearly cap?
3. What sort of person would be most likely to decline to volunteer based on fears about potential liability?

PART 3. LICENSING AND LIABILITY OF VOLUNTEER EMERGENCY MANAGEMENT WORKERS

Steps that a governmental organization might take to ensure coverage for affiliated volunteers are laid out below in Indiana Code 10-14-3:

TITLE 10. PUBLIC SAFETY
ARTICLE 14. EMERGENCY MANAGEMENT
CHAPTER 3. EMERGENCY MANAGEMENT AND DISASTER LAW

IC 10-14-3-3

Emergency Management Worker

Sec. 3. As used in this chapter, "emergency management worker" includes any…volunteer…of:

(1) the state;
(2) other:
   (A) states;
   (B) territories; or
   (C) possessions;
(3) the District of Columbia;
(4) the federal government;
(5) any neighboring country;
(6) any political subdivision of an entity described in subdivisions (1) through (5); or
(7) any agency or organization;

performing emergency management services at any place in Indiana subject to the order or control of, or under a request of, the state government or any political subdivision of the state. The term includes a volunteer health practitioner registered under IC 10-14-3.5.

---

[8] Id. 14503(a)(1), (2) & (3). The Act also excludes from liability protection any specific misconduct constituting a crime of violence or international terrorism, hate crime, sexual offense, civil rights violation, or which is caused by the influence of alcohol or drugs in violation of state law as well as volunteers performing services for groups responsible for federal hate crimes (e.g., crimes that manifest evidence of prejudice based on race, religion, sexual orientation, or ethnicity). Id. 14503(f) & 14505(4). (Federal hate crimes are defined at 28 U.S.C. 534 note.)

[9] *See*, e.g., Soler, et al., "The Ten Year Malpractice Experience of a Large Urban EMS System," *Annals of Emergency Medicine 14*: 982, 985 (1985); Goldberg, et al, "A Review of Prehospital Care Litigation in a Large Metropolitan EMS System," *Annals of Emergency Medicine, 19*:557, 558 (1990).

IC 10-14-3-15

Governmental functions; liability; emergency management workers:

Sec. 15. (a) Any function under this chapter and any other activity relating to emergency management is a governmental function. The state, any political subdivision, any other agencies of the state or political subdivision of the state, or, except in cases of willful misconduct, gross negligence, or bad faith, any emergency management worker complying with or reasonably attempting to comply with this chapter or any order or rule adopted under this chapter, or under any ordinance relating to blackout or other precautionary measures enacted by any political subdivision of the state, is not liable for the death of or injury to persons or for damage to property as a result of any such activity. This section does not affect the right of any person to receive:

(1) benefits to which the person would otherwise be entitled under:
    (A) this chapter;
    (B) the worker's compensation law (IC 22-3-2 through IC 22-3-6); or
    (C) any pension law; or
(2) any benefits or compensation under any federal law.

(b) Any requirement for a license to practice any professional, mechanical, or other skill does not apply to any authorized emergency management worker who, in the course of performing duties as an emergency management worker, practices a professional, mechanical, or other skill during a disaster emergency.

(c) A volunteer working as an authorized emergency management worker may be covered by the medical treatment and burial expense provisions of the worker's compensation law (IC 22-3-2 through IC 22-3-6) and the worker's occupational diseases law (IC 22-3-7). If compensability of the injury is an issue, the administrative procedures of IC 22-3-2 through IC 22-3-7 shall be used to determine the issue.

As the reader can discern, the most important factor in obtaining the statutory protection offered under Indiana's worker's compensation law is whether the individual was "an authorized emergency management worker." In practice, in order to gain this coverage, the volunteer must be rostered as qualified by either the state or local emergency management agency or a volunteer organization recognized by either of those two governmental entities.

One difficult issue is the licensing of volunteers who enter a jurisdiction from another state to assist in emergency response. As the above statute indicates, Indiana solves the problem by waiving the "license to practice any professional, mechanical, or other skill." The Emergency Management Assistance Compact ("EMAC") [See model legislation at: http://www.emacweb.org/index.php?option=com_content&view=article&id=155&Itemid=271] (enacted in Indiana as IC 10-14-6) does not mandate a way to deal with this matter. Rather, it states that: "such person shall be deemed licensed, certified, or permitted by the state requesting assistance to render aid involving such skill to meet a declared emergency or disaster, subject to such limitations and conditions as the governor of the Requesting State may prescribe by executive order or otherwise." This approach makes sense when viewed in the context of the history of such licensure, which has always been a matter for the states in which the licensee works. On the other hand, FEMA's push for interstate certification pursuant to the authority of the National Incident Management System may conflict with this example of interstate comity. For more information on FEMA's credentialing efforts, the reader may wish to consult its Resource Management link at: http://www.fema.gov/emergency/nims/Resource Mngmnt.shtm#item3.

The mandates of individual entities within states further complicate matters. Hospitals, for example, require physicians that use their facilities to have practice privileges. These privileges are matters imposed by hospitals pursuant to insurance and other guidance. Even within a single county, it may require significant negotiation to come to an agreement that allows physicians to practice at hospitals at which they are not credentialed during an emergency. This issue must be explored before a disaster event with risk management professionals at hospitals and other entities with similar restrictions.

PART 4. WHO IS A VOLUNTEER?

As the case below illustrates, organizations as well as individuals may be accorded volunteer status.

Northern Indiana Public Service Co. v. Sharp*

Court of Appeal of Indiana, Third District

732 N.E.2d 848 (2000)

FACTS AND PROCEDURAL HISTORY

The facts are set forth in *Sharp I* are as follows:

On November 27 and 28, 1990, an estimated seven inches of rain fell in an eight-hour period in Northwest Indiana. As a result, the Little Calamut River began overflowing its banks in Highland. The water crossed Tri-State's parking lot, flowed across Indianapolis Boulevard, and entered into the Wicker Park Manor subdivision.

During the initial stage of the flooding on November 27, Krooswyk Trucking & Excavating, Inc. was engaged to provide sand for flood control. An initial attempt was made to establish a sand barrier across Indianapolis Boulevard, but the water continued to wash the sand away.

At approximately 8:00 A.M. on November 28, Gerald Krooswyk, president of Krooswyk trucking, suggested to Highland officials that a dike be built across Tri-State's parking lot to save the subdivision from further flooding. The dike would consist of gravel dumped across Tri-State's property from the river's edge along the edge of Indianapolis Boulevard to a high point at the end of the parking lot. The proposed location for the dike passed directly under NIPSCO's energized overhead power lines, which also crossed Indianapolis Boulevard.

At approximately 10:30 A.M. on November 28th, Highland officials approved the dike location and Krooswyk trucks began dumping gravel to form the dike on the Tri-State parking lot. At approximately 1:10 P.M., Robert Sharp ["Robert"], a Krooswyk employee, began backing his truck onto the pile of gravel forming the dike. After [Robert] raised the truck bed, the truck became energized when electricity from the overhead wires arced. [Robert] was electrocuted.

On August 21, 1991, [Robert] filed suit against both Highland and NIPSCO. The complaint alleged that Highland's negligent acts and NIPSCO's negligent and/or reckless conduct caused [Robert]'s death.

Highland and NIPSCO filed separate motions for summary judgment, both asserting the immunity defense found in I.C. 10-4-1-8. After a hearing on the motions, the trial court granted summary judgment in favor of Highland and NIPSCO....

Sharp I, 665 N.E.2d at 613. In Sharp I we held as follows:

The trial court correctly determined that Highland was immune from liability under the Act [Indiana's Civil Defense and Disaster Law of 1975]. Accordingly, the grant of summary judgment in favor of Highland is correct. The grant of summary judgment is affirmed.

The trial court also correctly determined that, as a matter of law, NIPSCO was a civil defense and disaster worker under the Act. Accordingly, the [trial] court was correct in determining that NIPSCO was immune from liability for its allegedly negligent acts. To the extent that the court's grant of summary judgment dealt with this question of law, it is affirmed.

However, the trial court erred in making a determination on the question of willful misconduct, gross negligence, and bad faith. The

---

* Note to the reader: IC 10-4-1-5 cited in the Sharp case was later replaced by parts of IC 10-14-3-3 and IC 10-14-3-15, cited above.

question was not raised in NIPSCO's summary judgment motion. Furthermore, it is a question for the trier of fact. To the extent that the court's grant of summary judgment dealt with this issue, it is reversed.

Finally, the trial court correctly granted summary judgment in favor of Tri-State. The grant of summary judgment is affirmed.

A jury trial in the cause of Sharp versus NIPSCO was commenced on March 30, 1998. The facts, as further developed at trial and most favorable to the verdict, are as follows: Throughout the night of November 27th, Krooswyk Trucking & Excavating, Inc. ("Krooswyk Trucking") and employee Robert dumped sand to erect a barrier across Indianapolis Boulevard. (R. 1952-53, 1957.) At approximately 2:00 A.M., the Town began evacuating residents of the Wicker Park subdivision who were located east of Indianapolis Boulevard. (R. 809-10, 854.) Meanwhile, the sand barrier continued to be washed away by rising waters. (R. 1953.) Between 8:00 and 9:00 A.M. of the following morning, Gerald Krooswyk ("Krooswyk"), of Krooswyk Trucking, and Robert discussed the possible locations of an additional dike with town officials. (R. 1958-62.) The proposal to build an additional dike was approved by town officials between 10:30 and 11:00 A.M. (R. 1940, 1962, 1964.)

The first load of gravel for the new dike (hereinafter referred to as the "Gravel Dike") arrived at the flood scene shortly after 11:00 A.M. (R. 1962, 1964.) The Gravel Dike passed directly under NIPSCO's energized overhead lines. (R. 1964.) Before construction of the Gravel Dike began, Krooswyk Trucking did not discuss with the town officials shutting off the power to these overhead lines. (R. 1964.) At approximately the same time that the construction on the Gravel Dike commenced, NIPSCO's line supervisor, Gene Shayatovich ("Shayatovich"), arrived at the command center, which was set up on the Indianapolis Boulevard overpass bridge. (R. 902, 953.) The construction site where the Gravel Dike was being built could be seen from this command center. (R. 863.)

At this time, Shayatovich, Fire Chief Haas ("Fire Chief"), and town officials Mike Pipta ("Pipta") and William Cameon ("Cameon") discussed electrical concerns in the Wicker Park homes becoming flooded. (R. 863, 867, 902.) The Fire Chief had reports that firemen who were attempting to evacuate Wicker Park had felt a current while in the homes. (R. 865.) The Fire Chief was also concerned about the potential for NIPSCO's electricity igniting fuel from underground tanks. (R. 853.) Accordingly, the Fire Chief requested that NIPSCO shut off power to the entire flooded area, which included the area of the Gravel Dike. (R. 864, 922.) However, Pipta requested that NIPSCO leave the power on at the pump station because it was continuing to pump out flood water. (R. 866.) Additionally, Cameon testified that town officials wanted the street lights to be left on for security reasons. (R. 939-40.) The town officials and Shayotovich devised a plan whereby electric power to the Wicker Park homes would be cut, but service would continue to the pumping station and street lights. (R. 939-40, 1516-19, 1589, 1616, 1618, 1619.) At approximately 12:00 P.M., a NIPSCO line crew, consisting of Mr. Walters ("Walters") and Mr. Laws ("Laws") arrived at the command center. (R. 1106.) Traveling by boat, Walters and Laws began the process of de-energizing the cut-out fuses located in the flooded subdivision. (R. 1126-28.)

While NIPSCO workers de-energized the power to the subdivision, Krooswyk Trucking continued its construction of the Gravel Dike. Robert was in the process of dumping his second load of gravel when his truck bed came into contact with NIPSCO's overhead power lines. (R. 792-93.) Electricity traveled through the truck's metal bed to the tires, which began to smolder and smoke. (R. 796-97; 2061-62.) Robert jumped from the truck onto the wet stone surface, then into the water, which resulted in his electrocution. (R. 798-99; Supp. R. 45.)

At trial, NIPSCO moved for judgment on the evidence both at the end of Sharp's case-in-chief and again at the close of all the evidence. (R. 408-12, 527-28.) The trial court denied both motions. (R. 530.) Thereafter, NIPSCO filed a motion to correct errors, which the trial court denied. (R. 539-41, 62628.) This appeal followed.

DISCUSSION AND DECISION

I. WHETHER THE TRIAL COURT ADMITTED EVIDENCE THAT WAS PRECLUDED BY THIS COURT'S PREVIOUS JUDGMENT

NIPSCO contends that the trial court erred by permitting Sharp to raise issues of negligence that were not presented to this Court on Sharp's appeal. (Appellant's Brief at 17.) Specifically, NIPSCO alleges that it was error for the trial court to consider testimony about NIPSCO's policies regarding pre-emergency planning and training because Sharp had not raised this issue in response to NIPSCO's motion for summary judgment. (Appellant's Brief at 17.) NIPSCO relies on the doctrines of the law of the case and res judicata in support of its arguments.

A. The Doctrines of the Law and the Case and *Res Judicata*

The law of the case doctrine mandates that when an appellate court decides a legal issue, both the trial court and the court on appeal are bound by that determination in any subsequent appeal involving the same case and relevantly similar facts. . . . In *State v. Lewis*, our supreme court further noted that the law of the case doctrine "'merely expresses the practice of courts generally to refuse to reopen what has been decided, not a limit on their power." 543 N.E.2d 1116, 1118 (Ind. 1989) . . . Moreover, a trial court may, upon remand after reversal of an order of summary judgment, recognize such additional facts as are dispositive of the case and rule accordingly. Id.

The doctrine of *res judicata* consists of two concepts, claim preclusion and issue preclusion. . . . Claim preclusion applies where a final judgment on the merits has been rendered which acts as a complete bar to a subsequent action on the same issue or claim between those parties and their privies. Issue preclusion bars the subsequent relitigation of the same fact or issue where that fact or issue was necessarily adjudicated in a former suit and the same fact or issue is presented in a subsequent action. Where issue preclusion applies, the previous judgment is conclusive only regarding those issues actually litigated and determined therein. *Res judicata* provides that a judgment on the merits is an absolute bar to a subsequent action between the same parties on the same claim. . . .

B. Analysis

In the appeal taken from the trial court's grant of summary judgment in favor of NIPSCO, our court affirmed that portion of the trial court's determination that NIPSCO was a civil defense and disaster worker under Indiana's Civil Defense and Disaster Law of 1975 (the "Act"). Accordingly, we held that NIPSCO was immune from liability for its *allegedly* negligent acts. However, we specifically reversed that portion of the trial court's judgment as it may have applied to the questions of willful and wanton misconduct, gross negligence, and bad faith, as judgment on such issues had not been sought in NIPSCO's motion for summary judgment.

As between Sharp and NIPSCO, our previous ruling did not amount to a final judgment on the merits as it related to willful and wanton misconduct, gross negligence, and bad faith, and therefore, did not trigger the *res judicata* concept of claim preclusion. However, the *res judicata* concept of issue preclusion did serve to limit the issues available for further litigation. Specifically, the issue of whether NIPSCO was a civil defense and disaster worker under the Act had already been determined, and thus was not available for further litigation upon remand. Applying the law of the case doctrine renders the same result; namely, the NIPSCO was acting as a civil defense and disaster worker and therefore was not liable for its allegedly negligent acts. Nevertheless, neither issue preclusion nor the law of the case doctrine serve to limit the introduction of evidence which is relevant to a determination of NIPSCO's alleged willful misconduct, gross negligence, or bad faith, an issue expressly reserved by our prior opinion.

On remand, Sharp raised the issue of whether NIPSCO's lack of pre-emergency planning and training amounted to gross negligence or willful or wanton misconduct. In support of this contention, Sharp also elicited facts not present in *Sharp I* regarding NIPSCO's duties at the flood scene and NIPSCO's knowledge of the risks posed to Krooswyk Trucking employees. Contrary to NIPSCO's argument on appeal, we find that neither the new issue nor the additional facts offended the doctrines of law of the case or *res judicata*. Rather, the issue of NIPSCO's pre-emergency planning and training was impliedly reserved upon remand as was the solicitation of additional facts essential to determine NIPSCO's duties and knowledge of risks as they bore on NIPSCO's willful or wanton misconduct....

NIPSCO further contends that at trial Sharp simply "recharacterized" negligence evidence as evidence of gross negligence or willful or wanton misconduct, and in the process usurped NIPSCO's statutory immunity in order to revisit issues of negligence already adjudicated.

(Appellant's Brief at 17-18.) NIPSCO has misinterpreted, and accordingly misapplied, our previous ruling. In *Sharp I*, we held in part that "NIPSCO was immune from liability for its allegedly negligent acts...kr not that NIPSCO was not negligent. Therefore, neither Indiana Code section 10-4-1-8, nor the doctrines of law of the case or *res judicata* have been rendered ineffective simply because the jury was permitted to consider NIPSCO's pre-emergency planning and training, or lack thereof, or the additional facts raised at trial pertaining to NIPSCO's duty or knowledge of risks. Simply stated, the new issue was not previously adjudicated and the additional facts, even if repetitive, are necessary to determine whether NIPSCO's conduct amounted to gross negligence or willful and wanton misconduct.... Based on the foregoing, we find that the trial court did not err when it admitted evidence of NIPSCO's emergency planning and training procedures or evidence of NIPSCO's duties and knowledge at the time of the flood. Evidence relevant to the above referenced issues was both impliedly reserved and essential to determine whether NIPSCO's conduct amounted to gross negligence or willful and wanton misconduct.

II. WHETHER THE TRIAL COURT ABUSED ITS DISCRETION WHEN IT DENIED NIPSCO'S MOTION FOR JUDGMENT ON THE EVIDENCE

At the close of Sharp's case, and again at the end of trial, NIPSCO moved for judgment on the evidence. NIPSCO argued at trial, and now contends, that the trial court erred in failing to grant its motion for judgment on the evidence because Sharp "failed to carry her burden of proving that NIPSCO was guilty of willful and wanton misconduct or gross negligence." (R. 408.) In essence, NIPSCO's argument is twofold: (1) that Sharp failed to prove that NIPSCO had actual knowledge that Krooswyk Trucking was building the Gravel Dike in close proximity to NIPSCO's overhead power lines, and (2) that NIPSCO's pre-emergency planning and plans did not amount to willful and wanton misconduct or gross negligence.

A. Standard of Review Judgment on the Evidence

The purpose for judgment on the evidence is to test the sufficiency of the evidence.... The grant or denial of a motion for judgment on the evidence is within the broad discretion of the trial court and will be reversed only for an abuse of that discretion....

B. Negligence

In the instant case, for Sharp to recover on a theory of negligence, she must establish a duty on that part of NIPSCO to conform its conduct to a standard of care arising from its relationship with Robert, a failure of NIPSCO to conform its conduct to the requisite standard of care required by the relationship and an injury to Robert proximately caused by the breach.... [Whether a duty of care exists is a question of law for the courts to decide.... Here, as discussed infra, our prior decision serves to limit the liability of NIPSCO to those acts, if any, that amounted to willful or wanton misconduct or gross negligence....

C. Analysis – NIPSCO's Duty to Robert

We proceed by analyzing NIPSCO's duty to Robert, as determined by the relationship between NIPSCO and Robert, NIPSCO's foreseeability of the resulting harm to Robert, and the public policy concerns raised by the Act.

NIPSCO's Relationship to Robert

We have previously held that:

Companies engaged in the generation and distribution of electricity have a duty to exercise reasonable care to keep distribution and transmission lines safely insulated in places where the general public may come into contact with them.

Northern Indiana Pub. Serv. Co. v. East Chicago Sanitary Dist., 590 N.E.2d 1067, 1072 (Ind. Ct. App. 1992)... More generally, we have held that electric utilities have a duty to exercise such care as a person of reasonable prudence would use under like conditions and circumstances....

Here, where the Town of Highland was a disaster area and town officials restricted access to the flood area to emergency workers only, it was the Town of Highland, not NIPSCO, which ensured that the general public did not come into contact with the at-issue power lines. Moreover, it may be generally said that the Town of Highland "was in control of the [flood] situation." *See Sharp I*, 665 N.E.2d at 616. Therefore, while NIPSCO had a relationship with, and a duty to exercise reasonable care for, the general public of Highland, it is not reasonable to extend this relationship, and its corresponding duty, to Krooswyk Trucking and its employees. As Krooswyk Trucking was under contract with Highland, and Highland, not NIPSCO, was in control of the flood emergency, we find that there was no recognizable relationship between NIPSCO and Robert.

NIPSCO's Foreseeability of Harm to Robert

Beyond the relationship between NIPSCO and Robert, we look to the evidence at trial from which the jury may have inferred either NIPSCO's actual knowledge of the specific risks posed to Robert and/or NIPSCO's general knowledge of the risks associated with flood conditions, to further assess whether NIPSCO owed a duty of care to Robert.

(a) NIPSCO's Actual Knowledge

The evidence most favorable to Sharp reveals that the construction site where the Gravel Dike was being built could be seen from the command center, that NIPSCO employee Shayatovich was present in the command center, and that NIPSCO employees saw dump trucks passing the command center. However, the Record is devoid of any evidence which would indicate that NIPSCO representatives were ever advised by town officials, or Krooswyk Trucking, of the Gravel Dike being constructed, and that such dike crossed under NIPSCO power lines. It requires undue speculation to impute actual knowledge to NIPSCO employees of the construction of the Gravel Dike under the subject energized NIPSCO power lines from these general facts, and thus to infer that NIPSCO had actual knowledge of the risk of electrocution posed to Krooswyk Trucking employees. Moreover, the plans for construction of the Gravel Dike, as devised by Krooswyk Trucking and approved by town officials, called for dumping stone well away from the NIPSCO overhead power lines and then moving the stone to the location of the dike with a front end loader. (R. 1140-41, 1167-68, 1964-65.) However, contrary to these plans, Robert backed his dump truck up to the dike, failed to stay at least twenty feet from the power lines, and came into contact with the power lines. Thus, while it may be inferred from the plan devised by Krooswyk Trucking that it had actual knowledge of the risk posed by these power lines, such facts do not amount to substantial evidence supporting the inference that NIPSCO had actual knowledge of any specific risk posed to Robert by its power lines.

(b) NIPSCO's General Preparation and Training for Floods

Sharp presented expert testimony regarding the actions which NIPSCO employees should have taken had they been properly trained and prepared for a flood emergency. Assuming, without deciding, that the following excerpts of testimony offered by Sharp's expert, Dr. Feinberg, were admissible, we nevertheless find that such evidence does not lead to the conclusion that NIPSCO could have foreseen the harm to Robert.

Dr. Feinberg testified that NIPSCO's duties for pre-emergency planning and training and assessment of dangers at the scene of the flood should have included (1) a duty to make "safety determinations" at the scene of the flood, (2) a duty to properly train NIPSCO employees for flood emergencies, (3) a duty to be "proactive" in assessing flood dangers, (4) a duty to assess whether equipment which could reach power lines was going to be used, (5) a duty to put "up something like a saw horse ... with instructions to the operator" or "stantions [sic] on either [of the electrical lines] with flags" ... "something to warn people saying don't come any closer," and (6) a duty to supply a spotter. (R. 1256, 1261, 1264-66.)

We find that the above planning and training procedures, even if implemented, would not have made the harm to Robert foreseeable. We make this finding because, regardless of the general policies NIPSCO had or did not have in place, NIPSCO had no actual knowledge that a portion of the Gravel Dike was being constructed under power lines. Further, the plan approved by Highland called for the gravel to be dumped well away from the power lines. Additionally, the testimony of various town officials indicates that NIPSCO employees provided skilled assistance in furthering the purposes of the Act; namely, NIPSCO employees de-energized cutout fuses located in the flooded sub-division of

Wicker Park, aiding in the "prompt and efficient rescue" of Highland citizens. *See* Ind. Code § 10-4-1-2 (b) (2). Thus, the evidence does not reveal that NIPSCO's general preparation and training for floods, or lack thereof, gave rise to the foreseeability of the harm to Robert.

The Act and Public Policy

Lastly, we consider the public policy concerns raised by the Act. The Civil Defense and Disaster Law of 1975 exists in part for the following purposes:

(1) to provide for emergency management under a state emergency management agency;
(2) to create local emergency management departments and to authorize and direct disaster and emergency management functions in the political subdivision of the state;
  (b) It is further declared to be the purpose of this chapter and the policy of the state:
  (2) to prepare for prompt and efficient rescue, care, and treatment of persons victimized or threatened by disaster[.]

Ind. Code § 10-4-1-2. Moreover, we have held that "generally speaking, the Act is intended to facilitate rescue and remedial measures in response to a disaster." *Sharp I*, 665 N.E.2d at 614. Additionally, the Act limits the liability of those workers deemed to be "emergency management workers" as follows:

> Neither the state nor any political subdivision of the state, nor any other agencies of the state or political subdivision of the state, nor, except in cases of willful misconduct, gross negligence, or bad faith, any emergency management worker complying with or reasonably attempting to comply with this chapter or any order or rule adopted under this chapter..., shall be liable for the death of or injury to persons...as a result of any such activity.

Ind. Code § 10-4-1-8 (emphasis added). Here, we do not find that the legislature's intent to facilitate rescue and remedial measures during disasters is furthered by imposing a duty on NIPSCO. To the contrary, the chain-of command established by the Act would be jeopardized by such a duty, as that duty would elevate the role of

NIPSCO to being co-equal with that of the Town of Highland, an effect clearly not intended by the plain language of the Act.

D. Conclusion

Assessing NIPSCO's relationship to Robert, the foreseeability of Robert's electrocution, and the public policy as announced by the Act and *Sharp I*, we hold that Sharp failed to prove that NIPSCO owed a duty to Robert. Accordingly, Sharp has failed to satisfy the requisite duty element of her negligence claim. Consequently, the trial court erred in denying NIPSCO's motion for judgment on the evidence.[4]

Reversed with instructions to enter judgment in accordance with this opinion.

BAKER, J., and MATTINGLY, J., concur.

QUESTIONS

1. Contrast the protections afforded to corporations such as NIPSCO by Indiana Code 10-4-1-8 with the coverage they might receive under the federal Volunteer Protection Act.

2. What other organizations might receive protection from liability under Indiana Code 10-4-1-8?

3. Discuss why the protections afforded by Indiana Code 10-4-1-8 and the federal Volunteer Protection Act are different for corporate entities.

---

[4] Additionally, Sharp alleges that NIPSCO "directly contravened the Town's directive that it turnoff [sic] the power to the entire flood area," citing, in piecemeal fashion, various excerpts from the Fire Chief's testimony. (Appellee's Brief at 16-19, 39.) However, our independent review of the Record reveals that no such inference may be drawn from the testimony of the various town officials who testified on this matter at trial. Specifically, we note that the following testimony of Fire Chief Haas stating that "but for Mr. Pipta, NIPSCO would have shut down all the power in the area[,]" and the testimony of Mr. Pipta stating that "it was because [he] wanted to keep the pumps on that the power was still on." (R. 909, 1616.) Accordingly, we find the Record to be devoid of facts from which one could reasonably infer that NIPSCO failed to follow the Town of Highland's directives.

New York, N.Y., October 13, 2001 – New York Firefighters at the site of the World Trade Center. Photo by Andrea Booher/ FEMA News Photo.

Chapter 9

RECOVERY BY RESPONDERS: THE RESCUE DOCTRINE, THE "FIREMAN'S RULE," POST TRAUMATIC STRESS DISORDER, AND WORLD TRADE CENTER SITE LITIGATION

The rescue doctrine and the "fireman's rule" have served over the years as balances upon one another. As you read the case below, consider how their interaction is evolving. The fireman's rule does not apply only to firefighters.

Duties to responders continue to evolve: post traumatic stress disorder law recognizes the mental sacrifices that responders make, and lawyers advocated on their behalf in the aftermath of the September 11, 2001 attacks.

PART 1. THE RESCUE DOCTRINE AND THE FIREMAN'S RULE

Heck v. Robey

Supreme Court of Indiana 659

N.E.2d 498 (1995)

SELBY, J.

The question in this interlocutory appeal is whether Robey, a paramedic, may recover against Heck and Peabody Coal for injuries incurred during Robey's rescue of Heck. Heck and Peabody Coal moved for summary judgment, arguing among other things that the fireman's rule bars Robey from recovering. The trial court denied the motions for summary judgment, holding that genuine issues of material fact exist and that the fireman's rule does not extend to paramedics. The Court of Appeals accepted jurisdiction over this interlocutory appeal and reverse the trial court, holding that the fireman's rule applies and that Robey's claim was barred. We disagree and conclude that the fireman's rule does not, as a matter of law, bar Robey's recovery. Under the rescue doctrine Heck and Peabody Coal owed no duty to Robey except to abstain from positive wrongful acts. The trial court properly found that a genuine issue of material fact exists as to whether Heck engaged in positive wrongful acts, and thus did not err in rejecting Heck and Peabody Coal's argument that the fireman's rule bars Robey from recovering.

I. FACTS

James L. Robey (Robey) was a licensed paramedic and emergency medical technician employed by Warrick Emergency Medical Service.

On January 16, 1990, Robey and his partner, Tracy Kavanaugh (Kavanaugh), responded to a "911" call to extricate Heck from a ditched vehicle

143

at the Squaw Creek Mine and to provide emergency medical services. The Squaw Creek Mine is a joint venture owed in part by Peabody Coal. Lawrence Heck (Heck), a Squaw Creek Mine employee weighting nearly 200 pounds, had driven a company vehicle into a large, steep ditch while on the job. According to Robey, "the truck fit upside down in the ditch. No a whole lot of it was above the ditch." Robey alleges that Heck was intoxicated at the time of the accident, which Heck and Peabody Coal do not dispute for purposes of their motions and this appeal.

Robey was in charge of the rescue operation. When Robey and Kavanaugh could not get into Heck's vehicle, they summoned firefighters who removed the vehicle door with hydraulic tools to aid in Heck's rescue. Robey and Kavanaugh then removed Heck, placed him on a spine board, and carried him out of the ditch, up to the road. Heck allegedly flailed and kicked in a combative manner during the rescue, requiring Kavanaugh to hold Heck's arms and forcing Robey to do most of the pulling necessary to extricate Heck. Because of the position of the truck, both Robey and Kavanaugh were on their knees while trying to lift Heck. Heck was so combative that Robey called for a back-up ambulance and requested Valium from the hospital to calm Heck. Robey sustained back injuries for which he now brings this negligence action. Carol S. Robey, Robey's spouse, bring her negligence action for loss of Robey's services, society, and consortium.

II. STANDARD REVIEW

Summary judgment is proper only when there is no genuine issue of material fact and the moving party is entitled to judgment as a matter of law....

III. DISCUSSION

A. Overview

The fireman's rule provides that firemen responding in emergencies are owed only the duty of abstaining from positive wrongful acts. The Court of Appeals characterizes the fireman's rule as an exception to the rescue doctrine, which is discussed below.... We first recognized the fireman's rule in 1893. *Woodruff v. Bowen* (1893), Ind., 136 Ind. 431, 34 N.E. 1113. In *Woodruff*, we held that a landowner owed no duty to a firefighter responding to a fire on the landowner's property except to abstain from positive wrongful acts. *Woodruff*, 136 Ind. at 442. We established this narrow, limited duty to firefighters because of the impracticability and expense of keeping one's property in the safest of conditions at all times on the off-chance that a firefighter might be required to enter the property in an emergency. Thus, as in other jurisdictions, Indiana's fireman's rule was based originally upon premises liability and concerned only the legal question of duty. Since adopting *Woodruff*, we have not addressed the propriety of its application in cases not involving premises liability.

We now take the opportunity to examine how the fireman's rule has developed since we last addressed the rule over 100 years ago. Since that time, the Court of Appeals has expanded the fireman's rule, holding that the rule acts as a complete bar to recovery by public safety officers except in limited situations. In this case, the Court of Appeals did not address the applicability of the rescue doctrine, holding instead that the fireman's rule extends to paramedics and bars their recovery. This case presents the question of the duty owed to a professional rescuer, such as a paramedic, under the rescue doctrine, rather than the question of a property owner's duty to a professional rescuer injured by a defect in the property. We discuss the rescue doctrine and the fireman's rule in turn.

B. The Rescue Doctrine

Robey alleges that Heck and Peabody Coal owed him a duty based upon the rescue doctrine.

This Court first recognized the rescue doctrine in *Neal v. Home Builders, Inc.* (1953), 232 Ind. 160, 111 N.E.2d 280, holding that "'one who has, through his negligence, endangered the safety of another may be held liable for injuries sustained by a third person in attempting to save such other from injury.'"[2] ... As Justice Cardozo eloquently explains:

> Danger invites rescue. The cry of distress is the summons to relief. The law does not ignore these reactions of the mind in tracing conduct to its consequences. It recognizes them as normal. It places their effects within the range of the natural and the probable. The wrong that imperils life is a wrong to the imperiled victim; it is a wrong also to his rescuer. *Wagner v. International Ry. Co.*, 232 N.Y. 176, 133 N.E. 437 (N.Y. 1921).

Heck encounters that he cannot be liable because Neal suggests that the rescue doctrine applies only when there are three parties: a tortfeasor, a party injured as a result of the tortfeasor's negligence, and a rescuer of the party injured.... Thus, a person who injures himself while acting in a careless or reckless manner may owe a duty to his or her own rescuer; the duty stems from an implied invitation to rescue....

[T]he 'rescue doctrine' under any conception of it contemplates a voluntary act by a rescuer who in an emergency attempts a 'rescue' prompted by a spontaneous, humane motive to save human life, and which 'rescue' the rescuer had no duty to attempt in the sense of a legal obligation or in the sense of a duty fastened on him by virtue of his employment.... The rescue doctrine is designed to encourage and reward humanitarian acts.

It is undisputed that Robey responded to the accident in his capacity as a paramedic pursuant to a "911" call. Robey concedes that his actions were not voluntary, since he was "acting under a duty imposed on him as an employee..." (R. at 67-68). Back lack-of-voluntariness and the existence of this duty of employment would ordinarily suffice to defeat Robey's argument that Heck and Peabody Coal owed him a duty under the rescue doctrine. While we commend professional rescuers for their contributions to society, a professional rescue attempt stemming from a "911" call simply lacks the spontaneous and impulsive character that the rescue doctrine was designed to protect.[3]

Moreover, we have held that "only those who have a close proximity in time and distance to the party requiring assistance are within the class of potential rescuers." *Lambert*, 492 N.E.2d at 291. In one oft-cited case, Justice Cardozo suggested in dictum that continuity may be significant:

> We may assume, though we are not required to decide that peril and rescue must be in substance one transaction; that the sight of one must have aroused the impulse to the other; in short, that there must be unbroken continuity between the commission of the wrong and the effort to avert its consequences.... Robey was not at or near the site at the time of the accident, and his response came only after an emergency alert.

Thus, the rescue doctrine will not ordinarily create, for professional rescuers responding to emergency calls within the course of their employment, the same level of duty owed to lay rescuers present at or near the scene of an accident. It is true that Robey's arrival and actions at the scene of the rescue were not the result of a continuous transaction of spontaneous, impulsive, and gratuitous reaction to Heck's peril. However, Robey presented evidence creating a genuine issue of material fact as to whether Heck engaged in flailing and kicking in a combative manner during the rescue attempt, thus making

---

[2] The rescue doctrine is particularly significant historically, because at the time the doctrine gained acceptance, contributory negligence was an absolute bar to recovery. The rescue doctrine protected the rescuer's cause of action by precluding the rescuer from being found contributorily negligent for "voluntary placing himself in a perilous position to prevent another person from suffering serious injury or death...." Jeffrey F. Ghent, Annotation, Rescue Doctrine: Applicability and Application of Comparative Negligence Principles, 75 A.L.R. 4th 875, 876 (1990).

[3] Notably, we are deciding the scope of duty under the rescue doctrine as a matter of law based upon the relationship of the parties, the foreseeability of harm, and public policy, *see Webb v. Jarvis* (1991), Ind., 575 N.E.2d 992, 995, not based on assumption of risk. As we discussed below, the fireman's rule can no longer be based upon an assumption-of-risk rationale. See generally 1 Arthur Best, Comparative Negligence Law and Practice § 4.20[3] (1995); *see also Christensen v. Murphy*, 296 Ore. 610, 678 P.2d 1210 (Ore. 1984). But *see Fox v. Hawkins* (1992), Ind.App., 594 N.E.2d 493.

extrication more dangerous.[4] Thus, Heck's duty to Robey does not arise under the rescue doctrine, but rather from the relationship between Heck and Robey following the commencement of the rescue attempt, and imposes responsibility upon Heck for his conduct during the rescue.

C. The Fireman's Rule

We must now determine whether the fireman's rule constitutes an absolute bar to Robey's recovery, as held by the Court of Appeals. We conclude that it does not.

In *Woodruff v. Bowen* (1893), Ind., 136 Ind. 431, 34 N.E. 1113, we held that a landowner owed no duty to a firefighter responding to a fire on the landowner's property except to abstain from positive wrongful acts. Id. at 442. This Court stated that

"the owner of a building in a populous city does not owe it as a duty at common law, independent of any statute or ordinance, to keep such building safe for firemen or other officers, who, in a contingency, may enter the same under a license conferred by law. It seems to be settled, however, that such duty may be imposed either by statute or by an ordinance adopted for that purpose." 136 Ind. at 442-43.

Thus, the fireman's rule in this Court's jurisprudence was simply a shorthand way of characterizing the duty owed by a landowner to those coming onto the premises under a public duty during emergencies. Since that time, our Court of Appeals expanded the rule, adopting the following test.[5] "The rule basically provides that professionals, whose occupations by nature expose them to particular risks, may not hold another negligent for creating the situation to which they respond in their professional capacity." *Koehn v. Devereaux* (1986), Ind.App., 495 N.E.2d 211, 215. Courts have used several different rationales to support the fireman's rule. In this case, the Court of Appeals held that the fireman's rule "rests on three distinct but related theoretical pedestals: the law of premises liability, the defense of incurred risk, and the concerns of public policy," thus expanding the rationale behind the rule to include the additional justifications of assumption of risk and public policy. *Heck*, 630 N.E.2d at 1363 (emphasis added).

First, we note that premises liability is not at issue in this case. Second, we observe that when determining whether a common-law duty exists, we consider public policy. In the previous section, we held that, consonant with public policy, the rescue doctrine does not extend to an imperiled person's conduct after rescuers have arrived and are attempting rescue. We decline to take the inconsistent position that public policy favors holding that the defendants have complete immunity based solely upon the plaintiff's occupation. Moreover, the public policy concern of encouraging the public to seek assistance when it is needed is not impinged by this view.

The third rationale relied upon by the Court of Appeals in holding that Robey is based as a matter of law from asserting this negligence claim is incurred risk.[6] In 1942, the United States Supreme Court outlined the history of the assumption-of-risk defense:

Perhaps the nature of the present problem can best be seen against the background of one hundred years of master-servant tort doctrine.

---

[4] Robey argues that Heck engaged in willful and wanton conduct by operating a vehicle while intoxicated in violation of Ind. Code § 9-30-5-1. Even assuming arguendo that Heck violated the statutory prohibition against driving while intoxicated, we agree with the Court of Appeals that Ind. Code § 9-30-5-1 is intended to benefit the general public on the State's highways, not paramedics attempting to rescue intoxicated drivers. Therefore any violation of this statute alone would not constitute per se willful or wanton behavior.

[5] The Court of Appeals departed from the rule as stated in *Woodruff* when it barred firefighters injured off-premises from recovering based upon the rule. *Koehn v. Devereaux* (1986), Ind.App., 495 N.E.2d 211, 215. The Court of Appeals also expanded the doctrine to include police officers, *Koop v. Bailey* (1986), Ind.App., 502 N.E.2d 116, 117-18. The Court of Appeals recently declined to extend the fireman's rule to building inspectors. *Sam v. Wesley* (1995), Ind.App., 647 N.E.2d 382, 1995 WL 89485 at *2.

[6] In other jurisdictions, "incurred risk" is referred to as "assumption of risk" and this opinion reflects this practice when discussing case-law from other jurisdictions. In Indiana, there exist separate doctrines referred to as "assumption of risk" in contract cases, and "incurred risk" in non-contract cases. *See Whitebirch v. Sitller* (1991), Ind.App., 580 N.E.2d 262, 265 n.2. In Indiana assumption of risk, or an enforceable express consnet to hold another harmless and/or relieve them of duty, remains a complete bar to recovery.

Assumption of risk is a judicially created rule which was developed in response to the general impulse of common law courts at the beginning of this period to insulate the employer as much as possible from bearing the "human overhead" which is an inevitable part of the cost – to some-one – of the doing of industrialized business. The general purpose behind this development in the common law seems to have been to give maximum freedom to expanding industry. The assumption of risk doctrine for example was at-tributed by this Court to "a rule of public policy, inasmuch as an opposite doctrine would not only subject employers to unreasonable and often ru-inous responsibilities, thereby embarrassing all branches of business," but would also encourage carelessness on the part of the employee.

Tiller v. Atlantic Coast Line R. Co., 318 U.S. 54, 58-59, 87 L. Ed. 610, 63 S.

Ct. 444 (1942) (footnotes omitted). Out of these origins, assumption of risk expanded beyond master-servant relationships into other areas of tort law until it became a firmly established part of our jurisprudence.

Tort law changed significantly over the years, however, and in 1983 Indiana became the fortieth state to adopt either a comparative fault or com-parative negligence scheme.... As a comparative fault statute, the Indiana Comparative Fault Act (Act) eliminated contributory negligence as a complete defense, as well as other common-law defenses. The Act created a modified comparative fault scheme in which the plaintiff can recover only if his or her fault is less than or equal to "the fault of all persons whose fault proximately con-tributed to the complainant's damages."

Importantly, the Act provides in part:

> "Fault" includes any act or omission that is negli-gent, willful, wanton, or reckless toward the person or property of the actor others, but does not in-clude an intentional act. The term also includes un-reasonable assumption of risk not constituting an enforceable express consent, incurred risk, and un-reasonable failure to avoid an injury or to mitigate damages.

Ind. Code § 34-4-33-2 (emphasis added). Consequently, the complete defense of "incurred risk" no longer exists; it is subsumed by the concept of fault in our comparative fault scheme. As a component of fault, it is subject to the Act's apportionment scheme that reduces or eliminates the plaintiff's recovery depending on the degree of the plaintiff's fault.... Any rule that purports to effect an absolute defense based upon in-curred risk is contrary to our comparative fault scheme.[10]

The Court of Appeals held that Robey im-pliedly assume the risk of injury to the primary sense, based upon his choice of occupation. Ac-cording to the Court of Appeals, when the plaintiff incurs risk in the primary sense, the plaintiff re-lieves the defendant of any duty towards the plain-tiff, and no negligence action may lie. *Heck*, 630 N.E.2d at 1366. The Court of Appeals held that

> [a] paramedic knows the work involves certain risks and has, thereby, implicitly agreed to relieve negligent persons of responsibility when the risks lead to foreseeable harm. In essence, Robey "con-sented" to the harm risked or "waived" an applica-tion of the rescue doctrine.

Heck, 630 N.E.2d at 1366.... Under the Act, a plaintiff may relieve a defendant of what would otherwise be his or her duty to the plaintiff only by an express consent....

Neither Heck nor Peabody Coal allege that Robey expressly consented to bear the risks of the rescue operation by holding Heck and Peabody Coal harmless for any injury. We there-fore conclude that Heck and Peabody Coal have not demonstrated that Robey is barred as a matter of law from bringing this action. A defen-dant will prevail at summary judgment if the plaintiff expressly assumed the risks of the activ-ity and agreed to hold the defendant harmless (and that agreement is not unconscionable or contrary to public policy) or if the defendant oth-erwise had no duty to the plaintiff. The court, however, may not hold that the defendant had no duty to the plaintiff based upon incurred risk, as

---

[10] Of course, as a part of the fault, incurred risk may effectively reduce or eliminate a plaintiff's recovery, depending upon the percentage of fault attributed to the plaintiff by the fact-finder.

compared to express enforceable consent. As a result, we reject the Court of Appeals' conclusion that because Robey "implicitly agreed" to the risks of working as a professional rescuer, he was barred as a matter of law from recovering under the fireman's rule. While the fact-finder may determine that Robey incurred risk, the fact-finder must consider any incurred risk as fault for apportionment purposes under the Act.

The question of whether there is continued viability for the fireman's rule limiting the duty of care owed by the owner of urban premises, as stated in *Woodruff* is not directly presented in this case, and we decline to address it at this time.

IV. CONCLUSION

For the foregoing reasons we affirm the trial court's denial of summary judgment and remand for further proceedings consistent with this opinion.

SHEPARD, C. J., and DeBRULER, DICKSON, and SULLIVAN, J. J., concur.

QUESTIONS

1. Explain how the rescue doctrine and the fireman's rule apply to the activities of emergency responders.
2. Compare the level of duty owed to professional rescuers responding to emergency calls within the course of their employment with that owed to lay rescuers present at or near the scene of an accident.
3. The Indiana Supreme Court noted that public policy does not favor holding that the defendants have complete immunity based solely upon the plaintiff's occupation. How far does the defendants' potential liability extend under the fireman's rule after this decision?
4. The Court concludes that "the question of whether there is continued viability for the fireman's rule limiting the duty of care owed by the owner of urban premises is not directly presented in this case, and we decline to address it at this time." In the wake of *Heck*, how do you think the Court might address that issue in the future?
5. Discuss the policy arguments on both sides of the issue of whether emergency responders should be prohibited from seeking recovery from persons they respond to assist.
6. Consider the effect of insurance on the fireman's rule. Should insurance make a difference in whether an injured responder may recover?

PART 2: POST-TRAUMATIC STRESS DISORDER

Means v. Baltimore County, Maryland

689 A.2d 1238 (Md. App. 1997)

Maryland Court of Special Appeals

Opinion by Raker, J.

In this Workers' Compensation case, we must decide whether post-traumatic stress disorder (PTSD) unaccompanied by physical disease may be compensable as an occupational disease under the Maryland Workers' Compensation Act, now codified as Title 9 of the Labor and Employment Article of the Maryland Code (1991 Repl. Vol., 1996 Cum. Supp.). We shall hold that PTSD can be compensable as an occupational disease.

I.

Appellant Doreen Kay Means has been employed by Baltimore County since 1986. She was initially hired as a Certified Respiratory Therapist, also known as a paramedic, based at the Towson Fire Station. Her duties as a paramedic involved responding to emergency calls and rendering aid at the scenes of accidents and other emergencies. Means filed the workers' compensation claim at issue in this case in February, 1994. She claimed that she "was diagnosed as suffering post traumatic stress syndrome as a result of working a medic unit." Because Means's alleged PTSD is based on events occurring several years before the claim was filed, we turn now to a chronology of those events.

Means contends that the PTSD she allegedly suffered was caused by a particularly severe accident in 1987 involving a van carrying five teenagers. As the first medical personnel crew on the scene, she provided aid and declared the teenagers dead. A few days after this accident, Means responded to another emergency call, with equally serious injuries and fatalities.

Shortly after these incidents in March, 1987, Means was transferred, upon her request, to the Brooklandville Fire Station, a station with a reputation for receiving few emergency calls. After a year, Means was transferred to the Randallstown Fire Station where she remained until February, 1992.

Sometime prior to February, 1992, Means requested a demotion from paramedic to firefighter. In conjunction with the demotion to firefighter, she was transferred back to the Towson station. Although she had been demoted, she was on several occasions required to act as a paramedic at the Towson station. In 1992, Means was required to serve as the paramedic at a particularly gruesome motorcycle accident. The victim had not been wearing a helmet and his scalp had been torn away from his skull. After this accident, Means felt that she "woke up" and remembered the particularly traumatic accidents in 1987 when she had previously worked out of the Towson station.

After the motorcycle accident, Means frequently missed work and began seeing a psychiatrist and a therapist at the Psychological Services Section of the Baltimore County Police and Fire Departments. In her initial visit to the therapist on June 15, 1992, Means reported suffering from flashbacks of the van accident, headaches, crying spells, and difficulty concentrating. She reported that her return to the Towson station had "really upset her and brought back painful memories." In clinical intake notes dated June 17, 1992, the therapist treating Means noted that her "symptoms sound as though they could possibly be part of a post-traumatic reaction or disorder." The following notations appear at the conclusion of the clinical intake notes from Means's first meeting with the therapist:

INITIAL DIAGNOSIS (DSM-III-R)

Axis I R/O Post Traumatic Stress Disorder

Axis II Deferred

Axis III None noted.

The therapist concluded that "further evaluation is necessary to determine if client may be experiencing a post-traumatic reaction of delayed onset." Means remained under the therapists' care at the County's Psychological Services Section until October, 1992. Means was subsequently evaluated in July and October, 1995, by Dr. Joseph M. Eisenberg. Dr. Eisenberg wrote in his evaluation of Means that it was his "opinion that the initial diagnosis in 1992 should have been Post-Traumatic Stress Disorder, delayed onset." Means proffered that Dr. Eisenberg would testify that she suffered from PTSD caused by her employment as a paramedic.

Means filed a workers' compensation claim for PTSD in February, 1994, seeking compensation for 110 hours of missed work. She identified February 1, 1992, as the date of disablement, the same date as her transfer back to the Towson station. On January 6, 1995, the Workers' Compensation Commission held a hearing on Means's claim and concluded that she had not suffered an occupational disease arising out of and in the course of her employment. Means filed a petition for judicial review in the Circuit Court for Baltimore County. See § 9-737.

The County filed a motion for summary judgment. The County presented two arguments: (1) that as a matter of law, Means failed to establish that she suffered from PTSD; and (2) that as a

matter of law, PTSD may not form the basis of an occupational disease claim. The trial court granted the County's motion for summary judgment on the second ground. Means noted a timely appeal to the Court of Special Appeals, and we granted certiorari before consideration by that court. We shall reverse.

II.

A. In Maryland, workers' compensation encompasses two categories of compensable events: accidental personal injury and occupational diseases. §§ 9-501, 9-502; *Lovellette v. City of Baltimore*, 297 Md. 271, 279, 465 A.2d 1141, 1146 (1983). Section 9-101(b) defines "accidental personal injury" as follows:

(b) Accidental personal injury. – "Accidental personal injury" means:

(1) an accidental injury that arises out of and in the course of employment;

(2) an injury caused by a willful or negligent act of a third person directed against a covered employee in the course of the employment of the covered employee; or

(3) a disease or infection that naturally results from an accidental injury that arises out of and in the course of employment, including:

(i) an occupational disease; and

(ii) frostbite or sunstroke caused by a weather condition.

This Court has described accidental injuries as those that involve "the injury and destruction of tissue by the application of external force, such as a blow." *Foble v. Knefely* 6 A.2d 48, 53 (1939). Occupational disease is defined in § 9-101(g) of the Act as follows:

(g) Occupational disease. – "Occupational disease" means a disease contracted by a covered employee:

(1) as the result of and in the course of employment; and

(2) that causes the covered employee to become temporarily or permanently, partially or totally incapacitated.

While the Act does not further define "occupational disease," this Court has further delineated the term "as some ailment, disorder, or illness which is the expectable result of working under conditions naturally inherent in the employment and inseparable therefrom, and is ordinarily slow and insidious in its approach."

Foble, 6 A.2d at 53.

Not all diseases which meet this definition are compensable. Section 9-101(g) must be read in conjunction with § 9-502(d). Section 9-502 reads, in pertinent part:

(d) Limitation on liability. – An employer and insurer are liable to provide compensation . . . only if:

(1) the occupational disease that caused the death or disability:

(i) is due to the nature of an employment in which hazards of the occupational disease exist and the covered employee was employed before the date of disablement; or

(ii) has manifestations that are consistent with those known to result from exposure to a biological, chemical, or physical agent that is attributable to the type of employment in which the covered employee was employed before the date of disablement. . . .

The limitations imposed by § 9-502(d) seek to ensure that only those diseases directly caused by the employment be compensable. *Davis v. Dynacorp*, 647 A.2d 446, 451 (1994).

Occupational diseases have not always been compensable under the Act. The legislative history of the Act suggests that the General Assembly was reluctant to recognize occupational diseases as compensable under workers' compensation. *See*

Miller v. Western Electric Co., 528 A.2d 486, 490 (1987); *see* generally Thomas S. Cook, Workers' Compensation and Stress Claims: Remedial Intent and Restrictive Application, 62 NOTRE DAME L. REV. 879, 889-91 (1987) (discussing state legislatures' early and continuing reluctance regarding occupational disease claims). In 1939, however, the General Assembly recognized occupational disease "as a problem, like on-the-job accidental injury, that an industrial society had to address in a comprehensive fashion," and enacted Maryland's first occupational disease statute. Miller, 528 A.2d at 491; *see* 1939 Md. Laws ch. 465.

Chapter 465 of the Acts of 1939 enumerated thirty-four diseases that were compensable under the Act as occupational diseases. The statute required employers to compensate only for those thirty-four specified diseases and only when caused by the process or occupation specified. For example, asbestosis was compensable if arising out of "any process or occupation involving an exposure to or direct contact with asbestos dust." 1939 Md. Laws ch. 465, § 1, at 995. In 1951, the occupational disease statute was repealed and reenacted, 1951 Md. Laws ch. 289, § 1, at 752, replacing the schedule format with the more general definition of occupational disease that remains in effect today. With this statutory framework and history in mind, we now turn to examine the compensability of PTSD under Maryland's Act.

B. The compensability of work-related mental disabilities unaccompanied by physical illness has been a controversial topic in workers' compensation law over the past decade. Cook, *supra*, at 879. Workers' compensation claims based on mental injuries caused by mental stimuli have been coined "mental-mental" claims, in contrast to "physical-mental" 3 and "mental-physical" 4 claims. *See* A. LARSON, THE LAW OF WORKMEN'S COMPENSATION, § 42.20 (1996). Means makes a mental-mental claim – she alleges that a mental stimulus (the memory of the traumatic accidents) caused a mental injury (PTSD). A majority of the states have found mental-mental claims to be compensable under some circumstances.

This Court has recognized mental-mental claims to be compensable in the context of accidental injury. *Belcher v. T. Rowe Price*, 621 A.2d 872 (1993). In Belcher, the employee, a secretary for T. Rowe Price Foundation, worked on the top floor of an office building located next to a construction site. One morning a three-ton beam broke loose from its crane and crashed through the roof of T. Rowe Price's building, landing with a deafening noise only five feet from Belcher's desk. The power in the building went out, pipes and wires were ripped apart, and debris covered Belcher. Thereafter, Belcher experienced panic attacks, nightmares, and chest pains which were diagnosed as symptoms of PTSD. 621 A.2d at 874-75.

This Court held that Belcher was entitled to workers' compensation for her injuries because they resulted from an accidental personal injury under the terms of the Act. We reasoned that when a mental injury is precipitated by an "unexpected and unforeseen event that occurs suddenly or violently," 621 A.2d at 887 (quoting *Sparks v. Tulane Medical Ctr. Hosp. & Clinic*, 546 So. 2d 138, 147 (La. 1989)), a worker may recover for that mental injury if the injury is "capable of objective determination." 621 A.2d at 890.

One year later, this Court addressed the question of mental-mental claims in the context of occupational diseases. *Davis v. Dynacorp*, 647 A.2d 446 (1994). Davis was a computer operator employed by Dyncorp. Davis alleged that he was continually subjected to serious harassment by his coworkers. The sustained harassment, Davis contended, caused PTSD which prevented him from returning to work 647 A.2d at 447. In contrast to Belcher, Davis maintained that he was entitled to compensation because he suffered an occupational disease, not an accidental personal injury.

This Court held that Davis's claim did not constitute an occupational disease under §§ 9-101(g) and 9-502(d)(1)(i) because the alleged disease was not "due to the nature of an employment in which hazards of the occupational disease exist." § 9-502(d)(1)(i). We concluded that "nothing peculiar to Davis's duties as a computer operator…made him more susceptible to harassment than in any other kind of employment." Davis, 647 A.2d at 451.

Because Davis's particular claim could not constitute an occupational disease under the Act, we did not reach the issue of whether, as a matter of law, gradually resulting, purely mental diseases could ever be compensable occupational diseases. 647 A.2d at 452. In conclusion, however, we addressed the possibility that gradually resulting mental diseases may be compensable occupational diseases. Judge Chasanow, writing for the Court, observed:

> We are not willing to rule out the possibility that some gradually resulting, purely mental diseases could be compensable occupational diseases or that there may be circumstances where work-induced stress may result in a compensable occupational disease. Today, we merely hold that the mental disease resulting from the harassment encountered by Davis was not due to the nature of his employment.

647 A.2d at 452.

C. This case requires us to resolve the issue that we did not reach in Davis, i.e., whether, as a matter of law, PTSD should be excluded from compensable occupational diseases. We reach this issue because, unlike the occupation of computer operator in Davis, the occupation of paramedic is "an employment in which hazards of the occupational disease exist." § 9-502(d)(1)(i). We hold today that PTSD may be compensable as an occupational disease under the Workers' Compensation Act if the claimant can present sufficient evidence to meet the statutory requirements. *See* § 9-101(g) (disease must be contracted as the result of and in the course of employment and the disease must cause the employee to become incapacitated); § 9-502(d)(1)(i) (disease must be due to nature of an employment in which the hazards of the occupational disease exist).

In Davis, 336 Md. at 237, 647 A.2d at 451, we posed the question as follows:

> "The question becomes whether mental disease caused by his job harassment may be reasonably characterized as due to the general character of Davis's employment."

We conclude that Means's asserted PTSD may be reasonably characterized as due to the general character of her employment as a paramedic. Unlike the computer operator in Davis who divided his time between programming computers and reading manuals, Means's employment as a paramedic exposed her to events that could potentially cause PTSD.

We conclude that PTSD may be compatible with the general character of occupational disease. We have consistently described occupational disease as "some ailment, disorder, or illness which is the expectable result of working under conditions naturally inherent in the employment and inseparable therefrom, and is ordinarily slow and insidious in its approach." *Foble v. Knefely*, 6 A.2d 48, 53 (1939); *see also* Davis, 647 A.2d at 449; *Lovellette v. City of Baltimore*, 465 A.2d 1141, 1146 (1983). In the *American Psychiatric Association's Diagnostic and Statistical Manual (DSM-IV)*, a condition may only be diagnosed as PTSD after the symptoms have persisted for over one month. AMERICAN PSYCHIATRIC ASSOCIATION, QUICK REFERENCE TO THE DIAGNOSTIC CRITERIA FROM DSM-IV 211 (1994). Although the outbreak of symptoms may be experienced a few days to a few weeks after the trauma, symptoms may also be delayed. *See* H. Kaplan et al., *Synopsis of Psychiatry* 610 (7th ed. 1994). Based on these criteria, PTSD can be slow and insidious.

Although the structure and history of the occupational disease statutes reflect an intent by the legislature to treat occupational disease differently from accidental injury in some respects, in light of our holding in Belcher we see no sound reason to treat them differently in this regard. In Miller, we noted one basis for treating occupational disease and accidental injury differently:

> The problems of showing disability and causation simply appear less formidable in the [accidental injury] context. The injury, at least to the lay eye, is relatively easy to see and evaluate, and its connection to the employment is more readily apparent.

Miller, 528 A.2d at 492.

This quotation echoes the criticisms of those who oppose compensation for workers who suffer from PTSD arising out of their employment. This Court has addressed PTSD in other contexts, and we have concluded that expert testimony concerning PTSD is "as evidentiarily reliable as an opinion by an orthopedist who has been engaged only to testify ascribing a plaintiff's subjective complaints of low back pain to soft tissue injury resulting from an automobile accident." *State v. Allewalt*, 517 A.2d 741, 746 (1986).... Workers who suffer back pain or soft tissue injury as a result of accidents or diseases arising in the course of employment are not denied compensation due to the difficulty of verification. Judge Orth, writing for the Court in Belcher, observed:

> We have come to appreciate that a mind may be injured as well as a body maimed. A person's psychic trauma does not vary depending upon the type of legal action in which the harm is scrutinized.... The inability to work and the loss of earning power are the same.

Belcher, 621 A.2d at 886.

Other states that maintain a distinction between accidental injury and occupational disease in their workers' compensation statute similarly have held mental disorders to be compensable as occupational diseases, e.g., *City of Aurora v. Industrial Comm'n*, 710 P.2d 1122, 1123 (Colo. Ct. App. 1985); *Martinez v. University of California*, 601 P.2d 425, 426 (N.M. 1979); *Pulley v. City of Durham*, 468 S.E.2d 506, 510 (N.C. Ct. App. 1996); *James v. State Accident Ins. Fund*, 624 P.2d 565, 568 (Or. 1981); *Gatlin v. City of Knoxville*, 822 S.W.2d 587, 590 (Tenn. 1991); *see O'Loughlin v. Circle A. Constr.*, 112 Idaho 1048, 739 P.2d 347, 353 (Idaho 1987).

We stress that we do not today hold that Means's alleged PTSD is necessarily compensable as an occupational disease. We hold only that if the claimant can successfully prove that PTSD meets the statutory requirements, PTSD is not as a matter of law excluded from compensable occupational diseases and that the non-physical nature of Means's claim does not per se exclude her from coverage under the Act.[10] Means must prove that she contracted PTSD "as the result of and in the course of employment." § 9-101(g). Furthermore, she must prove that the mental illness she suffers is due to the nature of a paramedic's job and that employment as a paramedic entails the hazard of developing PTSD. § 9-502(d)(1)(i).

JUDGMENT OF THE CIRCUIT COURT FOR BALTIMORE COUNTY REVERSED. CASE REMANDED TO THAT COURT FOR FURTHER PROCEEDINGS CONSISTENT WITH THIS OPINION. COSTS TO BE PAID BY BALTIMORE COUNTY.

---

[10] We do not address today whether occupational disease based on mental illness shall be governed by special standards distinct from those applied in cases of physical disease. This issue has been neither briefed nor argued before this Court. We point out, however, that there are various tests, standards, and conditions for compensability among the many states that compensate mental-mental claims.

There are essentially four different standards that courts apply to determine which mental injuries will be compensable. Cf. LARSON, supra note 3, § 42.25. Some states treat mental injuries no differently than physical injuries or physical diseases and allow compensation as long as the mental injury arises out of the employment. See, e.g., *City of Aurora v. Industrial Comm'n*, 710 P.2d 1122, 1124 (Colo. Ct. App. 1985) (addressing occupational disease); *Hansen v. Von Duprin, Inc.*, 507 N.E.2d 573, 576 (Ind. 1987); Albanese's Case, 378 Mass. 14, 389 N.E.2d 83 (Mass. 1979); *Martinez v. University of California*, 93 N.M. 455, 601 P.2d 425 (N.M. 1979) (occupational disease).

Other states apply an objective test; that is, a mental injury is compensable if the average worker would have been harmed by the stressful conditions in the workplace. E.g., *State v. Cephas*, 637 A.2d 20, 27 (Del. 1994); *Sturgis v. District of Columbia Dept. of Employment Servs.*, 629 A.2d 547, 551 (D.C. 1993); *Goyden v. State Judiciary*, 256 N.J. Super. 438, 607 A.2d 651, 655 (N.J. Super. Ct. App. Div. 1991), aff'd per curiam, 128 N.J. 54, 607 A.2d 622 (N.J. 1992); *Wilson v. Workmen's Compensation Appeal Bd.*, 542 Pa. 614, 669 A.2d 338, 344 (Pa. 1996).

The third test requires that the claimant prove that the injurious mental stimulus was greater than the usual day-to-day stress experienced in the workplace. E.g., ALASKA STAT. § 23.30.395(17) (1996); ME. REV. STAT. ANN. tit. 39A, § 201(3) (1995); OR. REV. STAT. § 656.802 (1995); R.I. GEN. LAWS § 28-34-2 (1996); *Owens v. National Health Labs.*, 8 Ark. App. 92, 648 S.W.2d 829, 831 (Ark. Ct. App. 1983); *Dunlavey v. Economy Fire & Casualty Co.*, 526 N.W.2d 845, 853 (Iowa 1995); *Borden, Inc. v. Eskridge*, 604 So. 2d 1071, 1073-74 (Miss. 1991); Wilson, 669 A.2d at 344 (applying both the objective test and the abnormal working condition test); *Bedini v. Frost*, 678 A.2d 893, 894 (Vt. 1996); *Consolidated Freightways v. Drake*, 678 P.2d 874, 878 (Wyo. 1984).

Finally, some states allow recovery for purely mental injuries only when induced by a traumatic event, shock, or fright in the workplace. E.g., LA. REV. STAT. § 23:1021 (1996); N.M. STAT. ANN. § 52-1-24 (1996); UTAH CODE ANN. § 35-1-45.1 (1996); *Pathfinder Co. v. Industrial Comm'n*, 62 Ill. 2d 556, 343 N.E.2d 913, 917 (Ill. 1976); *Wolfe v. Sibley, Lindsay, & Curr Co.*, 36 N.Y.2d 505, 330 N.E.2d 603, 606, 369 N.Y.S.2d 637 (N.Y. 1975); *Gatlin v. City of Knoxville*, 822 S.W.2d 587, 590 (Tenn. 1991); *Hercules, Inc. v. Gunther*, 13 Va. App. 357, 412 S.E.2d 185, 189 (Va. Ct. App. 1991).

QUESTIONS

1. What are the signs and symptoms of PTSD?
2. Why do you think that recognition of PTSD as a compensable work-related injury has been such a relatively recent phenomenon?
3. What challenges do you see for Judges in deciding whether to award damages for PTSD?
4. Describe any barriers that might exist to a person with PTSD or other mental challenge brought on by working in the extremely stressful emergency services arena. Perform research to determine the support services that exist to assist such personnel, and describe the positive and negative and negative aspects of the different services.
5. In contrast to the *Means* case which was decided in Maryland, the State of Illinois is reluctant to award workers compensation for PTSD to emergency responders. That State reasons that PTSD claims are easy to fabricate. Illinois denied PTSD claims of a Paramedic whose patient died due to mechanical problems with the rig, reasoning

that she had had multiple other patients expire. See *Burney v. Jersey Cmty. Hosp.*, 04 IL.W.C. 41965, No. 06 I.W.C.C. 1168 (Ill. Workers' Comp. Comm'n 2006). The State also refused the PTSD claim of an Illinois firefighter who feared for his life after being trapped on the second floor by a flashover. The workers' compensation arbitrator said that entering a burning building is what firemen do and firefighters must expect these risks, adding that the danger of flashover is not greater than the danger to which all firefighters are exposed. See *Perry v. City of Peoria*, 01 I.I.C 0791, No. 98 IL.W.C. 28990 (Ill. Indus. Comm'n 2001). So, apparently, in Illinois, first responders cannot get compensation for PTSD if their claim arises from exposure to the kinds of risks that they face on a regular basis. To what extent, if any, might this information influence your consideration of working in Illinois if you were a well-credentialed emergency responder with multiple options?

PART 3: WORLD TRADE CENTER SITE LITIGATION

This case is an excellent vehicle for transition from the emergency response to the emergency management sections of the book. The case opens with a thorough description of the effects on the responders who worked the debris pile left in the aftermath of the attacks on the World Trade Center in New York on September 11, 2001. The main issue involved is what entities are responsible for.

In Re: World Trade Center Disaster Site Litigation

456 F. Supp. 2d 520 (SDNY 2006)

OPINION By: ALVIN K. HELLERSTEIN, United States District Judge.

OPINION DENYING AND GRANTING MOTIONS FOR JUDGMENT ON THE PLEADINGS AND FOR SUMMARY JUDGMENT

It took ten months to remove the debris that resulted when the terrorists crashed their hijacked airplanes into the Twin Towers of the World Trade Center on September 11, 2001. Thousands

of workers converged on the site, toiling day and night, seven days a week until they completed their jobs. They risked their lives from shifting debris, fires, smoke, and acrid and polluted air to complete their work in record time, in an extraordinary effort to close the gaping hole caused by the terrorists to the landscape and psyche of New York and the nation.

I consider in this Opinion the claims of approximately 3,000 of these workers, claiming permanent injury to their respiratory systems and their health and vitality, and a shortening of their lives. They claim that the City and its contractors, and other Defendants, were negligent in monitoring the air and assuring appropriate safety in the workplace, particularly in not providing adequate respiratory equipment, and assuring proper use thereof.

Defendants now move to dismiss these claims, contending that they are immune from suit pursuant to state and federal laws providing immunity for actions undertaken in response to a disaster created by an enemy attack on the state and nation. Plaintiffs argue that Defendants are not immune, particularly in light of Congress' clear contemplation, in the Air Transportation Safety and System Stabilization Act of 2001, that the City was exposed to numerous claims resulting from or relating to the terrorist-related aircraft crashes of September 11, 2001, and granting to the City a cap to limit its potential liability stemming from such claims. Furthermore, Plaintiffs argue, Congress again recognized the City's exposure to suits such as those at bar by granting a one billion dollar fund to the City to pay for the City's losses, liabilities and expenses, enabling the City to create a captive insurance fund to insure its exposure.

I discuss the various motions of the City and other Defendants in this Opinion and hold that the Defendants are benefited by limited immunity, limited according to time and activity, and that the issues are fact-intensive and cannot be decided on motion at this juncture. My conclusion also expresses some suggestions for the future progression of these cases, to enable the parties to begin discussions of settlements and to prepare for trial.

I. INTRODUCTION

The terrorist attacks of September 11, 2001 inflicted a gaping wound on the structure and spirit of New York City. But it did not defeat the City, nor its population. As the nation began to absorb the enormity of the devastation and loss of lives that resulted from the terrorist attacks, an army of responders – instrumentalities of federal, state and city governments, private contractors, and thousands of firemen, policemen, paramedics, and construction workers – descended on the site of the devastation in New York City, initially to participate in the desperate search for survivors and, after all hope of life had faded, to assist in the recovery of remains and the clearing of debris. Working night and day, seven days a week, overcoming intense heat, persistent fires, and noxious fumes, the work was done and the site was cleared, in just under ten months – record time.

The extraordinary efforts of the men and women who worked on the site took a toll. A few have died, with at least one of their deaths having been attributed to the poisons they breathed while looking for survivors and clearing the debris. Anthony DePalma, Debate Revives as 9/11 Dust is Called Fatal, N.Y. Times, April 14, 2006, at B2. Many others allege serious respiratory injuries, threatening to shorten their lives and afflict their remaining years. Anthony DePalma, Illness Persisting in 9/11 Workers, Big Study Finds, N.Y. Times, Sept. 5, 2006, at A6. A study released by doctors at Mount Sinai Medical Center shows that approximately 70 percent of the 10,000 workers who were tested reported that they suffer from new or substantially increased respiratory problems since September 11. Id. In all, more than 3,000 of these men and women have filed suit in this Court, and even more suits are likely as respiratory injuries continue to manifest themselves. Under procedures outlined in the New York General Municipal Law section 50-e, allowing for leave to serve notice of claims upon the City of New York outside of the prescribed 90-day period, hundreds of additional persons have gained leave by the New York Supreme Court also to file suits, adding to the lawsuits consolidated before me.

The main Defendant in these lawsuits is the City of New York. The City's Department of Design and Construction coordinated all the work on the site, drawing on the expertise of other City agencies, for example, the City Office of Emergency Management, and of the federal and state governments. The Port Authority of the States of New York and New Jersey, the owner of the World Trade Center site, also had a role, and it, too, is named as a Defendant. The site was divided into quadrants, and in each, a general contractor and numerous subcontractors undertook the work; they also are Defendants....

II. THE FACTUAL BACKGROUND

The devastation wrought by the terrorist attacks of September 11, 2001 was unimaginable. One and Two World Trade Center, once great symbols of the City's economic strength and vitality, came crashing down, consuming and entombing the remains of almost 3,000 people who were caught inside the buildings. Three World Trade Center, Four World Trade Center, and Six World Trade Center sustained extensive fire damage. Seven World Trade Center caught fire from the falling debris of the Twin Towers and, after burning unimpeded for several hours, also collapsed. See In re September 11 Property Damage and Bus. Loss Litig. (Aegis Ins. Serv. v. The Port Authority), 468 F. Supp. 2d 508, (S.D.N.Y. 2006) (hereinafter 7WTC). The shopping malls and parking lots beneath the World Trade Center complex sustained massive structural damage and partial collapse. The World Financial Center and the glass enclosure of its "Winter Garden," the Verizon Building at West and Vesey Streets, the Deutsche Bank Building at 90 West Street, the St. Nicholas Church at Barclay and West streets, and 125 Cedar Street also sustained extensive damage.

Rescue and recovery workers, and the contractors and government agencies overseeing their efforts, were faced with the daunting task of conducting an emergency response in a hellish setting: a smoldering pile of twisted and entangled shapes of rods and beams towering over twelve stories high and weighing more than one and half million tons, harboring fires within and emanating clouds of noxious dust and vapors. The fires took three months to go out, until December 2001. Conditions were further exasperated by the nature of the various substances present at the site. Thousands of tons of hazardous materials were released into the air of lower Manhattan, including asbestos, lead, mercury, cadmium, polychlorinated biphenyls ("PCBs"), benzene, and chromium, among others, creating a mixture of toxic gases and ultra-fine particles never before experienced on such a scale. As Plaintiffs' Master Complaint alleges:

> asbestos, lead, and mercury from such items as personal computers, main frame computers, and copy machines; mercury from florescent lights; plastics, polyvinyl chloride insulations of cables, nylon carpeting, and other materials containing and/or producing dioxins and other harmful materials when burned; benzene from the [**9] jet fuel and other petroleum products stored in the towers; [*527] lead ammunition from the on-site Secret Service shooting range; and arsenic, lead, mercury and chromium stored in the U.S. Customs Laboratory. (Pls.' Master Compl. P376, dated Aug. 19, 2005)

In the initial days following the attacks of September 11, the primary focus of all at the site was rescuing any potential survivors. In re World Trade Center Disaster Site Litig. (*Hickey v. Port Authority of N.Y. & N.J.*), 270 F. Supp. 2d 357, 372 (S.D.N.Y. 2003); aff'd in part, dismissed in part by *McNally v. The Port Authority*, 414 F.3d 352 (2d Cir. 2005) (hereinafter Hickey). Unfortunately, there were too few and, by September 29, 2001, all hope of finding lives having faded, the search for survivors closed, and efforts turned to "demolition of the ruined structures of the Towers, removal of thousands of tons of debris, and cleanup of the World Trade Center site[.]" Hickey, 270 F. Supp. at 372.

A. The Declarations of Emergency: The Immediate Government Response

In the aftermath of the attacks, government leaders at the local, state and federal levels took immediate action to secure physical assistance and funding for the recovery effort at the World Trade Center site. The Mayor of the City of New

York, the Governor of the State of New York, and the President of the United States all declared states of emergency, authorizing and directing government agencies and officials to undertake those measures necessary to assist the City of New York in its process of recovery.

Pursuant to the authority granted him under the Executive Law of New York, N.Y. Exec. Law § 24 (McKinney 2006), the Mayor of the City of New York, Rudolph W. Giuliani, issued a Mayoral Order on September 11, 2001, proclaiming a local state of emergency based on the danger to public safety posed by the attacks. In declaring a state of emergency, the Mayor directed "the Police, Fire and Health Commissioners and the Director of Emergency Management to take whatever steps are necessary to preserve the public safety and to render all required and available assistance to protect the security, well-being and health of the residents of the City." Proclamation of a State of Emergency, Mayor Rudolph W. Giuliani (September 11, 2001). In subsequent proclamations, and pursuant to Executive Law section 24(1)(g) allowing for suspension of local laws and regulations during states of emergency, the Mayor directed that local regulations governing the leasing of real property to the City be suspended so as to "permit the immediate leasing of office and other space for use by City agencies in order to continue to provide essential services and critical functions of the City." Proclamation of State of Emergency, Mayor Rudolph Giuliani (Sept. 14, 2001). The Proclamation of Emergency was renewed by Mayoral Order every five days, as mandated by the Executive Law, throughout the duration of the recovery and cleanup efforts at the World Trade Center site, through the end of June 2002. See e.g., Proclamation of a State of Emergency, Mayor Rudolph W. Giuliani (Sept. 11, 2001); Proclamation of State of Emergency, Mayor Michael R. Bloomberg (June 29, 2002).

A disaster emergency was also declared for the State of New York by Executive Order of Governor George E. Pataki on September 11, 2001, pursuant to the authority granted him under the New York State and Local Natural Disaster and Man-Made Disaster Preparedness Law ("Disaster Act"). N.Y. Exec. Law §§ 20-29-g (McKinney 2006). Noting the "unspeakable atrocities" that occurred in New York City, Washington D.C., and Pennsylvania, Governor Pataki "direct[ed] the implementation of the State Disaster Preparedness Plan and authorize[d]," various state agencies to take "all appropriate actions to assist in every way all persons killed or injured and their families, and protect state property and to assist those affected local governments and individuals in responding to and recovering from this disaster, and to provide such other assistance as necessary to protect the public health and safety[.]" Exec. Order No. 113 of Governor George E. Pataki (Sept. 11, 2001), N.Y. Comp. Codes R. & Regs. tit. 9, § 5.113 (2005).

On September 14, 2001, President George W. Bush, acting pursuant to the National Emergencies Act, 50 U.S.C. §§ 1601-1651 (2006), declared the existence of a national state of emergency "by reason of the terrorist attacks at the World Trade Center... and the Pentagon, and the continuing and immediate threat of further attacks on the United States." Proclamation No. 7463, 66 Fed. Reg. 48, 199 (Sept. 14, 2001). The declaration was deemed effective as of September 11, 2001. The declaration served also to activate the provisions of the Stafford Disaster Relief and Emergency Assistance Act ("Stafford Act"). 42 U.S.C. §§ 5121-5206 (2006). Pursuant to the Presidential declaration of a national emergency, the Director of the Federal Emergency Management Agency ("FEMA"), Joe M. Allbaugh, declared that a national emergency existed in the State of New York and, in the interest of ensuring the provision of federal assistance, authorized FEMA "to allocate from funds available for these purposes, such amounts as [are] necessary for Federal disaster assistance and administrative expenses." 66 Fed. Reg. 48,682 (Sept. 21, 2001).

B. The City Asserts Control and the Recovery Operation Commences

The City response began mere moments after the terrorist attacks on New York City. American Airlines Flight 11 crashed into One World Trade Center at 8:40 A.M. By 8:50 A.M. on September

11, the City, initially through the Fire Department, had established its Incident Command Post and had asserted control over the World Trade Center complex and the surrounding areas. The rescue and recovery efforts at the site were thereafter coordinated through the City Office of Emergency Management ("OEM"), with the Fire Department designated as the incident commander for the site, and with the City Department of Design and Construction ("DDC") assuming total control over all aspects of safety, construction, demolition, and cleanup activities at the site.

On September 12, 2001, the DDC set up a temporary command center at Public School IS 89 in lower Manhattan, immediately to the North of the World Trade Center site, and commenced daily meetings to organize rescue and recovery efforts. Of utmost concern to the DDC was securing the World Trade Center site and limiting access to the area. Together with other City agencies, including the OEM, the DDC established stringent protocols determining "not only who would have access to the site, but also how that access would take place and under what constraints." The City further enlisted the Port Authority of New York and New Jersey (the "Port Authority") to assist in maintaining the security of the perimeter and to report observed safety protocol discrepancies.

The City also engaged private contractors for the recovery effort. On September 15, 2001, FEMA confirmed that contracts could be awarded without need for competitive bidding under the emergency conditions existing after September 11. Requirements for competitive bidding having been waived, and pursuant to the Declarations of Emergency issued at the City, State and Federal levels, the DDC engaged Bovis Lend Lease, AMEC Construction Management, Tully Construction Company, and Turner Construction Company to "provide the work necessary for removal and demolition services." These four contractors were designated as the City's Primary Contractors and assumed lead roles in the recovery and cleanup efforts at the site. The efforts of the Primary Contractors were coordinated, and supervised, through the DDC at twice daily

meetings held at the temporary command center, and by numerous visits to the worksite. By September 14, 2001, the DDC had divided the site into four quadrants with a Primary Contractor assigned as a "construction manager" for each individual quadrant. The Primary Contractors acted as supervisors for their individual quadrants, with responsibility for enforcing applicable regulations and ensuring compliance.

Cognizant also of the need for additional space outside of the World Trade Center complex to which all debris from the site could be removed and where searches for evidence and human remains could be conducted, the City re-opened the Fresh Kills Landfill ("Fresh Kills") on Staten Island. As debris was cleared from the site, it was loaded onto barges and transferred by the Department of Sanitation to Fresh Kills for sorting and further inspection.

The DDC, with the assistance of the Port Authority engineers, regulated and controlled the removal of debris from the site, and all issues pertaining thereto. Global Positioning System devices were installed in all vehicles carrying debris from the site, allowing the City to improve the efficiency of the debris removal process. All trucks leaving the site were issued "Load Debris Tickets" containing information about the destination of the truck and its cargo. As debris was removed from the site, the DDC tracked and coordinated the following: "(a) FDNY/Rescue operations; (b) Structural concerns; (c) Progress of work; (d) Weather; (e) Trucking/Traffic operations-Manifests; (f) Safety concerns; (g) Manpower (contractor/CM Staff-site & home office); (h) Equipment on site; (i) Idle equipment; (j) Material deliveries/usage; and (k) Crane issues." The DDC further controlled all aspects relating to persons working at the site, from regulating shift changes and meal allowances, to payment and payroll.

In the initial days and weeks following September 11, the City and its Contractors, together with public utilities, worked also to restore essential services to the City. The September 11 attacks resulted in the immediate loss of power to all of lower Manhattan and in the destruction of critical components of the gas and steam infrastructure. The Con Edison substations, which had been located directly beneath World Trade Center Seven, were destroyed by fire and by the building's ultimate collapse, resulting in a critical disruption of services to Lower Manhattan. See 7WTC, 2006 U.S. Dist. LEXIS 749, 2006 WL 62019 at *7-10. Con Edison assumed sole responsibility for restoring electric, gas and steam services and related facilities that were damaged or destroyed due to the events of September 11. The Verizon Building, located at 140 West Street, also sustained severe structural damage, crippling the phone system. Other critical services, such as the transportation system running through the World Trade Center site, were also destroyed and disrupted.

C. The Development of Health and Safety Standards at the Site

Conditions at the World Trade Center site, particularly the hazards posed by the dust and contaminants that enveloped lower Manhattan for weeks following the attacks, posed significant dangers to the rescue and recovery workers. In the months following September 11, and continuing to the close of operations at the site in June of 2002, the Occupational Safety and Health Administration ("OSHA") reported levels of various contaminants, including dioxin and asbestos, in excess of OSHA's permissible exposure limits. The debris pile itself, containing what remained of two 110-story towers of concrete and steel, created its own volatile, unstable, and inherently dangerous worksite. Implementation and enforcement of viable and responsive health and safety standards was therefore essential. The workers at the site were presented with a dangerous environment, below and surrounding their work activities, threatening their health and safety.

By September 12, 2001, the City had established itself as the lead entity charged with the development and enforcement of health and safety standards at the site, and had instituted daily meetings with representatives of the Primary Contractors as well as with the FDNY, NYPD, OEM, and OSHA, to organize the rescue and recovery operation. These meetings would continue throughout the duration of the recovery effort. At the request of the DDC, Bechtel Environmental Safety & Health ("Bechtel") also began work at the site on September 12, 2001, assisting the City with monitoring compliance with health and safety standards.

Critical to any health and safety plan was the development of appropriate standards for the use of personal protective equipment ("PPE"), including respirators. Within hours of the collapse, the FDNY advised its employees that respirators should be worn at the site and placed an order for over 5,000 respirators and 10,000 cartridges to provide its employees with the necessary respiratory protection.[30] The FDNY also ordered adapters to convert 15,000 "Scott" facemasks, designed for use with self-contained breathing equipment, to be used instead with filter cartridges. The City was also focused on establishing appropriate PPE standards at the site and, by September 21, 2001, plans were in place to:

> Develop the job and site specific PPE requirements (DDC); Develop and distribute a list of required PPE to all site emergency response and workers, including posters for staging areas (DDC); Provide respirator fit-testing, maintenance and use information, and comprehensive technical support (NYCDOH); Aggressively promote minimum PPE use in all areas where highest exposures may occur. (NYCDOH, NYPD, FDNY, National Guard)

The DDC, together with the New York City Department of Health ("City DOH"), assumed primary responsibility for developing and enforcing PPE requirements, with each agency at various times proclaiming itself as the lead agency in charge of worker health and safety. As a general matter, the DDC assumed responsibility for City and contractor personnel, while the City DOH assumed responsibility for FDNY and NYPD personnel.

---

[30] However, it was not until September 28, 2001 that a written order was prepared and it was not until November 26, 2001 that all necessary bureaucratic authorization was given. On September 22, 2001, the FDNY complained of the delay, and admonished officials "OEM must develop a plan...to address overall use & respirator issue[s]."

On September 20 and 22, the City DOH issued criteria for minimum safety gear to be worn at the site. Similarly, on September 21, and again on October 19, the City DOH issued orders mandating the use of specific protective actions to be taken as personnel and vehicles left the site:

It is hereby ordered that all persons leaving the WTC site shall follow personal hygiene protocols, including but not limited to...removal or HEPA vacuuming of work clothes...

It is further ordered that all vehicles leaving the WTC site be spray washed[.]

(Pls.' J.A. Vol. 5, Ex. 64 (City DOH Order dated Sept. 21, 2001).) On October 22, 2001, the City DOH issued a directive specifically addressing the use of safety equipment and respirators:

Personal Protective Equipment Required in Debris Area

- Hardhat or helmet

***

- Respirator (half-face reusable) with P100/organic vapor/acid gas (OVAG) filter cartridges

(Pls.' J.A. Vol. 6, Ex. 97 (Health Bulletin dated Oct. 22, 2001).)

The DDC, together with Bechtel, also played an important role in establishing health and safety protocols, periodically issuing Environmental Health and Safety Bulletins outlining the applicable safety standards in force at the site. In a Bulletin issued in February 2002, the DDC, after consultation with OSHA representatives, announced its concurrence with the City DOH determination as to minimum respiratory protective equipment:

A half-face respirator with P-100, organic vapor, acid gas filters/cartridges is required within the confines of the slurry wall and for any activity or area outside the slurry wall that generates dust, fumes or vapors[.]

D. The Implementation and Enforcement of Health and Safety Standards

From as early as September 12, 2001, the DDC, working primarily with Bechtel, the lead contractor in charge of health and safety concerns at the site, and with the assistance of the Port Authority and the Primary Contractors, began conducting inspections in order to enforce compliance with applicable PPE requirements. Federal agencies, including the Environmental Protection Agency ("EPA") and OSHA, also participated in safety monitoring and assumed a leading role with respect to certain tasks relevant to health and safety monitoring.

By October of 2001, the City had distributed an initial version of its comprehensive Environmental Safety and Health Plan (the "ES&H Plan"), addressing all aspects of worker safety at the site. Numerous revisions to the Plan would follow. The ES&H Plan operated to define the "minimum acceptable requirements for ensuring workers' safety and health at the World Trade Center (WTC) Emergency Project" and established the hierarchy of responsibility and enforcement. Setting forth its general purpose and the DDC's lead role in its enforcement, the Plan expressly provided as follows:

This ES&H Plan is directly applicable to all work conducted by agency and prime contractor/subcontractor personnel engaged in any activity associated with the cleanup and recovery efforts on the WTC Emergency Project. The DDC has overall responsibility for the site's ES&H program.

(Pls.' J.A. Vol. 8, Ex. 149.)

The ES&H Plan provided that the DDC had lead responsibility for "Environmental Health and Safety Services," with the City's Primary Contractors also assuming responsibility for enforcing the terms of the Plan. Specifically, the Plan provided that "each prime contractor and their subcontractors are responsible for implementation, enforcement and compliance with all aspects of this plan." Site monitoring was also performed by other city, state and federal agencies. The results of such monitoring, and all tests setting forth rates of exposure were to be provided to the City DOH for presentation of a final report to the DDC. Further, the DDC alone had the authority to stop work in the event of workplace hazards.

Separately from the ES&H Plan, the DDC and FDNY, as co-incident commanders at the site, together with the Primary Contractors, four separate employer/employee associations, and

OSHA entered into the WTC Emergency Project Partnership Agreement (the "Partnership Agreement") on November 20, 2001, and which was subsequently revised on April 10, 2002. The Partnership Agreement affirmed "the value of working in a cooperative, focused and voluntary effort to ensure a safe and healthful environment for everyone involved," and memorialized the parties' commitment to "advance the goal of environmental health and safety" and to "share safety hazard data."

In accordance with the ES&H Plan and the Partnership Agreement, reports documenting the results of these safety compliance inspections were prepared throughout the duration of the recovery effort at the direction of the DDC. The reports prepared by the DDC, and reports and documents prepared by the Primary Contractors and other agencies at the site, highlight ongoing and persistent problems with enforcing compliance with applicable PPE requirements. Within the first month of initiating operations at the site, the Primary Contractors, AMEC, Tully, Bovis, and Turner, had all documented problems with PPE compliance and particularly with respirator use. Indeed, problems with respirator usage would pervade the entirety of the recovery operation at the site, with estimates prepared in January of 2002 showing compliance rates below 29 percent.

E. The Role of Federal Agencies

The enormity of the task necessitated the involvement of, and cooperation with, federal agencies. Although the City, through the DDC, assumed primary control over the site, several federal agencies, including FEMA, OSHA, the EPA and the United States Army Corps of Engineers ("Army Corps"), participated in the rescue and recovery effort. These various agencies would ultimately play an active role in the efforts at the World Trade Center, most particularly through their attendance at meetings addressing overall concerns of worker health and safety and through their assistance in developing and enforcing appropriate health and safety protocols responsive to such concerns.

1. The Activation of Federal Assistance

On September 14, 2001, President George W. Bush declared a National State of Emergency, thereby activating the Stafford Disaster Relief and Emergency Assistance Act ("Stafford Act"), 42 U.S.C. §§ 5121-5206 (2006). Activation of the Stafford Act by Declaration of National Emergency allowed for implementation of the course of federal assistance provided pursuant to the framework outlined in the Federal Response Plan ("FRP").

The FRP, an agreement among twenty-seven federal agencies, "establishes a process and structure for the systematic, coordinated, and effective delivery of federal assistance to address the consequences of any major disaster or emergency declared under the [Stafford Act]." Specifically, the FRP sets forth a "Basic Plan," presenting "the policies and concept of operations that guide how the Federal Government will assist disaster-stricken State and local governments." (Id. at 4.) The Basic Plan provides that, upon exhaustion of local resources and at the request of the affected local government, FEMA shall operate as the lead federal agency for coordinating an appropriate federal response, providing for both technical and financial assistance. (Id. at 7-8, 12.)

The FRP further coordinates the structure and nature of federal assistance by grouping the types of federal assistance most likely to be utilized by overwhelmed state and local governments into twelve separate Emergency Support Functions ("ESFs").[8] (Id. at 13.) Each individual ESF is headed by a primary agency "designated on the basis of its authorities, resources, and capabilities in the particular functional area," and assisted by one or more other federal agencies acting in a supporting capacity. (Id.) As the lead agency in charge of coordinating any federal response pursuant to a declaration of emergency, FEMA is authorized to activate "some or all of the ESFs, as necessary." (Id.)

Pursuant to activation of the FRP, and FEMA's subsequent activation of the relevant ESFs, OSHA, the EPA and the Army Corps each provided technical and physical assistance to the City of New York in their respective areas of expertise and authority. Federal financial assistance was also provided throughout the duration of the recovery effort with FEMA promising to cover the cost of all operations at the World Trade Center Site as well as at Fresh Kills Landfill. (Defs.' J.A., Vol. 2, Ex. S (Press Release, The White House, dated Sept. 18, 2001).)

2. The Role of the Occupational Safety and Health Administration

In keeping with its designation under the FRP, OSHA assumed the lead role for developing and enforcing respirator requirements at the site. Within a few days of the attacks, OSHA had begun working with the City and other local agencies to develop comprehensive and effective PPE requirements and overall worker health and safety protocols. OSHA advised the City that the use of P-100 filters, rather than standard combination filter/cartridges, would provide workers with adequate protections against the respiratory dangers presented at the site. OSHA also performed atmospheric monitoring throughout the rescue and recovery effort to establish the geographical boundaries within which respirator use would be required and in order to determine the level of respiratory protection needed to protect workers against contamination by the surrounding atmosphere.

On approximately September 20, 2001, at the request of the City DOH, OSHA assumed a role as the "lead agency for distributing, fitting, and training for respirators for the recovery of workers." OSHA took over the operation of the FDNY's respirator distribution staging area, ultimately becoming the "sole provider of respiratory protection equipment, fit-testing and training for new shift FDNY personnel." OSHA also distributed respirators to NYPD personnel.

Distribution of respirators and other safety equipment at the site was coordinated through plywood huts set up by OSHA at various points surrounding the World Trade Center project. OSHA ultimately distributed over 131,000 respirators. All individuals to whom OSHA distributed respirators received qualitative and quantitative fit-checks, and training for the proper use, storage,

and maintenance of respiratory equipment. OSHA also developed a 10-hour health and safety course focusing on the proper use of respirators. All individuals who worked in a supervisory role at the site, including employees of the Primary Contractors, were required to attend the course prior to being admitted to the site.

OSHA, however, did not assume direct supervisory power to assure that workers used respiratory equipment consistently and efficiently. DOH email ("Unfortunately, OSHA has taken an 'advisory' role to date.'"). OSHA's role was thus one of "assistance and consultation, **not** enforcement." (Even though OSHA deployed over 1,000 inspectors to the site to report safety violations to the various contractors and to document observed safety violations in weekly reports, the prevalence of violations supported by anecdotal evidence of substantial non-use of PPE suggests that safety standards were not uniformly and consistently enforced. Furthermore, it is not clear if OSHA inspectors had the authority to stop work at the site upon observing critical safety violations. Per the express terms of the ES&H Plan, the DDC retained exclusive stop work authority. Nevertheless, Walter Murray of Turner Construction testified that he believed that OSHA also had authority to stop work at the site. By the close of operations at the World Trade Center site, OSHA had identified more than 9,000 hazards as needing correction. Defendants ask me to presume that employers had corrected the problems pointed out to them, but it would be improper for me to do so on a Rule 12(c) motion. Clearly, problems with enforcing PPE requirements persisted and not OSHA, but the DDC and the private Contractors, were responsible to ensure that compliance at the site would be enforced.

3. The Role of the Environmental Protection Agency

Under the FRP, the EPA is designated as the lead agency responsible for the cleanup of sites contaminated by "hazardous materials release[s] caused by a catastrophic event." (National Contingency Plan, 40 C.F.R. § 300.130(i)). As such, the EPA, in conjunction with the City Department of

Environmental Protection ("City DEP"), assumed the lead role for "hazardous waste disposal" at the World Trade Center site. ("EPA Response to September 11: Oh My God, Look at that Plane.") The EPA further assumed "primary responsibility for monitoring the ambient air, water and drinking water and coordinating the sampling data for all the response agencies." (Hearing Before U.S. Senate Committee on Environment and Public Works, Feb. 11, 2001).) In accordance with its lead role in environmental monitoring at the site, the EPA also assumed responsibility for public dissemination of the results of environmental monitoring tests conducted at the site, regularly publishing such results on its website and at the tented area where workers ate their meals.

Environmental testing by the EPA was coordinated with OSHA and other City agencies through the use of multiple ambient air monitors at locations in and around the World Trade Center site and through the collection of data from pre-existing air monitors in the area. At the close of operations in June of 2002, OSHA had taken over 6,100 air samples and had turned over the results of these tests to the EPA for further evaluation and examination. Although other City agencies, including the City DOH and the City DEP, also conducted environmental monitoring at the site, the results of such monitoring were coordinated through the EPA and shared with the Environmental Assessment Group, a inter-agency group comprised of federal, state and City agency representatives. On the basis of these tests, both the EPA and the Environmental Assessment Group determined to adopt the rule that all workers be required to wear respirators at all times while working on the pile.

The EPA thus worked with City and federal agencies to develop adequate health and safety protocols, but limited its role to monitoring of air quality and removal of hazardous materials. Indeed, the EPA expressly acknowledged that it lacked "authority to enforce the worker health and safety policies for non-EPA/USCG employees." ("We have observed very inconsistent compliance with our recommendations, however, we do not have the authority to enforce the worker health and safety policies[.]")

4. The Role of the Army Corps of Engineers

On October 1, 2001, at the request of the City under ESF 3 (Public Works and Engineering) of the FRP, the Army Corps assumed control over the coordination, implementation, structure and enforcement of safety and health procedures and protocols at Fresh Kills. Although the Army Corps' role at Fresh Kills was limited as an initial matter to the development and implementation of health and safety standards for contractor personnel only, as of October 10, 2001, the scope of its role at Fresh Kills was expanded to include "the implementation of a comprehensive health and safety plan for all personnel working at the landfill site."

In accordance with its expanded role at the Fresh Kills Landfill, the Army Corps identified the following areas for inclusion in a comprehensive health and safety plan: "Establish baseline worker exposure levels for each job site activity; Develop job and site specific PPE requirements; Establish an ongoing worker exposure monitoring program; Issue daily work site air quality updates; Implement dust control measures; Establish the presence of health and safety team monitors." To assist in the development and enforcement of such a comprehensive plan, the Army Corps contracted with a private contractor, Phillips & Jordan, to act as the Construction Manager for the Fresh Kills site. Specifically, Phillips & Jordan assumed responsibility for the management of the forensic recovery operation at the site and for the enforcement of the site safety and health plan. The Army Corps further enlisted the assistance of a subcontractor, Evans Environmental & Geosciences, to develop an Environmental Safety and Health Plan tailored to the concerns presented by the recovery operations at Fresh Kills and to develop and implement necessary training and monitoring programs.

F. The Rescue and Recovery Effort Comes to a Close

From the time that the rescue and recovery operation began at the World Trade Center site in the moments following the September 11 attacks, to the close of operations in June of 2002, work at the site never ceased, continuing twenty-four hours a day, seven days a week, including holidays, with the exception only of Veteran's Day 2001. Despite the enormity of the task, however, work progressed at a rate that many could not have imagined and, as early as April of 2002, the transition of control over the site from the DDC to the Port Authority was being designed and implemented.

On May 10, 2002, control over Seven World Trade Center was returned to the Port Authority. The turnover of control as to the remainder of the World Trade Center complex followed shortly thereafter, on June 30, 2002, with the Port Authority once again assuming complete responsibility for the site. Although control has officially been returned to the Port Authority, work at the site continues to this day with efforts now turned to the completion of all steps necessary to rebuilding.

G. The Continuing Vitality of Applicable Safety Standards and Labor Laws

Throughout the duration of the rescue and recovery effort, the City agencies and Primary Contractors remained obligated to abide by applicable safety standards and labor laws.

Pursuant to the authority granted under Executive Law section 24, both the Governor of the State of New York and the Mayor of the City of New York had the capacity to suspend, or direct the suspension of, "any of its local laws, ordinances or regulations, or parts thereof subject to federal and state constitutional, statutory and regulatory limitations, which may prevent, hinder, or delay necessary action in coping with a disaster or recovery therefrom[.]" N.Y. Exec. Law § 24 (McKinney 2006). Both declined, however, to authorize wholesale suspension of applicable laws and regulations, instead authorizing only the suspension of the sign-in/sign-out procedures required by New York State Labor Law for the period from September 11, 2001 to October 14, 2001.

In the absence of any such suspension of applicable labor laws, the City expressly mandated that its Primary Contractors abide by and enforce all relevant laws, ordinances, and regulations. Specifically, draft contracts between the DDC and the contractors show that compliance with applicable safety standards was required. The DDS mandated compliance with the following:

a. New York State Uniform Fire Prevention and Building Code;
b. National Fire Prevention Association (NFPA) requirements;
c. National Electrical Code (NEC);
d. American National Standard – ANSI, A117.1 1986 or current edition (Accessibility and Usability by the Handicapped);
e. Occupational Safety and Health Administration (OSHA);
f. New York State Department of Labor Rules and Regulations;
g. New York State Energy Code;
h. Local Codes and Ordinances;
i. New York State Department of Health Requirements; and
j. New York State Department of Environmental Conservation

By their pleadings, however, Plaintiffs allege that "non-compliance with relevant statutes, regulations and ordinances was rampant," with workers being encouraged in a least some instances to forego filing of complaints for safety violations so as not to interfere with the recovery operation. Plaintiffs contend that this failure to enforce applicable safety standards and the failure to provide adequate and appropriate respiratory protection resulted in unprecedented exposure to various contaminants and toxins, and thus gave rise to the respiratory injuries that now plague so many of the rescue workers.

A. The Procedural Background

Indeed, just as the process of recovery and rebuilding began, those who participated in the efforts that made such recovery possible began to suffer from a host of respiratory ailments. In the months following September 11, and continuing until today, thousands of suits alleging respiratory injuries sustained as a result of violations of various state and federal safety laws and regulations were filed in New York State Court. To date, more than 3,000 cases alleging respiratory injuries have been filed, with thousands more in the offing. By their Master Complaint dated August 19, 2005, Plaintiffs asserted ten separate claims for recovery. Counts One and Two asserted claims pursuant to New York Labor Law. Counts Three and Four asserted claims pursuant to those provisions of the General Municipal Law allowing suits by injured and deceased firefighters and police officers and their representatives. Count Five stated a claim sounding in common law negligence. Counts Six and Seven asserted claims for medical monitoring and fear of cancer, respectively. Count Eight asserted a claim for fraud and misrepresentation. Count Nine asserted a claim for wrongful death. Finally, Count Ten asserted a claim on behalf of all derivative plaintiffs.

The Defendants removed the actions to federal court asserting jurisdiction under the Air Transportation Safety and System Stabilization Act ("ATSSSA" or "the Act"), 49 U.S.C. § 40101 note (2006). The Act provides a federal cause of action for actions for damages "arising out of" the terrorist-related aircraft crashes of September 11 and vests the District Court for the Southern District of New York with "original and exclusive jurisdiction over all actions brought for any claim (including any claim for loss of property, personal injury, or death) resulting from or relating to the terrorist-related aircraft crashes of September 11, 2001." ATSSSA § 408(b)(3). Motions to remand followed.

In determining the scope of federal jurisdiction provided under the Act, and noting the important concerns of federalism counseling against the imputation of a congressional intent to preempt an entire panoply of state law without a clearly expressed intent to provide such preemption, I held "that claims alleging respiratory injuries suffered at the World Trade Center site, up to and including September 29, 2001, [were] preempted by section 408 of the Act, and that claims incurred after that date or at different sites

[were] not preempted." Hickey, 270 F. Supp. 2d at 374. Cognizant, however, of the importance of a final determination as to the scope of my jurisdiction under the Act, I certified the order providing for federal jurisdiction for interlocutory appeal, 28 U.S.C. § 1292(b), and stayed the remand of cases not subject to federal jurisdiction pending review by the Court of Appeals. Id. at 381.

Proceeding on the assumption that federal jurisdiction existed as to all cases coordinated under 21 MC 100 absent a decision to the contrary by the Court of Appeals, I directed that the parties proceed on a limited, albeit extensive, course of discovery, focusing on Defendants' anticipated dispositive defense of immunity under state and federal law and with the aim of establishing a joint offer of proof, alleviating Plaintiffs of the burden of proving all factual averments. (*See* Case Management Order No. 3 ("CMO 3"), dated Feb. 7. 2005.) I ordered further that Plaintiffs file separate claims for each individual claimant, holding that the individual issues relevant to each claimant predominated over common issues. Thus, as the parties awaited decision by the Court of Appeals, the litigation continued to move forward.

As discovery relevant to the asserted defenses continued, the Court of Appeals delivered its decision. Although the Court concurred with my finding of jurisdiction as to respiratory injuries sustained at the World Trade Center site in the period between September 11 and September 29, 2001, and with my general proposition that, by enactment of the Act, Congress did not in fact intend to "displace the entire panoply of state law 'regulat[ing] the health and safety of the workplace,'" *McNally v. The Port Authority*, 414 F.3d 352, 379 (2d Cir. 2005), the Court expressed in dicta its disagreement with that portion of my Opinion remanding those actions asserting injuries arising at sites other than the World Trade Center and subsequent to September 29. See Id. at 380 ("We need not take the phrase 'relating to' to any metaphysical extreme in order to conclude that it encompasses the claims brought before the district court..., i.e., that airborne toxins and other contaminants emanating from the debris created by the crashes caused respiratory injuries to plaintiffs employed to sift, remove, transport, or dispose of that debris."). By Order of July 22, 2005, I adopted the reasoning of the Court of Appeals in McNally without prejudice to future submissions as to the extent of my jurisdiction pursuant to the Act.

Having adopted the reasoning of McNally, I turned to furthering the progress of the litigation. The parties appeared before me for a status conference on November 7, 2005 to address the status of limited discovery pursuant to CMO 3 and to address the practicability of adopting a joint offer of proof. At the conference it became readily apparent that any hope for a joint offer of proof had been dashed and I conceded that any further efforts in this regard would be futile. Although abandoning efforts to arrive at a joint offer of proof, I nevertheless acknowledged the dispositive nature of the immunity defenses and directed the parties to proceed, after completion of discovery, by motion. The parties subsequently completed the mandated course of pursuant to CMO 3 in early 2006, with Defendants then proceeding to make motions for judgment on the pleadings and motions for summary judgment on the basis of state and federal immunity.

B. The Pending Motions

The motions now pending before me concern whether Plaintiffs may proceed with their claims alleging respiratory injuries sustained at the World Trade Center site during the recovery and cleanup efforts following September 11 against the City and its Contractors (the "City Defendants"), [and] the Port Authority of New York and New Jersey (the "Port Authority")...The various Defendants assert immunity on the basis of state and federal statutory and common law immunity.

The City Defendants move for judgment on the pleadings and summary judgment. By their motion for judgment on the pleadings, the City Defendants assert immunity pursuant to the New York State Defense Emergency Act ("SDEA"), the New York Disaster Act ("Disaster Act"), and under state common law. By their motion for summary judgment, the City Defendants assert immunity pursuant to federal law.

...The Port Authority moves both for dismissal and for summary judgment on the basis of immunity pursuant to the SDEA and moves also for judgment on the pleadings asserting immunity under state common law...[T]he Port Authority move[s] for summary judgment on the ground of federal immunity.

IV. THE PREEMPTIVE EFFECT OF THE ATSSSA

The Air Transportation Safety and System Stabilization Act, 49 U.S.C. § 40101 note, sets forth the essential statutory framework for the consideration and treatment of claims arising from the terrorist attacks of September 11, 2001. Enacted following September 11, the Act was aimed at protecting the airline industry from potential economic collapse while simultaneously providing a simplified avenue for recovery by those immediate victims of the attacks who wished to take advantage of that method, and an exclusive federal remedy for all others having a cause of action. See Colaio v. Feinberg, 262 F. Supp. 2d 273 (S.D.N.Y. 2003); see also Hickey, 270 F. Supp. 2d at 362.

By amendments, the protections of the Act were extended to additional parties – of relevance here, the City of New York and the Port Authority of New York and New Jersey. ATSSSA §§ 408(a)(1), (3). The Act, as originally enacted, limited the liabilities of the airline industry defendants to the defendants' insurance coverages. The liability of the Port Authority in relation to its property interest in the World Trade Center was also limited to the aggregate of its liability insurance. ATSSSA § 408(a)(1). Separately, the liability of the City was limited to the higher of its liability insurance coverage or $ 350,000,000. ATSSSA § 408(a)(3).

As to plaintiffs, those who sustained injury as a direct result of the attacks, were given a special means of redress, in a specially authorized and appropriated Victim Compensation Fund ("VCF") or, alternatively, through litigation in the federal courts, specifically the United States District Court for the Southern District of New York. ATSSSA §§ 408(a), 408(b). Other plaintiffs, including those in the lawsuits which are the subject of this Opinion, were also granted rights of action arising from the ATSSSA but, since they incurred their injuries in the days and months following September 11, could not take advantage of the VCF and were limited to suit in the United States District Court for the Southern District of New York.

Plaintiffs argue that granting immunity to the Defendants, pursuant either to state or federal immunity doctrines, would contradict Congressional intent as expressed in the Act. Particularly, Plaintiffs argue, Congress appropriated one billion dollars to the City, enabling the City to establish a Captive Insurance Fund. Consolidated Appropriations Resolution, Pub. L. No. 108-7 (2003). Plaintiffs argue that Congress could not have legislated a $350 million ceiling on the City's aggregate liability and a $1 billion special litigation authorization without recognizing the City's exposure to the claims of those who engaged in the debris removal and cleanup work, for any other source of significant exposure is not apparent. This legislative purpose, Plaintiffs argue, preempts pre-existing state and federal law providing immunity to the City and its contractors, as well as to the other Defendants.

A. The Doctrine of Preemption

Determinations as to whether a particular federal law operates to preempt state law are "fundamentally a matter of Congress' intent." McNally, 414 F.3d at 371 (citing *English v. General Elec. Co.*, 496 U.S. 72, 78-79, 110 S. Ct. 2270, 110 L. Ed. 2d 65 (1990); *Hillsborough County v. Automated Med. Labs., Inc.*, 471 U.S. 707, 713, 105 S. Ct. 2371, 85 L. Ed. 2d 714 (1985)). The doctrine of preemption, rooted in the Supremacy Clause of the United States Constitution, U.S. Const. art. VI, cl. 2, provides that to the extent a state law conflicts with federal law, the contradictory state law is "without effect." Hickey, 270 F. Supp. 2d at 366. The Supreme Court has

recognized two general types of preemption, express and implied.

Congress can expressly preempt state law by explicitly providing for the displacement of state law. See *Hillsborough County v. Automated Med. Labs., Inc.*, 471 U.S. 707, 713, 105 S. Ct. 2371, 85 L. Ed. 2d 714 (1985). Even if Congress does not expressly preempt state law, state law may be impliedly preempted if Congress either 1) has legislated in a particular field so pervasively that no room is left for concurrent state legislation, see id.; or 2) has enacted federal laws that conflict with state laws, see *Geier v. American Honda Motor Co.*, 529 U.S. 861, 873-74, 120 S. Ct. 1913, 146 L. Ed. 2d 914 (2000). Id.

There are "no fixed meanings or shorthand rules to demarcate just where the traditional jurisdiction of state courts and application of state law must be ousted." Id. at 367. Courts are to adopt a fluid approach to setting the scope of a particular statute's preemptive effect.

Beginning with the presumption that Congress does not intend to displace state law, especially in traditional areas of state control..., courts go on to weigh whether the central motivations behind the federal law favor preemption – in other words, whether allowing state law to govern would undermine federal control or weaken the protections Congress intended to provide.

Id. As a general rule, preemption of areas traditionally subject to state regulation will not be found unless it is the "clear and manifest purpose of Congress." *CSX Transp. Inc. v. Easterwood*, 507 U.S. 658, 664, 113 S. Ct. 1732, 123 L. Ed. 2d 387 (1993) (quoting *Rice v. Santa Fe Elevator Corp.*, 331 U.S. 218, 67 S. Ct. 1146, 91 L. Ed. 1447 (1947)).

If, however, traditional state law principles contradict the purpose of a federal statute, preemption will be applied. In *Burnett v. Grattan*, 468 U.S. 42, 104 S. Ct. 2924, 82 L. Ed. 2d 36 (1984), the Supreme Court held that application of a restrictive administrative statute of limitations to bar litigation under the Civil Rights Act "ignore[d] the dominant characteristic of civil rights actions: they belong in court." Id. at 50 (citing *McDonald v. West Branch*, 466 U.S. 284,

104 S. Ct. 1799, 80 L. Ed. 2d 302 (1984)). Thus, the Court approved application of a more liberal statute of limitations. Id. at 55. The Court of Appeals for the Second Circuit reached a similar conclusion regarding the application of restrictive state statutes in a section 1983 action by a former university professor, noting that "[i]t would be anomalous for a federal court to apply a state policy restricting remedies against public officials to a federal statute that is designed to augment remedies against those officials, especially a federal statute that affords remedies for the protection of constitutional rights." *Pauk v. Board of Trustees of City Univ. of New York*, 654 F.2d 856, 862 (1981).

B. The Alleged Preemptive Effect of the ATSSSA and the Captive Insurance Fund

The Act expressly provides that those who sustained injury as a result of the September 11 attacks may gain redress from the VCF or, alternatively, seek it in traditional litigation, but only in the federal court for the Southern District of New York. Both manners of relief are, however, limited in several important respects. As an initial matter, eligibility for the VCF is limited to those who suffered physical injury or death as a result of the terrorist-related aircraft crashes and who were "on the planes or at the World Trade Center, the Pentagon, or the crash site at Shanksville, Pennsylvania at the time, or in the immediate aftermath, of September 11, 2001." Hickey, 270 F. Supp. 2d at 362 (quoting ATSSSA § 405(c)). Further, upon the submission of a claim under the VCF, individuals "waive the right to file a civil action...in any Federal or State court for damages sustained as a result of" the September 11 attacks. ATSSSA § 405.

To the extent persons who were injured as a result of September 11 chose not to participate in, or were not eligible for, the VCF, section 408 of the Act, as amended in November of 2001, establishes a "federal cause of action" as the "exclusive remedy for damages arising out of" the September 11 attacks. ATSSSA § 408(b)(1). The section further vests the United States District Court for the Southern District of New York

"with original and exclusive jurisdiction" over all claims "resulting from or relating to the terrorist-related aircraft crashes[.]" ATSSSA § 408(b)(3). Creation of a federal cause of action, however, did not serve to preclude application of state law. Instead, Congress expressly provided that, in actions brought pursuant to the Act, the "substantive law for decision...shall be derived from the law...of the state in which the crash occurred unless such law is inconsistent with or preempted by federal law." ATSSSA § 408(b)(2).

Section 408(a) operates also to limit the potential liability of defendants likely to be named in actions resulting from or relating to the September 11 attacks (other than the terrorists themselves or those who conspired with, or aided and abetted them). Thus, the liability of any "air carrier, aircraft manufacturer, airport sponsor, or person with a property interest in the World Trade Center" may not exceed "the limits of [their] liability insurance coverage." ATSSSA § 408(a)(1). The potential liability of the City of New York is also limited. By amendment to the Act, passed at the urging of the City, the Act expressly provides that "[l]iability for all claims...against the City of New York shall not exceed the greater of the City's insurance coverage or $ 350,000,000." ATSSSA § 408(a)(2). In addition, further to provide financial protection to the City of New York and its Contractors, another Act of Congress, passed one year later, appropriated 1 billion dollars in federal funding to "establish a captive insurance company...for claims arising from debris removal[.]" Consolidated Appropriations Resolution, 2003, Pub. L. No. 108-7 (2003). The Captive Policy insures the City and its Contractors against potential liability, as well as associated costs and expenses, and affords the City and its Contractors fully-funded financial protection against all potential claims. WTC Captive Insurance Company Liability Insurance Policy, § 2.04 ("Defense of Suits and Related Defense Costs") (the "Captive Insurance Fund").

C. Discussion

As noted earlier, I begin "with the presumption that Congress does not intend to displace state law[.]" Hickey, 270 F. Supp. 2d at 366. Pre-emption will not be found unless it is the "clear and manifest purpose of Congress." CSX Transp. Inc., 507 U.S. at 664 (quoting *Rice v. Santa Fe Elevator Corp.*, 331 U.S. 218, 67 S. Ct. 1146, 91 L. Ed. 1447 (1947)).

Here, neither the plain language of the statute, nor its purpose, evince a Congressional intent to supplant the application of state law principles. To the contrary, Congress expressly provided that the "substantive law for decision" in any suit relating to the September 11 attacks "shall be derived from the law...of the state in which the crash occurred," unless "inconsistent with or preempted by federal law." ATSSSA § 408(b)(2). Application of such state substantive law is not "inconsistent" with federal law as embodied in the Act. See 7WTC, 2006 U.S. Dist. LEXIS 749, 2006 WL 62019 at *6-8 (granting immunity to the City of New York pursuant to the New York State Defense Emergency Act as to claims asserted by Consolidated Edison's insurers for damage resulting from the collapse of Seven World Trade Center). Nothing in the Act or its legislative history suggests that defenses against potential lawsuits are prohibited. It is Plaintiffs' burden to show that Congress intended to remove reliance on such defenses insofar as they directly contradict Congressional intent, see *Silkwood v. Kerr-McGee Corp.*, 464 U.S. 238, 255, 104 S. Ct. 615, 78 L. Ed. 2d 443 (1984), and Plaintiffs are unable to meet such a burden.

The limited purpose of section 408 was to provide an efficient and rational means for the adjudication of claims relating to or resulting from September 11. It was not intended to guarantee compensation to those injured by the September 11 attacks. The possibility that compensation might ultimately be denied to those who declined to, or were ineligible to, participate in the Fund was contemplated by Congress and expressed by Senator John McCain during the Congressional debate: "To ensure that the victims and families of victims who were physically injured or killed on September 11th are compensated even if courts determine that the airlines and any other potential corporate defendants are not liable for the harm...the [Act] also creates a

victims' compensation fund." 147 Cong. Rec. S9589091 at S9594 (Sept. 21, 2001 (statement of Sen. McCain); see also Colaio, 262 F. Supp. 2d 273.

Thus, I decline to extend the scope of federal preemption under the Act to preclude application of otherwise available state law immunity defenses. I further decline to bar application of federal immunity doctrines as contradictory to the alleged compensatory purpose of the Act. The preemptive effect of the Act serves to displace, "not the substantive standards governing liability, but only the state-law damages remedies," McNally, 414 F.3d at 380, in the sense of placing a monetary limit to those remedies. Risks are inherent in the very nature of litigation, even where claims are brought pursuant to express Act of Congress. In re September 11 Litig., Opinion and Order Regulating Testimony at Depositions, 236 F.R.D. 164 (S.D.N.Y. 2006). Although "[i]t would be anomalous for a federal court to apply a state policy restricting remedies," Pauk, 654 F.2d at 862, to a statute designed to expand remedies, the ATSSSA is not such a statute. Nor is it the role of the federal courts to eliminate entirely the risks inherent to litigation. By enactment of the ATSSSA, Congress intended to create a federal forum for the adjudication of claims arising out of the September 11 attacks, to consolidate and rationalize all such suits in a single forum, and to protect defendants against a multiplicity of actions in multiple forums. To infer from the Act that all Plaintiffs should be entitled to compensation would run counter to the otherwise clearly expressed intent of Congress. It is an inference that I decline to make.

Having thus ruled, I concede a sense of disquiet. I have not come across federal legislation of the type we have here – providing a limit to how much aggregate liability the City can incur – except in bankruptcy or insolvency practice, and there is no suggestion of that here. Nor, except perhaps in the case of atomic energy disasters, have I encountered a contribution from the federal fisc to create a litigation defense fund, and even in that case there is a significant difference. Here, the City was given one billion dollars to assure against liability, loss or expense, while in the case of atomic energy disaster, there is a provision only of federal insurance. 42 U.S.C. § 2210 (2005).

What, one may ask, was the City, and Congress, contemplating if not the kinds of suits as those over which I preside? Clearly, provisions of insurance do not concede liability, as Defendants stress, but never have we seen provisions as those we have here.

Provisionally, and subject to further consideration as these cases progress, the case for preemption has not been made. But neither is there to be a blank check to enrich lawyers with endless stratagems of motions and delays. Congress legislated for the public welfare, and that translates in this instance to speedy proceedings towards the merits, to distinguish between cases proving injuries arising from the terrorist-related aircraft crashes and from the conditions resulting from those crashes, and cases lacking proof of such a right to recover, and to fashion remedies appropriate to those showing a right to relief. These are the goals that I set for the parties, and for myself in presiding over these cases, to bring these cases to a place where they can be settled or tried as speedily as the interest of justice allows.

V. THE MOTIONS FOR JUDGMENT ON THE PLEADINGS – STATE IMMUNITY

The various Defendants move for judgment on the pleadings on the basis of various state statutes and doctrines of common law providing for immunity, and assert that liability may not attach for actions taken in response to the terrorist attacks of September 11. Specifically, the Defendants, in various groupings, assert immunity under the New York State Defense Emergency Act, the New York State and Local Natural Disaster and Man-Made Disaster Preparedness Law, and New York common law.

A. The Standard of Review

The standard employed in reviewing a motion for judgment on the pleadings pursuant to Rule 12(c), Fed. R. Civ. P., is the same as that employed for motions brought pursuant to Rule 12(b)(6), Fed. R. Civ. P.

A Rule 12(b)(6) motion requires the court to determine if the plaintiff has stated a legally sufficient claim. A motion to dismiss under Rule 12(b)(6) may be granted only if "it appears beyond doubt that the plaintiff can prove no set of facts in support of his claim which would entitle him to relief." *Conley v. Gibson*, 355 U.S. 41, 45-46, 78 S. Ct. 99, 2 L. Ed. 2d 80 (1957); *Branum v. Clark*, 927 F.2d 698, 705 (2d Cir. 1991). The court's function is "not to assay the weight of the evidence which might be offered in support" of the complaint, but "merely to assess the legal feasibility" of the complaint. *Geisler v. Petrocelli*, 616 F.2d 636, 639 (2d Cir. 1980). In evaluating whether plaintiff may ultimately prevail, the court must take the facts alleged in the complaint as true and draw all reasonable inferences in favor of the plaintiff. See *Jackson Nat'l Life Ins. Co. v. Merrill Lynch & Co.*, 32 F.3d 697, 699-700 (2d Cir. 1994). If matters outside of the pleadings are considered by the court, the motion shall be treated as motion for summary judgment, Fed. R. Civ. P. 56, and all parties should be given opportunity to submit relevant materials. Fed. R. Civ. P. 12(c).

B. The New York State Defense Emergency Act

Enacted in 1951, at the height of the Cold War, the New York State Defense Emergency Act (the "SDEA") provides for a comprehensive response to attacks upon the United States, and the State of New York, by coordinating the private and public sectors to "make possible the recovery of the people and the rehabilitation of the economic and social life of the state following any such attack." N.Y. Unconsol. Law SDEA § 9102-a (McKinney 2006). . In the interest of ensuring that public and private entities will work aggressively to prepare for and to respond to attacks, the SDEA provides immunity for actions taken "in good faith carrying out, complying with or attempting to comply with" any law or order requiring such a unified response and relating to "civil defense." SDEA § 9193. See also 7WTC, 2006 U.S. Dist. LEXIS 749, 2006 WL 62019 at *6.

The City Defendants, the Port Authority, the Silverstein Defendants, and Con Ed, join in asserting immunity pursuant to the SDEA for actions taken in the wake of the September 11 attacks. Defendants assert that all actions they took in response to the attacks were taken in good faith, to comply with the Declarations of Emergency issued by the President, the Governor, and the Mayor, that all actions related to the "civil defense," and that their actions thus fall within the express grant of immunity provided by the SDEA.

1. The Immunity Provision of the SDEA

The immunity provision of the SDEA provides for protection to various entities – government, individuals, partnerships, and corporations – who, in good faith, are engaged in activities in preparation for and responsive to attacks on the State, pursuant to the law or to duly promulgated rules, regulations or orders. The protection extends to preparations against attacks, see 7WTC, 2006 U.S. Dist. LEXIS 749, 2006 WL 62019 at *6-*8, and to "activities...following attacks, including, as part of the definition of "civil defense," essential debris clearance, immediately essential repairs of damaged facilities, and the restoration of essential services. SDEA § 9103(5). The SDEA provides that parties engaged in such civil defense activities "shall not be liable for any injury or death to persons or damage to property as the result thereof."
SDEA § 9193(1).

2. The Continued Vitality of the SDEA

Plaintiffs argue that the SDEA was enacted during the height of the Cold War as the nation braced for what seemed to be inevitable nuclear attack, and is no longer viable. Plaintiffs assert further that the SDEA was effectively repealed by subsequent passage of the New York State and Local Natural Disaster and Man-Made Disaster Act, N.Y. Exec. Law §§ 20-29-g (McKinney 2006) (the "Disaster Act"). Plaintiffs' arguments are without merit and run counter to the plain language and express purpose of both the SDEA and the Disaster Act. The dangers to our nation, state and city that led to the passage of the SDEA are dangers that exist today. In 1947, we were concerned with the warlike stance of the Soviet Union and international communism. Today, we are concerned with international terrorism. Prime Minister Nikita S. Khruschev at the United Nations threatened to bury us, pounding his shoe on the lectern for emphasis. The terrorists who sent their bombers into the World Trade Center buried almost three thousand of us, and there are too many threats from too many sources to allow us to forget our concerns in sanguine obliviousness.

Enactment of the SDEA was responsive to a perceived threat "of atomic conflict with communist nations and the concomitant need for a comprehensive plan to ensure the survival of the State's citizens in the event of foreign attack." *Fitzgibbon v. County of Nassau*, 147 A.D.2d 40, 541 N.Y.S.2d 845, 847 (2d Dep't 1989). As I observed in an earlier decision, the statute "is not limited to a particular time or a particular threat." 7 WTC, 2006 U.S. Dist. LEXIS 749, 2006 WL 62019, at *6. I ruled that the legislative history and plain language of the statute show that it was based upon "a broad notion of 'enemy attack' using any kind of weapon capable of inflicting mass injury." Id. Thus, the statute gives broad meaning to the term "attack," defining it as:

> [a]ny attack, actual or imminent, or series of attacks by an enemy or a foreign nation upon the United States causing, or which may cause, substantial damage or injury to civilian property or persons in the United States in any manner by sabotage or by the use of bombs, shellfire, or nuclear, radiological, chemical, bacteriological, or biological means or other weapons or processes.

SDEA § 9103(2). By its plain language the SDEA "is not limited to a nuclear attack or particular enemy." Daly, 793 N.Y.S.2d at 716. Here, it is clear that the hijacking and subsequent crashes of American Airlines Flight 11 and United Airlines Flight 175 into the Twin Towers, resulting in the destruction of a once great financial and business center and in the loss of thousands of lives, constitute an "attack" as contemplated under the SDEA. The SDEA remains viable legislation.

With regard to the Disaster Act, as enacted in 1951, its application was limited to natural disasters such as "flood, drought, tidal wave, fire, hurricane, earthquake, windstorm, or other storm, landslide, or other catastrophe," and not disasters resulting from "an enemy attack as defined in the New York state defense emergency act." N.Y. Executive Laws Art. 2 § 10 (repealed). Almost thirty years later, in 1978, the definitions were expanded to include disasters resulting from "man-made causes." N.Y. Cons. Laws, Executive Laws Art. 2(b)§ 20(2)(a). But nothing in the plain language of the statute or in its legislative history suggests an intent to override the immunities provided by the SDEA. The two statutes must be read together, harmoniously. See *J.E.M. AG Supply, Inc. v. Pioneer Hi-Bred Int'l, Inc.*, 534 U.S. 124, 143-44 (2001) ("[W]hen two statutes are capable of coexistence, it is the duty of the courts, absent a clearly expressed congressional intention to the contrary, to regard each as effective.") (quoting *Morton v. Mancari*, 417 U.S. 535 (1974)) (internal quotation marks omitted). Thus, I hold that the SDEA remains a valid basis for asserting immunity...

3. Qualifying Laws Under the SDEA

The grant of immunity provided by the SDEA is limited to actions taken "in good faith carrying out, complying with, or attempting to comply with" "any law, any rule, regulation or order duly promulgated or issued pursuant to [the SDEA]" and "relating to civil defense[.]" SDEA § 9193(1). All agree that neither the Mayor nor the Governor of New York invoked any provisions of the SDEA in their various declarations of emergency following September 11. But that does not mean that the governmental entities, and the persons, firms and corporations that were engaged in debris removal and cleanup of the World Trade Center site did not act pursuant to the laws, regulations and orders issued by the President, Governor and Mayor to organize the carrying out of such activities.

The SDEA defines "law" broadly to include "[a] general or special statute, law, city, or village charter, local law, ordinance, resolution, rule, regulation, order or rule of common law." SDEA § 9103(17). The Executive Orders issued by the Mayor of the City of New York and the Governor of the State of New York are precisely such laws. By their executive orders, the Mayor and the Governor ordered the City and State civil defense agencies to engage in civil defense activities. "Civil Defense Activities" are defined by the statute to include immediately essential repairs and restoration of essential services. The City and State agencies acted also pursuant to the common law, for when an emergent disaster threatens society as a whole, the doctrine of *salus populi supreme lex* (the welfare of the people is the highest law) requires the government to act, enlisting persons, firms and corporations in the private sector to eliminate the threat to society and restore society's ability to function. *See* Matter of Cheesebrough, 78 N.Y. 232 (1879). *Salus populi* means "that society has a right that corresponds to the right of self-preservation in the individual, and it rests upon necessity because there can be no effective government without it." Daly, 793 N.Y.S.2d at 721-22 (quoting 20 N.Y. Jur. 2d. Const. Law § 192). *Salus Populi* and the SDEA coincide, for both encourage immediate action to preserve society. *See* id. at 718.

Plaintiffs argue that the Contractor Defendants, since they were engaged by private contracts and since they acted, not pursuant to the Mayor's and Governor's Executive Orders, but pursuant to their contracts, were not "carrying out, complying with or attempting to comply with any law, any rule, regulation or order duly promulgated or issued pursuant to [the SDEA]." SDEA § 9193(1). Plaintiffs cite *Abbott v. Page Airways* 245 N.E.2d 388 (N.Y. 1969), where the New York Court of Appeals declined to extend SDEA immunity to a helicopter company that was engaged by the Director of the Civil Defense Office of Monroe County, New York, pursuant to contract, to hover over a riotous demonstration in the City of Rochester to enable the Director to survey the disorder. The chartered helicopter crashed, killing four people, injuring several others, and causing extensive property damage. Id. The helicopter company claimed immunity pursuant to the SDEA. The New York Court of Appeals held, however, that the helicopter company was not entitled to immunity from suit pursuant to the SDEA, as the defendant was "doing nothing more...than engaging in its regular course of business of providing air transportation for hire." Id. at 391. "[T]he policy of New York State," the Court of Appeals held, "has been to reduce rather than increase the obstacles to recovery of damages for negligently caused injury or death." Id.

Plaintiffs contend that the Contractor Defendants were paid for their services, and thus were "doing nothing more...than engaging in [the] regular course" of their respective businesses. Id. However, one cannot compare the havoc caused by the September 11 attacks, and the concomitant need for private assistance with the riot in Rochester in the mid-1960s. The declarations of emergency following the September 11 attacks authorized the City's remedial actions and its call for the assistance and cooperation of scores of private entities and thousands of workers. In contrast, the helicopter company in Abbott was involved in the emergency response in the most limited and peripheral sense. To deny immunity here based on the mere fact of payment for services rendered, without considering the circumstances presented by the September 11 attacks, would run counter to the plain language of the SDEA, expressly [**82] contemplating participation by private actors and providing an intentionally expansive definition of legal authorities. Plaintiffs' argument is more properly addressed to the question whether Defendants were undertaking civil defense activities in good faith, and it is to this argument that I now turn.

4. Civil Defense Activities and the Requirement of Good Faith

A. Civil Defense Activities

The grant of immunity under the SDEA is limited to "civil defense" "activities and measures designed or undertaken...to minimize the effects upon the civil population caused or which would be caused by an attack." SDEA § 9193. Where the activities are taken in response to an attack, civil defense measures are defined to include, among a wide range of measures, "essential debris clearance," "immediately essential emergency repair or restoration of damaged vital facilities," "means and methods for the recovery and rehabilitation of the state," and "the restoration of essential community services, industrial and manufacturing capacity, and commercial and financial activities in the state." Specifically, civil defense activities responsive to an attack include: activities for fire fighting; rescue, emergency medical, health and sanitation services; monitoring for radiation and other specific hazards of special weapons; decontamination procedures; unexploded bomb reconnaissance; essential debris clearance; emergency welfare measures; immediately essential emergency repair or restoration of damaged vital facilities; the implementation of the means and methods for the recovery and rehabilitation of the state; effective utilization of all persons and materials; care and shelter for those made homeless; distribution of stockpiled food, water, medical supplies, machinery and other equipment; the preservation of raw materials; the restoration of essential community services, industrial and manufacturing capacity, and commercial and financial activities in the state; and the resumption of educational programs[].
SDEA § 9103(5).

"[T]he framers [of the SDEA] undoubtedly anticipated that the various civil defense functions contemplated by the Act would be undertaken *during the rush of emergency.*" Fitzgibbon, 541 N.Y.S.2d at 849 (emphasis added). Thus, in Fitzgibbon, the court excluded routine patrol functions from the immunity provisions of the Act. The court held that although directing traffic in the event of an attack would be covered by SDEA immunity, the SDEA would not grant immunity to the actions of an auxiliary police officer engaged in otherwise routine patrol functions. Id. at 849-50. The court reasoned that if a defendant was engaged in conduct otherwise ordinary and routine to its business, the mere presence of an emergency condition would not operate to mandate the extension of SDEA immunity to otherwise routine conduct. *See* Abbott, 245 N.E.2d 388 *see* also 7 WTC, 2006 U.S. Dist. LEXIS 749, 2006 WL 62019, at *6 (distinguishing between "routine conduct for which there is no immunity under the [S]DEA, and conduct that a municipality performed while engaged in a civil defense function"). Thus, the statute defines as a civil defense activity, not all debris clearance, but only "essential" debris clearance; not all repair or restoration of damaged vital facilities, but only that which is "immediately essential" and which constitutes "emergency" repair or restoration; and not the restoration of all community services, but only those which are "essential."

B. The Requirement of Good Faith

Civil defense actions pursuant to the SDEA must also be taken in "good faith carrying out, complying with or attempting to comply with" the legal authorities enumerated therein in order for immunity to attach. SDEA § 9193(1). In this respect, Defendants again urge that the relevant inquiry is not *how* the Defendants acted, but rather *why* the Defendants acted. According to Defendants, the good faith requirement refers only to Defendants' compliance, or attempted compliance, with a civil defense law within the meaning of the SDEA. Moreover, Defendants assert that, regardless of the proper form of inquiry, Plaintiffs have in any event failed to make a showing of bad faith sufficient to defeat the pending motion for judgment on the pleadings.

To date, no cases, either federal or state, have directly addressed the nature of the SDEA's good faith requirement. The courts have, however, interpreted the parallel good faith provision of the SDEA's precursor statute, the War Emergency Act ("WEA"), in a series of cases addressing claims arising from injuries sustained during the mandatory blackouts of the 1940s. Defining good faith as "an honesty of intention," *Smith v. Town of Orangetown*, 57 F. Supp. 52, 55-57 (S.D.N.Y. 1944), the courts have focused their inquiry on whether the individual defendant's alleged negligence resulted from a good faith attempt to comply with an order or regulation cognizable under the WEA.

Thus, in Smith, the District Court affirmed the jury's grant of immunity to a police officer who, while driving during a blackout, drove into a group of soldiers, killing one and wounding several. Id. at 53. The analysis adopted by the trial court required a two part inquiry: first, whether the defendant had acted negligently, and, if so, whether the defendant had been pursuing his duties in good faith at the time of the alleged negligence. Id. at 54. So long as the defendant was acting in a good faith attempt to comply with his duties, liability for negligence would not attach. Id.

The grant of immunity under the WEA, however, was not absolute. In *Jones v. Gray*, 267 A.D. 242, 45 N.Y.S.2d 519 (App. Div. 1943), a case factually similar to Smith, an air raid warden, while allegedly responding to a blackout, crashed his car into another vehicle, killing several people. Id. at 520. The defendant argued that he was immune from liability pursuant to the WEA. The court rejected this defense finding that the warden was not "in good faith...attempting to perform his duties as an air warden at the time of the collision," but instead was on what could only be characterized as a fateful "joy ride." Id. at 523.

Good faith thus may not be inferred simply from the fact that, at the time of the allegedly negligent acts, the Defendants were acting in a manner responsive to a declaration of emergency. Nor may cases granting immunity for isolated incidents of negligence be extended as to allow for the wholesale grant of immunity to [*553] incidents of negligence that occurred, not at isolated intervals, but rather continually over a period of several months. Although the question of why the Defendants acted may ultimately prove critical to determining immunity under the SDEA, the interests of justice mandate also a consideration of *how* the Defendants acted. Special consideration must also be given where the obligation of good faith runs, not to third parties potentially injured by the attacks, but instead to those directly engaged, for all intents and purposes as employees, in the recovery effort. Those who engage others in their service owe definite and specific obligations which may not easily be abandoned.

C. Discussion

Defendants argue that as long as they were attempting to comply with a legal authority "relating to civil defense" and doing so "in good faith," they should be entitled to SDEA immunity, and should not be accountable for any deficiencies or negligence in how they carried out their activities, even to the workmen to whom they would otherwise owe a duty. Under this interpretation of the SDEA, immunity should extend to all of Defendants' activities that were responsive to the civil defense emergency created by the attacks, from beginning to end, as they concerned the implementation of "means and methods for the recovery and rehabilitation" of the City. SDEA § 9103(5). Thus, there should not be an inquiry into whether or when an emergency condition existed or ceased to exist, or whether the specific actions undertaken were themselves in good faith. The plain language of the statute, however, considered together with the unique and important public policies implicated by the instant litigation, counsel against adoption of Defendants' arguments. The proper analysis requires consideration of potential temporal limitations on the grant of immunity as well as a consideration of the obligation of good faith that extends beyond mere implementation.

Critically, the statute extends immunity, not to all activities following an emergency, but limits immunity only to those activities that are, in themselves, "essential," or "immediately essential," and "emergency" in nature. The few cases are in accord, limiting the statute and generally finding liability....

The limitation of immunity to acts undertaken in the context of an emergency is essential to ensure "the least possible interference with the existing division of the powers of the government and the least possible infringement of the liberties of the people[.]" SDEA § 9102. Limitations on the right to seek redress for injuries sustained must be strictly limited, extended only where necessary to restore the ability of society again to begin functioning. Defendants argue that immunity is necessary to encourage companies to volunteer their efforts; the fear of lawsuits,

Defendants argue, otherwise will cause them to hold back. But individual workers also are essential to the response effort, and those who claim injury are the very individuals who, without thought of self, rushed to the aid of the City and their fallen comrades. Their efforts also must be encouraged, for their fear of injury without redress can cause such volunteers also to hold back. A delicate balance has to be struck, one that encourages both companies and individuals to come forward to clear the effects of the blows to society.

These two competing interests, namely the need to allow for an immediate and effective response to an attack on the state as against the need to ensure persons injured the right of redress, may be considered as existing along a spectrum. Where the emergency condition predominates, the interest of protecting those engaged to assist in the emergency response must necessarily be of less concern. In the rush of an emergency, the ability and capacity to adequately implement and effect necessary safety procedures is greatly reduced. The SDEA's immunity provision operates to ensure that fear of liability will not operate to dissuade government and private entities from responding to a disaster, even in the absence of otherwise mandated safety protocols and procedures. However, as the emergency condition fades, as the rights and obligations of persons and entities engaged in the response effort become regulated by contract, as procedures and protocols are implemented to protect against potential dangers, the need for immunity diminishes and the obligations and duties otherwise imposed once again must be protected.

There is no bright-line demarcation between that which is emergent and that which is done in the more normal course. In a jurisdictional setting, I proposed to create a two-week period after September 11, to September 29, 2001, at which time, pursuant to order of the Mayor, efforts turned from rescue of victims to clearance of debris. *See* Hickey, 270 F. Supp. 2d at 374 (holding that causes of action of workers claiming respiratory injury arose under the Air Transportation Safety and System Stabilization Act before September 29, 2001, and under New York

State law thereafter). The Court of Appeals, however, disapproved of any such bright-line and ruled that claims throughout the period of the rescue and recovery effort were covered by the Act. *See* McNally, 414 F.3d at 380-81. The New York Supreme Court applied the bright-line rule of Hickey to claims under the SDEA, but the questionable efficacy of Hickey casts doubt also on the decision of the New York Supreme Court. Daly, 793 N.Y.S.2d at 719.

That an emergency existed for some time following September 11 is without question. But whether the emergency lasted for days, or weeks, or months, and in connection with which precise activities, are fact-intensive questions, not possible to answer in connection with a Rule 12 motion addressed to the pleadings. Clearly, the desperate search for survivors constituted an unprecedented setting, and nothing about what took place in connection with such a search could be considered ordinary or routine – at least, so it would seem. But at some point after the attacks, the emergency conditions subsided, the extraordinary settled into a routine, and the Defendants' argument for immunity loses cogency under the teachings of the New York cases.

The difficulty of discerning the precise point at which the emergency condition ceased to exist is further complicated by the SDEA's parallel requirement of good faith. The SDEA mandates that all actions pursuant to the Act be undertaken "in good faith carrying out, complying with or attempting to comply with" relevant legal authorities SDEA § 9193(1). The factual record before me shows clearly that, in responding to the September 11 attacks, Defendants endeavored to develop a viable health and safety plan for workers at the site. As Defendants put it, the efforts to ensure worker health and safety were extensive and included: (1) employing health and safety experts; (2) participating in daily health and safety meetings; (3) implementing health and safety protocols, plans, and agreements; (4) adopting a clear and consistent respiratory protection policy, devised by experts and based on proven monitoring and analytical techniques; (5) sharing environmental information with intergovernmental agencies; (6) consulting

with top experts in the selection of respirators; (7) overcoming Herculean challenges in acquiring and distributing respirators to workers from the outset; and (8) consistently making an effort to ensure worker compliance with respirator use directives.

The pleadings, however, show also that there were critical lapses in the enforcement of safety standards and in the dissemination of vital information about the safety of the air at Ground Zero to those most affected, the workers themselves. As of January 2002, compliance rates for respirator usage fell below twenty-nine percent, and Plaintiffs allege that "[r]espirator fit testing done at and around the WTC site was illusory, at best," and, in any event, "did not meet OSHA standards."

Accepting all allegations alleged in the complaint as true, as I am required to do on a motion pursuant to Rule 12(c), Fed. R. Civ. P., I cannot determine at this stage that Defendants' actions all were in good faith, nor which acts might have been, and which appear not to have been, conducted in good faith. There is no standard in case law to help me determine if good faith should be considered according to the motives with which actions were undertaken, in which case good faith likely would be found, or if good faith should be considered in the context of the particular acts as to which the Defendants allegedly were negligent, in which case standards of good faith would tend to merge into standards of negligence. Regardless of Defendants' efforts and motives to develop safety standards, if the standards were not implemented or enforced in a systematic and reasonable manner, liability well may attach. It may not be sufficient for Defendants to assert simply that they were acting in a good faith attempt to comply with the various declarations of emergency, without considering also what they did, and did not, do. The cases over which I preside involve, not third parties claiming injury because of conduct of an emergency provider, but by the emergency providers themselves, based on allegations that those in charge of their respective workplaces did not provide the training and equipment available and necessary and appropriate to allow them to do their work and to protect them against the dangers incident to their workplaces. Whether Plaintiffs will be able to make a showing of bad faith is a question of fact for the jury that may not be summarily determined at this stage in the litigation. *See* Smith, 57 F. Supp. at 55 T]he existence of good faith is an issue for the jury to decide in any case[.]").

All of this teaches that one has to be cautious in considering a Rule 12(c) motion for judgment on the pleadings on the basis of a fact-laden defense. Defendants' motion accepts Plaintiffs' allegations of respiratory injury as fully proved, and so they must be in the context of a Rule 12(c) motion for judgment on the pleadings. Fed. R. Civ. P. 12(c). A valid affirmative defense defeats even a well-pleaded claim, and the SDEA certainly is a valid defense, but the scope and extent of applicability of the defense is unclear, and a judge must be hesitant to apply such a defense beyond its narrow parameters to a valid claim of injury.

C. The Argument of Immunity Under the New York Disaster Act

The New York State and Local Natural Disaster and Man-Made Disaster Preparedness Law (the "Disaster Act"), N.Y. Exec. Law §§ 20-29-g (McKinney 2006), provides another basis for the City Defendants' arguments of immunity.

1. The Limited Scope of the Disaster Act's Application

The Disaster Act provides that, upon the "occurrence or imminent threat of widespread or severe damage, injury, or loss of life or property, resulting from any man-made or natural causes," Disaster Act § 20(2)(a), the "chief executive of any political subdivision is... authorized and empowered to and shall use any and all facilities, equipment, supplies, personnel and other resources of his political subdivision in such manner as may be necessary or appropriate to cope with the disaster or any emergency resulting therefrom." Disaster Act § 25(1). The Disaster Act provides also for immunity for discretionary functions performed by a political subdivision's officers

and employees: "A political subdivision shall not be liable for any claim based upon the exercise or performance or the failure to exercise or perform a discretionary function or duty on the part of any officer or employee in carrying out" its provisions. Disaster Act § 25(5). Political subdivisions are given immunity for those of its acts responding to a disaster that: (1) constitute discretionary functions or duties; and (2) are performed by its officers or employees.

Thus, the Disaster Act protects political subdivisions, and immunizes then against lawsuits based on discretionary acts of their political officers and employees. There also are additional limitations relating to the Governor's powers to suspend laws, but not to suspend those laws that safeguard the public health and welfare. Section 29-a of the Disaster Act grants the Governor of the State of New York the authority to "suspend specific provisions of any statute, local law, ordinance, or orders, rules or regulations... during a state disaster emergency, if compliance with such provisions would prevent, hinder, or delay action necessary to cope with the disaster." Disaster Act § 29-a(1). By implication, the obligations of the government to act consistently with statutory or regulatory duties remain in effect unless the Governor orders the suspension of such laws. Furthermore, the Governor may not suspend a law which safeguards public health and welfare and which is not "reasonably necessary" to the responsive effort.

No suspension [of law] shall be made which does not safeguard the health and welfare of the public and which is not reasonably necessary to the disaster effort.

Disaster Act § 29-a(2)(b).

The rationale of the Disaster Act is clear. Political subdivisions of the State are to act to preserve the health and welfare of the public, but not at the expense of the public's health and welfare. The firemen, policemen and construction workers worked at the site to further the health and welfare of the public, but the political subdivisions that regulated their work were not at liberty to sacrifice their health and welfare, except possibly if there was no other way to combat the emergency effects of the disaster. And, significantly, Governor Pataki did not exercise his authority under section 29-a of the Disaster Act to suspend the protections of the New York Labor Law governing the health and welfare of workers at construction sites, presumably because "compliance with such provisions [did not] prevent, hinder, or delay action necessary to cope with the disaster." Disaster Act § 29-a(1). It follows that the provisions of the New York Labor Law relevant to worker safety remained in effect throughout the duration of the rescue and recovery effort, and that it was incumbent on those responsible for supervising the site to comply with those laws.

Thus, the City Defendants cannot argue cogently that they should be immunized for their decisions and conduct in putting out the fires and removing the debris from the rubble of Towers One, Two and Seven no matter what they did. As I ruled in relation to the SDEA, specific actions have to be evaluated according to time, place and necessity. There is likely to be a setting where immunity should be upheld, but the decisions cannot be made on motion, without a complete record. There is nothing in the Disaster Act that extends immunity beyond that provided by the SDEA, except perhaps if the Governor had found the need to suspend the New York Labor Law and other regulatory protections of the workplace.

The City Defendants argue that a liberal extension of immunity is important to encourage public actors to respond fully to a public disaster, for "[a] public officer, haunted by the specter of a lawsuit, may well be subject to the twin tendencies of procrastination and compromise to the detriment of the proper performance of his duties." *Rottkamp v. Young*, 21 A.D.2d 373, 249 N.Y.S.2d 330, PIN (App. Div. 1964). But large numbers of working people, as well as City officials, must work together to restore a society struggling to recover from a disaster. Extending too large an immunity to official actions, at the expense of the health and welfare of working people, will discourage one group of people while encouraging another group of people. "Neither the City nor any other entity has discretion to violate an applicable statute," Daly, 793

N.Y.S.2d at 721, least of all the strong policy of New York protective of worker safety. Hickey, 270 F. Supp. 2d 357. The Disaster Act should not be read to create blanket immunity.

Like the SDEA, the Disaster Act's grant of immunity is limited to actions that are emergent in their own quality, and not only because those actions were intended to alleviate a previous emergency condition. The Disaster Act authorizes actions upon the "occurrence or imminent threat of" a disaster, Disaster Act § 20(2)(a), and which are "necessary or appropriate to cope with the disaster or any emergency resulting therefrom." Disaster Act § 25(1). Immunity is thus provided, not necessarily to any actions taken in consequence of a disaster, but instead only to those actions which are necessary to cope with the disaster. Immunity is to be conferred sparingly, and is not to be a cloak excusing all accountability. Abbott, 23 N.Y.2d 502, 245 N.E.2d 388, 297 N.Y.S.2d 713. The issue is fact intensive, and fixing the precise point when emergency efforts became routine is difficult. A proper resolution of the issue requires a properly developed factual record. *See* McNally, 414 F.3d 352.

For all these reasons, I decline to grant immunity under the Disaster Act beyond that which is available under the SDEA.

2. The Extension of Disaster Act Immunity to Non-Government Actors

The Disaster Act expressly limits its grant of immunity to actions taken by political subdivisions and their employees and officers. The City Defendants argue, however, that the grant of immunity should be extended to the private contractors whom the City engaged, and that doing so would be consistent with the policy of the Disaster Act and basic principles of common law. The argument, insofar as it argues for an immunity wider than that available under the SDEA, is without merit.

The immunity provision of the Disaster Act, section 25(5), offers protection only to "political subdivision[s]," and only when their officers or employees "exercise or perform" discretionary functions or duties." *See* also Disaster Act § 29-b(2), (3) (granting immunity to local civil defense forces operating under the direction and command of the city civil defense director as authorized by the governor). Had the legislature intended for immunity to extend to private actors, it could easily have so provided. I decline to legislate that which the legislature did not provide.

D. New York State Common Law Immunity

The City Defendants, together with the Port Authority and Con Edison, seek immunity also on the basis of state common law, arguing that they performed uniquely governmental functions in the aftermath of the September 11 attacks. In general, although the common law does provide immunity for acts that are "completely sovereign in nature and completely foreign to any activity which could be carried out by a private person," *Williams v. State*, 90 A.D.2d 861, 456 N.Y.S.2d 491, 493 (App. Div. 1982), there are important limitations. Common law immunity is limited to governmental acts that are discretionary in nature. Furthermore, the common

law does not extend immunity to private entities that are sued for negligent or willful behavior.... Important public policy considerations require that the immunity granted to private entities acting at the behest of government entities be limited. Royal Ins. Co., 918 F. Supp. at 659-60.

The City and Port Authority argue that all acts performed in response to the September 11 attacks were discretionary and unique to governmental entities, including the exercise of traditional police powers in the interest of public health and safety. My task, however, is not to evaluate claims in their generality, but to "scrutinize specific claims," for "[i]t is the specific act or omission out of which the injury is claimed to have arisen and the capacity in which that act or failure to act occurred" that determines governmental immunity. In re September 11 Litigation, 280 F. Supp. 2d at 303 (quoting *Weiner v. Metro. Transp. Auth.*, 55 N.Y.2d 175, 433 N.E.2d 124, 127, 448 N.Y.S.2d 141 (1982). The issue cannot be decided in the context of a motion for judgment on the pleadings. It requires a proper, and fully developed, factual record.

VI. THE MOTIONS FOR SUMMARY JUDGMENT – FEDERAL IMMUNITY

Defendants argue that they are entitled to immunity also under federal law, pursuant to the Stafford Act and under principles of federal derivative immunity. The Defendants assert that their actions after September 11 mainly were at the express direction of federal government agencies and that immunity should therefore extend to them insofar as they acted at the behest of the federal government. The Defendants assert further that to impose liability would run counter to basic principles of law to the extent that the federal government is the real party in interest. Defendants move for summary judgment, after full discovery relevant to their defense of federal immunity.

A. Standard of Review

Summary judgment is warranted if the "pleadings, depositions, answers to interrogatories,

and admissions on file, together with the affidavits... show that there is no genuine issue as to any material fact and that the moving party is entitled to judgment as a matter of law." Fed. R. Civ. P. 56(c). A "genuine issue" of "material fact" exists "if the evidence is such that a reasonable jury could return a verdict for the nonmoving party." *Anderson v. Liberty Lobby, Inc.*, 477 U.S. 242, 248, 106 S. Ct. 2505, 91 L. Ed. 2d 202 (1986). Although all facts and inferences therefrom are to be construed in favor of the party opposing the motion, see *Harlen Assocs. v. Village of Mineola*, 273 F.3d 494, 498 (2d Cir. 2001), the non-moving party must raise more than just "metaphysical doubt as to the material facts," *Matsushita Elec. Indus. Co. v. Zenith Radio Corp.*, 475 U.S. 574, 586, 106 S. Ct. 1348 (1986). "[M]ere speculation and conjecture is insufficient to preclude the granting of the motion." Harlen, 273 F.3d at 499. "If the evidence is merely colorable or is not significantly probative, summary judgment may be granted." Anderson, 477 U.S. at 249-50 (citations omitted).

B. Derivative Federal Immunity

The practicalities of modern governance have led courts, over the past sixty years, to extend the immunity traditionally afforded to the federal government for actions taken in furtherance of its government functions to private entities hired to facilitate the government in the implementation of its programs and goals. *See Yearsley v. W.A. Ross Const. Co.*, 309 U.S. 18, 60 S. Ct. 413, 84 L. Ed. 554 (1940). The extension of such immunity stems from the premise that, "[t]o insulate the United States from its discretionary decisions, but not to do likewise when the United States enters into contracts with others to execute the will of the United States 'makes little sense.'" *Richland-Lexington Airport District v. Atlas Properties, Inc.*, 854 F. Supp. 400 (D.S.C. 1994) (quoting *Boyle v. United Techs. Corp.*, 487 U.S. 500, 511-512, 108 S. Ct. 2510, 101 L. Ed. 2d 442 (1988)). Indeed, the primary purpose of the defense "is to prevent the contractor from being held liable when the government is actually at fault" but is otherwise immune from liability. *Trevino v. General Dynamics Corp.*, 865 F.2d 1474, 1478 (5th Cir. 1989).

Defendants, the City of New York and its Contractors as well as the Port Authority and the WTC Defendants, urge that under the principles first enunciated in Yearsley, 309 U.S. 18, 60 S. Ct. 413, 84 L. Ed. 554, and further developed in Boyle, 487 U.S. at 505, and Richland-Lexington, 854 F. Supp. 400, they are entitled to immunity for actions taken in the aftermath of the September 11 attacks to the limited extent that such actions were controlled by and undertaken pursuant to the direction of federal agencies, namely the Army Corps, OSHA and the EPA. They contend that to impose liability on Defendants for actions that were undertaken at the direction of federal agencies would work a manifest injustice and would severely hinder the ability of the government to provide an effective and immediate response to future disasters.

1. The Relevant Case Law

The Supreme Court first recognized the doctrine of derivative immunity for private contractors, the so-called government contractor defense, in Yearsley, 309 U.S. 18, 60 S. Ct. 413, 84 L. Ed. 554, dismissing a suit by landowners for damage to their property arising from work performed by a private contractor pursuant to a contract with the federal government. The landowners alleged that the contractor's negligence in constructing dikes in the Missouri River at the direction of the federal government led to the erosion of ninety-five acres of private property. Id. The Court held that the interests of justice required that federal immunity be extended to private contractors where: (1) the contractor was working pursuant to the authorization and direction of the federal government; and (2) the acts complained of fell within the scope of such government directives. Id. at 21. Under the rule set forth in Yearsley, so long as, "[the] authority to carry out the project was validly conferred, that is, if what was done was within the constitutional power of Congress, there is no liability on the part of the contractor for executing" the will of the federal government. Id. Thus, liability may attach only where the contractor exceeded the scope of his authorization, or where the authority itself was not validly conferred. Id.

The purpose and scope of the government contractor defense was clarified and expanded in Boyle. 487 U.S. 500, 108 S. Ct. 2510, 101 L. Ed. 2d 442. There, pursuant to specifications provided by the United States, a private manufacturer had constructed a military helicopter with an allegedly defective escape hatch. Id. at 502-503. As a result of the defective hatch, the co-pilot was killed when he was unable to open the hatch after the helicopter crashed into the water. Id. A jury had returned a verdict in favor of the executor of the co-pilot's estate on the basis of state law. Although remanding for further consideration, the Supreme Court held that principles of government immunity applied with equal force to the case of a private manufacturer acting pursuant to precise government specifications. Id. Rejecting the argument that federal immunity is necessarily limited to actions taken by an official in the course of performing his or her duties as a federal employee, the Court instead framed the essential issue as the "uniquely federal interest... in getting the Government's work done." Id. at 505. Such unique federal interest, in turn, mandated the displacement of state law to the extent that application of state law principles would give rise to a "significant conflict" with an "identifiable 'federal policy or interest.'" Id. at 507 (quoting *Wallis v. Pan American Petroleum Corp.*, 384 U.S. 63, 69, 86 S. Ct. 1301, 16 L. Ed. 2d 369 (1966)).

The Court made clear, however, that a federal interest alone is insufficient to displace state law. Id. at 507. State law is instead displaced only where it presents a significant conflict with federal policy, such that "the application of state law would 'frustrate specific objectives' of 'federal legislation.'" Id. Although the conflict need not be as absolute as that which is required under ordinary preemption analysis where Congress has legislated in an area traditionally governed by state law, "conflict there must be." Id. at 508. In the absence of a conflict, state law obligations remain in effect. Thus, if:

> the United States contracts for the purchase and installation of an air-conditioning unit, specifying the cooling capacity but not the precise manner of construction, a state law imposing upon the manufacturer of such units a duty to include a certain safety feature would not be a duty identical to anything promised the Government, but neither would it be contrary. The contractor could comply with both its contractual obligations and the state-prescribed duty of care. No one suggests that state law would generally be pre-empted in this context. Id. at 509.

In the case presented in Boyle, however, the Court recognized the potential for the finding of a "significant conflict." Id. at 511. Neither the federal government nor its officers could be sued for designing an unsafe and defective helicopter because of an inadequate escape hatch, for the Federal Tort Claims Act ("FTCA") precluded the imposition of liability for "the exercise or performance or the failure to exercise or perform a discretionary function or duty." 28 U.S.C. § 2680(a).

The Court in Boyle therefore reasoned that, if the federal government could not be sued for performing discretionary acts, neither could a private contractor, and any state law providing for such liability would necessarily give rise to an inherent conflict. On the basis of such a significant conflict, Boyle held, the imposition of liability pursuant to state law must fail where: "(1) the United States approved reasonably precise specifications; (2) the equipment conformed to those specifications; and (3) the supplier warned the United States about the dangers in the use of the equipment that were known to the supplier but not to the United States." Boyle, 487 U.S. at 512.

Boyle ruled that there was a unique federal interest implicated in the securing of military procurement contracts. The rule of Boyle, however, is more expansive and extends to other fields of strong federal interest. The dispositive issue is "whether there is a uniquely federal interest in the subject matter of the contract." Richland-Lexington, 854 F. Supp. at 422. Thus, in Richland-Lexington, the District Court for South Carolina extended federal immunity to a private contractor hired by the Environmental Protection Agency to cleanup and stockpile contaminated soil. Id. 423-24. Applying Boyle, the court ruled that the actions of the EPA were discretionary in nature, and that the discretionary function exception of the FTCA operated to prevent it and its employees and officials from being sued. Id. at 423. The court then applied Boyle's three-prong test, and concluded that state law conflicted insofar as it operated to impose liability on a private contractor hired by the EPA and therefore was displaced. First, the court found that the EPA had approved the site for cleanup, had determined the best method for cleanup, and had hired the defendant contractor to excavate and remove contaminated soil from the site according to that method. Second, the court found that the defendant contractor had performed pursuant to performance specifications issued by the EPA and under the EPA's supervision. Third, the court found that the defendant contractor had not been aware of any dangers with respect to the cleanup activity as to which the EPA was not also aware. Id. at 423-24.

The court concluded that state tort law was displaced and the defendant contractor was therefore immune from suit, sharing in the immunity traditionally granted the federal government. Id. at 424.

The crucial element is that the government contractor acted in compliance with "reasonably precise specifications" approved by the United States. Boyle, 487 U.S. at 512. The very essence of the defense is to "prevent the contractor from being held liable when the government is actually at fault," Trevino, 865 F.2d at 1478, for "[w]hen a contractor acts under the authority and direction of the United States, it shares in the immunity enjoyed by the Government." Zinck, 690 F. Supp. at 1333 (citing Yearsley, 309 U.S. 18, 60 S. Ct. 413, 84 L. Ed. 554). If, however, the private actor was acting independently of precise government directions and approvals, the defense does not apply. Trevino, 865 F.2d at 1480; see also *Tate v. Boeing Helicopters*, 55 F.3d 1150 (6th Cir. 1995). Furthermore, the government must supervise and control the contractor's actions, for if it does not, or if it fails to exercise supervisory judgment as to the particularities of the project, state law allowing lawsuits for negligence and federal policy providing for immunity are not in conflict and displacement of state law may not be warranted. Tate, 55 F.3d at 1154. In the latter case, the discretionary functions of the government are not implicated and the government contractor defense as enunciated in Yearsley and Boyle may not apply. Trevino, 865 F.2d at 1480.

When the government merely accepts, without any substantive review or evaluation, decisions made by a government contractor, then the contractor, not the government, is exercising discretion. A rubber stamp is not a discretionary [*563] function; therefore, a rubber [**119] stamp is not "approval" under Boyle. Id. Thus, derivative immunity arises where the government: (a) approves in its discretion reasonably precise specifications, (b) supervises and controls the implementation of those specifications, and (c) the contractor is not aware of reasons not known to the government why the application is unsafe or unreasonable. Id. Immunity will not attach, however, where the private contractor acts, or

knows material facts, independently of the federal agency.

2. Application to the Rescue and Recovery Efforts at Ground Zero

Defendants claim that federal agencies assumed exclusive control over three distinct areas of the rescue and recovery effort following September 11, and that they are therefore entitled to share in the immunity enjoyed by the federal government with respect to all claims arising from those areas. Specifically, Defendants claim federal control as follows: (1) the Army Corps assumed control over the design, implementation and enforcement of environmental health and safety monitoring at Fresh Kills; (2) OSHA assumed the lead role for respirator distribution, fit-testing, training and use at and around Ground Zero; and (3) the EPA assumed lead responsibility for environmental monitoring and hazardous waste removal. The Defendants thus argue that they are entitled to derivative immunity from all claims relating to the health and safety protocols in effect at Fresh Kills; claims relating to respirator distribution, fit-testing and training at and around Ground Zero; and all claims arising from air quality monitoring and hazardous materials removal conducted at the World Trade Center site.

The first step in the Boyle analysis is to identify the unique federal interest. That lies in the Federal Response Plan ("FRP") which expressly provides for "a process and structure for the systematic, coordinated, and effective delivery of Federal assistance to address the consequences of any major disaster or emergency[.]" (FRP, Ex. L at 1.) The Presidential Declaration of a National State of Emergency on September 14, 2001 activated the FRP, leading the Army Corps of Engineers, OSHA and the EPA to become involved in the recovery efforts at Ground Zero.

The next step is to identify if a substantial conflict exists between federal policy and application of state law, as "[c]onflict there must be." Boyle, 487 U.S. at 508. Pursuant to the Stafford Act, a federal agency or employee acting in accordance with the FRP "shall not be liable for any claim based upon the exercise or performance of or the failure to exercise or perform a discretionary function or duty[.]" FRP at 8; 42 U.S.C. § 5148. To the extent the actions challenged were discretionary in nature, allowing for "'second-guessing' of these judgments...through state tort suits... would produce the same effect sought to be avoided" by the Stafford Act and FRP exemption. Boyle, 487 U.S. at 511. It would make little sense to immunize the federal agencies and officers from liability for decisions made as to protocols to be followed at the site while at the same time imposing liability on the City and other Defendants simply for abiding by such federally-determined protocols.

The question, however, is whether the same or different conduct is implicated, that is, whether the contractor is merely carrying out that which federal officers chose to do in the exercise of their discretion, under the supervision and control of federal officers, or whether the contractor did something materially different or additional for which they are being sued. Here, to the extent that the relevant federal [*564] agencies did not exercise oversight over Defendants' actions, federal immunity will not operate to protect Defendants from liability by suspending application of state law. Tate, 55 F.3d at 1154.

On October 1, 2001, and at the request of the City, the Army Corps assumed limited control over operations at Fresh Kills Landfill and, on October 10, 2001, assumed responsibility for "the implementation of a comprehensive health and safety plan for all personnel working at the landfill site." (Defs.' J.A. Vol. 5, Ex. BJ (Amendment to 3-COE, Oct. 10, 2001).) It engaged Phillips & Jordan to develop and implement the various protocols and procedures for a comprehensive scheme for managing health and safety at the landfill, including dust control measures and monitoring compliance with protocols by workers. (Defs.' J.A. Vol. 5, Ex. BK (Phillips & Jordan Disaster Recovery Group: Experience).)

However, there has been no showing that the Army Corps exercised a level of control and supervision over the operations at the Fresh Kills worksite sufficient to warrant the extension of federal immunity to the City and other Defendants. To the contrary, the record suggests that the City continued to exercise an independent degree of supervisory control over operations. It was the DDC, and not the Army Corps, that designated Fresh Kills as one of the zones of the rescue and recovery effort. It was the City, not the Army Corps, that commissioned the Environmental Health and Safety Plan ("ES&H"), first distributed in October of 2001, to address the implementation of health and safety protocols within the four primary zones constituting the former site of the World Trade Center and surrounding areas, and which addressed itself to the issues of worker safety at Fresh Kills, incorporating the more extensive recommendations by Phillips & Jordan and placing such recommendations on file with the DDC.

The mere fact that a federal agency, rather than a private contractor, assumed responsibility for operations within the Fresh Kills zone does not require that immunity automatically attach. Although the Plaintiffs cannot "second-guess" the reasonableness of the Army Corps' recommendations, the Plaintiffs are not precluded from attacking other aspects of the operations which were outside the direction of the Army Corps, directly or through Phillips & Jordan, or which were not part of the work that the Army Corps supervised and controlled.[22] Indeed, the clear understanding was that the Army Corps, far from assuming authoritative control over operations at Fresh Kills, worked "on an as required basis" as the City of New York requested its assistance.

As to OHSA, within days of the September 11 attacks, and at the request of the City DOH, OSHA became the "lead agency for distributing, fitting, and training for respirators for the recovery

---

[22] It should be noted that nothing in the FRP, pursuant to which the Army Corps assumed its role at Fresh Kills, supposes that reliance on federal assistance and resources serves to divest a local authority of control. (See FRP, Ex L ("The combined emergency management authorities, policies, procedures, and resources of local, State, and Federal governments as well as voluntary disaster relief organizations...constitute a national disaster response framework for providing assistance following a major disaster or emergency.")).

of workers." (Defs.' J.A., Vol. 2, Ex. Q (Clark Dep.) at 87:19-88:6; Pls.' J.A. Vol. 4, Ex. 56.) OSHA selected the type of respirators to be distributed, and coordinated the distribution of respirators at huts set up throughout the World Trade Center site, providing respirator-fit and -use training, including a mandatory 10-hour health and safety course for all individuals working in a supervisory role at the site. Further, to ensure compliance with established standards, OSHA deployed over 1,000 inspectors to the site to supervise the work of the Primary Contractors and to document violations.

And yet, despite its designated position as the lead agency for respirator training and use, the record clearly shows that OHSA continued to work in an advisory capacity, providing assistance only as needed and requested by the City. Throughout the recovery effort, the City DOH continued to act as the agency responsible for the development of job and site-specific PPE requirements, including establishing relevant respiratory requirements. The ES&H Plan clearly sets out the DDC, with the assistance of other City agencies and the Primary Contractors, as having primary responsibility for environmental safety and health at the site. Moreover, OSHA expressly declined to assume an enforcement role at the site, instead limiting its inspectors to reporting any observed violations to the relevant Primary Contractor. Enforcement was thus left to City agencies and the Primary Contractors. The cooperative relationship between OSHA and City agencies was memorialized in the WTC Emergency Project Partnership Agreement, affirming "the value of working in a cooperative, focused and voluntary effort to ensure a safe and healthful environment for everyone involved.

The EPA also played an important and vital role in the rescue and recovery efforts at Ground Zero. Like OSHA, the EPA assumed its role at the site pursuant to a direct request by the City and pursuant to the authority granted the agency under the FRP. Specifically, the EPA assumed lead responsibility for the removal of hazardous materials at the site and for monitoring and reporting on atmospheric conditions. Air monitoring at the site was coordinated with OSHA and

relevant City agencies through the use of air monitors stationed at locations in and around the World Trade Center site. At the same time, the City, through its own agencies, continued to conduct its own tests, independent of federal air quality monitoring. Both the City and the EPA shared their data with the inter-agency Environmental Assessment Group. Determinations as to respirator requirements on the basis of test results, in turn, were made, not by the EPA separately, but rather by the Environmental Assessment Group as a collective group.

The record thus shows that the City never abandoned its overall responsibility for worker health and safety. The EPA expressly acknowledged that it lacked the authority to enforce the worker health and safety policies for non-EPA employees. Moreover, even though the EPA had assumed lead responsibility for the removal of toxic waste at the site, it in fact exercised its authority in conjunction with the City DEP. ("The Environmental Protection Agency (EPA), in conjunction wit the City Department of Environmental Conservation, (DEP) is responsible for recovering, handling and disposing hazardous wastes at the disaster site.").) Thus, like the Army Corps and OSHA, the EPA operated at the site in an advisory capacity only, never divesting the City of its authority or its duty to protect those working at its behest.

Defendants urge that active participation of the federal agencies in developing protocols for health and safety suffices to extend immunity. But this is not so. The essential purpose of the derivative immunity doctrine is to "prevent the contractor from being held liable when the government is actually at fault[.]" Trevino, 865 F.2d at 1478. If the private actor acted in compliance with "reasonably precise specifications" established by the federal agency, and under the supervision and control of the federal agency, immunity is extended. Boyle, 487 U.S. at 512. But immunity is not extended where the private actor acts independently, or outside of, or in addition to, the government's specifications. Moreover, the safety of the workplace is heavily regulated by New York state labor law, a subject that I developed in Hickey, and to which I return in the next section.

Unless adherence to the requirements of the labor law conflicts with federally-developed protocols, the state regulation is not ousted by the federal interest. Id. at 509.

Here, the record does not show exclusive federal control. The record instead presents a picture of cooperation and collaboration, with federal agencies providing assistance pursuant to the request of the City and expressly declining to assume an enforcement role, deferring instead to the City agencies and the Primary Contractors. The City and the Primary Contractors it engaged exercised supervisory responsibility for worker health and safety at the site and for ensuring compliance with applicable safety standards. The exercise of discretionary functions by the federal agencies did not conflict with a continuing obligation by Defendants to abide by the mandatory laws regulating the health and safety of the workplace.

It is beyond any doubt that the City and other Defendants relied upon the assistance and expertise of the federal agencies. To the extent that reliance, and adoption of federal standards and protocols is shown, and the Defendants' conduct is tantamount to actions by the federal authority, the Defendants enjoy the same immunity as would be conferred on discretionary acts and decisions of federal officers and employees. At this point, however, the record is not sufficiently clear to enable the court to demark the boundary between federally instructed discretionary decisions, and those made by the various Defendants. There are material, triable issues of fact that will have to be resolved. At his point in the pre-trial proceedings, the Motions for Summary Judgment are denied.

C. Stafford Act Immunity

The various Defendants separately assert immunity pursuant to the Stafford Act. 42 U.S.C. § 5148 (2006). The Stafford Act provides that the federal government, "shall not be liable for any claim based upon the exercise or performance of or the failure to exercise or perform a discretionary function on the part of a Federal agency or an employee of the Federal Government in carrying out the provisions" of the FRP. 42 U.S.C. § 5148. I decline, however, to extend the grant of immunity provided by the Stafford Act to non-federal actors beyond the limits of federal common law, for to do so would extend the Act beyond its plain terms.

Nothing in the plain language of the Stafford Act provides that immunity should be extended to non-federal actors. The mere fact that non-federal actors may participate in response and recovery efforts does not in and of itself require that immunity be extended to actors not included in the statutory provision of immunity. Indeed, "this court is not at liberty, because it thinks the provisions [of a statute] inconsistent or illogical, to rewrite them in order to bring them into harmony with its views as to the underlying purpose of Congress." *Helvering v. New York Trust Co.*, 292 U.S. 455, 472 (1934). That the Stafford Act's immunity provision should not be judicially rewritten is further underscored by the doctrine of derivative federal immunity otherwise available (and previously discussed).

Immunity provisions are to be interpreted narrowly. Abbott, 23 N.Y.2d 502, 245 N.E.2d 388. Our system of justice is premised on accountability, save for specific exceptions based on statute or fundamental common law principles of necessity. Defendants urge that courts should encourage private actors to enlist in recovery efforts from mass disasters, but the same policy has to be sensitive to the individual workers who risk their lives. The job of restoring society cannot be based on a system rewarding businesses, but being indifferent to the health and welfare of working people.

D. Other Bases for Federal Immunity

The Port Authority and the World Trade Center Defendants, as well as Con Ed, present additional arguments for immunity under federal law.

The Port Authority and the World Trade Center (Silverstein) Defendants argue that the federal government's promise to pay 100 percent of all costs associated with the rescue and recovery efforts at Ground Zero makes the federal government the real party in interest, thus mandating the dismissal of all claims as against them. They cite to the Presidential Declaration of National Emergency on September 14, 2001, by which President Bush ordered FEMA to pay all costs associated with the efforts taken in response to the September 11 attacks. Such promise of payment, however, does not operate to suspend ordinary rules of rights and obligations running between a tortfeasor and his alleged victim. The question properly before me by these pending

motions is whether Defendants are in some manner entitled to immunity for acts undertaken in response to the September 11 attacks. That Defendants may ultimately seek indemnification from the federal government on the basis of FEMA's alleged obligation to pay 100 percent of costs is irrelevant to this larger question.

ConEd's additional argument, additional to arguments of derivative governmental immunity and the Stafford Act, is based on its legal responsibility to restore energy to lower Manhattan....

VIII. CONCLUSION

For the reasons discussed in the Opinion, the motions by the City and by the Port Authority to dismiss the complaints against them, and against the contractors engaged by the City are denied; the motions by ConEd and the World Trade Center Defendants are granted, and the complaints against them are dismissed.

The state and federal statutes that provide immunity protect the remaining Defendants against suit, but the precise scope and extent of the immunity varies according to date, place and activity. As I indicated in the prior sections of this Opinion, the fact-intensive nature of the issue makes its resolution unsuitable for resolution by motion. Discovery, additional proceedings, and a more extensive factual record, and perhaps a trial, will be required.

The number of these cases present unusual complexities in these pre-trial proceedings. Plaintiffs, we have been told, worked for various contractors, on various dates, and in different portions of the World Trade Center site. Many contracting companies worked at the site, performing varying functions, each under separate contract, and each subject to varying supervision by different government inspectors, working for different federal, state and City agencies, and by the primary contractors and higher-level subcon-

tractors. Although a considerable amount of discovery has taken place in connection with the pending motions, there is much more to come – much, much more.

The needs of the parties, and the public interest, do not permit the leisurely and expensive progression of proceedings that are often characteristic of complex litigation. If even a minority of the Plaintiffs suffered serious injuries to their respiratory tracts arising from the acrid air of September 11, their claims deserve to be heard when a recovery could make a difference to their lives. Conversely, if complaints lack merit, or if defenses prove to be valid, defendants are entitled to the earliest possible release. The availability of a one billion dollar fund authorized by Congress should not serve as encouragement to lengthen and complicate these proceedings. The scar to the public interest needs to be cleansed, speedily, in good time....

SO ORDERED.

Dated: New York, New York

October 17, 2006

ALVIN K. HELLERSTEIN

United States District Judge

QUESTIONS

1. Emergency responders knew the job was dangerous when they took it. In fact for some, the element of danger adds a bit of spice to the work. As a general matter, what is your opinion of emergency responders suing their employers? Is it fair? Why or why not? Is it fair in this case? What if anything makes this case different?

2. As the Second Edition of this book goes to press, Congress has agreed to compensate those who were exposed to substances while working the site of the 9-11 attacks on the twin towers. Congress at this point has excluded coverage for cancer. Perform research to discover the current status of coverage for cancer among WTC site responders. Discuss whether cancer, particularly cancer associated with lung disorders, should be covered. Perform research to discover why Congress did not cover such cancers and make that knowledge a part of your discussion.

3. List the federal and state agencies involved in the clean-up of the twin towers site and further disposal of debris as well as human remains. Describe their roles as well as the amount and type of control they had over various aspects of the process.

4. Review the lists prepared in response to question 3. Clearly, the amount and type of control various agencies exerted was very important in the Court's evaluation of who might be responsible for injuries to the responders. Reviewing your lists, the case, and rereading the materials on HAZMT events, who do you think bears responsibility for injuries to the responders? Explain your answer.

5. Reread the materials on immunities in the case. What are the public policy arguments in favor of immunities in a case like this one?

6. Recall and review the *Buttram* case. Discuss the similarities and differences between the *Buttram* and *WTC* cases.

Section II

EMERGENCY MANAGEMENT LAW

Don't think it won't happen just because it hasn't happened yet.

The Road and the Sky
Jackson Browne

Goshen, N.Y., September 22, 2011 – Governor Cuomo announces $2.4 million from the state's Agricultural and Community Recovery Fund (ACRF) will be distributed to farmers to help New York's agricultural communities from storm damages. Photo by Elissa Jun/FEMA.

Chapter 10

POWERS OF GOVERNORS

Gubernatorial powers differ widely among the states. Consider the relative strengths of the Governors of Indiana and New Jersey while reading the materials below.

PART 1. STATE EMERGENCY MANAGEMENT STATUTE

Indiana Code 10-14-3-11 (2010)

Governor – Direction and control of department – Powers and duties.

(a) The governor has general direction and control of the agency and is responsible for carrying out this chapter. In the event of disaster or emergency beyond local control, the governor may assume direct operational control over all or any part of the emergency management functions within Indiana.

(b) In performing the governor's duties under this chapter, the governor may do the following:

  (1) Make, amend, and rescind the necessary orders, rules, and regulations to carry out this chapter with due consideration of the plans of the federal government.

  (2) Cooperate with the President of the United States and the heads of the armed forces, the Federal Emergency Management Agency, and the officers and agencies of other states in matters pertaining to emergency management and disaster preparedness, response, and recovery of the state and nation. In cooperating under this subdivision, the governor may take any measures that the governor considers proper to carry into effect any request of the President of the United States and the appropriate federal officers and agencies for any emergency management action, including the direction or control of disaster preparations, including the following:

  (A) Mobilizing emergency management forces and other tests and exercises.

  (B) Providing warnings and signals for drills, actual emergencies, or disasters.

  (C) Shutting off water mains, gas mains, and electric power connections and suspending any other utility service.

  (D) Conducting civilians and the movement and cessation of movement of pedestrians and vehicular traffic during, before, and after drills, actual emergencies, or other disasters.

  (E) Holding public meetings or gatherings.

  (F) Evacuating and receiving the civilian population.

  (3) Take any action and give any direction to state and local law enforcement officers and agencies as may be reasonable and necessary for securing compliance with this chapter and with any orders, rules, and regulations made under this chapter.

  (4) Employ any measure and give any direction to the state department of health or local boards of health as is reasonably

191

necessary for securing compliance with this chapter or with the findings or recommendations of the state department of health or local boards of health because of conditions arising from actual or threatened:

    (A) national security emergencies; or

    (B) manmade or natural disasters or emergencies.

(5) Use the services and facilities of existing officers, agencies of the state, and of political subdivisions. All officers and agencies of the state and of political subdivisions shall cooperate with and extend services and facilities to the governor as the governor may request.

(6) Establish agencies and offices and appoint executive, technical, clerical, and other personnel necessary to carry out this chapter, including the appointment of full-time state and area directors.

INDIANA CODE 10-14-3-12 (2010)

Declaration of disaster emergency – Powers and duties of governor.

(a) The governor shall declare a disaster emergency by executive order or proclamation if the governor determines that a disaster has occurred or that the occurrence or the threat of a disaster is imminent. The state of disaster emergency continues until the governor:

(1) determines that the threat or danger has passed or the disaster has been dealt with to the extent that emergency conditions no longer exist; and

(2) terminates the state of disaster emergency by executive order or proclamation.

A state of disaster emergency may not continue for longer than thirty (30) days unless the state of disaster emergency is renewed by the governor. The general assembly, by concurrent resolution, may terminate a state of disaster emergency at any time. If the general assembly terminates a state of disaster emergency under this subsection, the governor shall issue an executive order or proclamation ending the state of disaster emergency. All executive orders or proclamations issued under this subsection must indicate the nature of the disaster, the area or areas threatened, and the conditions which have brought the disaster about or that make possible termination of the state of disaster emergency. An executive order or proclamation under this subsection shall be disseminated promptly by means calculated to bring the order's or proclamation's contents to the attention of the general public. Unless the circumstances attendant upon the disaster prevent or impede, an executive order or proclamation shall be promptly filed with the secretary of state and with the clerk of the city or town affected or with the clerk of the circuit court.

(b) An executive order or proclamation of a state of disaster emergency:

(1) activates the disaster response and recovery aspects of the state, local, and interjurisdictional disaster emergency plans applicable to the affected political subdivision or area; and

(2) is authority for:

    (A) deployment and use of any forces to which the plan or plans apply; and

    (B) use or distribution of any supplies, equipment, materials, and facilities assembled, stockpiled, or arranged to be made available under this chapter or under any other law relating to disaster emergencies.

(c) During the continuance of any state of disaster emergency, the governor is commander-in-chief of the organized and unorganized militia and of all other forces available for emergency duty. To the greatest extent practicable, the governor shall delegate or assign command authority by prior arrangement embodied in appropriate executive orders or regulations. This section does not restrict the governor's authority to delegate or assign command authority by orders issued at the time of the disaster emergency.

(d) In addition to the governor's other powers, the governor may do the following while the state of emergency exists:

(1) Suspend the provisions of any regulatory statute prescribing the procedures for conduct of state business, or the orders, rules, or regulations of any state agency if strict compliance with any of these provisions would in any way prevent, hinder, or delay necessary action in coping with the emergency.

(2) Use all available resources of the state government and of each political subdivision of the state reasonably necessary to cope with the disaster emergency.

(3) Transfer the direction, personnel, or functions of state departments and agencies or units for performing or facilitating emergency services.

(4) Subject to any applicable requirements for compensation under section 31 [IC 10-14-3-31] of this chapter, commandeer or use any private property if the governor finds this action necessary to cope with the disaster emergency.

(5) Assist in the evacuation of all or part of the population from any stricken or threatened area in Indiana if the governor considers this action necessary for the preservation of life or other disaster mitigation, response, or recovery.

(6) Prescribe routes, modes of transportation, and destinations in connection with evacuation.

(7) Control ingress to and egress from a disaster area, the movement of persons within the area, and the occupancy of premises in the area.

(8) Suspend or limit the sale, dispensing, or transportation of alcoholic beverages, firearms, explosives, and combustibles.

(9) Make provision for the availability and use of temporary emergency housing.

(10) Allow persons who:

(A) are registered as volunteer health practitioners by an approved registration system under IC 10-14-3.5; or

(B) hold a license to practice:

(i) medicine;

(ii) dentistry;

(iii) pharmacy;

(iv) nursing;

(v) engineering;

(vi) veterinary medicine;

(vii) mortuary service; and

(viii) similar other professions as may be specified by the governor; to practice their respective profession in Indiana during the period of the state of emergency if the state in which a person's license was issued has a mutual aid compact for emergency management with Indiana.

(11) Give specific authority to allocate drugs, foodstuffs, and other essential materials and services.

PART 2. GUBERNATORIAL EXECUTIVE ORDERS

Michael S. Herman[*]

Copyright © 1999 Rutgers School of Law – Camden
Rutgers Law Journal

30 Rutgers L. J. 987 (1999)
Reprinted by permission

I. INTRODUCTION

It is said that the New Jersey Governor is, by Constitutional authority, one of the most powerful governors in the nation: That power is readily apparent in the executive orders promulgated by New Jersey governors. Executive orders function as legal, policy, and political tools which have been used with increasing frequency by New Jersey governors, particularly in the past two decades.[2] Use of that tool raises political, legal, and constitutional questions.

II. NATURE OF EXECUTIVE ORDERS

Executive orders are both a strong and fragile Governor's tool. Like statutes, court decisions, and regulations, executive orders are legally binding law which the Governor can create or revoke by the mere stroke of a pen. So long as the Governor is acting within her authority, she may issue or repeal an executive order without the procedural or other safeguards that other types of law require. An executive order need not follow a defined process like a piece of legislation, comply with administrative procedures like a rule or regulation, or follow stare decisis as the courts must. Thus, the nature of executive orders gives a great deal of power to a single individual.

Nevertheless, executive orders are also very fragile because they can be overturned at any time by the authorizing Governor, a future Governor, or the Legislature[21] and quite often are. Usually, when a new Governor comes into office, she issues an executive order rescinding the orders of previous Governors. Governors often repeal their own executive orders when deemed necessary or when conditions have changed, such as the end of a state of emergency....

III. STRUCTURE OF EXECUTIVE ORDERS

Executive orders have many of the same attributes as other types of law, but are also unique in many respects. They are most often constructed of three main parts. The first part contains clauses beginning with "whereas," which help establish the purpose of the order. Like purpose clauses in legislation, they help to justify the order and explain what the order is intended to accomplish. For example, in an order declaring a state of emergency, the whereas clauses may

[*] J.D., Rutgers School of Law-Camden, 1990; M.A., Eagleton Institute of Politics, Rutgers University, 1988. The author is currently an attorney with the Federal Emergency Management Agency and served as Policy Advisor and Special Assistant to Governor Jim Florio from 1990 to 1994. The views expressed by the author are his own and do not necessarily reflect the views of any employer, current or previous. The author would like to express his appreciation for the assistance and guidance of Robert F. Williams, Distinguished Professor of Law, Rutgers School of Law-Camden, Jack M. Sabatino, Associate Dean, Rutgers School of Law-Camden, and Alan Rosenthal, Professor of Public Policy, Eagleton Institute of Politics, Rutgers University, as well as the invaluable research assistance of Jason Cohen, Rutgers School of Law-Camden, Class of 1998. (Reprinted by permission (1999). *Rutgers Law Journal 30*:4, pp. 987–991.

outline the scope and magnitude of the emergency that the order addresses. Along with detailing the actions that the Governor plans to take, those clauses permit the Governor to express a particular concern or interest in the matter. Whereas clauses may also set forth the factual predicates upon which the Governor's legal authority is based, such as prison overcrowding.

Conversely, the Governor may use this section of an executive order to make broad statements about issues where little or no significant substantive action is required or which might not even be within the Governor's authority. In those orders, the clauses serve the same function as legislative or executive proclamations. Such orders include those requiring flags to be flown at half-staff, which generally contain a long section of whereas clauses recognizing the important contributions of a deceased. However, the action required by those orders is limited to flying flags in memorial for a specified number of days.

The second main section of an executive order makes a broad statement invoking the Governor's powers. Usually, this section cites the general authority of the Governor granted by the constitution and laws of the state. However, occasionally this section may cite a specific authority given to the Governor. This is where the Governor cites and declares her powers to issue the mandate of the executive order. This section may also contain other statements, such as those used in an order declaring a state of emergency.

The final section of an executive order contains the substance of the order. This typically consists of one or more clauses ordering the action deemed necessary. The length of this section depends upon the nature and complexity of the order itself and may range from a single paragraph to a number of pages.

IV. TYPES OF EXECUTIVE ORDERS

Despite the provisions for a strong governor in the 1947 New Jersey Constitution and the remarks of that convention's delegates, the New Jersey Constitution does not explicitly authorize gubernatorial executive orders. Nevertheless, executive orders have become a well-used tool of New Jersey governors.

There are two methods of categorizing executive orders. One classifies them by purpose; the other classifies them by origin of authority (if any). Although this section of the article focuses mainly upon the latter method, some discussion of the former is helpful at the outset.

A. Orders Categorized by Purpose

Executive orders may serve two broad categories of purpose. The most common and well-known orders are those which relate to the Governor's constitutional duty to faithfully execute the laws and thus are derived from the Governor's role as head of the executive branch. This category includes orders which direct the administration of law or the performance of certain acts and duties by executive branch officers and agencies. This category also includes orders for ceremonial and formal functions....

Executive orders may also have substantive legal effects. That type of order is a relatively new phenomenon, little discussed except within the past thirty years....

B. Orders Categorized by their Source of Authority

The authority for executive orders in New Jersey is derived from a number of different sources including the New Jersey Constitution, state legislation, federal legislation, the Governor's emergency powers, and judicial orders and decisions.

1. Constitutional Authority

Although there is no specific provision in the New Jersey Constitution which empowers the Governor to issue executive orders, there are clearly constitutional provisions from which the Governor's authority to issue executive orders may be implied. For example, article V, section I, paragraph 11 provides that the Governor must execute the laws. That section has been interpreted as giving the Governor a number of implied powers, including the authority to promulgate executive orders....

Article V, section I, paragraph 12 of the New Jersey Constitution is also relevant to gubernatorial executive orders because it empowers the Governor to convene the Legislature or Senate alone when the public interest requires, and appoints the Governor as Commander-in-Chief of all military and naval forces of the state. At least one executive order has been used to call a special session. The Governor's power as Commander-in-Chief has been used at least twice to issue orders concerning the state's military and naval forces.

2. Specific Statutory Authority

The least controversial executive orders issued in New Jersey are probably those specifically authorized by legislation. These orders can be broken down into two categories: those authorized by legislation and those intended to implement legislation.

(a) Executive Orders Authorized by Legislation

...The Right to Know Law,[77] which permits public inspection of public records, has been the source of authority for a number of executive orders, including three orders issued by Governor Hughes shortly after the statute was passed in 1963. A number of cases have been decided under the Right to Know Law and its implementing executive orders. In those cases, the courts have recognized the legal validity of such orders in establishing what records may be made available to the public....

(b) Executive Orders that Implement Legislation

All of New Jersey's recent Governors have issued executive orders that directly implement legislation....

4. Authority Under Emergency Powers

Some of the Governor's broadest powers to issue executive orders are authorized by the emergency powers granted by the legislature. Those powers are derived from New Jersey's Disaster Control Act.[108] The Act was passed in 1941 and empowered the Governor to assist the federal government in the war effort. In 1942, the Act was expanded to give the Governor broader powers in providing for civilian defense. In 1949, the Act was expanded to include any emergency resulting from natural causes such as natural disasters, and in 1953, it was expanded to include emergencies which arise from "unnatural causes.[109]

The purpose of the Act is extremely broad:

> The purpose of this act is to provide for the health, safety, and welfare of the people of the State of New Jersey and to aid in the prevention of damage to and the destruction of property during any emergency as herein defined by prescribing a course of conduct for the civilian population of this State during such emergency and by centralizing control of all civilian activitieFs having to do with such emergency under the Governor and for that purpose to give to the Governor control over such resources of the State Government and of each an every political subdivision thereof as may be necessary to cope with any condition that shall arise out of such emergency and to invest the Governor with all other power convenient or necessary to effectuate such purpose.[110]

In addition to the broad powers spelled out by the Act's purpose, the Act also grants "Emergency powers to the Governor":

> The Governor is authorized to utilize and employ all the available resources of the State Government and of each and every political subdivision of this State, whether of men, properties or instrumentalities, and to commandeer and utilize any personal services and privately owned property necessary to avoid or protect against any emergency subject to the future payment of the reasonable value of such services and privately owned property as hereinafter in this act provided.[111]

Any doubt about the Governor's sweeping emergency powers is clearly erased by the Act's provision which grants the Governor the power to promulgate orders, rules, and regulations:

In order to accomplish the purposes of this act, the Governor is empowered to make such orders, rules and regulations as may be necessary adequately to meet the various problems presented by any emergency and from time to time to amend or rescind such orders, rules and regulations, including among others the following subjects:

i. On any matter that may be necessary to protect the health, safety and welfare of the people or that will aid in the prevention of loss to and destruction of property.

j. Such other matters whatsoever as are or may become necessary in the fair, impartial, stringent and comprehensive administration of this act.[112]

That section also provides that the Governor's orders, rules, and regulations shall be promulgated by proclamation and are binding on all political subdivisions, public agencies, public officials, and public employees of the state.[113]

(a) Judicial Interpretation of Gubernatorial Emergency Powers and Orders

The extent to which the Governor can promulgate orders under New Jersey's Disaster Control Act was decided by the New Jersey Supreme Court in two cases: *Worthington v. Fauver*[114] and *County of Gloucester v. State*.[115] Both cases dealt with executive orders which declared states of emergency due to prison overcrowding and authorized the Commissioner of Corrections to house state prisoners in county correctional facilities. Although those decisions reached opposite results and *Gloucester* partially overturned *Worthington*, both demonstrate the great deal of deference courts give to the Governor's authority to issue gubernatorial executive orders that declare states of emergency.

In *Worthington*, the court examined Governor Byrne's Executive Orders Nos. 106 and 108, when a county subject to the orders challenged them. The court found that prison overcrowding constituted an "emergency" under the Disaster Control Act and explicitly rejected the County's argument that prison overcrowding is not an "unusual incident," as required under the Act's definition of "disaster."[120] However, the court did put some limitations on the Governor's power under the Disaster Control Act. Although the

court found that the Governor had power under the Act to address prison overcrowding on a temporary emergency basis, the court found that the Act did not grant the Governor permanent authority to do so.[121]

The court also determined that the emergency executive orders did not violate the principle of separation of powers under article III, section I of the New Jersey Constitution. The court addressed three issues in reaching that conclusion. First, it determined that the orders did not represent a usurpation of legislative power by the executive branch. The court found that the emergency powers must be authorized and that the Disaster Control Act was such a delegation. Second, the court determined that the enabling Act was not an unconstitutional delegation of power by the Legislature. The court found that the Legislature must have concluded such a delegation would not threaten its powers when the Act was passed, because the Legislature specifically delegated the power to use county resources in emergencies. Finally, the court determined that the Legislature's delegation of power and the executive implementation of emergency orders did not unlawfully encroach upon the sphere of the judicial branch. That issue was raised because prior court decisions on prison overcrowding were part of the impetus for the executive orders.

In *Gloucester*, the court picked up where it left off in *Worthington*. By the time the court heard *Gloucester*, an additional fourteen executive orders had been issued, extending the state of emergency for nearly twelve years. In *Worthington*, the court recognized that the Governor had the power to execute those orders, but explicitly declined to address the number of times the orders could be renewed. Rather, it simply recognized that the action could not be permanent.[129] After twelve years, the court found this issue could no longer go unaddressed, and announced:

We left for another day the question of how often the Governor could renew Executive Order No. 106, or the length of time during which he could continue to exercise these extraordinary powers, before we would conclude that he had exceeded his statutory authority.

That day has come.[130]

The court found that, after almost twelve years, prison overcrowding was no longer an "emergency" under the Disaster Control Act and that a longterm solution to the problem could not be addressed by gubernatorial executive orders issued under the Disaster Control Act.[132] However, the court found the issue could be addressed by an executive order if the Legislature chose to declare a "continuing emergency" and explicitly gave the Governor the authority to address the matter by executive order.[133] The court also delayed the effective date of its order for one year to allow legislative and executive action on the matter.[134]

In 1994, as that one-year deadline approached, the Legislature passed an act that granted the Governor authority to issue executive orders dealing with prison overcrowding for two years or until the Legislature determined the emergency no longer existed.[135] On the same day the law went into effect, Governor Whitman issued Executive Order No. 16, exercising the authority delegated to her by the Act. As the statute was about to expire, the Legislature again passed an act granting those powers to the Governor for an additional two years, and Governor Whitman issued Executive Order No. 48.

While *Worthington* and *Gloucester* appear to reach opposite results, they both demonstrate the deference the courts give to Governors to promulgate executive orders. In *Worthington*, the court gave the Governor very broad discretion to issue executive orders under the Disaster Control Act. In *Gloucester*, the court found that the limits of that discretionary power had been breached because the Governor had exceeded his authority under the Disaster Control Act after twelve years. Nevertheless, the court held that the Governor could continue to issue such orders if specifically authorized by the Legislature, and the Legislature approved of that finding when it passed legislation authorizing the Governor to issue further executive orders on prison overcrowding....

VI. JUDICIAL AVOIDANCE OF INVALIDATING EXECUTIVE ORDERS

In all cases where the courts have been asked to address gubernatorial executive orders in New Jersey, one trend is clear: the courts have consistently avoided dealing directly with the issue of the Governor's authority to promulgate executive orders and thus, with the possibility of invalidating of those orders. This judicial avoidance has been accomplished by several means. New Jersey courts have denied challenges to executive orders on jurisdictional grounds by finding the issues moot or not yet ripe. The courts have also dealt with the subject matter of challenged orders either without addressing the orders themselves or by upholding the orders' actions on other grounds. While there may be a greater judicial willingness to address these challenges of late, it appears that this trend occurred only after the courts were left with no other alternative by the legislative and executive branches....

VIII. CONCLUSION

Continued use of executive orders by the executive branch has serious implications for our government and society. Our government depends upon a separation of powers and a system of checks and balances in order to function as intended by the framers of the state and federal constitutions.[209] If the Governor is permitted to continue exercising inherently legislative functions, that delicate balance may fail.

Moreover, the accumulation of legislative power by the Governor is inherently undemocratic. Of the three branches of government, the Legislature is the most representative body of the people. The Governor is elected to four-year terms and is a lame duck in her second term, thus unanswerable to the people at least in an electoral sense. That fact alone creates the possibility that an executive could make decisions and pass laws without regard to the will or interest of the people.

Because the courts are reluctant to check the Governor's power as currently used to issue executive orders, and, because it is unrealistic to assume the Governor will check her own power, the legislative branch alone is left to check that power. Ironically, it may be that the Legislature's inaction caused the increased use of executive orders in the first place.

However, it is also possible that the Legislature and citizenry wish the Governor to have such power. Because the 1947 Constitution was drafted primarily to strengthen the Office of the Governor,[210] at a minimum, more public dialogue on the subject, and possibly explicit legislative action, may be required to clarify this delegation of power.

QUESTIONS

1. Compare the emergency powers of the Governors of Indiana and New Jersey.
2. Which Governor has more options regarding courses of action in an emergency? Explain your conclusion.
3. What are the shortcomings and advantages of a state system that provides for a "strong" Governor?
4. What are the shortcomings and advantages of a system that provides for a "weak" Governor?
5. Describe any differences that may result in planning for emergency situations under the two types of Governor. Why might such differences occur?
6. Discuss any challenges that attorney advisor to the Governor of Indiana might face in drafting a Declaration of Emergency.
7. What issues do you see that might be present in a state such as New Jersey where an emergency can have a lengthy duration?
8. Consider the New Jersey prison overcrowding emergency described in the article above. How might such an emergency be handled under the Indiana statutory scheme? When evaluating this question, be sure to keep in mind the Indiana requirements for emergency declaration duration and renewal.

NOTES

1. See Jack M. Sabatino, Assertion and Self-Restraint: The Exercise of Governmental Powers Distributed Under the New Jersey Constitution, *29 Rutgers L.J.* 799, 825 (1998) (noting deferential posture New Jersey courts have generally adopted toward gubernatorial executive orders).
2. This trend may have eased off slightly with Governor Whitman. Unlike her two predecessors, both houses of the Legislature have been controlled by her party for her entire term to date, and she and the majority appear to agree on basic principles. However, even with that accord, Governor Whitman issued 91 orders in her first term. *See* table infra Part II for a more detailed analysis of orders issued by governors under the Constitution of 1947.

21. It appears that the only way the courts can strike down an executive order is by showing that it is unconstitutional or without legal authority.
77. NJ. Stat. Ann. § § 47:1A-1 to -4 (West 1989). Title 47, section 1A-2 of the New Jersey Statutes provides in relevant part:
    Except as otherwise provided in this act or by any other statute, resolution of either or both houses of the Legislature, executive order of the Governor, rule of court, any federal law, regulation or order, or by any regulation promulgated under the authority of any statute or executive order of the Governor, all records which are required by law to be made, maintained or kept on file by any board, body, agency, department, commission or official of the State or of

any political subdivision thereof or by any public board, body, commission or authority created pursuant to law by the State or any of its political subdivisions, or by any official acting for or on behalf thereof (each of which is hereinafter referred to as the "custodian" thereof) shall, for the purposes of this act, be deemed to be public records.
NJ. Stat. Ann. § 47:1A-2 (West Supp. 1999) (emphasis added).

108. L. 1941, ch. 393, § § 1-3 (codified at NJ. Stat. Ann. app. A. § 9-30 to -32 (West 1992)).

109. *See Worthington v. Fauver*, 88 NJ. 183, 192-93, 440 A.2d 1128, 1132-33 (1982), for a history of the evolution of the Disaster Control Act. One state constitutional provision deals with emergencies. Article IV, section VI, paragraph 4 of the New Jersey Constitution provides:
The Legislature, in order to insure continuity of State, county and local governmental operations in periods of emergency resulting from disasters caused by enemy attack, shall have the power and the immediate and continuing duty by legislation (1) to provide, prior to the occurrence of the emergency, for prompt and temporary succession to the powers and duties of public offices, of whatever nature and whether filled by election or appointment, the incumbents of which may become unavailable for carrying on the powers and duties of such offices, and (2) to adopt such other measures as may be necessary and proper for insuring the continuity of governmental operations. In the exercise of the powers hereby conferred the Legislature shall in all respects conform to the requirements of this Constitution except to the extent that in the judgment of the Legislature to do so would be impracticable or would admit of undue delay.
NJ. Const. art. IV, § VI, P 4. This provision, which became effective December 7, 1961, only applies to enemy attacks and appears to be the result of concerns about nuclear attack during the cold war. *See Williams*, New Jersey Constitution, *supra* note 36, at 71-72.

110. NJ. Stat. Ann. app. A. § 9-33 (West 1992).

111. Id. at § 9-34.

112. Id. at § 9-45.

113. Id.

114. 88 NJ. 183, 440 A.2d 1128 (1982).

115. 132 NJ. 141, 623 A.2d 763 (1993).

120. Id. at 194, 440 A.2d at 1137. The County argued that overcrowding was not "unusual" because it had been foreseen as far back as 1977 and that it was not an "incident" because it was not sudden and unforeseen. Both arguments were rejected as too narrow an interpretation of the statute. Id.; *see also Worthington v. Fauver*, 180 NJ. Super. 368, 379, 434 A.2d 1134, 1140 (App. Div. 1981) (Joelson, J., dissenting) (asserting that the emerging order was within the Governor's statutory powers); NJ. Stat. Ann. app. A. § 9-33.1(1) (defining "disaster").

121. *Worthington*, 88 NJ. at 202-04; 440 A.2d at 1137-38. However, while the Governor did not have authority to issue orders on a permanent basis, the court stated that the Legislature could grant the Governor that authority. Id. at 203, 440 A.2d at 1138. The Legislature did just that when the court later invalidated similar executive orders in *County of Gloucester v. State*, 132 NJ. 141, 623 A.2d 763 (1993). *See infra* notes 129-38.

129. Id. at 204, 440 A.2d at 1138-39.

130. *County of Gloucester v. State*, 132 NJ. 141, 149, 623 A.2d 763, 767 (1993) (quoting *Worthington*, 88 NJ. 183, 203-04, 440 A.2d 1128, 1138 (1982)).

131. Id. at 150, 623 A.2d at 768.

132. Id. at 153, 623 A.2d at 769.

133. Id.

134. Id.

135. Corrections–Prison Crowding–Emergency Act, ch. 12, § 1-4, 1994 NJ. Sess. Law Serv. ch. 1 (West). That Act provides: "There is hereby declared a continuing state of emergency in connection with the crowding in State prisons and other penal and correctional institutions of the New Jersey Department of Corrections." Id. at ch. 12, § 2.

209. Article III, paragraph 1 of the New Jersey Constitution provides:
The powers of the government shall be divided among three distinct branches, the legislative, executive, and judicial. No person or persons belonging to or constituting one branch shall exercise any of the powers properly belonging to either of the others, except as expressly provided in this Constitution. NJ. Const. art. III, P 1 . . .

210. *See supra* notes 18-20 and accompanying text.

State and local emergency managers train together on legal responsibilities. David Barrabee photo.

Chapter 11

STATE AND LOCAL RESPONSIBILITIES

The following statute sets out one state's requirements for a statewide emergency operations plan (EOP). Consider whether there are any hazards that the plan is not required to address.

INDIANA CODE 10-14-3-9 (2010)

Emergency Operations Plan

(a) The agency shall prepare and maintain a current state emergency operations plan. The plan may provide for the following:
  (1) Prevention and minimization of injury and damage caused by disaster.
  (2) Prompt and effective response to disaster.
  (3) Emergency relief.
  (4) Identification of areas particularly vulnerable to disaster.
  (5) Recommendations for:
    (A) zoning;
    (B) building;
    (C) other land use controls;
    (D) safety measures for securing mobile homes or other nonpermanent or semipermanent structures; and
    (E) other preventive and preparedness measures designed to eliminate or reduce disaster or its impact that must be disseminated to both the fire prevention and building safety commission and local authorities.
  (6) Assistance to local officials in designing local emergency action plans.
  (7) Authorization and procedures for the erection or other construction of temporary works designed to protect against or mitigate danger, damage, or loss from flood, conflagration, or other disaster.
  (8) Preparation and distribution to the appropriate state and local officials of state catalogs of federal, state, and private assistance programs.
  (9) Organization of manpower and chains of command.
  (10) Coordination of federal, state, and local disaster activities.
  (11) Coordination of the state disaster plan with the disaster plans of the federal government.
  (12) Other necessary matters.
(b) The agency shall take an integral part in the development and revision of local and interjurisdictional disaster plans prepared under section 17 [IC 10-14-3-17] of this chapter. The agency shall employ or otherwise secure the services of professional and technical personnel capable of providing expert assistance to political subdivisions, a political subdivision's disaster agencies, and interjurisdictional planning and disaster agencies. These personnel:
  (1) shall consult with subdivisions and government agencies on a regularly scheduled basis;
  (2) shall make field examinations of the areas, circumstances, and conditions to which

particular local and interjurisdictional disaster plans are intended to apply; and

(3) may suggest revisions.

(c) In preparing and revising the state disaster plan, the agency shall seek the advice and assistance of local government, business, labor, industry, agriculture, civic and volunteer organizations, and community leaders. In advising local and interjurisdictional agencies, the agency shall encourage local and interjurisdictional agencies to seek advice from the sources specified in this subsection.

(d) The state disaster plan or any part of the plan may be incorporated in rules of the agency or by executive orders.

(e) The agency shall do the following:

(1) Determine requirements of the state and political subdivisions for food, clothing, and other necessities in the event of an emergency.

(2) Procure and pre-position supplies, medicines, materials, and equipment.

(3) Adopt standards and requirements for local and interjurisdictional disaster plans.

(4) Provide for mobile support units.

(5) Assist political subdivisions, political subdivisions' disaster agencies, and interjurisdictional disaster agencies to establish and operate training programs and public information programs.

(6) Make surveys of industries, resources, and facilities in Indiana, both public and private, necessary to carry out this chapter.

(7) Plan and make arrangements for the availability and use of any private facilities, services, and property, and if necessary and if the private facilities, services, or property is used, provide for payment for the use under agreed upon terms and conditions.

(8) Establish a register of persons with types of training and skills important in emergency prevention, preparedness, response, and recovery.

(9) Establish a register of mobile and construction equipment and temporary housing available for use in a disaster emergency.

(10) Prepare, for issuance by the governor, executive orders, proclamations, and regulations necessary or appropriate in coping with disaster.

(11) Cooperate with the federal government and any public or private agency or entity in achieving any purpose of this chapter and in implementing programs for disaster prevention, preparation, response, and recovery.

(12) Do other things necessary, incidental, or appropriate to implement this chapter.

(f) The agency shall ascertain the rapid and efficient communications that exist in times of disaster emergencies. The agency shall consider the desirability of supplementing these communications resources or of integrating these resources into a comprehensive intrastate or state-federal telecommunications or other communications system or network. In studying the character and feasibility of any system, the agency shall evaluate the possibility of multipurpose use of the system for general state and local governmental purposes. The agency shall make appropriate recommendations to the governor.

(g) The agency shall develop a statewide mutual aid program to implement the statewide mutual aid agreement.

B. THE STRUCTURE AND RESPONSIBILITIES OF STATE AND LOCAL EMERGENCY MANAGEMENT AUTHORITIES

William C. Nicholson

Emergency management is a young and growing field. With the creation of the Federal Emergency Management Agency (FEMA) in 1978, the all-hazards approach to emergency management was born, replacing the nuclear attack image of civil defense. Since then, emergency management has evolved as an international discipline featuring ever-widening fields of research, necessary in a world rapidly becoming more technologically complex and dangerous.

As an all-hazards discipline, emergency management is responsible for protection from all potential dangers, through the entire disaster life cycle. As described by FEMA, the disaster life cycle sums up the process through which emergency managers prepare for emergencies and disasters, respond to them when they occur, help people and institutions recover from them, mitigate their effects, reduce the risk of loss, and prevent disasters such as catastrophic, uncontrolled wildfires from occurring.

State and local emergency management agencies (SEMA and local EMAs) conduct and coordinate these activities in the cycle of disaster. Local EMAs are the "point of the spear" in first direct contact with the emergency or disaster event.

Requirements for Emergency Management Performance

Planning, training, and exercising requirements for local EMAs are found in the Performance Partnership Agreement (PPA) and the Cooperative Agreement (CA) with the Federal Emergency Management Agency (FEMA). These agreements describe the conditions which must be met before the federal government, through SEMA, will release emergency management funds to local units of government. The PPA, which forms a strategic plan, is revised every five years.

The PPA provides long-term guidance to the state agency and FEMA. The CA is an agreement reached every year that sets specific goals for EMA in the federal fiscal year, which runs from October 1 through September 30. The responsibilities of Local EMAs take the form of "Outputs," whose fulfillment SEMA oversees and reports to FEMA. SEMA co-operates with Local EMAs to reduce these outputs to specific tasks, which are called "Compliance Requirements."

Some counties in various states may not choose to comply with the terms of the PPA and CA, and hence do not receive federal funds. The fact that the vast majority of counties are in compliance with these documents, however, may be found by a court to establish a standard of care, creating potential legal liability for those units of government with a noncompliant EMA.

Roles of State Emergency Response Commissions (SERC) and Local Emergency Planning Committee (LEPC)

Despite the description of emergency management as an all-hazards discipline, state and local units of government are required by the federal Emergency Planning and Right to Know Act (EPCRA) to split off an important part of emergency management to another entity. EPCRA is part of Title III of the Superfund Amendment and Reauthorization Act of 1986 (SARA Title III).[1] EPCRA requires planning for releases of extremely hazardous substances (EHS) to be assured by a State Emergency Response Agency (SERC).[2] The SERC in turn establishes emergency planning districts and supervises Local Emergency Planning Committees (LEPCs).[3] LEPCs are responsible for the actual planning for EHS releases.[4]

The LEPC has specific roles that are mandated by law, some of which overlap the broad responsibilities of Local EMA. LEPCs are part of the SERC, and therefore have the status of a state agency. While Local EMA makes reports to the State Emergency Management Agency (SEMA), the EMA is not a smaller unit of SEMA, but rather is a part of local government. These differences in organization can potentially have a profound effect on the interrelationship between the LEPC and the SERC, as well as between the LEPC and the Local EMA.

---

[1] 42 USC §§ 11001-11050 (2010). Indiana parallel sections to EPCRA are cited herein.
[2] 42 USC § 11001(a) (2010).
[3] 42 USC § 11001(b), (c) (2010).
[4] 42 USC § 11003 (a) (2010).

SERC and LEPC Interface

In most states, the requirements of EPCRA have been enacted into state law.[5] Most states have taken the further step of enacting into state law their own additional implementing legislation. These additional laws cover the SERC's organization and specific duties of the SERC and LEPCs.[6]

The role of the SERC is fourfold.[7] First, the SERC encourages and supports the development of emergency planning efforts to provide information about potential chemical hazards. Second, the SERC assists the state in complying with the requirements of SARA. Third, the SERC designs and supervises the operation of emergency planning districts, which are each served by an LEPC.[8] Fourth, the SERC gathers and distributes information needed for effective emergency response planning.

LEPCs have significant responsibilities as well.[9] First, in every state, the LEPC must satisfy the requirements of SARA.[10] Some states require more. In Indiana, for example, the following are required by law: The LEPC must prepare and submit, at least yearly, a roster of committee members for the SERC's approval. The LEPC must meet at least quarterly. The LEPC must submit the fiscal report required under Indiana Code 6-6-10-8. It is important to note that, in accordance with Indiana Code 6-6-10-7(a), a LEPC cannot receive its funds unless these four requirements have been met.[11]

Indiana's SERC exercises its oversight duties regarding LEPCs activities through specific requirements for financial reporting. The report required under Indiana Code 6-6-10-8 is the LEPC's yearly fiscal report to the Indiana SERC of monies expended in the preceding year origi-nating in the emergency planning and right to know fund. Money from the fund may be spent on seven categories:

1. preparing and updating the SARA Title III plan;
2. establishing and implementing procedures for dealing with public information requests;
3. training for emergency response planning, information management, and hazardous materials ("HAZMAT") incident response;
4. equipping a HAZMAT response team which provides response throughout the LEPC's district if the equipment is consistent with team training;
5. purchasing communications gear for the LEPC's administrative use;
6. paying LEPC members a $20 meeting stipend; and
7. paying for SARA Title III risk communication, chemical accident related, and accident prevention projects submitted to and approved by the IERC.

As the preceding discussion indicates, the SERC is the loving yet potentially firm parent that oversees the activities of LEPCs. The legal relationship between the Indiana SERC and LEPCs has, in fact, led the Indiana Attorney General to opine that LEPCs are state agencies.

LEPC/SEMA Interface

In different states, varying agencies may fulfill administrative support responsibilities for the SERC. Usually, the state Department of Environmental Management or Protection assists the SERC. In Indiana, SEMA fulfills that support role. This was a change from the assistance previously being given by the Department of

---

[5] Indiana Code 13-25-2 (2010).
[6] Indiana Code 13-25-1 (2010).
[7] Indiana Code 13-25-1-6(a) (2010).
[8] 42 USC § 11001(b), (c) (2010).
[9] Indiana Code 13-25-1-6(b) (2010).
[10] 42 USC §§ 11001-11050 (2010).
[11] In addition, the Indiana SERC may withhold funds from an LEPC under Indiana Code 6-6-109 for failing to do any of seven requirements annually: failure to submit the fiscal report, failure to provide proof of the published legal notice required by SARA, failure to submit an updated plan, failure to submit current LEPC bylaws, failure to present evidence of a compliant exercise, failure to provide a current roster, and failure to submit minutes of quarterly LEPC meetings.

Environmental Management. It must be noted here that Indiana SEMA's role under the law is only of a supportive nature: it provides a home for the personnel who carry out tasks directed by the Indiana SERC. SEMA does not supervise LEPCs. The SERC supervises LEPCs, as mandated by the federal EPCRA statute.

SEMA and Local EMA Interface

In contrast to the LEPCs, Local EMAs are clearly part of local government.[12] As described above, the LEPC has specific roles that are mandated by law. Some of these legal duties overlap the broad responsibilities of the county EMA. Both are tasked with the preparedness phase of emergency management: planning, training, and exercising responsibilities. Frequent common membership of the LEPC and the county's EMA Advisory Council also has the potential to create conflicts.

Planning Responsibilities

The basic emergency management organization is Local EMA. Under Indiana Code 10-14-3-17, for example, the county EMA is required to "...prepare and keep current a local disaster emergency plan for its area." As previously mentioned, the LEPC's planning requirement under EPCRA is specifically limited to SARA Title III chemical discharges.[13]

Under the all-hazards approach to emergency management, potential emergencies involving HAZMAT must be anticipated. The Local EMA's emergency operations plan ("EOP") must include an appendix for response to HAZMAT incidents which is prepared and kept current. Potential HAZMAT emergencies include the extremely hazardous materials covered by SARA Title III, as well as a variety of other substances. In Indiana, this mandate is dealt with through

planning Compliance Requirements for the Local EMA. These include submission of the local EOP, in whole, including the SARA Title III plan as a hazard-specific appendix.

The overlapping planning responsibilities of LEPCs and EMAs may, from time to time, result in some duplication of effort, and even disagreements as to the proper approaches to the nine planning elements[14] required of LEPCs. In order to integrate the planning responsibilities of EMAs with those of LEPCs, the Indiana SERC's approach, by policy, was to mandate that the SARA Title III plan should be a hazard-specific appendix to the EOP. The SARA Title III plan, although an appendix to the EMA's EOP, remains the responsibility of the LEPC. The SARA Title III plan must be updated on a yearly basis.[15]

As an appendix, the SARA Title III plan should take advantage of the remainder of the EOP, referring such matters as community notification, evacuation, training, and exercising to the appropriate portions of the EOP, rather than re-inventing the wheel.

Training Responsibilities

Training and exercising are matters for which LEPCs may legitimately expend funds. LEPCs are required to include training in their plans.[16]

The PPA sets the following objectives for training: develop and deliver training to individuals and groups with key emergency management responsibilities, focusing on capability shortfalls in relation to identified risks. Such training is to be conducted at all levels of government and the private sector, and will include "on the job" training for EMA managers from unaffected areas during actual emergencies.

The CA contains specific requirements for every year that move states, and their local units of government, closer to the goals specified in

---

[12] Under Indiana Code 10-14-3-7 (2010), the legislature found that local EMAs are needed to preserve the lives and property of the state. The statute further states that "Each county shall maintain...a county emergency management organization."
[13] 42 USC § 11003 (c) (2010), Indiana Code 13-25-2-5 (2010).
[14] 42 USC § 11003 (c) (2010).
[15] 42 USC § 11003 (a) (2010).
[16] 42 USC § 11003 (c)(8) (2010).

the PPA. The training specifications of the CA set objectives for training. The following items were required objectives in a recent Indiana CA: continue to train the state forward response team, provide assistance to state agencies in development of SOPs, conduct earthquake training of state and local agencies, continue to deliver training through SEMA and the state's Public Safety Training Institute (PSTI), develop new training through SEMA and PSTI, provide an aggressive public awareness program, and conduct SARA Title III training. Required training Compliance Requirements for Local EMAs included: one SEMA training course annually for each EMA staff member for whom reimbursement for salaries was received from the federal government.

Exercising Responsibilities

EPCRA requires that the LEPC's plan must be exercised at least once annually.[17] SERCs typically permit this exercise to be a tabletop, functional, or full-scale exercise. SERCs frequently require advance notification and brief description of the exercise to be conducted so that SERC staff may monitor the exercises.

The EPCRA statute specifically refers to an annual "exercise."[18] Due to this fact, SERCs that permit an actual event to be used to substitute for an exercise often impose specific requirements on that event for it to qualify as the required exercise. Typically, an actual SARA Title III release must meet certain threshold tests before it may qualify to take the place of the yearly exercise. These requirements may include, among others: the number and type of responding units and the type and quantity of chemical involved.

A PPA sets the exercise objectives like the following examples: (1) establish a comprehensive exercise program to test and evaluate all aspects of the state's emergency management system, including Radiological Emergency Planning (REP), Chemical Stockpile Emergency Preparedness Planning (CSEPP), Earthquake (EQ), HAZMAT,

Terrorism, etc.; (2) conduct statewide tabletop exercises that integrate federal, state, and local response forces; and (3) emphasize evaluation and correction of identified deficiencies.

The CA requires exercises like the following during a typical federal fiscal year: participation in federally developed exercises that evaluate federal response planning concepts, monitor all exercises and develop a process to correct deficits, and require that all participants in the federal reimbursement program meet the exercise requirements set by FEMA including correction of deficiencies. Required exercise Compliance Requirements for Local EMAs normally include: one "real world" event or one full-scale exercise or one functional exercise or two tabletop exercises. All events or exercises must meet set criteria.

Working together, Local EMAs and LEPCs can unite their requirements in a combined exercise that will satisfy both the SERC and SEMA.

Common Membership Issues

Some states engender community support for the Local EMA through the vehicle of a legislatively mandated body, such as the County Emergency Management Advisory Council in Indiana. While this approach can build valuable relationships, it may also create potential legal difficulties. LEPCs and EMA Advisory Councils frequently have members in common, particularly in rural areas. The Local EMA director is almost always an LEPC member, and may even be the chairman of the LEPC. Sometimes, this situation may tend to blur the lines of responsibility between the EMA and the LEPC. As indicated above, LEPCs and EMAs are parts of different governmental entities. These diverging lines of authority create legal requirements for separate control, separate finances, and separate meetings.

Separate Control

The two groups must be separately controlled due to the legal lines of authority above the LEPC

---

[17] 42 USC § 11003 (c)(9) (2010).
[18] Id.

and the Local EMA. The twelve (12) membership categories for the LEPC specified at EPCRA Section 301(c) must be included in the roster submitted to the IERC for approval at least annually by February 14. The six (6) membership categories for the County Emergency Management Advisory Council, which oversees local EMA activities, are specified at Indiana Code 10-14-3-17(c).

Separate Finances

Financing sources and responsibilities are also different for the two agencies. The Local EMA's budget is part of the local unit of government's budget. If the unit is in compliance with FEMA's requirements, partial reimbursement will be provided, as discussed above. The local unit of government retains control over the EMA's budget due to that entity's being a part of local government.

The sources of LEPC finances vary from state to state. Some states impose fees on all entities that transport EHS through the state. Others impose filing fees on Tier II filers. In Indiana, LEPC funds are distributed through county government. Other states distribute the funds directly from the SERC to the LEPC. The Indiana statute requires that the county fiscal body "shall" appropriate the funds requested by the LEPC in a compliant spending plan.[19] The intent and effect of this law is that the county act as administrator of the funds, but that it have no say as to the actual expenditure of the funds.

The funds derived from the Indiana Local Emergency Planning and Right to Know Fund may, by law, be expended only on LEPC activities. The reason for this limitation is that the funds come from SARA Title III fees, and the funds are earmarked for expenditure on preparing for response to releases of the substances generating the funds. Although some county leaders may question this straightforward limitation, it derives from both federal and state statute. Other states that distribute funds directly

to their LEPCs without going through county government avoid the difficulties that the Indiana system has sometimes generated.

Separate Meetings

Frequently, given the common membership of LEPCs and EMA Advisory Councils, meetings will be scheduled on the same night for both the EMA Advisory Council and the LEPC. Often at such meetings, several different subjects will be discussed. Such matters may be LEPC or EMA issues, or even issues that involve both entities. Legally, the matters must be separated and handled in different meetings. One group's meeting must be held and adjourned prior to dealing with the matters from the perspective of the other entity.

Over the years, the requirement of state law that LEPCs meet every three months has proven burdensome, particularly in rural jurisdictions. This has resulted in some members leaving the groups. A new law went into effect on July 1, 2010, allowing meetings to take place once every six months.[31]

A Different Approach

As previously mentioned, the LEPC has specific roles that are mandated by law, some of which overlap the broad responsibilities of Local EMA. This has led some observers to believe that the division between the two entities is artificial and not helpful. North Carolina took a different approach to allocating the responsibilities of the LEPC and Local EMA Advisory Council. In that state, the LEPC is specifically mandated by state law to provide guidance to the Local EMA. This means that, in North Carolina, the LEPC performs the role of the Local EMA Advisory Council. Therefore, in North Carolina, no parallel organizations are required. The LEPC's duties under state law are similar to those described above. The LEPC also has the local advisory obligations discussed above. The multitude

---

[19] Indiana Code 6-6-10-7(d) (2010).
[31] Indiana Code 13-25-1-6(b)(3) (2010).

of duplicative meetings that led the Indiana legislature to cut LEPC meetings from 4 yearly to 2 does not exist. In fact, since the two are united, in North Carolina the LEPC is likely to meet far more often than 4 times a year. Monthly meetings of the group are commonplace in North Carolina.

CONCLUSION

The common goals of saving lives and protecting property unite Local EMAs and LEPCs, just as they unite SEMA and the SERC. All involved must, however, work with both mutual respect and a clear understanding of the legal constraints on Local EMAs and LEPCs.

Local EMAs and LEPCs can combine resources in appropriate ways to save funds and prevent repetitive activity. These steps must be taken with care, however, since the two agencies are units of different levels of government. Working carefully together, Local EMAs and LEPCs can combine forces to help assure that the best possible steps have been taken to plan, train and exercise for all hazards, including potential SARA Title III Extra Hazardous Substances releases.

———————————————

QUESTIONS

1. Why do communities engage in emergency planning?
2. What does "all hazards" mean to you?
3. Discuss how the contents of a local plan might vary from those in a state plan such as that required by the statute above.
4. Where might a rural community find qualified persons to serve on the LEPC or County Emergency Management Advisory Council? Consider the same issue for an urban area.
5. As the article indicates, planning involves much more than putting words on paper. Describe additional planning activities that

LEPCs and County EMA Councils must undertake.
6. How might the LEPCs' and County EMA Councils' activities be accomplished with fewer procedural requirements?
7. Discuss the advantages and disadvantages of the Indiana and North Carolina approaches to addressing the planning requirements described above.
8. Describe the planning arrangements you would decree to address the responsibilities of state and local governments if you were starting with a clean slate.

New Orleans, LA, September 4, 2005 – Mayor Nagin, Mike Brown, Secretary Chertoff. Jocelyn Augustino/FEMA.

Nashville, TN, May 10, 2010 – Resident describes flooding to Napolitano and Fugate. David Fine/FEMA.

Spencer, SD, June 5, 1998 – James Lee Witt with disaster victims. Andrea Booher/FEMA.

Washington, D.C., October 1, 2001 – FEMA Director Allbaugh introduces President Bush. Greg Schaler/FEMA.

Chapter 12

FEDERAL EMERGENCY MANAGEMENT: HISTORY, EVOLUTION, AND CHALLENGES

William C. Nicholson

Emergency management on the federal level is a creature of constant change. Since the first enactment to support emergency response in 1803, evolution has been the continuing theme of federal emergency management legislation and underlying assumptions and goals. The current structure for federal response is the National Response Framework [2008] replacing the National Response Plan [2004] and other federal plans with which it interacts as appropriate. Yet, with creation of the Department of Homeland Security, perhaps the greatest changes are yet to come.

I. A BRIEF HISTORY OF FEDERAL EMERGENCY MANAGEMENT: THE ORIGINS[1]

The Federal Emergency Management Agency – a part of the Department of Homeland Security as of March 1, 2003 – is responsible for what are termed the four phases of emergency management: mitigation, preparedness, response, and recovery. Its roots date back to the Congressional Act of 1803, which is generally thought of as the first piece of disaster legislation. The law provided assistance to a New Hampshire town after a devastating fire. In the next century, legislation responded on a case-by-case basis to a variety of over 100 disasters, including hurricanes, earthquakes, floods, and other natural disasters.

The precursor national structure for emergency organization was created in response to the so-called "Great War." In fact, Congress created the Council of National Defense (CND) on August 29, 1916, prior to America's entry into the First World War. President Wilson established the Advisory Commission to the Council of National Defense (ACCND) on October 11, 1916. The CND put in place a State Council Section

on April 6, 1917. The State Council Chairmen requested all Governors to create State Councils of Defense, and by April 9, 1918, these had been established in every state. By the date of the Armistice ending the First World War on November 11, 1918, 182,000 local units had been created.

The advent of the Armistice led to rapid the dissolution of State and Local Defense Councils after November 11, 1918. In response to this rapid dissolution, the Council of National Defense requested State and Local Councils to stay intact on December 12, 1918.

As time went by, a federal approach to problems exceeding the capacity of states to respond to disasters including those generated by natural events became more general. During the 1930s, the Reconstruction Finance Corporation was authorized to generate disaster loans for repair and reconstruction of various public facilities following a large earthquake. This precedent later broadened to cover other types of disasters. By 1934, the Bureau of Public Roads began under federal authority to provide funding for highways and bridges harmed by natural disasters. Under the 1941 Flood Control Act, the U.S. Army Corps of Engineers was given increased authority to put in place flood control projects in a variety of waterways. This one-at-a-time arrangement for provision of disaster assistance was not consistent. The result of this experience was legislation requiring greatly increased collaboration between the various federal departments. The President was authorized to organize these executive branch activities.

During the 1940s, the Second World War led to revival of structures fashioned during the Great War, as well as creation of new organizations. The Council of National Defense that Congress created in 1916 was reactivated on May 29, 1940. A new Office of Emergency Management was established in the Executive Office of the President by Executive Order (EO 8248) on September 8, 1939, not coincidentally one week after the invasion of Poland by Nazi Germany. President Roosevelt asked Congress for $150 million for Civil Defense on February 24, 1940. In December, 1940, the Model Uniform State Civil Defense Law was sent to the states. Executive Order 8757 established the Office for Civilian Defense in Office of Emergency Management on May 20, 1941. Emergency Medical Services was established on July 5, 1941. The familiar official Civil Defense Insignia was adopted on July 24, 1941. Japanese attacked Pearl Harbor on December 7, 1941. In a repeat of the aftermath of the First World War's desire for a return to normalcy, the administration abolished the Office for Civilian Defense on June 30, 1945. The atomic bomb fell on Nagasaki on August 9, 1945. Shortly after, World War 2 ended.

Almost without delay, the Cold War began. On March 27, 1948, the Secretary of Defense, created by the National Security Act of 1947, established the Office of Civil Defense Planning. A year and a half later, on September 23, 1949, President Truman announced that Russians had exploded their own atomic bomb. The Federal government began to send Civil Defense Planning Advisory Bulletins to Governors on October 5, 1949. Radiological monitoring and medical and health courses sponsored by federal government begin on March 27, 1950.

On September 30, 1950 Congress passed a law "To authorize Federal assistance to states and local governments in major disasters and for other purposes," which vested in the President authority to coordinate activities of all federal agencies in providing disaster assistance.[2] To support the civilian side of this effort, EO 10186 created the Federal Civil Defense Administration (FCDA) on December 1, 1950. The vital nature of widespread mutual aid, and the necessity that it be based on well-written identical documents was quickly recognized. The FCDA submitted a model interstate mutual aid agreement to all Governors on December 12, 1950, a mere eleven days after that organization's creation! On the international front, Canada and the United States effected a civil defense mutual aid agreement on March 27, 1951.

The Federal Civil Defense Act of 1950[3] established the Federal Civil Defense Administration (FCDA) as independent agency on January 12, 1951. FCDA announced the availability of matching federal funds for construction of fallout shelters in critical target areas on February 12, 1951. This was the first of many opportunities for state and local governments to augment their preparedness efforts through use of federal matching funds. EO 10611 created the Civil Defense Coordinating Board to ensure participation by all federal agencies in the national civil defense structure on May 11, 1955.

On August 8, 1958, Congress passed a law that clarified responsibility for Civil Defense, vesting it jointly on Federal, State and local governments, and authorizing contribution of funds to states and locals for (a) personnel and administrative expenses and (b) students attending Civil Defense schools.[4]

During the 1950s, there were relatively few disasters compared with later decades. During the 1960s and early 1970s, in contrast, enormous disasters occurred in different parts of the nation. These called for very large response and recovery endeavors undertaken by the Federal Disaster Assistance Administration, which was located within the Department of Housing and Urban

Development. Natural disasters included Hurricane Carla, which struck in 1962, Hurricane Betsy in 1965, Hurricane Camille in 1969, and Hurricane Agnes in 1972. Huge earthquakes included the Alaskan Earthquake in 1964 and the 1971 San Fernando Earthquake in Southern California. These happenings focused national consideration on the prevalence of natural disasters and resulted in significantly increased legislation. The National Flood Insurance Act,[5] passed in 1968, created new flood protections for homeowners. The Disaster Relief Act, enacted in 1974, [Public Law 93-288]created the mechanism for Presidential disaster declarations.

Emergency and disaster activities were still, however, not united. Adding to the hazards associated with natural disasters, the dangers posed by nuclear power plants and the transportation of hazardous materials resulted in the involvement of more than 100 federal agencies in some feature of disasters, hazards, and emergencies. At the state and local level, many programs and policies similar to those of the federal government compounded the intricacy of federal disaster relief efforts. Concerned by these duplications of effort, the National Governor's Association worked to lessen the number of agencies with which state and local governments were required to work. The Governors requested President Jimmy Carter, himself a former state chief executive, to unify federal emergency functions. The theory was the STATES would have "one-stop shopping"!

II. CREATION OF FEMA

President Carter's Reorganization Plan No. 3 of 1978,[6] issued pursuant to 5 USC §§ 901 et seq. (1978), put together many previously separated federal disaster responsibilities into the newly created Federal Emergency Management Agency (FEMA).[7] FEMA absorbed a number of agencies, including the Federal Insurance Administration, the National Fire Protection and Control Administration, the Federal Preparedness Agency of the General Services Administration, and the Federal Disaster's Assistance Administration activities from HUD. The Defense Department's Defense Civil Preparedness Agency also transferred to the new agency.

President Carter named John Macy to be the first director for FEMA. Macy accentuated the similarities between preparedness for natural hazards and that needed for civil defense measures. This was the time when FEMA began to develop the Integrated Emergency Management System with an all-hazards approach that continues to the present day. It should be noted that a series of classified directives were issued concerning the federal civil defense effort, the last being NSDD-66 in 1992, all designed to dictate changed directions in civil defense policy. But in reality from 1980 on civil defense was largely a state grant program not a factor in the strategic nuclear balance. This system includes direction, control, and warning systems that are common to all varieties of emergency from minor ordinary natural events to the supreme calamity – war.

A multitude of remarkable tests confronted the new agency during its early life. These challenges highlighted the complexity of all-hazards emergency management. Among the first disasters and emergencies faced by FEMA were the toxic waste problems at Love Canal, the Cuban refugee crisis, and the release of radiation from the Three Mile Island nuclear power plant.

In 1994, Public Law 103-337 repealed the federal Civil Defense Act of 1950 and incorporated some portions of that Act into the Robert T. Stafford Disaster Relief and Emergency Act.[8] This Congressional action continued emergency management law's evolution. The 1989 Loma Prieta Earthquake and Hurricane Andrew in 1992 continued the national focus of attention on FEMA. In 1993, President Clinton nominated James Lee Witt, the former emergency management director for Arkansas, to be the new FEMA director. Witt's experience as a state emergency manager stood him in good stead as he took the reins at FEMA. James Lee Witt's appointment to President Clinton's cabinet brought emergency management to the public's attention in a new way. Based on his background as a state director, he brought about wide changes in the way FEMA did business. Ordered by President Clinton to focus on FEMA's natural disaster programs, functions, and activities, Witt reformed disaster relief and recovery operations and brought about a strong new emphasis on the mitigation and preparedness phases of emergency management. He also emphasized to agency employees the vital nature of customer service. Under Witt, FEMA successfully handled a wide variety of natural disasters.

He also led FEMA to expand the range of disasters managed by the agency, such as coordination of the response to the crash of TWA flight 880 off Long Island, New York on July 17, 1996. With the end of the Cold War, James Lee Witt was able to change the allocation of FEMA's limited resources from traditional civil defense activities into more innovative disaster relief, recovery, and mitigation programs. Strangely perhaps just before Witt's confirmation by the Senate the first WTC bombing occurred to which FEMA responded. Again in 1995 the bombing of the Murrah Building in Oklahoma City provided a geographically confined attack but again FEMA led the response.

III. THE SEPTEMBER 11, 2001 ATTACKS – TERRORISM TAKES CENTER STAGE

The terrorist attacks of September 11, 2001, just 9 months into the first George W. Bush Administration, brought natural preparedness and homeland security to the top of the agency's priority list. The challenges of leading response to and recovery from the terrorist attacks put the agency to the test in unparalleled ways. FEMA synchronized its work to support and partner with the newly established White House Office of Homeland Security.[9] FEMA's Office of National Preparedness became responsible for working, training, and equipping the nation's first responders to deal with nuclear, biological, chemical, and ordinance weapons of mass destruction. But

in passage of the Homeland Security Act of 2002 in November 2002 FEMA lost its name and status as an independent Executive Branch Agency when it became part of DHS on March 1, 2003. It new title was the Emergency Preparedness and Response Directorate.

The first edition of this book confidently stated at this point that "FEMA's 'all-hazards' approach to disasters will continue to apply to homeland security issues, as all federal plans will be consolidated by the department into an all-hazard plan." While this is nominally true, the reality has proven to be more problematic for FEMA and the citizens it serves.

IV. HURRICANE KATRINA AND CALLS FOR FEMA'S INDEPENDENCE

After President Bush's election, he replaced James Lee Witt as head of FEMA with Joseph Allbaugh, the former national campaign manager of Bush-Cheney 2000. Allbaugh, who had no experience in emergency management, appointed his longtime friend Michael Brown to be General Counsel of FEMA. In September 2001, Allbaugh named Brown, who also had no previous emergency management background, to be his deputy. In 2002, just as FEMA became part of DHS, Allbaugh left federal service and Brown replaced him as Under Secretary of DHS and

FEMA Director. Allbaugh went on to become a Washington lobbyist, and in 2005 he was described as "the man to see if you want a contract in Iraq, or a piece of the action on homeland security, or, apparently, a shot at rebuilding New Orleans."[10]

As indicated in the section on NIMS and the NRP/NRF in Chapter 6, even before DHS was created, there were tensions between its terrorism and all hazards missions. Some warned that tiny FEMA would be swallowed by the mega-agency and lose its identity and missions. In the period immediately following DHS' creation, that certainly appeared to be the case. "It's a competing balance inside the Department [of Homeland Security]," former DHS Undersecretary and FEMA head Michael Brown stated in 2004.[11] "The department has two missions. One is to prevent terrorism. The other is to prepare the country for all hazards. My job is to convince and show and lead by example that the all-hazard approach fits into their terrorism prevention."[12] Clearly, the focus at DHS was how all hazards should "fit into" a single hazard – terrorism prevention.

Many of those in senior positions at DHS commented on the amount of time they spent testifying before various Congressional committees. This is due to the fact that when DHS was created, the committees involved did not give up their jurisdiction to a single oversight committee. Instead, 108 committees have oversight. With responsibility so parceled out, it may perhaps be the case that accountability is lacking. The erosion of Congress' oversight function for DHS to the executive branch is a natural result of the way the Homeland Security Act of 2002 set up the committee structure. It is one of an ongoing litany of ways in which the separation of powers set in place by the founders of the United States has blurred since the nation's founding.

The Bush administration's priorities were reflected in terrorism-focused funding choices. President Bush's 2005 budget proposal would have transferred, reduced, or eliminated grants aimed at supporting local and state "all-hazards" efforts, focused on floods, wildfires, and blizzards, into programs focused solely on terrorism. Emergency managers in state and local governments questioned funding decisions that manifest a "myopic focus on terrorism."[13] The Bush administration tried to slash funding for Emergency Management Performance Grants (EMPGs),[14] as well as limiting the portion of the money that could be spent on personnel. The administration proposed a fiscal year 2006 EMPG budget cut from $180 million to $170 million.[15] In response, the International Association of Emergency Managers ("IAEM"),[16] at its mid-year meeting on February 27, 2005, adopted a resolution to increase

EMPG funding from $180 million to $280 million and keep it as a separate account rather than commingling it with antiterrorism funds.[17] EMPG funding continued to grow until 2011, although more slowly than IAEM would prefer.

EMPG Funding – FY 2002 through FY 2008

FY 2002	$134,693,410
FY 2003	$170,312,798
FY 2004	$204,710,257
FY 2005	$173,828,342
FY 2006	$177,655,500
FY 2007	$244,000,000
FY 2008	$291,450,000[18]

This trend continued through 2010. In 2009, the EMPG was $306,022,500.[19] The 2010 EMPG was $329.8 billion.[20] In FY 2011, EMPG funding fell a bit to $329,040,400.[21]

In 2004, former FEMA Director James Lee Witt expressed a cautionary view of FEMA's status, which contained an implied warning for the future: "[FEMA] has been buried beneath a massive bureaucracy whose main and seemingly only focus is fighting terrorism and while that is absolutely critical, it should not be at the expense of preparing for and responding to natural disasters."[22]

Witt's concerns were borne out in the aftermath of Hurricane Katrina in 2005. FEMA's performance during the Katrina response was generally poor. "It is clear the federal government in general and the Department of Homeland Security (DHS) in particular were not prepared to respond to the catastrophic effects of Hurricane Katrina."[23] The House Committee investigating Katrina found that the tension between all-hazards and terrorism missions was a factor in FEMA's problems after Katrina. "In particular, the decline in preparedness has been seen as a result of the separation of the preparedness function from FEMA, the drain of long-term professional staff along with their institutional knowledge and expertise, and the diminished readiness of FEMA's national emergency response teams."[24]

Following FEMA's poor response to Hurricane Katrina in 2005, many voices were raised calling for FEMA to be "fixed." Some called for the military to continue to take the lead role in response, as President Bush tasked it to do after FEMA's Katrina failure.[25] Others renewed pressure for FEMA to resume its status as an independent agency. The need for an independent FEMA had been expressed during consideration of the Homeland Security Act of 2002. Congressman Oberstar, in a harrowing foreshadowing of the Katrina debacle, stated during the debate on the HS Act:

> This is July 2002. Let us fast forward to July 2003. The majority has prevailed. FEMA is a box in the mammoth bureaucracy of the Department of Homeland Security. Flood waters are swirling around your city. You call for help. You get the Department of Homeland Security. The switchboard sends your call to the Under Secretary's office which looks up "disaster" on their organizational chart and sends you to the Congressional Liaison Office, which then promises to get a message back to you in 24 hours. Eventually, they find FEMA, by which time you are stranded on the roof of your house waving a white handkerchief and screaming for help. FEMA, the word comes back, sorry, is looking for suspected terrorists some place and will get back to you as soon as we can.[26]

Congress decided that the situation could be repaired with FEMA remaining in DHS. They adopted the Post-Katrina Emergency Management Reform Act of 2006,[27] effective fully March 31, 2007, that addressed many concerns about FEMA performance expressed after the hurricane. It returned Preparedness to FEMA, including: United States Fire Administration (USFA); Office of Grants and Training (G&T); Chemical Stockpile Emergency Preparedness Division (CSEP); Radiological Emergency Preparedness Program (REPP); and Office of National Capital Region Coordination (NCRC). Additional headquarters positions created at FEMA by the Post-Katrina Act include a Disability Coordinator, residing in the FEMA Office of Equal Rights, a Small State and Rural Advocate, a Law Enforcement Advisor to the Administrator and a National Advisory Council.

The Act specifically excluded certain elements of the Preparedness Directorate from transfer to FEMA. The legacy Preparedness Directorate was renamed the National Protection and Programs Directorate (NPPD). NPPD includes the following offices: Office of the Undersecretary; Office of Infrastructure Protection; Office of Cyber Security and Communications; Office of Risk Management and Analysis; and Office of Intergovernmental Programs. Infrastructure protection as well as risk management and analysis are functions that could be argued to be important parts of traditional preparedness. IAEM continues to lobby for return of all Preparedness Directorate assets to FEMA.[28]

The continuing tensions between the "all hazards" and terrorism missions led some to push for FEMA's separation from DHS in the immediate aftermath of President Obama's election in 2008. On February 25, 2009, Congressman James L. Oberstar, a Minnesota Democrat and chairman of the House Transportation and Infrastructure Committee, introduced H.R. 1174, the "FEMA Independence Act of 2009" whose major goal was to remove FEMA from DHS.[29] Other authoritative voices also called for FEMA to resume its independent status.[30] In contrast, Representative Bennie G. Thompson, chairman of the House Homeland Security Committee, has been a long-time supporter of keeping FEMA in DHS.[31]

In February 2009, the DHS Inspector General weighed in on the topic with a report entitled "FEMA: In or Out?"[32] The report made two major points supporting FEMA remaining a part of DHS: (1) It takes years for a complex organization to develop; and (2) Success depends on leadership more than structure. In the end, President Obama decided not to spend the political capital that would be needed to give FEMA independent cabinet status. Instead, he appointed W. Craig Fugate to be FEMA Administrator.[33] The former Emergency Management Director for the State of Florida, Fugate has years of experience in dealing with some of the most major disasters in the history of the United States. Taking this step was an implicit endorsement of the DHS Inspector General's report, as Fugate is generally believed by the emergency management community to be among the best emergency management leaders in the nation.[34]

V. FUNDING ISSUES

Historically, disaster relief funding has not been a political issue.

> Disaster assistance is an almost perfect political currency. It serves humanitarian purposes that only the cynical academic could question. It is largely funded out of supplemental appropriation and thus **DOES NOT OFFICIALLY ADD TO THE BUDGET DEFICIT** [emphasis added] . . . And it is extremely difficult to pinpoint exactly how much money the federal government spends on disasters. In the words of one FEMA staff member (who will remain anonymous) disaster assistance is the last big source of pork barrel in the federal government.[35]

Concerns about the size of the national debt and budget deficits in 2011 changed that view. Following the devastating Joplin, MO tornado outbreaks in May, 2011, House Majority Leader Eric Cantor (R-VA) announced that relief aid would be available but must be offset by cuts in the federal budget.[36] In the aftermath of Hurricane Irene, which struck the east coast in late August 2011, FEMA's reserve fund grew close to exhaustion. Mr. Cantor said that areas hit by Irene would get help, but insisted that funds to cover the payments would have to come from somewhere else in the budget. As the quote above indicates, Mr. Cantor's politicization of disaster relief contrasts with how such assistance has historically been treated.[37] In response, FEMA put work "on hold" for recovery projects in Joplin, MO, which was rebuilding from the tornadoes that flattened the city earlier in 2011, as well as six states in the deep south.[38]

In contrast to Majority Leader Cantor's desire to limit relief funding, FEMA Administrator Fugate emphasizes funding as an element of national unity, "In this country, Americans have always come to the aid of other Americans in a crisis and disaster," Fugate said on NBC's "Today" show on August 31, 2011. "We look at these large-scale disasters as something that's hard to budget for. This is a question that's best left for the appropriators and people who deal with these issues."[39]

Funding will be a continuing problem. As mentioned above, following a history of steady increases, the FY 2011 Emergency Management Performance Grant (EMPG) was reduced compared with that of the previous year. This may be the harbinger of a series of declines in funding level for the EMPG.

VI. FEMA'S FUTURE

Clearly, the decision following Hurricane Katrina was that the eggs would remain scrambled and that FEMA would stay within DHS. The Post-Katrina Emergency Management Reform Act addresses many of the concerns that arose in the wake of Katrina, but the Bush administration chose to ignore some of its requirements, and there is no reason to believe that future administrations might not do the same. FEMA's future – and that of emergency management at all levels of government – will continue to rely upon the personal relationships of the individuals who are in charge of the White House, DHS, and FEMA. This is always the case, given that these agencies are part of the Executive branch of government.

The fact that structurally FEMA continues to be under DHS means that the problems observed in the aftermath of Hurricane Katrina could manifest themselves following future disasters. FEMA is a small and relatively nimble subagency submerged in a large and clumsy bureaucracy. Al-Qaeda talks about cyber-terrorism and home-grown terrorism, and yet the DHS is still devoting vast sums of money to fighting the 9-11 hijackers. FEMA appears to be better at lessons learned, yet this expertise appears not to be respected within DHS as a whole.

The DHS agency structure dictates how funding is distributed and prioritized. DHS is and will continue to be a law enforcement-focused organization. When a funding choice comes, terrorism prevention will likely continue to be preferred over mitigation of natural hazards. Despite its demonstrated expertise in the field of all hazards emergency management, FEMA's will most likely be but one voice whose volume will wax and wane depending on the internal politics at DHS.

QUESTIONS

1. Why do we need federal emergency management, if we do at all?
2. How has federal emergency management changed over the years since its creation?
3. Perform research to find out how FEMA handled Hurricane Katrina to and how it performed before, during, and after Hurricane Irene in August 2011. Compare FEMA's actions in the two events. In your discussion, be sure to include the following elements:
   • type of event;
   • severity of event;
   • personalities involved;
   • assets available;
   • funding issues; and
   • relevant changes of law.
4. Divide the class into two groups – "FEMA in" and "FEMA out" – debate the pros and cons of FEMA being a part of DHS or being an independent agency. Include all aspects of the debate, including overall costs and benefits. Have each member of each group perform independent research and bring in an article supporting her/his view.
5. The author views the Bush Administration's failure to abide by the Post-Katrina Emergency Management Reform Act's requirements as reflecting a broader issue in the federal government. The Executive branch assumes that Legislative branch will abdicate its Constitutional responsibilities to act as a co-equal with the President and the Courts. The author posits that the failure of Congress to hold President Bush's feet to the legal fire demonstrates the correctness of this view. Do you agree with this view? If not, why not?
6. Discuss the implications of the view discussed in questions for future executive actions in emergency situations. You may wish to look at the National Emergencies Act (50 USC 1651-1651).
7. Some politicians believe that all governmental doings should privatized. Divide the class into two groups and discuss the issue of whether FEMA's activities should continue or be performed by the private sector.
8. The quote emphasized in the text states that **"disaster assistance...DOES NOT OFFICIALLY ADD TO THE BUDGET DEFICIT."** This statement from Platt's 1999 book reflects the traditional view of disaster assistance. Review the paragraph in which the quote occurs and answer the following questions:
   • Why has this been the prevailing view?
   • Why has it changed?
   • State and defend your view of the matter.
9. The closing sentence states "Despite its demonstrated expertise in the field of all hazards emergency management, FEMA's will most likely be but one voice whose volume will wax and wane depending on the internal politics at DHS." Do you agree or not? Explain your answer.

Appendix A

PRESIDENTIAL POLICY DIRECTIVE/PPD-8

THE WHITE HOUSE
WASHINGTON

March 30, 2011

SUBJECT: National Preparedness

This directive is aimed at strengthening the security and resilience of the United States through systematic preparation for the threats that pose the greatest risk to the security of the Nation, including acts of terrorism, cyber attacks, pandemics, and catastrophic natural disasters. Our national preparedness is the shared responsibility of all levels of government, the private and nonprofit sectors, and individual citizens. Everyone can contribute to safeguarding the Nation from harm. As such, while this directive is intended to galvanize action by the Federal Government, it is also aimed at facilitating an integrated, all-of-Nation, capabilities-based approach to preparedness.

Therefore, I hereby direct the development of a national preparedness goal that identifies the core capabilities necessary for preparedness and a national preparedness system to guide activities that will enable the Nation to achieve the goal. The system will allow the Nation to track the progress of our ability to build and improve the capabilities necessary to prevent, protect against, mitigate the effects of, respond to, and recover from those threats that pose the greatest risk to the security of the Nation.

The Assistant to the President for Homeland Security and Counterterrorism shall coordinate the interagency development of an implementation plan for completing the national preparedness goal and national preparedness system. The implementation plan shall be submitted to me within 60 days from the date of this directive, and shall assign departmental responsibilities and delivery timelines for the development of the national planning frameworks and associated interagency operational plans described below.

National Preparedness Goal

Within 180 days from the date of this directive, the Secretary of Homeland Security shall develop and submit the national preparedness goal to me, through the Assistant to the President for Homeland Security and Counterterrorism. The Secretary shall coordinate this effort with other executive departments and agencies, and consult with State, local, tribal, and territorial governments, the private and nonprofit sectors, and the public. The national preparedness goal shall be informed by the risk of specific threats and vulnerabilities – taking into account regional variations – and include concrete, measurable, and prioritized objectives to mitigate that risk. The national preparedness goal shall define the core capabilities necessary to prepare for the specific types of incidents that pose the greatest risk to the security of the Nation, and shall emphasize actions aimed at achieving an integrated, layered, and all-of-Nation preparedness approach that optimizes the use of available resources.

The national preparedness goal shall reflect the policy direction outlined in the National Security Strategy [May 2010], applicable Presidential Policy Directives, Homeland Security Presidential Directives, National Security Presidential Directives, and national strategies, as well as guidance from the Interagency Policy Committee process. The goal shall be reviewed regularly to evaluate consistency with these policies, evolving conditions, and the National Incident Management System.

National Preparedness System

The national preparedness system shall be an integrated set of guidance, programs, and processes that will enable the Nation to meet the national preparedness goal. Within 240 days from the date of this directive, the Secretary of Homeland Security shall develop and submit a description of the national preparedness system to me, through the Assistant to the President for Homeland

Security and Counterterrorism. The Secretary shall coordinate this effort with other executive departments and agencies, and consult with State, local, tribal, and territorial governments, the private and nonprofit sectors, and the public.

The national preparedness system shall be designed to help guide the domestic efforts of all levels of government, the private and nonprofit sectors, and the public to build and sustain the capabilities outlined in the national preparedness goal. The national preparedness system shall include guidance for planning, organization, equipment, training, and exercises to build and maintain domestic capabilities. It shall provide an all-of-Nation approach for building and sustaining a cycle of preparedness activities over time.

The national preparedness system shall include a series of integrated national planning frameworks, covering prevention, protection, mitigation, response, and recovery. The frameworks shall be built upon scalable, flexible, and adaptable coordinating structures to align key roles and responsibilities to deliver the necessary capabilities. The frameworks shall be coordinated under a unified system with a common terminology and approach, built around basic plans that support the all-hazards approach to preparedness and functional or incident annexes to describe any unique requirements for particular threats or scenarios, as needed. Each framework shall describe how actions taken in the framework are coordinated with relevant actions described in the other frameworks across the preparedness spectrum.

The national preparedness system shall include an interagency operational plan to support each national planning framework. Each interagency operational plan shall include a more detailed concept of operations; description of critical tasks and responsibilities; detailed resource, personnel, and sourcing requirements; and specific provisions for the rapid integration of resources and personnel.

All executive departments and agencies with roles in the national planning frameworks shall develop department-level operational plans to support the interagency operational plans, as needed. Each national planning framework shall include

guidance to support corresponding planning for State, local, tribal, and territorial governments.

The national preparedness system shall include resource guidance, such as arrangements enabling the ability to share personnel. It shall provide equipment guidance aimed at nationwide interoperability, and shall provide guidance for national training and exercise programs, to facilitate our ability to build and sustain the capabilities defined in the national preparedness goal and evaluate progress toward meeting the goal.

The national preparedness system shall include recommendations and guidance to support preparedness planning for businesses, communities, families, and individuals.

The national preparedness system shall include a comprehensive approach to assess national preparedness that uses consistent methodology to measure the operational readiness of national capabilities at the time of assessment, with clear, objective and quantifiable performance measures, against the target capability levels identified in the national preparedness goal.

Building and Sustaining Preparedness

The Secretary of Homeland Security shall coordinate a comprehensive campaign to build and sustain national preparedness, including public outreach and community-based and private-sector programs to enhance national resilience, the provision of Federal financial assistance, preparedness efforts by the Federal Government, and national research and development efforts.

National Preparedness Report

Within 1 year from the date of this directive, the Secretary of Homeland Security shall submit the first national preparedness report based on the national preparedness goal to me, through the Assistant to the President for Homeland Security and Counterterrorism. The Secretary shall coordinate this effort with other executive departments and agencies and consult with State, local, tribal, and territorial governments, the private and nonprofit sectors, and the public. The Secretary shall submit the report annually in sufficient time to allow it to inform the preparation of my Administration's budget.

Roles and Responsibilities

The Assistant to the President for Homeland Security and Counterterrorism shall periodically review progress toward achieving the national preparedness goal.

The Secretary of Homeland Security is responsible for coordinating the domestic all-hazards preparedness efforts of all executive departments and agencies, in consultation with State, local, tribal, and territorial governments, nongovernmental organizations, private-sector partners, and the general public, and for developing the national preparedness goal.

The heads of all executive departments and agencies with roles in prevention, protection, mitigation, response, and recovery are responsible for national preparedness efforts, including department-specific operational plans, as needed, consistent with their statutory roles and responsibilities.

Nothing in this directive is intended to alter or impede the ability to carry out the authorities of executive departments and agencies to perform their responsibilities under law and consistent with applicable legal authorities and other Presidential guidance. This directive shall be implemented consistent with relevant authorities, including the Post-Katrina Emergency Management Reform Act of 2006 and its assignment of responsibilities with respect to the Administrator of the Federal Emergency Management Agency.

Nothing in this directive is intended to interfere with the authority of the Attorney General or Director of the Federal Bureau of Investigation with regard to the direction, conduct, control, planning, organization, equipment, training, exercises, or other activities concerning domestic counterterrorism, intelligence, and law enforcement activities.

Nothing in this directive shall limit the authority of the Secretary of Defense with regard to the command and control, planning, organization, equipment, training, exercises, employment, or other activities of Department of Defense forces, or the allocation of Department of Defense resources.

If resolution on a particular matter called for in this directive cannot be reached between or among executive departments and agencies, the matter shall be referred to me through the Assistant to the President for Homeland Security and Counterterrorism.

This directive replaces Homeland Security Presidential Directive (HSPD)-8 (National Preparedness), issued December 17, 2003, and HSPD-8 Annex I (National Planning), issued December 4, 2007, which are hereby rescinded, except for paragraph 44 of HSPD-8 Annex I. Individual plans developed under HSPD-8 and Annex I remain in effect until rescinded or otherwise replaced.

Definitions

For the purposes of this directive:

(a) The term "national preparedness" refers to the actions taken to plan, organize, equip, train, and exercise to build and sustain the capabilities necessary to prevent, protect against, mitigate the effects of, respond to, and recover from those threats that pose the greatest risk to the security of the Nation.

(b) The term "security" refers to the protection of the Nation and its people, vital interests, and way of life.

(c) The term "resilience" refers to the ability to adapt to changing conditions and withstand and rapidly recover from disruption due to emergencies.

(d) The term "prevention" refers to those capabilities necessary to avoid, prevent, or stop a threatened or actual act of terrorism. Prevention capabilities include, but are not limited to, information sharing and warning; domestic counterterrorism; and preventing the acquisition or use of weapons of mass destruction (WMD). For purposes of the prevention framework called for in this directive, the term "prevention" refers to preventing imminent threats.

(e) The term "protection" refers to those capabilities necessary to secure the homeland against acts of terrorism and manmade or natural disasters. Protection capabilities include, but are not limited to, defense against WMD threats; defense of agriculture and food; critical infrastructure protection; protection of key leadership and events; border security; maritime security; transportation security; immigration security; and cybersecurity.

(f) The term "mitigation" refers to those capabilities necessary to reduce loss of life and property by lessening the impact of disasters. Mitigation capabilities include, but are not limited to, community-wide risk reduction projects; efforts to improve the resilience of critical infrastructure and key resource lifelines; risk reduction for specific vulnerabilities from natural hazards or acts of terrorism; and initiatives to reduce future risks after a disaster has occurred.

(g) The term "response" refers to those capabilities necessary to save lives, protect property and the environment, and meet basic human needs after an incident has occurred.

(h) The term "recovery" refers to those capabilities necessary to assist communities affected by an incident to recover effectively, including, but not limited to, rebuilding infrastructure systems; providing adequate interim and long-term housing for survivors; restoring health, social, and community services; promoting economic development; and restoring natural and cultural resources.

BARACK OBAMA

QUESTIONS ON PRESIDENTIAL POLICY DIRECTIVE/PPD-8 – NATIONAL PREPAREDNESS

Read PPD 8 and answer the following questions.

1. PPD 8 talks about "shared responsibility." Between whom is the responsibility shared, and does shared responsibility also mean shared control? If so, how does that work?
2. What is a National Preparedness "Goal" and a National Preparedness "System?"
3. What tasks does PPD 8 set out that must be accomplished?
4. Find and examine the National Preparedness Goal, First Edition issued September 30, 2011. Is it realistic? What aspects of it have you seen accomplished?
5. Find and examine the Final Disaster Recovery Framework Released September 30, 2011. To whom is this document directed? Discuss the resources needed to support the Framework and from where they will be taken.
6. Discuss how to build and accomplish a National Campaign to Build and Sustain Preparedness.
7. How could you as an individual, your class, your school or organization, your city or town, and your state contribute to a National Campaign to Build and Sustain Preparedness?

8. PRESIDENTIAL POLICY DIRECTIVE (PPD) 8 "National Preparedness" reflects current Congressional limitations, requiring an economic framework for planning.

We will analyze current performance against our intended capabilities, the defined targets, and associated performance measures. This analysis will enable us to individually and collectively determine necessary resource levels, inform resource allocation plans, and guide Federal preparedness assistance. Budget implications across the preparedness enterprise cannot be assessed without this detailed and specific information. This approach will allow for annual adjustments based on updated priorities and our resource posture.[32]

What will it cost to implement PPD 8? As you have examined PPD8 and supporting materials, discuss the cost estimates you have or have not encountered.

9. Given the limited resources available for emergency management, what priority do you believe that fulfilling the requirements

---

[32] National Preparedness Goal, p. 19.

of PPD 8 should have as assets are allocated? Defend your answer.

10. How would you do things differently if you were in charge?

NOTES

1. The author wishes to acknowledge the valuable assistance of William R. Cumming, who served for over 20 years as a staff attorney at FEMA. His knowledge of FEMA policy and history is without parallel. His Vacation Lane Blog is a great resource for anyone wishing an independent view of the matters discussed in this chapter. Many of the documents referenced can be tracked on the home page of that blog at: http://vlg338.blogspot.com
2. Public Law 875, 81st Congress
3. Public Law 920, 81st Congress
4. Public Law 85-606
5. 42 U.S.C. §§ 4001 et seq. (2010).
6. 43 FR 41943, 92 Stat. 3788 (1978).
7. *See* Executive Order 2127 of March 31, 1978 "Federal Emergency Management Agency," 3 CFR, 1979 Comp., p. 376 and Executive Order 12148 of July 20, 1979, as amended, "Federal Emergency Management," 3 CFR, 1979 Comp., p. 412 implementing the reorganization.
8. 42 U.S.C. §§ 5121-5206 (2010).
9. Established by Executive Order 13228 of October 8, 2001 "Establishing the Office of Homeland Security and the Homeland Security Council."
10. Timothy Noah, "Joe Allbaugh, Disaster Pimp" Slate (September 7, 2005). Found on line at http://slate.msn.com/toolbar.aspx?action=read&id=2125756
11. Robert Block, *Identity Crisis - Hurricane Tests Emergency Agency At Time of Ferment: Now Under Homeland Security, FEMA Has Lost Clout, Managers on Ground Say: Terrorist With 145 MPH Winds*, WALL ST. J., Aug. 16, 2004.
12. *Id.*
   According to [DHS Under Secretary Michael] Brown and other insiders, a quiet battle is under way within the Homeland Security Department. On one side are former law-enforcement officials, advocating secrecy, tight security and intelligence as the key to minimizing the trauma of any terrorist attack. On the other are firefighters and emergency managers who emphasize collaboration, information sharing, public awareness and mitigation efforts to reduce the impact of disasters.
13. Shaun Waterman, Analysis: *Fear of Being Eclipsed by Terror*, UNITED PRESS INT'L, Mar. 19, 2004,

http://www.upi.com/inc/view.php?StoryID=20040319-020617-1820r ("[E]mergency managers over the country...working in state and local governments to plan and prepare their communities for the worst, feel their budgets and in some cases their very existence being squeezed by what some say is a myopic focus on terrorism.").
14. Karin Fischer, *Anti-Terrorism Focus Could Hurt State Cuts Would Affect Ability to Respond to Natural Disasters,* CHARLESTON DAILY MAIL, Feb. 27, 2004, at 1C. ("State officials, however, are alarmed at reductions in grants that help pay for the salaries of emergency services staff. The appropriation for the Emergency Management Performance Grants would decrease by nine million dollars, and states would face limitations on the share that would go to pay personnel.").
15. INT'L ASS'N EMERGENCY MANAGERS, EMERGENCY MANAGEMENT PERFORMANCE GRANT (EMPG) FUNDING, http://www.iaem.com/resources/ADVOCACY/documents/IAEM2005positions.doc [hereinafter IAEM].
16. *See* http://www.iaem.com for more information on IAEM.
17. IAEM, *supra* note 294.
18. FEMA, "Emergency Management Performance Grants" found on line at http://www.fema.gov/emergency/empg/empg.shtm
19. http://www.fema.gov/government/grant/empg/index2009.shtm
20. http://www.fema.gov/government/grant/empg/index10.shtm
21. http://www.fema.gov/government/grant/empg/
22. James Lee Witt, President, James Lee Witt Assocs., Former Director of FEMA, Statement before Subcommittee on National Security, Emerging Threats and International Resources and the Subcommittee on Energy Policy, Natural Resources and Regulatory Affairs (Mar. 24, 2004), *available at* http://www.all-hands.net/Article592.html.
23. *See* "A Failure of Initiative: The Final Report of the Select Bipartisan Committee to Investigate the Preparation for and Response to Hurricane Katrina" found online at http://www.gpoaccess.gov/serialset/creports/pdf/hr109-377/fema.pdf

24. *Id.*

25. "Let FEMA Be FEMA" Washington Post, Tuesday, October 4, 2005, found online at http://www.washingtonpost.com/wp-dyn/content/article/2005/10/03/AR2005100301485.html

26. Transportation and Infrastructure Committee, Press release, Chairman Oberstar's opening comments on "An Independent FEMA: Restoring the Nation's Capabilities for Effective Emergency Management and Disaster Response" (May 14, 2009). Found online at: http://transportation.house.gov/News/PRArticle.aspx?NewsID=918

27. Post-Katrina Emergency Management Reform Act of 2006 found on line at http://frwebgate.access.gpo.gov/cgi-bin/getdoc.cgi?dbname=109_cong_bills&docid=f:s3721is.txt.pdf

28. IAEM Government Affairs Release, March 3, 2009. Found on line at archhttp://www.iaem.com/Committees/GovernmentAffairs/documents/IAEM-USACongressionalPrioritiesList030909.pdf

29. Transportation and Infrastructure Committee, Press release, Chairman Oberstar's opening comments on "An Independent FEMA: Restoring the Nation's Capabilities for Effective Emergency Management and Disaster Response" (May 14, 2009). Found on line at: http://transportation.house.gov/News/PRArticle.aspx?NewsID=918

30. See, e.g., http://www.hsaj.org/?fullarticle=6.1.1

31. "The department cannot sustain another massive reorganization five years into its initial development. This entity of twenty-two merged agencies needs time to build upon its improvements rather than prematurely dismantling it." Daniel Fowler, "Thompson Takes On Oberstar's Effort to Dislodge FEMA From DHS" CQ Homeland Security (December 22, 2008). Found on line at: http://www.apcointl.org/new/commcenter911/downloads/US%20Presidential%20Transition%20Weekly%20Report%201-2-09.pdf

32. DHS Office of Inspector General "FEMA: In or Out?" found on line at http://www.dhs.gov/xoig/assets/mgmtrpts/OIG_09-25_Feb09.pdf

33. Fugate had previously refused an offer by President Bush to replace Michael Brown as FEMA head. M.J. Stephey, "FEMA Chief W. Craig Fugate" TIME (Friday, Mar. 06, 2009) Found on line at: http://www.time.com/time/nation/article/0,8599,1883485,00.html

34. "Craig Fugate is widely considered to be among the very best emergency managers in the country, and Florida is being looked upon as one of the very best prepared states," Kathleen Tierney, director of the Natural Hazards Center at the University of Colorado at Boulder. Quoted in "Editorial: Dangerous Politics" Gainesville Sun (May 12, 2009). Found on line at: http://www.gainesville.com/article/20090512/OPINION01/905121002 "We are pleased that an experienced and nationally recognized professional emergency manager has been tapped to lead the nation's emergency management agency. We believe that this move is a strong and clear statement of the importance that the Administration places on having a fully functional FEMA led by top-notch professionals. Craig Fugate's decades of service in the field of emergency management began as a first responder, and then emergency management director in Alachua County, Florida. His demonstrated performance at the county level led to his selection as a senior member of the Florida Division of Emergency Management (FDEM). Craig's innovative approaches to all facets of emergency management and dedicated support to local and county emergency management led to his appointment by two governors as director, FDEM. In this position, he earned national recognition for creative programs, including his partnerships with the private sector to maximize its contributions to response operations and his embracing technology, through social networking and other innovations." Russ Decker, CEM, President IAEM, Letter of Support for Nomination of Craig Fugate to be FEMA Administrator (March 19, 2009). Found on line at: http://www.iaem.com/publications/news/documents/IAEMCraigFugateFEMA19March2009.pdf

35. Rutherford H. Platt, *Disasters and Democracy, 66* (Island Press 1999).

36. http://www.cbsnews.com/stories/2011/05/29/ftn/main20067183.shtml

37. Indeed, as recently as 2004 Speaker Cantor himself voted against a bill to offset disaster relief aid with budget cuts. http://www.huffingtonpost.com/2011/08/31/cantor-disaster-relief-hurricane-irene-fema_n_943280.html

38. http://www.businessweek.com/ap/financialnews/D9PDVT881.htm

39. http://www.politico.com/news/stories/0811/62395.html

Tifton, Ga., April 29, 2009 – As FEMA representatives sign in, State Public Assistance (SPA) Grants Manager Marty Itzkowitz directs registration at today's Public Assistance (PA) Applicants Briefing. Approximately 150 representatives from 43 counties are here to learn how FEMA can help their area recover from March severe storms and flooding. George Armstrong/FEMA.

Chapter 13

FEDERAL EMERGENCY MANAGEMENT GRANTS*

William C. Nicholson

I. WORKING WITH FEMA: GRANTS MANAGEMENT PROCESSES AND PROCEDURES

The attorney advising a client regarding FEMA is in a somewhat different position than his or her colleague providing advice on relations with other agencies, such as EPA. FEMA's major role is issuing, administering and managing grants, as contrasted with other federal agencies that regulate state and local governments as well as businesses within the states. When FEMA distributes funds, whether under a program such as the Chemical Stockpile Emergency Preparedness Program (CSEPP) or pursuant to an individual assistance program, such funds flow from FEMA to the state involved in the form of a grant. The state then distributes these funds to local units of government or individuals that are their ultimate recipients, who are termed subgrantees or subrecipients. The state has the responsibility, under 44 CFR Sec. 13.37, for administering these grants and assuring that subgrantees utilize grant funds in a manner that complies with state and federal law. Office of Management and Budget (OMB) Circular A-87 requires that "Governmental units assume responsibility for administering federal funds in a manner consistent with underlying agreements, program objectives, and the terms and conditions of the federal award."

A. GUIDANCE FOR GRANT ADMINISTRATORS

A number of tools provide guidance for the grant administrator. They include:

Office of Management and Budget (OMB) Circulars
(All circulars can be found at: http://www.whitehouse.gov/omb/circulars)
A-87 "Cost Principles for State, Local, and Indian Tribal Governments" A-102 "Grants and Cooperative Agreements with State and Local Governments"

A133/150 Audits

Treasury Guidance
Treasury Financial Manual
Cash Management Improvement Act

FEMA Guidance
44 CFR Part 13: Administrative Requirements.
44 CFR Part 206: Programs
Robert T. Stafford Disaster Relief and Emergency Act 42 USC §§ 5121-5206 (2012).

---

* Much of this material may be found at www.fema.gov

Headquarters Guidance
Office of Financial Management SOP Reconciling Financial Systems
Regional Guidance

Other Guidance
Federal Grants Handbook

B. STATE APPLICATION REQUIREMENTS FOR DISASTER GRANT FILE

1. Disaster Notice and Amendments: This section includes the Governor's formal letter requesting a federal disaster declaration. It is submitted to the FEMA Regional Office, forwarded by them to FEMA Headquarters, and the sent by Headquarters to the President. It also includes the President's letter approving the declaration and any amendments made by the state or FEMA.
2. FEMA-State Agreement and Modifications: Signed early in the disaster, this includes eligible geographical areas, the cost share, the particular grant programs activated, and designates the Governor's Authorized Representative (GAR).
3. Grant Application (FEMA Form 424) with Assurance: This is the application for federal assistance that must be completed by the state.
4. Grant Administration Plan: State must submit a revised plan annually.
5. Miscellaneous correspondence.

C. STATE APPLICATION REQUIREMENTS FOR A NON-DISASTER GRANT

The state must submit a completed Form SF 424 Grant Application. If reimbursement of indirect costs is desired, the state must reach an Indirect Cost Agreement with FEMA. A FEMA Form 2020 Budget Information¬NonConstruction Programs and a Budget Narrative must also be provided, as must a Program Narrative Statement that includes a statewide overview, performance measures and completion time frames. A FEMA Form 20-16 Summary Sheet for Assurances and Certifications with Assurances attachments must also be present.

D. REPORTING/RECONCILING REQUIREMENTS

Grantee Requirements
The Grantee must provide quarterly Financial Status Reports (FEMA Form 20-10) and Federal Cash Transaction Reports (PMS-272). Performance Reports must also be provided on a regular basis, which is frequently required to be quarterly as well.

FEMA Requirements
FEMA must perform a Cash on Hand Analysis quarterly and analyze trends, as well as a Cost Share Analysis and Systems Reconciliation.

E. POST-AWARD CHANGES REQUIRING FEMA APPROVAL

Any "change in scope or objectives" such as a change in approved performance measures requires FEMA prior approval, pursuant to 44 CFR 13.30(d)(1). The state must submit written justification for the change and changes to action items that would result. The approval process for such changes is the same as the pre-award process.

Cumulative transfers among direct cost budget categories that will exceed 10 percent of the total approved budget (including both federal and state share amounts) must be approved by FEMA under 44 CFR 13.30(c)(1)(ii). The approved revised budget will serve as the basis for calculating the 10 percent when applying for additional approvals for future revisions.

Requests for extension of the performance period for up to 12 months will be considered by the FEMA Regional Offices, but will not be granted automatically. Justification for such a request must be provided, including statements of current activity status, an explanation of why the extension is needed, the amount of funding required to finish the activity, and an estimated completion date. Work must be in progress on the approved scope of work during the original performance period to qualify for an extension of time. Further, all quarterly reports must be up to date.

From time to time, however, FEMA recognizes that one of the many burdens placed on overworked state and local emergency management personnel is spending the funds appropriated for them in a timely manner. If the problem is widespread enough, the agency may issue Bulletins similar in content to Grant Programs Directorate Information Bulletin No. 379, dated

February 17, 2012, in which Assistant Administrator Elizabeth M. Harman of the Grant Programs Directorate issued Guidance to State Administrative Agencies to Expedite the Expenditure of Certain DHS/FEMA Grant Funding. The Bulletin contains a commitment to provide grantees with additional flexibility to accelerate the spending of remaining FY 2007–FY2011 DHS/FEMA grant funds (including formula grant programs), consistent with existing laws, regulations and programmatic objectives1. These measures will apply also to FY 2012 grant awards[.]

It covers a wide variety of grants, including mitigation, EOC construction, HSGP, UASI, and others. Opportunities exist for: (A) reprioritization of grant funds; (B) waiver to existing program requirements such as personnel cost caps and/or program match requirements; (C) expansion of efforts that build and sustain capabilities; and (D) extending an award's period of performance based on extenuating circumstances.

Some may argue that an inability to spend funds is the mark of a program that is underfunded on the personnel side. Such persons will doubtless welcome the above-mentioned waiver on personnel cost caps.

F. CLOSEOUT-WINDING UP THE GRANT

Closing disasters in a timely and efficient manner is a high priority for FEMA. Another important Closeout goal is reduction of unliquidated obligations. The overall end desired is improved financial integrity. Within 90 days of the expiration or termination of a grant, the Grantee must submit the following reports: the Final Financial Status Report (FEMA Form 20-10); the Final Request for Payment (SF-270) if applicable; the Final Progress Report; an Invention Disclosure if applicable; a Federally Owned Property Report. To the extent that value remains in things that have been purchased with grant funds, such items generally are given to state or local emergency management for use in purposes consistent with the purpose of the original grant.

G. EXPENSE LIMITATIONS THAT MAY CAUSE DIFFICULTIES

Travel expenses generally require pre-approval. Concern that grant monies not be frittered away is the policy concern underlying this requirement. When permission is granted for travel, the amounts allowed must equal per diems of the governmental unit involved consistent with a written policy. If the unit has no written policy, per diems must follow federal allowances. Further, the circumstances under which such permission is granted are very limited. As a general rule, travel and food expenses are only permitted for events that directly relate to the grant's purpose, or as A-87 puts it, "Travel expenses incurred specifically to carry out the award." Advertising and public relations costs are not permitted, unless directly related to the grant's purpose or specifically required by the Federal award and then only as a direct cost. Costs of alcoholic beverages are unallowable. Similarly, entertainment costs, including amusement, diversion, and social activities and any costs directly associated with such costs (such as tickets to shows or sports events, meals, lodging, rentals, transportation, and gratuities) are unallowable.

Compensation for individual employees must be reasonable for the services rendered and conform to the established policy of the governmental unit consistently applied to both federal and non-federal activities. Appointments of personnel must be made in accordance with a governmental unit's laws and rules and meet merit system or other requirements of applicable federal law. Reasonable fringe benefits are allowable, so long as they are consistent between federal and non-federal employees. Special rules apply to pensions and severance pay.

Occasionally, the grantee may, in the process of exercising oversight, come upon improper expenditures by a subgrantee. Several difficulties might flow from such a discovery. The subgrantee may not have funds to cover repayment of improper payouts. Should the grantee discover that a subgrantee has made mistakes of a magnitude that they cannot be addressed within the subgrantee's assets, rapid consultation with the grantor agency is highly advisable. Of course, any misuse of funds springing from apparent bad faith must be quickly addressed. Since funds flow through the grantee (state government) to the subgrantee (local government), violations of both state and federal law may be involved. The grantee must be careful in this situation to take all steps mandated by state law upon discovery of violations. For example, state law may require notification of state auditors within a given period of time following discovery of impropriety. Although the grantee's discovery of improprieties could lead to the difficulties described above, both grantee and subgrantee are best served in such a situation by making good faith efforts to correct problems promptly upon their discovery.

H. AUDITS

From time to time, FEMA conducts audits grantee oversight of subgrantees. Discovery by FEMA of unaddressed offenses may result in significant penalties for the offending entities. If FEMA discovers less serious mistakes, the only result might be recommendations by the auditors. Failure to follow these recommendations can lead to continued fault being found. FEMA expects grantees to exercise proper oversight.

Circular A-133 "Compliance and Reporting Responsibilities" makes all grantees responsible for:

1. identifying all Federal awards received and expended and their respective programs;
2. maintaining internal control to assure compliance with regulation, laws and grant contractual provisions that could have a material effect on each Federal program administered;

3. complying with relevant laws, regulations, and grant contractual provisions;
4. preparing financial statements, including the schedule of expenditures of Federal awards;
5. ensuring that audits are properly conducted and reports submitted on time; and
6. following up on audit findings, and taking necessary corrective action.

Grantees have always performed most of these tasks, but the 1996 revision made them responsible for the follow-up on audit findings, which includes preparing a summary schedule of any prior audit findings and taking the necessary steps to resolve current year findings based on a corrective action plan.

Only a grantee that expends $300,000 or more in federal funds must have a single audit performed. Grantees spending $300,000 or more under one federal program may elect to have a program-specific audit performed, according to the circular. Those grantees that spend less than $300,000 in federal awards for a particular year are exempt from Federal audit requirements, although their records must be available for review or "reduced scope" audits.

Pass-through entities must monitor sub recipients, regardless of amount expended. State and local governments are given wide latitude in deciding how best to monitor their subgrantees. Subrecipient monitoring should not be limited to audit work and may include site visits, document review, and reporting requirements. Due diligence by grantees must include periodic review of all expenditures by subgrantees as well as coordination of yearly budgets. Discovery of irregularities should result in heightened oversight, including in more frequent examination of disbursements.

II. FINDING FEDERAL FUNDS BEFORE A DISASTER: PREDISASTER MITIGATION GRANTS AND PREPAREDNESS (NON-DISASTER) GRANTS

The trend in emergency management is toward devoting greater efforts to mitigation. Such programs as Project Impact have proven the effectiveness of enlisting community partnerships to work together with emergency management at the federal, state, and local levels to leverage limited funds and create safer communities. These programs must be administered in accord with the guidelines discussed in the preceding section.

One of the most visible examples of the trend toward mitigation is the Disaster Mitigation Act of 2000 (DMA). The DMA amended the Stafford Act (42 U.S.C. 5121 et seq.) to establish the Pre-Disaster Mitigation (PDM) Program and streamline administration of disaster relief. PDM benefitted from a first year $250,000,000 appropriation. By FFY 2009, the appropriation had shrunk to $100,000,000. The President's Federal Fiscal Year (FFY) 2012 budget proposal included just $36 million under the National Pre-Disaster Mitigation Fund. The DMA gives a high priority to mitigation of hazards at the local level, emphasizing risk assessment, implementation of loss reduction measures, and ensuring that critical services and facilities survive a disaster. With the unified efforts of economic incentives, awareness and education, and federal support promoted by the DMA, state and local governments (including Indian Tribes) are able to form effective community-based partnerships, implement effective hazard mitigation measures, leverage additional non-federal resources, and commit to long-term hazard mitigation efforts.

FEMA provides PDM grant funding to states as grantees. The grantee states pass on the funds to local units of government (subgrantees). The local units of government then utilize DMA assistance to support effective partnerships, improve the community's risk assessment, and establish mitigation priorities and a community hazard mitigation plan to reduce the risk from natural hazards. Section 203 of the Stafford Act (as amended by § 102 of the DMA) authorizes the PDM Program. During Federal Fiscal Year (FFY) 2002, $25 million was provided for the PDM Program. FEMA provides a Program Overview as well as yearly Program Guidance and National Ranking Factors on its Pre-Disaster Mitigation (PDM) Program website. http://www.fema.gov/government/grant/pdm/index.shtm

Panels composed of representatives from FEMA, State, Territories, local governments, federally recognized Indian Tribal governments, and other federal agencies will peer evaluate project and planning sub-applications on the basis of qualitative factors. The 2012 National Evaluation Panel Planning Factors and the 2012 National Evaluation Panel Project Factors will be available in the FEMA library.

FEMA and HUD are in partnership regarding the principles set forth in the HUD Sustainable Housing and Communities initiative and will utilize information from the PDM project and planning sub-applications to guide future opportunities for program collaboration. FEMA supports the HUD program goals for strategic local approaches to sustainable development by combining hazard mitigation objectives with the community development objectives. The community development objectives support regional planning efforts that integrate housing and transportation decisions, and increase state, regional, and local capacity to incorporate livability, sustainability, and social equity values into land use plans, zoning and infrastructure investments.

FEMA will note sustainability principles that are included in the PDM planning and project sub-applications. In other words, an application that incorporates the HUD principles will be more likely to receive favorable treatment as it competes for funding. For more information regarding the HUD Sustainable Housing and Communities initiative, please visit http://portal.hud.gov/portal/page/portal/HUD/program_offices/sustainable_housing_communities.

As specified in § 203(f), PDM total funds per State will be no less than $500,000 or 1 percent of appropriated funds, whichever is less. The remainder of the funds was allocated based on each state's percentage of national population, using most recent (2010) Census data as the basis for the division. Annual funding for each State, including local grants, may not exceed 15 percent of the annual appropriation. Not more than 10 percent of grant may be used to promote "mitigation technologies." The federal-local cost-share will be 75-25. (The statute specifies 90-10 for "small impoverished communities.") At least 25 percent of total eligible costs must be provided by nonfederal sources. All contributions, both cash and in-kind, are acceptable as part of the non-federal matching share.

Criteria for state and local hazard mitigation planning are established by 44 CFR Part 201. Countywide or multi-jurisdictional plans may make sense, since many hazards are better addressed by comprehensive hazard evaluation.

For updates on the Pre-Disaster Mitigation Program, the Hazard Mitigation Grant Program (HMGP), and the Flood Mitigation Assistance (FMA) Program, check the following website: http://www.fema.gov/government/grant/hma/index.shtm

Some emergency managers expressed concern that PDM Grants might be used for terrorism preparedness rather than mitigation of natural hazards as Congress had originally intended. The emergency manager with limited resources who wishes to maximize the opportunity to bring in grant funds may wish to consult availability of FEMA's Preparedness (Non-Disaster) Grants. FEMA provides state and local governments with preparedness program funding in the form of Non-Disaster Grants to enhance the capacity of state and local emergency responders to prevent, respond to, and recover from a weapons of mass destruction terrorism incident involving chemical, biological, radiological, nuclear, and explosive devices and cyber attacks. http://www.fema.gov/government grant/nondisaster.shtm. On August 24, 2011, FEMA announced more than $2.1 billion in these grants for FY 2011, down $780 million from FY 2010. http://www.fema.gov/news/ newsrelease.fema?id=57371 As budgets tighten, the necessity of procuring grant funds today that may be gone tomorrow becomes ever more self-evident.

QUESTIONS

1. What is the attorney's role in the grants management process on the local, state, and federal levels?
2. What proactive steps might the attorney take to ensure compliance with grant requirements?
3. List what other stakeholders might be of help in the grants management process, and describe the assets they could bring to the process.

4. How can an attorney assist in preparing FEMA grants?
5. How might attorneys make competitive grant proposals more likely to succeed than those of other jurisdictions?
6. Perform research to find the currently available Preparedness (Non-Disaster) Grants as well as PDM grants. Assign each class member one of the programs and have him/her report on the amounts available

and how to procure the grant. Discuss the relative merits of the grants.

7. In his discussion of FEMA, the author states that the "Prevention" phase created by DHS in the aftermath of 9/11 should be included as part of "Mitigation" which is one of the four traditional phases of emergency management. Do you agree, or do you believe it should be a separate fifth phase? Defend your viewpoint. How is this issue relevant to the grants process?

8. Discuss the relative merits of investing in mitigating against natural hazards vs. mitigating against terrorism.

The author (standing) works legal issues with local emergency managers.

Chapter 14

DIFFICULTIES IN MITIGATING
LEGAL EXPOSURE

OBTAINING COMPETENT LEGAL ADVICE:
CHALLENGES FOR EMERGENCY MANAGERS
AND ATTORNEYS

William C. Nicholson

I. INTRODUCTION

Emergency management is a creature of law.[1] On the federal, state, and local levels, legal enactments generate the formal structure, daily responsibilities, and powers of emergency management organizations.[2] The federal authority to regulate all phases of disasters and emergencies is rooted in the Constitution.[3] State enactments concerning this authority resemble one another.[4] They typically grant significant powers to Governors to deal with emergency events,[5] as well as establish and detail responsibilities for state[6] and local[7] emergency management organizations.

Under state laws, the major responsibility for emergency management rests with the local unit of government.[8] Experience has proven that "lawyering for a [local unit of government] during a crisis is rarely planned or anticipated. Yet, emergency and disaster conditions will require the prudent involvement of the [local government] attorney."[9] Local government attorneys (LGAs) and local emergency management coordinators (EMCs) are responsible for understanding and applying emergency management laws with the goals of maximizing public safety and procuring the greatest possible amount of financial support from other levels of government. For example, post-disaster debris removal may be eligible for federal reimbursement only if significant legal requirements are followed.[10]

Despite the requirement for understanding and compliance with legal standards, rural and less wealthy emergency management organizations, in particular, face significant challenges in obtaining competent legal advice.[11] As will be seen, however, even their wealthier fellows face similar problems. There are several reasons for this situation. Historical precedent, ingrained attitudes, and funding challenges together decrease the amount of competent legal advice available to local EMCs. This paper examines how well local units of government are fulfilling the mandate to incorporate legal counsel in the emergency management team.

This article was originally published as William C. Nicholson, "Obtaining Competent Legal Advice: Challenges for Local Emergency Managers and Attorneys," California Western Law Review Natural Disaster Law Issue, CALIFORNIA WESTERN LAW REVIEW, Vol. 46, No. 2, 343-368 (2010). Reprinted by permission.

II. THE REQUIREMENT FOR COMPETENT LEGAL COUNSEL
AND THE CONSIDERATION OF LEGAL ISSUES

Emergency management (EM) is a legal entity, and without law it would not exist in its current form. Yet "when disasters do strike, it immediately becomes obvious that the legal issues involved in local government disaster planning are some of the most misunderstood and confusing aspects of the entire process of disaster preparation and recovery."[12] Emergency management has four phases: (1) mitigation, the lessening or avoidance of a hazard; (2) preparedness, including planning, training, and exercising; (3) response, referring to actions taken in the immediate aftermath of an event to deal with its effects; and (4) recovery, bringing circumstances back to at least the status they had prior to the emergency event.[13] During all four phases of EM, laws delineate the proper ways for local government to exercise EM powers.[14] LGAs and EMCs must comprehend and employ these laws if they are to protect public safety and acquire maximum economic assistance from other levels of government. "Attorneys and their clients...need to educate and train one another regarding their respective roles. Clients need to learn, in advance of an emergency or other crisis, about the inherent ambiguities of law, and attorneys need to learn from clients what kinds of legal advice can be optimally utilized by the client."[15]

On April 29, 2004, the American National Standards Institute (ANSI) recommended to the 9/11 Commission that the National Fire Protection Association (NFPA) standard, "NFPA 1600 Recommended Practices for Disaster Management,"[16] be established as the national preparedness standard.[17] On July 22, 2004, the 9/11 Commission formally endorsed NFPA 1600, stating that it was encouraged by Department of Homeland Security (DHS) Secretary Ridge's praise for it and specifying its preference that its adoption be promoted by DHS.[18] NFPA 1600 sets up a shared set of norms for disaster management, emergency management, and business continuity programs.[19] One vital aspect of NFPA 1600 is its requirement that all EM and business continuity programs comply with all relevant laws, policies and industry practice.[20] The National Emergency Management Association (NEMA) endorsed NFPA 1600 as an appropriate standard for emergency management – as early as 1998 NEMA passed a resolution signaling its support of NFPA 1600.[21] The standard, along with other existing documents like the Capability Assessment for Readiness (CAR),[22] provides the foundation for the Emergency Management Accreditation Program (EMAP).[23] EMAP is supported by a large number of important players in emergency management,[24] including the Federal Emergency Management Agency (FEMA).[25] Although the program is voluntary, the fact that it is endorsed by such a wide variety of authorities means that EMAP is well on its way to becoming the de facto standard for EM in the United States. As more programs become accredited under the standard, it is more likely that a court will hold all EM to the norm.[26] Accepted industry practices frequently move from de facto to de jure acceptance through common law adoption in the courts.[27] Failure to comply with a commonly accepted industry standard like EMAP will likely be admitted into evidence in a trial alleging negligent failure to prepare for or respond to an emergency.[28] Clearly, evidence of not performing to the standards set by EMAP could lead to a finding of liability.[29]

EMAP requires EM organizations to comply and keep up with changes in relevant laws and authorities:

4.2: Laws and Authorities

OVERVIEW

Laws and authorities refer to the legal underpinning for the program. Federal, state, tribal and local statutes and implementing regulations establish legal authority for development and maintenance of the emergency management program and organization and define the emergency powers, authorities, and responsibilities of the chief executive and the program coordinator. These principles serve as the foundation for the program and its activities.

4.2.1 The emergency management program shall comply with applicable legislation, regulations, directives and policies. Legal authorities provide flexibility and responsiveness to execute emergency management activities in disaster and non-emergency situation [sic]. The emergency management program and the program's responsibilities are established in state and local law. Legal provisions identify the fundamental authorities for the program, planning, funding mechanisms and continuity of government.

4.2.2 The program has established and maintains a process for identifying and addressing proposed legislative and regulatory changes.[30]

The National Response Framework (NRF) lists a number of requirements for local officials. Among them is "understanding and implementing laws and regulations that support emergency management and response."[31] Regarding plans, the NRF states that they are acceptable only if "consistent with applicable laws."[32]

The National Incident Management System (NIMS) also recognizes the importance of knowing and complying with applicable laws:

To better serve their constituents, elected and appointed officials should do the following: ...Understand laws and regulations in their jurisdictions that pertain to emergency management and incident response.[33]

During the response phase of emergency management, NIMS also supports attorney membership in the Incident Management team:

(4) Additional Command Staff

Additional Command Staff positions may also be necessary, depending on the nature and location(s) of the incident or specific requirements established by Incident Command. For example, a legal counsel might be assigned to the Planning Section as a technical specialist or directly to the Command Staff to advise Incident Command on legal matters, such as emergency proclamations, the legality of evacuation orders, isolation and quarantine orders, and legal rights and restrictions pertaining to media access.[34]

Further, "the incident objectives and strategy must conform to the legal obligations and management objectives of all affected agencies, and may need to include specific issues relevant to critical infrastructure."[35]

The obligations to comply with existing law and monitor new law mean that the local emergency manager must have access to competent legal counsel on a continuing basis. As the foregoing discussion indicates, the legal issues to which LGAs may need to respond include a wide variety of emergency-management-specific matters of which they may be completely without knowledge.[36] As a member of the Command Staff during a response, the attorney must be prepared to answer legal questions in real time.[37] This temporal pressure stands in stark contrast to the usual situation, where lawyers contemplate issues with access to extensive legal resources and time to perform research with the goal of reaching the absolute, nuanced, best legal response.[38] Often, power outages make the electronic research tools and other support upon which an attorney might otherwise rely unavailable.[39]

A great deal of money is often at stake when local legal counsel advise regarding federal requirements. Issues like federal payment for debris clearance and reimbursement of disaster-related costs require strong, knowledgeable advocacy. Legal counsel will find that this process can be painful, but it is always professional.[40]

FEMA's Comprehensive Preparedness Guide (CPG) states that "as a public document, an EOP...cites its legal basis."[41] The CPG also requires an "Authorities and References" section, which "provides the legal basis for emergency operations and activities."[42] The "Concept of Operations" section must "describe how legal questions/issues are resolved as a result of preparedness, response, or recovery actions, including what liability protection is available to responders."[43]

FEMA's Guide for All Hazards Emergency Operations Planning (SLG 101) requires that, at the outset, one must "review local and/or State laws, rules, regulations, executive orders, etc., that may be considered enabling legislation," and "review Federal regulatory requirements."[44] Under "Identify Hazards," SLG 101 requires that the planner "begin with a list of hazards that concern emergency management in your jurisdiction. Laws...can help define the universe of hazards which the planning team should address in the all-hazard EOP."[45] Under "Authorities and References," SLG 101 requires that "the Basic Plan should indicate the legal basis for emergency operations and activities. Laws, statutes, ordinances, executive orders, regulations, and formal agreements relevant to emergencies should be listed."[46] A Legal Advisor or equivalent is found under "Resource Management," which describes the position as follows:

Legal Advisor

• When notified of an emergency, reports to the EOC or other location as specified by the Resource Manager.
• Advises Supply Coordinator and Procurement Team on contracts and questions of administrative law.[47]

In terms of actual compliance with legal requirements, the Nationwide Plan Review: Phase 2 Report found the following regarding compliance (Note: S = Sufficient, PS = Partly Sufficient, NS = Not Sufficient):[48]

States			Urban Areas		
S	PS	NS	S	PS	NS
68%	32%	0%	68%	32%	0%

At both the state and large city levels, about one-third of the plans reviewed were only partly satisfactory regarding coverage of enabling legislation. The review did not address rural or less populated areas which, one might reasonably hypothesize, would have fewer resources and, hence, lower compliance. The Nationwide Plan Review: Phase 2 Report flatly states that "if a plan['s]...laws and authorities are inadequate, it

cannot be assessed as fully adequate, feasible, or acceptable."[49] Similarly, the Congressional Research Service found that twenty-two out of fifty-one state-level pandemic influenza plans lacked discussion of isolation and quarantine laws, and, shockingly, only two of fifty-one plans discussed procedures for judicial review of due process protections.[50]

Unfortunately, on the local level, the lack of a comprehensive treatment of legal issues is a common failing of emergency operations plans.[51] For local governments, failure to have competent legal advice regarding EM law may result in significant liability. This is because local government plays the vital first response role with regard to disasters, and its leaders must make decisions that may have major legal consequences.

Irrespective of the significant, and apparently concurrent, role that the State maintains in the local emergency management function, it is the local government that is on the front line of any disaster. Thus, the local government attorney should become intimately familiar with the local comprehensive emergency plan and with the extent and limitations of authority that a local government may exercise during an impending or realized disaster.[52]

One article discussing legal response duties in the wake of Hurricane Katrina makes the point that [a] lawyer might attempt to argue that the failure to craft a disaster preparedness plan is not negligent because that type of precaution is not common in the legal profession. However, doing what is customary (i.e., not preparing a disaster plan) does not preclude a finding of liability based on a cost-benefit analysis.[53]

The American Bar Association (ABA) has made efforts to assist practitioners with planning for emergencies and disasters.[54] After the September 11, 2001 attacks, the ABA created a webpage with guidance for local bar associations desiring to improve their resilience to disasters.[55] Still, such endeavors are in their infancy – even emergency planning, a matter that has been the subject of considerable litigation, did not until fairly recently receive significant attention in legal publications.[56]

III. ANECDOTAL INFORMATION REGARDING LOCAL LEGAL COUNSEL AND LOCAL EMERGENCY MANAGEMENT COORDINATORS

Since being appointed as General Counsel to the Indiana State Emergency Management Agency in 1995, the author has personally observed and worked with many local government attorneys and local emergency managers. He has noted varying levels of understanding and competence regarding knowledge and application of EM law by these two groups.

Informal conversations with these individuals, as well as state and federal training authorities, reveal a number of beliefs regarding their knowledge and understanding of EM law. Anecdotal discussions indicate widespread ignorance of this aspect of the law among both groups. They reportedly have disagreed over application of the law in emergency situations. Interviews with emergency managers in particular reveal that potential liabilities are one of their greatest sources of concern. Rural units of government in particular suffer from disconnects between LGAs and EMCs.

The relationship between LGAs and EMCs is often tenuous at best. On the attorney side, this situation results from a number of factors. Those who are a one-person operation often must handle all legal work for their jurisdiction, including prosecuting felonies, misdemeanors, and traffic court, as well as providing all legal advice on the total universe of legal issues faced by local government. Providing good advice under these circumstances is a "real challenge."[57]

Many communities do not have full-time law departments, but rather have part-time[58] lawyers doing their legal work.[59] In many states, the majority of local legal counsel are part-time. In Iowa, for example, fifty-seven are full-time, and forty-two are part-time.[60] In Mississippi, fifty-seven are full-time, and forty-two are part-time.[61] North Carolina has seventy-five part-time and twenty-five full-time county attorneys.[62] In Florida, in contrast, forty-three counties have full-time LGAs, and twenty-four of sixty-seven counties use outside counsel.[63]

Part-time local counsel often earn less per hour for legal work done for the government than they do for their private clients.[64] While some are elected (as in Iowa for example),[65] others are appointees. These individuals get the job for a variety of reasons,[66] meaning that they may not come into the job with a background in government law.[67] As appointed rather than elected officials, some may not hold their position for a long period of time, and thus not have the opportunity to build competence in this complex area of law. Some LGAs do not believe they receive the respect to which their position and status entitle them.

LGAs' knowledge of EM law varies greatly, with only a handful having a detailed understanding of EM law. Few have had any formal training in EM law, with a major reason being that Continuing Legal Education (CLE) in EM law is rarely available to LGAs.[68] Even if such training was available, their local government employers do not have funds available to pay for training in EM law.[69] Few meet regularly with the local EMC, and some may meet the EMC only in the immediate aftermath of a disaster. They are more likely to be called on during mitigation than response, where they provide input on such issues as revising building and fire codes.

Their offices do not typically have a plan for disaster operations – indeed, they do not anticipate being a member of the resource management team during a disaster or emergency. Many do not make an effort when speaking to non-lawyers to use everyday language to explain legal matters, which complicates matters for the EMC who is trying to understand them. Generally, neither they nor the units of government they serve fully understand the potential liabilities for failure to fully integrate the legal piece into emergency management. LGAs may be unaware of existing resources, such as the ABA's "Checklist for State and Local Government Attorneys to Help Prepare for Disasters."[70]

Local emergency management coordinators also face many challenges that exacerbate the difficulty of obtaining competent legal advice and reflect those faced by LGAs.[71] Although emergency management is often referred to as "in the process of professionalization," its progress is far from universal.[72] The incomplete nature of that professionalization is part of the reason that legal standards are sometimes not understood or observed. "Current concerns about legal liability arising out of failure to prepare for known hazards, however, may force public officials to pay greater attention to risks to public health and safety."[73]

Fortunately, there is a structure parallel to that of EMAP to foster and certify the professional growth of individual emergency managers. The "gold standards" for emergency managers are the Certified Emergency Manager (CEM) and Associate Emergency Manager (AEM) credentials, which are administered by the International Association of Emergency Managers (IAEM).[74] These are both nationally and internationally recognized and accepted credentials. The CEM requires the applicant to achieve specific standards in the areas described below:

- Emergency management experience. Three years by date of application. Comprehensive experience must include participation in a full-scale exercise or actual disaster. Three professional references. Including current supervisor.
- Education. A four-year baccalaureate degree in any subject area.
- Training. 100 contact hours in emergency management training and 100 hours in general management training. Note: No more than 25% of hours can be in any one topic.
- Contributions to the profession. Six separate contributions in areas such as professional membership, speaking, publishing articles, serving on volunteer boards or committees and other areas beyond the scope of the emergency management job requirements.
- Comprehensive emergency management essay. Real-life scenarios are provided, and response must demonstrate knowledge, skills and abilities as listed in the essay instructions.
- Multiple-choice examination. Candidates sit for the 100-question exam after their initial application and the other requirements are satisfied. The exam is a maximum of two (2) hours. A pamphlet is available further describing format and sources.
- Three References. Including a reference from the candidate's current supervisor.[75]

EMAP recommends that agencies over a certain size employ personnel who have earned the CEM/AEM.[76]

The current situation, however, is that many EMCs are not full-time employees, and some are volunteers. In Kansas, for example, according to Major General Tod Bunting, Kansas Adjutant General and top emergency management official, in 2007 "just thirty-eight of the state's 105 counties boasted a full-time emergency manager."[77] Often, the EMC is "an individual who must often juggle several different roles, ranging in [sic] from law enforcement to county administrative duties."[78] Sometimes, counties find themselves so economically challenged that they must share an EMC.[79] In Indiana, of ninety-two counties, forty-six to forty-seven have full-time EMCs. About forty-five Indiana EMCs are part-time, and one is still a volunteer.[80]

In North Carolina, there are thirty-two full-time EM coordinators, while sixty-eight counties have coordinators who also spend part of their time on fire, EMS, law enforcement, or related duties. In order to receive Emergency Management Performance Grant funds, they must spend at least 50% of their time on EM duties. Base salary for all North Carolina EM coordinators ranges from $20,753 to $115,661. Eleven jurisdictions report having Certified Emergency Manager (CEM) credentialed staff. None reported being EMAP certified currently, although two are in the process of applying for EMAP certification and three are currently in the process of preparing for applications for EMAP certification.[81] The North Carolina Department of Crime Control and Public Safety's Division of Emergency Management gained EMAP certification in October 2008, and is working with local jurisdictions to support their efforts to follow suit.[82]

Most EMCs had a previous career in one of the emergency response agencies: law enforcement, fire service, or emergency medical services.

This experience forms their approach to EM duties, which can either help or hinder them. Some are retirees and their overall energy level may be less than it once was. Some do not believe that they receive the respect to which their position and status entitle them. Often, they are poorly paid.[83] Like the LGAs, their knowledge of EM law varies greatly, with few having a detailed understanding of EM law. Similarly, few have had any formal training in EM law. Like many Americans, they may dislike attorneys, regarding them as conceited and arrogant. Overall, many believe that attorneys are a hindrance rather than an asset, and do not regard legal counsel as an important part of the EM team. Despite these beliefs, however, many EMCs are very worried about potential liability risks, as are the jurisdictions that employ them.[84][su'.']

The anecdotal evidence suggests that the more rural and economically challenged a jurisdiction is, the more likely it is to have an LGA and an EMC who correspond to the descriptions above. Of course, the plural of anecdote is not data. One major shortcoming of the anecdotal approach is that the consistency needed for scientifically valid evaluation of the information provided is lacking. For example, the definition of "full-time" used by Duane Davis, Emergency Management Director, Jackson County, Indiana and President of the Emergency Management Alliance of Indiana, includes individuals who do not work for forty hours per week as EMCs,[85] while the Executive Assistant to the Director for Government Liaison, North Carolina Emergency Management's definition includes only full-time forty-hours-per-week EMCs.[86] There is a great need for scientifically valid research to fill the gap in basic knowledge regarding how LGAs and EMCs interact, what they know about EM law, where and how they learned it, and their attitudes toward the matter.

IV. EXISTING RESEARCH ON THE RELATIONSHIP BETWEEN LGAS AND EMCS

The relationship between LGAs and EMCs has been the subject of very little research. Indeed, the majority of research on LGAs in general[87] consists of anecdotal statements similar to those discussed above.[88]

The only study encountered addressing the interface between LGAs and EMCs is from a chapter in a 1990 book that compared twelve different municipal departments in a survey. It found that the municipal attorney's office was the least likely to participate in emergency planning activities.[89] The anecdotal evidence suggests that the situation has changed very little in the intervening nineteen years.

One surprising common element that LGAs report is the lack of planning for operations during a disaster by local law departments. Several articles criticize this lack of planning prior to the September 11, 2001 terrorist attacks. Howard D. Swanson wrote a law review article based on his experience as City Attorney for Grand Forks, North Dakota during and after the Red River floods of Spring 1997.[90] He noted that "inconsistent with the growth of EM programs on the

various governmental levels has been the apparent lack of involvement by city attorneys."[91] Similarly, Roger A. Nowadsky found that local legal offices were not proactive in addressing true potential crises.[92] Even large cities have not been immune to the failure to adequately plan for legal issues during emergencies. In 1993, the Law Department for the Port Authority in New York and New Jersey was unprepared for the assault on the World Trade Center complex – it had no specific plan in place to deal with a disaster like a terrorist attack.[93]

Proactive lawyering does take place, but it typically occurs in the context of ensuring that legal organizations like the courts continue to function after a disaster. Some parts of the legal system are better prepared than others to ensure continuity of operations. For example, judges for the United States Court of Appeals for the Fifth Circuit had already planned for a disaster similar to Hurricane Katrina, which led to their ability to continue functioning after that event.[94] Preparation for continuity of operations is a far different matter, however, than putting a structure in place that will allow close cooperation between LGAs and EMCs. The literature suggests that law firms take business continuity precautions, but for the most part, discussion of the attorney's role concentrates on what is to be done after the disaster rather than on proactive lawyering.[95]

Ernest Abbott, who served as Chief Counsel for FEMA during the Clinton administration, recently observed, "I would guess that Katrina has led to more involvement of legal departments than 9/11. After all, 9/11 only hit three places, whereas Katrina hit almost every state in the nation, whether through actual contact or by evacuation."[96] Any increase in involvement by LGAs is apparently incremental. One article discussing legal response duties after Katrina makes the point that "a disaster preparedness plan...is not common in the legal profession."[97]

In contrast to local government, FEMA has significant legal resources.[98] The Office of Field Counsel within the DHS/FEMA Office of the Chief Counsel currently has twenty-five field attorneys, who are termed "Disaster Assistance Employees" (DAE). Further, there are twelve attorneys at recovery offices who are dedicated to specific disasters. DAE are on-call intermittent employees, and they are the backbone of providing disaster assistance. They allow FEMA to surge up for all programs. The relationship between FEMA attorneys and the program people who perform emergency management activities has evolved. In the 1970s, if the field staff did not like legal advice, they simply ignored it. "The interface is always an issue, and you need to stay on top of it."[99]

V. TRAINING FOR LGAS AND EMCS ON LEGAL ISSUES IN EMERGENCY MANAGEMENT

Among those who work closely with LGAs and EMCs, there is general agreement that their cooperation and knowledge of EM law needs improvements. Again, the basis for this consensus is the sum of anecdotal information. Some attempts have been made to address the lack of legal knowledge. Past federal government legal education offerings have focused on federal issues, without being of much help to state and local attorneys.[100]

A workshop on March 5–6, 2009, titled "The Law and Catastrophic Disasters: Legal Issues in the Aftermath," provided a relatively thorough treatment of legal issues affecting state and local EM following a radiological event.[101] This tabletop exercise was very helpful, but as with much other training, it focused on response and recovery, rather than addressing the role of legal counsel during all phases of emergency management.[102]

There have been few EM law education offerings at the state and local levels. North Carolina is one exception. A CLE program entitled "Disaster Emergency Law and the Role of the Bar" took place on December 4, 2008.[103] Despite the event's title, however, perusal of the materials provided to participants indicates that EM law

was addressed only in passing. The topics included Public Health Preparedness and Legal Authorities, Workplace Disaster Preparedness, and Ethical Considerations and the Role of the Bar in Emergencies.[104] The North Carolina Emergency Management Association (NCEMA), the professional alliance of local EMCs, presents helpful talks on legal developments from time to time during its spring and fall conferences.[105] They are useful to those who attend, but not all EMCs belong to NCEMA. The cost of dues, travel, registration fees, and time away from other employment are all factors that may prevent some, particularly part-time EMCs, from taking advantage of NCEMA's offerings. Some online CLE offerings have been made available that deal specifically with emergency management law, but they may only be accessible for limited periods of time.[106]

The author has taught courses specifically directed to LGAs and local emergency managers in Indiana, Missouri, and New Jersey. In 2000[107] and 2001,[108] he presented two-day courses in Indiana to local emergency managers and legal counsel on "Legal Issues for Emergency Managers." In Missouri, he gave a workshop on the topic in 2007, entitled "Legal Issues in Homeland Security Emergency and Disaster Law: Working Together for Litigation Mitigation,"[109] and presented it again to the Missouri State Emergency Management Agency on March 13, 2009.[110] He offered a workshop on a similar topic in New Jersey in 2003.[111] In these educational presentations and other speeches, he emphasizes the need for LGAs and EMCs to work closely together in all phases of emergency management.[112]

One frustration for the author has been the lack of participation by LGAs. When he asks those LGAs in attendance their thoughts on why more of their compatriots do not join them at the events, they mention the lack of CLE credits, the lack of funds for training or transportation to training, the general lack of understanding among local leaders of the legal issues' importance to emergency management, and the lack of connectedness between LGAs' offices and emergency management. These opinions are again examples of anecdotal evidence regarding understanding of EM law, challenges facing LGAs, and the relations between LGAs and EMCs. Clearly, more education in legal issues is needed, as is greater support for LGAs and EMCs who need to obtain that knowledge.

VI. CONCLUSION

State and local governments vary in the quality of legal guidance afforded to their EMCs. Particularly troubling is the Nationwide Plan Review: Phase 2 Report's finding that plans for about one-third of states and large urban areas were only "partly satisfactory" in outlining enabling laws for EM.[113] These are the best-financed units of government in the nation below the federal level. It is disquieting that many do not know their enabling laws. After all, that information may be researched at leisure. The quality of legal advice given in the heat of emergency response must, perforce, be significantly lower for those unfamiliar with the laws regarding which they are advising.

If legal counsel is to partner with local emergency management, leadership is required.

It has to come from the top if attorneys are to be involved. This has to be on a jurisdiction-by-jurisdiction basis. I think that organizations like the National Governors' Association, the National League of Cities, and so on need presentations on the importance of involving attorneys at the state and local levels at all times. They need to be involved in the emergency operations plans and so on.[114]

Over time, local officials are growing to understand potential legal liabilities and the possible effects on their political futures of failure to address these risks. It may be that the personal consequences will motivate them to pay greater attention to emergency management legal hazards.[115] The jurisdiction that employs a CEM as its emergency management coordinator and is EMAP certified is far less likely to experience unexpected liability. Despite this trend toward professionalization, many local jurisdictions have neither full-time attorneys nor full-time emergency coordinators.

Rural units of government face the greatest challenges in obtaining competent legal advice on emergency management issues. They lack the funding of larger entities, and may often find themselves with LGAs and EMCs who lack experience in their posts and are short on knowledge of legal issues.

This situation must be addressed in order to maximize protection from potential liability for both local units of government and their emergency management employees. The solution must embrace scientifically based research to support the anecdotal tales of troubles with emergency management legal issues. When such research has been completed, documented results can be presented in proper forums, and widespread backing for additional education will likely arise.

Similarly, to be effective, EM legal education must have support from governmental authorities at all levels. EM education from the federal government through FEMA must incorporate legal issues. At a minimum, FEMA should create a "Legal Issues in Emergency Management" course that addresses actual concerns found at the state and local levels, rather than solely federal issues, which rarely arise for local EMCs. Federal and state bar associations and law schools have an important role to play as approvers, promoters, and providers of continuing legal education.

Only when the full dimensions of the legal challenges for emergency management have been fully explored and addressed by responsive educational offerings will the universal liability risk to local jurisdictions be mitigated. Only when all local jurisdictions have competent legal advice for emergency management on an ongoing basis will this important hazard be properly addressed.

Legal Topics

For related research and practice materials, see the following legal topics: Governments – Local, Governments – Finance, Public Health & Welfare, Law – Social Services Emergency Services, Torts – Negligence Proof, Custom – Business Customs.

QUESTIONS

1. What does the article say are the biggest challenges facing emergency managers and lawyers in working together? Which of them strikes you as the hardest to overcome? Why?
2. What personal issues do you think might come up between the two groups as lawyers and EMC's work together?
3. How can EMCs and LGAs lessen the friction between themselves?

4. What avenues do you suggest for making education in EM law more commonly available?
5. How can EMCs and LGAs manage to be available at the same time for EM legal education programs?
6. How can EMCs and LGAs work together to encourage development and dissemination of EM legal education that is suitable for both groups to attend together?

NOTES

1. "Laws help create the infrastructure through which emergencies are detected, prevented, declared, and addressed." James G. Hodge, Jr. & Evan D. Anderson, Principles and Practice of Legal Triage During Public Health Emergencies, 64 N.Y.U. Ann. Surv. Am. L. 249, 250 (2008).
2. Examples of emergency management law on the federal level include the Robert T. Stafford Act, Pub. L. No. 93-288, 88 Stat. 143 (1974), which provides the pattern for federal aid to states and localities and the Homeland Security Act of 2002, Pub. Law No. 107-296, 116 Stat. 2135 (2002), which establishes standards for local emergency responders and emergency management, among other requirements. On the state level, enactments such as Ind. Code 10-14-1 et seq. create the structure and delineate the functions of state and local emergency management.

On the local level, many cities and counties have enacted laws that explain with more detail the authorities granted to them under state statutes.
3. Lloyd Burton, The Constitutional Roots of All-Hazards Policy, Management, and Law, 5 J. of Homeland Sec. and Emergency Mgmt. 4 (2008).
4. Keith Bea, L. Cheryl Runyon & Kae M. Warnock, Congr. Res. Serv. Gov. and Fin. Division, Emergency Management and Homeland Security Statutory Authorities in the States, District of Columbia, and Insular Areas: A Summary 4 (2004).
5. See, e.g., Ind. Code § 10-14-3-11 (2009) (Governor – Direction and Control of Department – Powers and Duties) ("In the event of disaster or emergency beyond local control, the governor may assume direct operational control over...

emergency management functions within Indiana.... In performing the governor's duties under this chapter, the governor may do the following. ...").

6. See, e.g., Ind. Code § 10-14-2-4 (2009) (Functions of Agency). ("The agency shall do the following: (1) Coordinate the state's emergency plans; (2) Serve as the coordinating agency for all state efforts for preparedness for, response to, mitigation of, and recovery from emergencies and disasters; (3) Administer this article and IC 16-31; (4) Perform duties assigned to the agency by the governor.").

7. See, e.g., Ind. Code § 10-14-3-17 (2009) (Political Subdivisions – County, Local and Interjurisdictional Organizations – Organization – Powers and Duties).

8. See Ind. Code sections supra notes 5-7; Chris Green, Part Time Disaster Planners Common – Many Kansas Counties Do Without A Full Time Emergency Manager, HutchNews, July 29, 2007, http://www.hutchnews.com/Todaystop/Emergency.

9. Howard D. Swanson, The Delicate Art of Practicing Municipal Law Under Conditions of Hell and High Water, 76 N.D. L. Rev. 487, 500 (2000).

10. See, e.g., Georgia Emergency Management Agency, Debris Management Plan: Roles and Responsibilities 3 (2006). In the event of a Presidential disaster declaration, local governments may receive reimbursement, subject to cost-share provisions, for the cost they incur for emergency clearance of debris from roadways and other public access facilities, and for the costs of removal and disposal of debris that poses an immediate threat to life, public health and safety. To be eligible for reimbursement under the Public Assistance Program, contracts for debris removal must meet rules for Federal grants, which mean they are subject to the Common Rule specifying uniform administrative requirements for grants to states and local governments. FEMA's common rule provisions can be found in 44 CFR Part 13, and specific subsections, such as 13.36, describe procurement and other requirements. Public Assistance applicants should comply with their own procurement procedures in accordance with applicable State and local laws and regulations, provided that they conform to applicable Federal laws and standards identified in Part 13. Id.

11. See William C. Nicholson, Seeking Consensus on Homeland Security Standards: Adopting the National Response Plan and the National Incident Management System, 12 Widener L. Rev. 491, 491-559, 524-525 (2006).

12. Joseph G. Jarret & Michele L. Lieberman, Symposium, When the Wind Blows: The Role of the Local Government Attorney Before, During, and in the Aftermath of a Disaster, 36 Stetson L. Rev. 293, 294 (2007).

13. Keith Bea, Congr. Res. Serv., Federal Stafford Act Disaster Assistance: Presidential Declarations, Eligible Activities, and Funding (2005).

14. See William C. Nicholson, Emergency Management Legal Issues, in Emergency Management: Principles and Practice for Local Government 237 (Kathleen Tierney & William Waugh eds., 2d. ed.) (2007).

15. Demetrios L. Kouzoukas, The National Action Agenda for Public Health Legal Preparedness: Public Health Emergency Legal Preparedness – Legal Practitioner Perspectives, 36 J.L. Med. & Ethics 18, 20 (2008).

16. National Fire Protection Association, NFPA 1600 Standard on Disaster/Emergency Management and Business Continuity Programs (2004 ed.), available at http://www.preparednessllc.com/resources/nfpa_1600.html (follow "NFPA 1600 – 2004 Edition" hyperlink).

17. ANSI, "9-11 Commission Presented with Recommendation on Emergency Preparedness," https://www.ansi.org/news_publications/news_story.aspx?menuid=7&articleid=670.

18. National Commission on Terrorist Attacks Upon the United States, The 9/11 Commission Report 398 (2004), available at http://www.gpoaccess.gov/911.

19. For information on ANSI, see American National Standards Institute Homepage, http://www.ansi.org (last visited Mar. 9, 2010).

20. National Fire Protection Association, NFPA 1600 § 5.2 (2007).

21. National Emergency Management Association, NFPA 1600 Standard Resolution, available at http://www.nemaweb.org/?335.

22. The Capability Assessment for Readiness (CAR) is actually the "common root" for NFPA 1600 and EMAP. Paul Rasch, EMAP Standards Subcomm. Chair and EMAP Technical Comm. Member, remarks at The New Emergency Management Standards from the Emergency Management Accreditation Program (EMAP), EIIP

Virtual Forum Presentation (August 22, 2007) (transcripts available at http://www.emforum .org/vforum/lc070822.htm). FEMA created the CAR, which, previous to EMAP, served as its state emergency management assessment tool. See Press Release, FEMA, Emergency Management Assessments Completed in Nine States Under Joint FEMA/EMAP Initiative (July 23, 2003), http://www.fema.gov/news/newsrelease .fema?id=3712.

23. FEMA Press Release, supra note 22.

24. Many organizations collaborated on and supported the development of EMAP, including: The National Emergency Management Association (NEMA), International Association of Emergency Managers (IAEM), U.S. Department of Homeland Security Emergency Preparedness & Response Directorate (EPR/FEMA), U.S. Department of Justice Office of Justice Programs, U.S. Department of Transportation, National Governors Association, National League of Cities, Council of State Governments (CSG), National Conference of State Legislatures, National Association of Counties, individual states, and others.
EMAP, EMAPonline.org, EMAP History (follow "What is EMAP?" hyperlink, then follow "EMAP History" hyperlink).

25. FEMA Press Release, supra note 22.

26. "Where relevant to the case and upon a proper evidentiary foundation, safety standards promulgated by organizations such as ANSI may be admitted to show an accepted standard of care, the violation of which may be regarded as evidence of negligence." Kent Vill. Assocs. Joint Venture v. Smith, 657 A.2d 330, 337 (Md. Ct. Spec. App. 1995).

27. Indeed, custom and usage within an industry need not be complete or general where improved safety standards, which EMAP provides for emergency management, are involved. See T.J. Hooper v. N. Barge Corp., 60 F.2d 737 (2d Cir. 1932), cert. denied sub nom E. Transp. Co. v. N. Barge Corp., 287 U.S. 662 (1932).

28. "Evidence of compliance or noncompliance with an industry standard of care is almost always admissible evidence on whether the defendant exercised due care.…" David G. Owen, Proving Negligence in Modern Products Liability Litigation, 36 Ariz. St. L.J. 1003, 1023 (2004).

29. There are political costs, and, for local officials, potential legal costs that might be exacted if they fail to prepare for and respond adequately to a disaster. A means of addressing the risk of legal liability and mitigating potential political costs is adherence to accepted national standards. In emergency management…at the programmatic level, National Fire Protection Association (NFPA) 1600 has been acknowledged by Congress, the 9-11 Commission, and other bodies as the accepted international standards for emergency management programs. The Emergency Management Accreditation Program (EMAP) operationalizes and expands the NFPA standards for state and local emergency management programs. The EMAP standards affirm that emergency management programs include the public agencies, nongovernmental organizations, and businesses that constitute the capabilities of states and communities to deal with disasters. [This] program[] provides [a] benchmark[] for professional emergency managers and emergency management programs to ensure that they have the tools to manage risks and to deal with disasters.
William L. Waugh, Jr., The Political Costs of Failure in the Katrina and Rita Disasters, 604 Annals of The Am. Acad. of Political and Soc. Sci. 10, 22-23 (2006).

30. Emergency Management Accreditation Program, EMAP Standard § 4.2 (2007), available at http://www.emaponline.org (follow "What is EMAP" hyperlink; then follow "The EMAP Standard" hyperlink; follow the EMAP Standard image hyperlink).

31. FEMA, National Response Framework 16 (2008), available at http://www.fema.gov/pdf/ emergency/nrf/nrf-core.pdf.

32. Id. at 74.

33. FEMA, National Incident Management System 14 (2008), available at http://www.fema.gov/ pdf/emergency/nims/NIMS_core.pdf.

34. Id. at 95.

35. Id. at 122.

36. It is incumbent upon the attorney to learn, in advance, what will be expected of him or her in a disaster scenario. It is not at all uncommon for attorneys to be thrust into roles or given responsibilities that are not only outside of their job descriptions, but consist of areas of the law that are alien to them.
Joseph G. Jarret & Michele L. Lieberman, supra note 12, at 312.

37. The legal environment during emergencies is always in flux, necessitating constant review.

This requires regular communications with law and policymakers at all levels of government, along with private-sector actors, to ensure the legality of specific actions. Public health legal practitioners have to gather reliable, accurate information from public health practitioners or others in the field to understand prevalent public health needs of populations affected by an emergency. This information can be used to prioritize legal issues that may facilitate or impede public health efforts in advance or as they arise. Public health practitioners, healthcare workers, and others may face a plethora of legal issues, both actual and perceived. The capacity to classify quickly and respond to these issues in real time is at the core of legal triage.
James G. Hodge, Jr. & Evan D. Anderson, Principles and Practice of Legal Triage During Public Health Emergencies, 64 N.Y.U. Ann. Surv. Am. L. 249, 273-74 (2008) (emphasis added).

38. William C. Nicholson, An Essential Team: Local Emergency Managers and Legal Counsel, 37 Pub. Mgr. 81, 83 (2008).

39. "Power outages that accompany most disasters deny attorneys the use of electronic research, the ability to confer with colleagues, access to computer databases, and the use of other helpful resources." Joseph G. Jarret & Michele L. Lieberman, supra note 12, at 312.

40. "I have no particular recollections of friction between FEMA and state/local attorneys. Robust debate can be anticipated when millions of dollars may be at stake. There is of course, an appeal process when an applicant is dissatisfied with a specific FEMA decision." Telephone Interview with Mary Ellen Martinet, Associate Chief Counsel for Field Counsel, DHS/FEMA, & Michael Hirsch, former Associate Chief Counsel, DHS/FEMA. (Jan. 8, 2010) (Comments of Mary Ellen Martinet).

41. FEMA, Developing and Maintaining State, Territorial, Tribal, and Local Government Emergency Plans (CPG 101) 5-1 (2009), available at http://www.fema.gov/pdf/about/divisions/npd/cpg_101_layout.pdf.

42. Id. at 6-5.

43. Id. at C-8.

44. FEMA, Guide for All Hazards Emergency Operations Planning (SLG 101) 2-3 (1996).

45. Id. at 2-6.

46. Id. at 4-16.

47. Id. at 5-H-14.

48. FEMA, Nationwide Plan Review: Phase 2 Report 13 (2006).

49. Id. at 62.

50. Sarah A. Lister & Holly Stockdale, Congr. Res. Serv., Pandemic Influenza, An Analysis of State Preparedness and Response Plans 16 (2007) available at http://www.fas.org/sgp/crs/homesec/RL34190.pdf.

51. William C. Nicholson, Litigation Mitigation: Proactive Risk Management in the Wake of the West Warwick Club Fire, 1 J. Emergency Mgmt. 14, 17-18 (2003).

52. Joseph G. Jarret & Michele L. Lieberman, supra note 12, at 299.

53. Brenna G. Nava, Comment, Hurricane Katrina: The Duties and Responsibilities of an Attorney in the Wake of a Natural Disaster, 37 St. Mary's L.J. 1153, 1170-01 (2006).

54. The ABA provides a "Disaster Law Resources" webpage focused on law firm business continuity. ABAnet, http://www.abanet.org/disaster/council.html (last visited Mar. 10, 2010).

55. ABAnet, Member Assistance, https://www.abanet.org/barserv/disaster/member_assistance.html (last visited Mar. 14, 2010).

56. See generally Denis Binder., Emergency Action Plans: A Legal and Practical Blueprint, 63 U. Pitt. L. Rev. 791 (2002).

57. Telephone Interview with Corwin R. Ritchie, Executive Director, Iowa County Attorneys Association (Mar. 12, 2010).

58. "Part-time" means that the attorney is paid for fewer than forty hours weekly on government matters. Like many other attorneys, the part-timers are usually paid at an hourly rate. Telephone Conversation with P. C. McLaurin, Jr., Extension Professor and Leader, Center for Governmental Training and Technology, Mississippi State University Extension Service, contact person for the Mississippi Association of County Board Attorneys (Mar. 12, 2010).

59. Telephone Interview with Corwin R. Ritchie, supra note 57.

60. Id.

61. "The overwhelming majority are part-time. In Mississippi, approximately seventy-five County Board Attorneys are part-time, with six or seven being full-time." Telephone Conversation with P.C. McLaurin, Jr., supra note 58.

62. Telephone Interview with James B. Blackburn, Legislative Counsel, North Carolina Association of County Commissioners (Mar. 12, 2010).

63. Communication with Gail Ricks, Assistant to Virginia Delegal, Esq., General Counsel, Florida Association of Counties (Mar. 22, 2010).

64. "As a general rule, those who are part-time are paid at a lower hourly rate than they receive for their work for private clients." Telephone Conversation with P.C. McLaurin, Jr., supra note 58.

65. Telephone Interview with Corwin R. Ritchie, supra note 57.

66. "The attorney for the County Board of Supervisors is appointed to the position by the Board. They [sic] may be the most qualified or a friend – which includes political supporter – of the County Board of Supervisors members." Telephone Conversation with P.C. McLaurin, Jr., supra note 58.

67. Id.

68. For specific information on these programs, see infra Part V.

69. Local governments are cutting into their core missions due to economic shortfalls. See, e.g., Flager County Government, County Trims Budget, Emergency Services Positions Cut, http://www.flaglercounty.org/pages.php?PB=2&NS=255 (last visited Mar.12, 2010) ("Adjustments to Flagler County's budget for next year continue. Tuesday County Administrator Craig Coffey announced the impending layoff of Director of Emergency Services Nathan McCollum. Emergency Services encompasses the county's Emergency Management and Fire Rescue departments."); see also Julia Patterson, King County's Budget Crisis: Where Do We Go From Here?, Kent Reporter, Jan 23, 2010, available at http://www.pnwlocalnews.com/south_king/ken/opinion/82404507.html?utm_source=feedburner&utm_medium=feed&utm_campaign=Feed%3A+southkingopinion+(Opinion+-+South+King+County ("King County is facing a budget crisis of incomparable magnitude, primarily due to the worldwide economic recession. I have the challenging task of balancing a budget that will severely cut critical functions of government, or work with my colleagues to ask voters if they are willing to support a tax increase for criminal justice, health, and human services programs.").

70. American Bar Association, ABAnet.org, State and Local Government Law Section, "Checklist for State and Local Government Attorneys to Help Prepare for Possible Disaster," http://www.abanet.org/statelocal/checlist406.pdf (last visited March 10, 2010).

71. "The capabilities of state and local agencies, however, are still very uneven. Some are highly professional and very capable, and others clearly are not. Capabilities are largely determined by the level of experience with hazards and disasters and by funding levels." William L. Waugh, Living with Hazards and Dealing with Disasters: An Introduction to Emergency Management 14 (M.E. Sharpe, Inc. 2000).

72. "Yet, even as participants noted some level of professional status or movement forward as a field[,] they commented (often at great length) on ongoing challenges such as fragmentation, a lack of professional identity, a diminished valuation of emergency management, and the field's cyclical status flux based on other's decisions outside the emergency management community." Carol L. Cwiak, Strategies for Success: The Role of Power and Dependence in the Emergency Management Professionalization Process 55 (October 29, 2009) (Ph.D. dissertation), available at http://www.ndsu.edu/emgt/graduate/alumni.

73. Waugh, supra note 71, at 52.

74. See International Association of Emergency Managers, CEM(R) FAQ, http://www.iaem.com/certification/generalinfo/cem.htm (last visited Apr. 10, 2010).

75. Id.

76. Id.

77. Green, supra note 8.

78. Id.

79. Id.

80. Telephone Interview with Duane Davis, Emergency Management Director, Jackson County, IN, President Emergency Management Alliance of Ind. (Mar. 15, 2010).

81. Communication with Brenda Jones, Executive Assistant to the Director, Government Liaison, N.C. Emergency Management (March 31, 2010).

82. Interview with Douglas Hoell, Director, Division of Emergency Management, North Carolina Department of Crime Control and Public Safety (Mar. 3, 2009).

83. "According to data compiled by the Kansas Association of Counties in 2006, full-time emergency managers in smaller counties could make as little as $30,000 to $40,000 a year while larger counties can pay out $70,000 to $90,000 annually. Part-time managers in a couple counties

made as little as $7 or $8 an hour in 2006." Green, supra note 8.

84. "Current concerns about legal liability arising out of failure to prepare for known hazards, however, may force public officials to pay greater attention to risks to public health and safety." Waugh, supra note 71, at 52.

85. "Some Directors who are full-time preside over 9-11 call centers as well. There is a trend to separating these roles due to the increased workload, as emergency management and administering a 9-11 center both require a full-time commitment to do the job right." Davis did not have information available on how many EMCs in Indiana work forty hours per week for emergency management. Telephone Interview with Duane Davis, supra note 80.

86. Communication with Brenda Jones, supra note 81.

87. Nicholson, supra note 38, at 81-83.

88. In contrast, with regard to emergency management, Dr. Carol Cwiak's Ph.D. dissertation masterfully includes many anecdotal statements along with her well-grounded empirical evaluation of the current and future status of professionalization in that discipline. Cwiak, supra note 72.

89. See J.D. Kartez & Mark K. Lindell, Adaptive Planning for Community Disaster Response, in Cities and Disaster, North American Studies in Emergency Management 28 (Richard T. Sylves & William L. Waugh, Jr. eds., 1990).

90. Swanson, supra note 9.

91. Id. at 488.

92. Roger A. Nowadsky, Lawyering Your Municipality Through a Natural Disaster or Emergency, 27 Urb. Law 9, 9 (1995).

93. Jeffrey S. Green & Ira Tripathi, Coping with Chaos: The World Trade Center Bombing And Recovery Effort, 27 Urb. Law. 41, 46 (1995).

94. John Council, Katrina Kicks the Fifth Circuit out of Marble Courthouse in the Big Easy, Tex. Law., Sept. 5, 2005, at 5, available at 9/5/2005 TEXLAW 5 (Westlaw).

95. See, e.g., Nava, supra note 53, at 1178.

96. Telephone Interview with Ernest Abbott, Former FEMA General Counsel and Principal in FEMA Law Associates (Feb. 7, 2009).

97. Nava, supra note 53, at 1170-71.

98. Telephone Interview with Mary Ellen Martinet & Michael Hirsch, supra note 40.

99. Id. (comments of Michael Hirsch).

100. FEMA, through the Emergency Management Institute, taught a class to state and FEMA attorneys September 9-10, 1998 entitled "Course E709: Expediting Disaster Response and Recovery Pursuant to the Stafford Act." The course focused on the federal side of emergency law. Subsequently, FEMA has worked to educate state-level attorneys through the National Emergency Management Association (NEMA) Legal Attorneys Committee during their twice-yearly meetings. Telephone Interview with Tamara S. Little, Assistant Attorney General, State of Ohio, NEMA Legal Attorneys Committee Chair (Mar. 21, 2002).

101. The event was designed and coordinated by Ernest Abbott, Former FEMA General Counsel and Principal in FEMA Law Associates. See generally FEMA Law, E-Newsletter, http://www.fema-law.com/E-Newsletters/Issue10Feb09.pdf. This program was presented in Washington, D.C., in conjunction with the NEMA meeting. The association with NEMA, the national organization composed of state EM directors, ensured attendance by state attorneys. The costs of travel, food, lodging, and registration resulted in few local attorneys attending the event.

102. See CAILAW, The Law and Catastrophic Disasters: Legal Issues in the Aftermath, available at http://www.cailaw.org/Brochures_2009/Disaster Management.pdf.

103. The North Carolina Bar Association presented this event at the N.C. Bar Center in Cary, N.C. on Dec. 4, 2008.

104. North Carolina Bar Association Foundation, Disaster Emergency Law and the Role of the Bar (2008).

105. See, e.g., on Wednesday, Mar. 17, 2010 from 8:30–9:25 a.m., a program entitled "Liability Issues in Emergency Management" was offered at the Spring Conference of the N.C. Emergency Management Association; see Conference Schedule, https://ncema.renci.org/_layouts/Renci/NCEMAConference/Schedule.aspx (last viewed Mar. 10, 2010).

106. See, e.g., West Legal Ed Center, Online CLE Course, "Administrative Law and Emergency Management: Katrina and Beyond II," http://westlegaledcenter.com/program_guide/course_detail.jsf?courseId=3739400.

107. William C. Nicholson, Lecture at the Indiana SEMA Course: Legal Issues for Emergency Managers (August 12-13, 2000).

108. William C. Nicholson, Lecture at the Indiana SEMA Course: Legal Issues for Emergency Managers (June 16-17, 2001).

109. William C. Nicholson, Lecture at the Missouri Emergency Manager's Training: Legal Issues in Homeland Security Emergency and Disaster Law – Working Together for Litigation Mitigation (Mar. 16, 2007).

110. Id. (Mar. 13, 2009).

111. William C. Nicholson, Remarks at the Annual Meeting Seminar of the New Jersey Emergency Management Association: Legal Issues for New Jersey Emergency Management Coordinators (November 19, 2003).

112. See, e.g., William C. Nicholson, Plenary Speech, Emergency Management Alliance of Indiana Conference: Emergency Management and Legal Attorneys – Partnering for Safety and Litigation Mitigation (October 18, 2002).

113. FEMA, Nationwide Plan Review: Phase 2 Report 13 (2006).

114. Telephone Interview with Mary Ellen Martinet & Michael Hirsch, supra note 40 (comments of Michael Hirsch).

115. "The legal liability of local officials for failure to prepare reasonably for disaster and the political liability of elected chief executives at all levels when they have not invested appropriately in emergency management offices and programs are encouraging innovation and reform." Waugh, supra note 71, at 193.

Spanish Grant, Texas, April 15, 2010 – An employee of the Seaquest company, which is responsible for tearing down homes in the Federal buy out program waters down a home in Spanish Grant, TX. which is being torn down being sponsored by FEMA and the State of Texas. Residents who chose to participate in the buyout program are paid fair market value of their homes and the home is demolished. The local and State governments will determine the final use of the land. Photo by Patsy Lynch/FEMA.

Chapter 15

LEGAL STEPS FOR MITIGATION

Mitigation focuses on breaking the cycle of disaster damage, reconstruction, and repeated damage. It includes "prevention" which DHS considers to be a fifth phase of emergency management.

PART 1. QUANTIFYING LEGAL RISK

William C. Nicholson and Lucien Canton*

Background

For the past sixteen years, Nicholson has worked with emergency managers to help them lower their risk of liability. The goal of "litigation mitigation," however, has proven to be an elusive one. This Chapter's goal is to move closer to universal protection from liability – not through expanded immunities that would excuse bad actors, but by means of a tool that can accurately identify and measure exposure to this common risk. As an experienced Director of Emergency Services for the City of San Francisco and senior FEMA employee, Canton understands the dangers that exposure to liability poses for jurisdictions. His practical experience provides a "reality check" as we develop this tool.

I. INTRODUCTION

Many laws require that competent legal advice be rendered on an ongoing basis to emergency managers. The legal history of emergency response and emergency management is replete with instances where the lack of such counsel has been a key factor in exposing jurisdictions to danger of liability. In some instances, that danger has evolved into actual legal emergencies, including lawsuits and resultant money judgments against jurisdictions.

For years, some commentators have cautioned about the hazard of legal liability, even going so far as to term it a "universal risk." Emergency

* Lucien G. Canton, CEM is an independent management consultant specializing in strategic planning for crisis. He is the former Director of Emergency Services for San Francisco CA where he coordinated the emergency management program and served as policy advisor to the Mayor on emergency management and Homeland Security issues. Prior to his appointment, Mr. Canton served as an Emergency Management Programs Specialist and Chief of the Hazard Mitigation Branch for FEMA Region IX where he assisted in responding to over eighteen major disasters. Mr. Canton is a Certified Emergency Manager, a Certified Protection Professional, and a Certified Business Continuity Professional. He is also certified as an assessor for the Emergency Management Accreditation Program. He holds a Master of Business Administration in International Management degree and is the author of *Emergency Management: Concepts and Strategies for Effective Programs* and of numerous articles on emergency management and security topics. Mr. Canton's book *Emergency Management: Concepts and Strategies for Effective Programs* ISBN: 978-0-471-73487-1 (2006) is widely regarded as the best available guide for creating and maintaining a top quality emergency management program. Although it is aimed at practicing emergency managers, William Nicholson and others have used it with great success as a textbook in college-level courses in the subject. Mr. Canton may be reached through his web site: http://luciencanton.com/

255

managers and the lawyers who counsel them generally agree with this view.

The idea of quantifying legal risk is not new. Insurance companies routinely perform this task, as do publicly held corporations. With regard to asbestos, for example, accountants and actuaries routinely estimate potential liabilities from firm- and industry-wide data on type of use, years of exposure, and the state of employee industrial protections as awareness of risk improved. Their approaches are usually proprietary and not available for use by the general public. Further, they do not evaluate emergency management liability risks.

There are difficulties in quantifying legal risks. Like all risk assessment, it is not an exact science. Our goal is to put together an approach that is similar to that used by local emergency managers in measuring other risks so that they understand that the risk is there and that the only way to reduce it is by obtaining regular, competent legal advice.

The challenge faced by emergency managers regarding planning for liability risks is simply this – historically, there has been no paradigm for measuring liability risk in this area. This paper proposes the bare bones of such a paradigm.

II. LIABILITY RISK MATRIX

The following matrix proposes to evaluate liability risk based on a jurisdiction's experience over a 5 year period. Of course, the greater the amount of information available, the better such a risk analysis will be – a 10-year period would be better.

One important thing to keep in mind when evaluating liability risk is that it is a function of both preparedness and external factors, like any other risk. Greater preparedness causes the risk to diminish. External factors, particularly a measurement of how much litigation the jurisdiction has experienced, must also be taken into account. Another thing to bear in mind is that litigation against emergency management is likely to increase in the aftermath of disasters. During the disaster, injuries and property damage have likely occurred, and these elements are like blood in the water to plaintiffs' attorneys. Note that the matrix rewards:

- Frequent involvement of a specific attorney.
- Experience shows that these two factors are vital in obtaining competent legal advice.

Ongoing, frequent exposure to these matters likewise results in appreciation of the variety of legal issues specific to emergency management. Knowing the law will result in far better advice during emergent events, when prompt, correct input on legal matters is vital to prevent liability exposure.

Further, performance of specific tasks relating to subjects on which emergency management needs input is also rewarded. These include: review of the variety of plans required by NFPA 1600 and EMAP; review of Mutual Aid Agreements (MAAs); as well as participation in training and exercising. Other specific tasks may be included as needed.

III. FACTORS IN THE LIABILITY RISK MATRIX

The factors included in the Matrix are those which experience has shown to be important in reducing exposure to liability.

A. History of Litigation

Different jurisdictions have different histories in terms of lawsuits. Urban areas are commonly

believed to be more litigious than rural locations. Rather than rely on accepted wisdom, however, this approach keys in on the jurisdiction's actual history of suits. The overall number of suits gives an idea of the litigiousness of the populace, while a separate tally of suits based on emergency management issues reflects the level of lawsuit danger specific to the department. Canton notes that his sense is that while jurisdictions are sued pretty regularly, the actual possibility of an EM related action is pretty low barring an actual event.

Some larger jurisdictions may protest that they are frequently sued, and thus the Matrix calculates them as at a high risk for liability, regardless of mitigation steps they might take. This result reflects the fact that these jurisdictions are at a high risk of liability, not that the Matrix is incorrect.

B. Specific Attorney Assigned to Emergency Management

If the same lawyer is involved with the subject, he/she gets to know the area of law as well as the client agency's personnel. Since this is an area of law that is both complex and unique in many ways, learning emergency management law before a disaster strikes is vital. The attorney must give advice "on the fly" during an event – there is no time for leisurely consultation of multiple sources to find the most nuanced and exact answer. After disaster strikes, many resources may be unavailable – the net may be inaccessible, libraries could be flooded out, and electricity may be off. Legal periodicals often emphasize the importance of knowing the client's business for successful representation.

C. Meetings Between Attorney and Emergency Manager

While the ideal might be for the attorney to participate in the monthly meetings of the Emergency Management Advisory Council (or whatever the local group with this responsibility is called), in practice that will probably not be possible. Canton notes that, when he was Director of Emergency Services for the City of San Francisco from 1996–2004, "My attorney participated in semi-annual exercises and planning sessions as appropriate but didn't make our monthly planning meeting. Also, there really wasn't that much business to discuss. Since he billed his time to my department, monthly meetings might be a cost prohibitive for smaller jurisdictions."

D. Specific Emergency Management-related Tasks Performed by Attorneys

Specific issues must be addressed through attorney input on an "as needed" basis. As the preceding discussion makes clear, high on the list for attorney attention are emergency plans. No less important are the other matters listed below. NFPA 1600 requires the following plans:

- Strategy
- Prevention
- Mitigation
- Emergency operations/response
- Business continuity, and recovery
- Review Mutual Aid Agreements for legal content and sufficiency
- Participate in training and exercising

Canton notes that participation in exercises is absolutely vital in integrating the attorney into emergency management. "This where we really sorted out their needs (e.g., computer, templates, reference files, etc.) and it makes them a part of the team."

IV. REQUIREMENTS FOR LEGAL ADVICE IN EMERGENCY MANAGEMENT

There are a number of specific matters that need an attorney's input. Generally, they are matters that have legal consequences for the jurisdiction. Often, legal input is specifically required.

A. NFPA 1600

National Fire Protection Association (NFPA) standard, "NFPA 1600 Recommended Practices for Disaster Management,"[33] establishes a shared

---

[33] NFPA 1600 Standard on Disaster/Emergency Management and Business Continuity Programs (2004), http://www.nasttpo.org/NFPA1600.htm

set of standards for disaster management, emergency management and business continuity programs. NFPA 1600 mandates that all EM and business continuity programs comply with all relevant laws, policies and industry practice. To do so, a lawyer's advice is necessary. Section 4.2 of the Emergency Management Accreditation Program (EMAP) similarly requires EM organizations to comply and keep up with changes in relevant laws and authorities.

4.2: Laws and Authorities

Overview

Laws and authorities refer to the legal underpinning for the program. Federal, state, tribal and local statutes and implementing regulations establish legal authority for development and maintenance of the program and organization and define the emergency powers, authorities, and responsibilities of the chief executive and the program coordinator. These principles serve as the foundation for the program and its activities.

4.2.1 The emergency management program shall comply with applicable legislation, regulations, directives and policies. Legal authorities provide flexibility and responsiveness to execute emergency management activities in disaster and non-emergency situation. [sic] The emergency management program and the program's responsibilities are established in state and local law. Legal provisions identify the fundamental authorities for the program, planning, funding mechanisms and continuity of government.

4.2.2 The program has established and maintains a process for identifying and addressing proposed legislative and regulatory changes.[34]

An attorney's advice is necessary to perform this important task.

B. National Response System and National Response Framework

The National Response Framework (NRF) requires local officials to "[u]nderstand...and implement...laws and regulations that support emergency management and response."[35] The NRF states that plans are acceptable only if "... consistent with applicable laws."[36] The National Incident Management System (NIMS) also recognizes the importance of knowing and complying with applicable laws:

> To better serve their constituents, elected and appointed officials should do the following:...Understand laws and regulations in their jurisdictions that pertain to emergency management and incident response.[37]

During emergency management's response phase, NIMS advocates for attorney membership in the Incident Management team.

(4) Additional Command Staff

Additional Command Staff positions may also be necessary, depending on the nature and location(s) of the incident or specific requirements established by Incident Command. For example, a legal counsel might be assigned to the Planning Section as a technical specialist or directly to the Command Staff to advise Incident Command on legal matters, such as emergency proclamations, the legality of evacuation and quarantine orders, and legal rights and restrictions pertaining to media access.[38]

Also, "[t]he incident objectives and strategy must conform to the legal obligations and management objectives of all affected agencies, and may need to include specific issues relevant to critical infrastructure."[39] The duty to comply with

---

[34] EMERGENCY MANAGEMENT ACCREDITATION PROGRAM, EMAP STANDARD § 4.2 (2007) available at: http://www.emaponline .org/ (follow "What is EMAP" hyperlink; then follow "The EMAP Standard" hyperlink; click on the EMAP Standard image) (last visited February 28, 2010).

[35] FEMA, NATIONAL RESPONSE FRAMEWORK 16 (2008), available at http://www.fema.gov/pdf/emergency/nrf/nrf-core.pdf.

[36] *Id.* at 74.

[37] FEMA, NATIONAL INCIDENT MANAGEMENT SYSTEM 14 (2008), available at http://www.fema.gov/pdf/emergency/nims/ NIMS_core.pdf.

[38] *Id.* at 95

[39] *Id.* at 122.

existing law and monitor new law means that the local emergency manager must have access to the advice of competent legal counsel.

C. Guide for All Hazards Emergency Operations Planning (SLG 101)

FEMA's Guide for All Hazards Emergency Operations Planning (SLG 101) (superseded by CPG 101 Version 2, discussed below) requires at that the first stage of planning is to "[r]eview local and/or State laws, rules, regulations, executive orders, etc., that may be considered enabling legislation. Review Federal regulatory requirements."[40] Under "Identifying Hazards," SLG 101 the planner must "[b]egin with a list of hazards that concern emergency management in your jurisdiction. Laws... can help define the universe of hazards which the planning team should address in the all-hazard EOP."[41] Under "Authorities and References," SLG 101 directs that "[t]he Basic Plan should indicate the legal basis for emergency operations and activities. Laws, statutes, ordinances, executive orders, regulations, and formal agreements relevant to emergencies should be listed."[42] Under Resource Management, a Legal Advisor or equivalent position is described as:

Legal Advisor:
1. When notified of an emergency, reports to the EOC or other location as specified by the Resource Manager.
2. Advises Supply Coordinator and Procurement Team on contracts and questions of administrative law.[43]

The Nationwide Plan Review Phase 2 Report evaluated compliance with legal requirements.

(Note: S = Sufficient, PS = Partly Sufficient, NS = Not Sufficient):[44]

	States			Urban Areas		
	S	PS	NS	S	PS	NS
Does the Plan outline appropriate local, State, and Federal laws, rules, regulations, executive orders, agreements, etc., that may be considered enabling legislation (per *SLG 101* page 2-3) for catastrophic incidents?	68%	32%	0%	68%	31%	1%

At both the state and large city levels, just one-third of the plans reviewed were only "partly satisfactory" as regards enabling legislation. The review did not speak to the status of rural or less populated regions which have fewer resources and, very likely, much lower compliance. The Nationwide Plan Review Phase 2 Report warns that "[i]f a plan['s]... laws and authorities are inadequate, it cannot be assessed as fully adequate, feasible, or acceptable."[45] The Congressional Research Service likewise found that 22 out of 51 state-level pandemic influenza plans needed discussion of isolation and quarantine laws, and, surprisingly, only 2 of 51 plans discussed procedures for judicial review of due process protections.[46]

Unfortunately, lack of complete treatment of legal issues is a common shortcoming of local emergency operations plans.[47] For local governments, not having competent legal advice regarding EM law may result in major liability. This is true due to the fact that local government plays the vital first response role in disasters, and its leaders must make choices with major legal consequences.

Irrespective of the significant, and apparently concurrent, role that the State maintains in the

---

[40] FEMA, Guide for All Hazards Emergency Operations Planning (SLG 101)2-3 (1996).
[41] *Id.* at 2-6.
[42] *Id.* at 4-16.
[43] *Id.* at 5-H-14.
[44] FEMA, Nationwide Plan Review Phase 2 Report 13 (2006).
[45] *Id.* at 62.
[46] Sarah A. Lister, Holly Stockdale, CONGRESSIONAL RESEARCH SERVICE, *Pandemic Influenza, An Analysis of State Preparedness and Response Plans*, Order Code 34190 (2007) available at http://www.fas.org/sgp/crs/homesec/RL34190.pdf
[47] William C. Nicholson, "Litigation Mitigation: Proactive Risk Management in the Wake of the West Warwick Club Fire." 1 J. OF EMERGENCY MANAGEMENT, 14, 17–18 (2003).

local emergency management function, it is the local government that is on the front line of any disaster. Thus, the local government attorney should become intimately familiar with the local comprehensive emergency plan and with the extent and limitations of authority that a local government may exercise during an impending or realized disaster.[48]

Regarding legal response duties following Hurricane Katrina, one article makes the point that:

> [a] lawyer might attempt to argue that the failure to craft a disaster preparedness plan is not negligent because that type of precaution is not common in the legal profession. However, doing what is customary (i.e., not preparing a disaster plan) does not preclude a finding of liability based on a cost-benefit analysis.[49]

Emergency planning has been the subject of considerable litigation. Still, this vital matter did not until fairly recently has not received the attention in legal publications that it is due.[50]

D. Comprehensive Preparedness Guide 101 Version 2.0 (November 2010) (CPG 101)

FEMA's latest planning guidance reinforces the requirements for legal advice on an ongoing basis to a greater extent that even the previous documents cited above. FEMA's Comprehensive Preparedness Guide 101 Version 2.0 (CPG 101)[51] issued November 2010 is titled "Developing and Maintaining Emergency Operations Plans." CPG 101 is consistent with other FEMA guidance – and was "designed to complement the

use of... guides [like that for the REPP annex] where required by law or regulation."[52] Planners must respect Constitutional requirements and guidance promulgated to ensure their enforcement. The Guide notes that Planners should ensure compliance with the requirements of Title VI of the Civil Rights Act of 1964, Executive Order 13166, the Americans with Disabilities Act, Section 504 of the Rehabilitation Act, and other Federal, state, or local laws and anti-discrimination laws.[53]

From the beginning, a lawyer must be involved in the planning process, as the Guide requires that "[a]s a public document, the EOP... states its legal basis...."[54] Similarly, according to the National Response Framework, plans are acceptable only if "... consistent with applicable laws." CPG 101 mandates an "Authorities and References" section.

This section provides the legal basis for emergency operations and activities. This section of the plan includes:

- Lists of laws, statutes, ordinances, executive orders, regulations, and formal agreements relevant to emergencies (e.g., MAAs)
- Specification of the extent and limits of the emergency authorities granted to the senior official, including the conditions under which these authorities become effective and when they would be terminated
- Pre-delegation of emergency authorities (i.e., enabling measures sufficient to ensure that specific emergency-related authorities can be exercised by the elected or appointed leadership or their designated successors)

---

[48] Joseph G. Jarret and Michele L. Lieberman, "Local Government Law Symposium: Article: "When the Wind Blows": The Role of the Local Government Attorney Before, During, and in the Aftermath of a Disaster" 36 Stetson L. Rev. 293, 298 (Winter, 2007).

[49] Brenna G. Nava, Comment: *Hurricane Katrina: The Duties and Responsibilities of an Attorney in the Wake of a Natural Disaster*, 37 St. Mary's L.J., 1153, 1170-01(2006).

[50] *See generally* Denis Binder., *Emergency Action Plans: A Legal and Practical Blueprint*, 63 U. Pitt. L. Rev. 791(2002).

[51] Comprehensive Preparedness Guide 101 Version 2.0 Found on line at: http://www.fema.gov/pdf/about/divisions/npd/CPG_101_V2.pdf

[52] CPG 101 at Intro-1.

[53] CPG 101 footnote 4.

[54] FEMA, Developing and Maintaining State, Territorial, Tribal, and Local Government Emergency Plans: Comprehensive Preparedness Guide 5-1 (2009), Mar. available at http://www.fema.gov/pdf/about/divisions/npd/cpg_101_layout.pdf (last visited February 28, 2010).

- Provisions for COOP and COG (e.g., the succession of decision-making authority and operational control) to ensure that critical emergency functions can be performed.[55]

In the "Concept of Operations" section, the plan must "Describe how legal questions/issues are resolved as a result of preparedness, response, or recovery actions, including what liability protection is available to responders."[56]

In the Guide's discussion of the planning process, the issue of constraints and restraints addresses legal issues thusly:

> A constraint is something that must be done ("must do"), while a restraint is something that prohibits action ("must not do"). They may be caused by a law, regulation, or management directive;[57]

Analysis of the restraining or constraining effect of laws and regulations must clearly be performed by legal counsel.

Under "Incorporating Individuals with Access and Functional Needs" in Preparedness, the Guide asks "Does the plan include a definition for "individuals with disabilities and others with access and functional needs," consistent with all applicable laws?"[58]

After the plan is completed, it must be properly disseminated, or promulgated. Under CPG 101, the plan "promulgation process should be based in a specific statute, law, or ordinance."[59] Again, a matter on which a lawyer needs to advise.

The Guide makes the point that "Plan maintenance is also critical to the continued utility of the plans an organization has developed. A number of operations have had setbacks due to...outdated laws."[60] "Teams should also consider reviewing and updating the plan after the following events.... The enactment of new or amended laws or ordinances."[61] Failure to do so may expose a jurisdiction to liability as its plan relies on outdated laws that no longer provide the protection they once did.

Clearly, for an emergency operations plan to be compliant with CPG 101, legal counsel ought to be closely involved in its development and maintenance. The attorney must review all plans for legal content and sufficiency on an ongoing basis. As CPG 101 states, "*Acceptability*. A plan is acceptable if it meets the requirements driven by a threat or incident, meets decision maker and public cost and time limitations, and is **consistent with the law**."[62]

V. USING THE MATRIX

The factors discussed must be totaled over a five-year period in order to get a time-based perspective on their frequency. The amount of liability risk may be quantified for different tasks performed by the attorney. A set value is assigned to each task, then the impact of each task on the total amount of liability risk is calculated, based on the following matrix (see p. 262). The total score reflects the level of liability risk: Low, Medium, or High.

So, how does our jurisdiction perform under this matrix? With a score of 77, it is in the medium danger range. Note that this number could be readily reduced by greater involvement of the attorney in emergency management. Reviewing plans every year (–50) changes this into minor exposure (–73).

---

[55] *Id.* at 6-5.
[56] *Id.* at C-8.
[57] CPG 101 page 4-11
[58] CPG 101 page 4-20
[59] CPG 101 page 4-25
[60] CPG 101 page 4-26
[61] CPG 101 page 4-26
[62] CPG 101 page 4-17

Liability Risk Matrix

Minor	**Medium**	**Major**
0 and below	0 – 100	over 100

Event	Points
Jurisdiction sued	+ 2 (each instance)
Jurisdiction sued (EM-related)	+ 5 (each instance)
Specific attorney assigned to EM	– 5 (each year)
Specific attorney not assigned to EM	+ 5 (each year)
Bi-annual meetings w/ attorney assigned to EM	– 5 (each instance)
No bi-annual meetings w/ attorney assigned to EM	+ 5 (each month w/o meeting)
Specific EM-related tasks by attorney	– 5 (each instance/as needed)
	+ 5 (w/o each task/when needed)

Review all plans for legal content and sufficiency

(points per plan for all required by law – strategy, prevention, mitigation, emergency operations/response, business continuity, and recovery)

Review MAAs for legal content and sufficiency

Participate in training and exercising

Practical Application of the Matrix

So, based on a 5-year period for a hypothetical jurisdiction, the following might apply:

Jurisdiction sued 6 times	12
EM sued twice	10
Specific lawyer assigned to EM	– 25
Lawyer meets average of 2× yearly w/ EM	$– 10 \times 5 = -50$
Lawyer reviews 5 plans twice	$– 25 \times 2 = - 50$
5 plans not reviewed for 3 years	$50 \times 3 = 150$
Three MAAs each reviewed 2 times	– 30
Three MAAs each not reviewed 3 times	$30 \times 3 = 90$
Participate in training and exercising 1×/year	$– 10 \times 3 = -30$
Overall score	77

VI. CONCLUSION

Emergency managers often express concern regarding potential liability risks. Generally, however, they find themselves unable to answer the important question, "How MUCH risk does the jurisdiction face?" The Liability Risk Matrix is a tool designed to answer that question. It quantifies liability risks so that the local emergency management director may measure their severity rank them among other risks facing the jurisdiction.

What is needed at this point is field testing of the Liability Risk Matrix. The input of emergency management directors who provide real-world experience will assist in fine-tuning the Matrix. The eventual goal is to create a tool that is easy to use and reliable. The end product hopefully will have a significant impact in lowering liability risks for emergency management – finally achieving the elusive goal of "litigation mitigation."

Of course, every jurisdiction must prioritize its expenditures. One would need to be impaired not to perceive the downward trend of income for many jurisdictions is accompanied by increasing populations of people needing services of all kinds. A request for additional allocation of expensive attorney assets is unlikely to meet with success unless it is well reasoned out and lucidly presented. The emergency manager thinking on how to obtain additional expensive resources has to explain the nature of mitigation, and how it applies to the litigation process.

"Avoid litigation at all costs" said Abraham Lincoln, and that advice is even truer for a unit of government than it is for an individual. The amounts saved through lawsuits avoided may be astronomical. This means that not only do you have to make sure you have legal advice for emergency management, but that you have taken active steps to add to the lawyer's viewpoint that of an emergency manager! Hopefully, you will intrigue the lawyer enough that he or she will get in the habit of sticking his or her head into all the offices he or she advises just to see what might develop into a problem and what pro-active legal steps might avoid it. Now THAT is the spirit of LITIGATION MITIGATION!

QUESTIONS

1. Divide the class into two groups to debate the issue of whether liability risk can be measured.
2. If liability risk can be measured, is the approach taken by the authors a valid one? Why or why not?
3. What additional factors might you suggest to take into account in measuring legal exposure?
4. What obstacles prevent measuring liability risk in your jurisdiction?
5. How would you go about incorporating liability risk into the planning process?
6. The introductory material mentions that this chapter includes what DHS refers to as the "Prevention" phase in Mitigation. What is your opinion of having 4 vs. 5 phases of emergency management? What are the advantages and disadvantages of breaking Prevention out as a separate entity?

PART 2: MITIGATION CASE

As you read this case, consider what circumstances, if any, justify the government in interfering with a person's ability to use property as that person may see fit.

ROBERTA GOVE VS ZONING BOARD OF APPEALS

SUPREME JUDICIAL COURT OF MASSACHUSETTS

444 Mass. 754; 831 N.E.2d 865; 2005 Mass. LEXIS 423

April 4, 2005, Argued

July 26, 2005, Decided

OPINION

MARSHALL, C.J.

Roberta Gove owns "lot 93," an undeveloped parcel of land within a "coastal conservancy district" (conservancy district) in Chatham. In 1998, Ann and Donald J. Grenier agreed to buy lot 93 from Gove, contingent on regulatory approval for the construction of a single-family house on the property. Because Chatham prohibits construction of new residences in the conservancy district, the zoning board of appeals of Chatham (board) denied the Greniers a building permit. Gove and the Greniers sought relief in the Superior Court on statutory and constitutional grounds, contending that the prohibition against residential construction on lot 93 did not substantially further a legitimate State interest and that the board had effected a taking of lot 93, without compensation, in violation of the Fifth and Fourteenth Amendments to the United States Constitution and art. 10 of the Massachusetts Declaration of Rights. After a two-day bench trial, a judge in the Superior Court ruled in favor of the defendants on all counts. The Appeals Court affirmed. *Grenier v. Zoning Bd. of Appeals of Chatham*, 62 Mass. App. Ct. 62, 814 N.E.2d 1154 (2004). We granted Gove's application for further appellate review, and now affirm the judgment of the Superior Court.

1. Background. Lot 93 is located in the Little Beach section of Chatham, nearly all of which was acquired by Gove's parents (the Horne family) in 1926. In time, members of the family developed a motel, marina, rental "cottage colonies," and a number of single-family houses in Little Beach. The family also sold several lots for development. In 1975, that portion of Little Beach still owned by the Horne family was divided, by the terms of the will of Gove's mother, among Gove and her three brothers. Gove received several lots outright and sixteen other lots in fractional ownership, to be shared with her brothers. Gove also obtained title to at least two cottages. Gove continues to own one cottage in Little Beach; she sold a second in 1996.

Little Beach is part of a narrow, low-lying peninsula, bounded by Chatham Harbor and Stage Harbor, at the extreme southeastern corner of Cape Cod. In recent years, a "breach" has formed in the barrier island that long separated Chatham Harbor from the open ocean. The breach, which is widening, lies directly across the harbor from Little Beach, and a land surveyor familiar with the area testified that Little Beach is now "wide open to the Atlantic Ocean" and prone to northeasterly storm tides. Chatham is known for its vulnerability to storms, and, according to an expert retained by Gove and the Greniers, in recent years Chatham has "as a direct result of the breach" experienced a "significant erosion problem," including "houses falling into the sea." The same expert testified that, since the appearance of the breach, "there had been a significant rise in the mean high water [near lot 93] along Chatham Harbor." The record indicates that virtually no development has occurred in Little Beach since 1980.

Even before the breach developed, Little Beach was prone to inundation by seawater. Gove testified that the area was flooded by hurricanes in 1938, 1944, and 1954, and by a significant offshore ocean storm in 1991. None of these storms struck Chatham directly, but in the 1944 hurricane, Stage Harbor experienced a storm surge some nine feet above sea level. The 1954 hurricane damaged buildings and flooded roads in Little Beach. The 1991 storm flooded the area around lot 93 to a depth of between seven and nine feet above sea level, placing most, if not all, of the parcel underwater. The 1944, 1954, and 1991 storms, while significant, were less severe than the hypothetical "hundred year storm"[6] used for planning purposes, which is projected to flood the area to a depth of ten feet. According to another expert called by Gove and the Greniers, during storms, roads in Little Beach can become so flooded as to be impassable even to emergency vehicles, and access to the area requires "other emergency response methods," such as "helicopters or boats." The same expert conceded that, in an "extreme" event, the area could be flooded for four days, and that, in "more severe events" than a hundred year storm, storm surge flooding in Little Beach would exceed ten feet.

Lot 93 itself consists of approximately 1.8 acres. The lot is within approximately 500 feet of both Stage Harbor and Chatham Harbor and, according to one expert, is susceptible to coastal flooding "from both the front side and the backside of [the] property." The lot is bisected by a tidal creek, which is prone to flooding as well.

---

[6] "A hundred year storm" is, as the term suggests, a statistical approximation of the most severe storm likely to occur in one century.

The highest point on the property is 8.7 feet above sea level, and much of the property is less than four feet above sea level and technically a "wetland." According to a 1998 map issued by the Federal Emergency Management Agency, lot 93 lies entirely within flood hazard "Zone A," an area defined by its vulnerability to "significant flooding" in "hundred year storms." Lot 93 also lies immediately outside "Zone V" where, during a hundred year storm, significant flooding with wave action can be expected.

Gove inherited lot 93 in 1975, when residential development was permitted on the parcel. In 1985, however, the town placed all of the land within the hundred year coastal flood plain, including lot 93, into the conservancy district. The stated purposes of the conservancy district include maintaining the ground water supply, protecting coastal areas, protecting public health and safety, reducing the risk to people and property from "extreme high tides and the rising sea level," and conserving natural resources. The town zoning officer testified that the conservancy district serves to mitigate the "total public safety problem" of coastal flooding, and was specifically intended to protect both residents and public safety personnel.

The bylaw governing the conservancy district bars without exception the construction of new residential dwellings. The bylaw does allow specified nonresidential uses, either as of right or by special permit.[7] The zoning officer testified that the nonresidential uses are less likely to create a danger in the event of a flood than are residential structures, in part because structures "ancillary" to homes "tend to break off" in storms and "do a lot of collateral damage to other structures and property," whereas such damage is less likely when nonresidential structures, normally more firmly anchored to the ground, are built.

In the years before the zoning regulations were amended to restrict development in the hundred year flood zone, Gove attempted to sell lot 93.

She listed the lot and another she owned with a local broker but "had no offers" on the properties, and she withdrew them from the market. Gove further testified that, whatever its value before the breach developed, lot 93's worth had "plummeted" as a result of the breach and the property "had no value...whatsoever" in the early 1990s. By the late 1990s, property in the area had gained value, she said, but the land, she clarified, was still most attractive to those who "have lived in the area" and were unswayed by frequent media reports of storm damage in Chatham.

In 1998, the Greniers contracted to purchase lot 93 from Gove for $192,000, contingent on their ability to obtain permits for a home and a septic system on the site. The Greniers proposed to develop a house on lot 93 on land between 5.3 and 7.0 feet in elevation. They proposed to construct the home raised on pilings, so that the level of the first floor would be above the level of a hundred year flood.

A zoning officer denied the Greniers a permit to build a house on the property. The board upheld the decision of the zoning officer. Gove and the Greniers then filed one suit against the selectmen and board and another against the conservation commission of Chatham. A Superior Court judge consolidated the actions, and the parties agreed to a bifurcated trial in which all claims, except for the issue of compensation, would first be tried before a judge, with issues of compensation then tried, if necessary, to a jury.

After a two-day trial during which both parties presented expert testimony, the Superior Court judge found "it is undisputed that [lot 93] lies in the flood plain and that its potential flooding would adversely affect the surrounding area." He found insufficient evidence to support Gove's takings claim, and concluded that she and the Greniers had failed to carry their burden of demonstrating that the board's decision was "legally untenable," "an abuse of discretion, or was arbitrary or capricious."

---

[7] Uses allowed as of right include fishing and harvesting activities, shellfishing, outdoor recreation, the installation of floats, maintenance of roadways, installation of utilities, agricultural activities, dredging for navigational purposes, and construction and maintenance of public boat launches and public beaches. By special permit, additional uses are allowed, including the construction of "catwalks, piers, ramps, stairs, unpaved trails, boathouses, boat shelters, [and] roadside stands"; structures for marinas and boatyards; driveways and roadways; and private boat launching ramps.

2. Discussion. At trial, Gove attempted to prove that the board had effected a taking of lot 93 by subjecting the property to land use regulation in a manner that failed substantially to advance legitimate State interests. *Lopes v. Peabody*, 417 Mass. 299, 303-304, 629 N.E.2d 1312 (1994) (*Lopes*). See *Nectow v. Cambridge*, 277 U.S. 183, 72 L. Ed. 842, 48 S. Ct. 447 (1928) (invalidating irrational application of zoning ordinance). Gove also contended that the town had deprived her of any beneficial use of lot 93, see *Lucas v. South Carolina Coastal Council*, 505 U.S. 1003, 120 L. Ed. 2d 798, 112 S. Ct. 2886 (1992) (*Lucas*), and disrupted her reasonable expectation of developing the property. See *Penn Cent. Transp. Co. v. New York City*, 438 U.S. 104, 57 L. Ed. 2d 631, 98 S. Ct. 2646 (1978) (*Penn Central*).

We discuss Gove's theories in turn.

a. Legitimate State interests. Relying on *Lopes*, Gove argues first that the zoning regulations, as applied to lot 93, failed substantially to advance legitimate State interests. See *Greenfield Country Estates Tenants Ass'n v. Deep*, 423 Mass. 81, 86, 666 N.E.2d 988 (1996). *Lopes* followed the Supreme Court's holding in *Agins v. Tiburon*, 447 U.S. 255, 260, 65 L. Ed. 2d 106, 100 S. Ct. 2138 (1980) (*Agins*), that "the application of a general zoning law to particular property effects a taking if the ordinance does not substantially advance legitimate State interests." See *Lopes, supra* at 305. This term, however, the United States Supreme Court reconsidered the validity of the *Agins* "substantially advances State interests" standard "as a freestanding takings test," and concluded that "this formula prescribes an inquiry in the nature of a due process, not a takings test, and that it has no proper place in our takings jurisprudence." *Lingle v. Chevron U.S.A. Inc.*, 161 L. Ed. 2d 876, 125 S. Ct. 2074, 2083 (2005) (*Lingle*).[11]

In practical effect, *Lingle* renders a zoning ordinance valid under the United States Constitution unless its application bears no "reasonable relation to the State's legitimate purpose." *Exxon Corp. v. Governor of Maryland*, 437 U.S. 117, 125, 57 L. Ed. 2d 91, 98 S. Ct. 2207 (1978). See *Lingle, supra* at 2085; *id.* at 2087 (Kennedy, J., concurring) (discussing nature of due process review). This highly deferential test neither involves "heightened scrutiny," *id.* at 2085, nor allows a court to question the "wisdom" of an ordinance. *Exxon Corp. v. Governor of Maryland, supra* at 124. See *Ferguson v. Skrupa*, 372 U.S. 726, 730-732, 10 L. Ed. 2d 93, 83 S. Ct. 1028 (1963); *Nectow v. Cambridge, supra* at 188; *Euclid v. Ambler Realty Co.*, 272 U.S. 365, 395, 71 L. Ed. 303, 47 S. Ct. 114, 4 Ohio Law Abs. 816 (1926). To the extent that *Lopes* conflicts with *Lingle*, our earlier decision is, of course, overruled.

In this case, the evidence clearly establishes a reasonable relationship between the prohibition against residential development on lot 93 and legitimate State interests.[13] Gove offered no testimony meaningfully questioning the conservancy district's reasonable relationship to the protection of rescue workers and residents, the effectiveness of the town's resources to respond to natural disasters, and the preservation of neighboring property.[14] Having addressed Gove's due process concerns, we turn now to consider her takings claim.

---

[11] As the Lingle Court explained: "The Takings Clause presupposes that the government has acted in pursuit of a valid public purpose.... Conversely, if a government action is found to be impermissible – for instance because it fails to meet the 'public use' requirement or is so arbitrary as to violate due process – that is the end of the inquiry. No amount of compensation can authorize such action" (citations omitted). *Lingle v. Chevron U.S.A. Inc.*, 161 L. Ed. 2d 876, 125 S. Ct. 2074, 2084 (2005).

[13] In addition to evidence of potential danger to rescue workers, an expert for Gove and the Greniers testified that in an especially severe storm, the proposed house "could certainly be picked up off its foundation and floated" away, potentially damaging neighboring homes. There was other evidence on which the judge could base a finding that a house on lot 93 would pose a danger to surrounding structures, including that the stairs required to reach the raised home, "tend to break off" in storms and "do a lot of collateral damage to other structures and property."

[14] Gove points to the testimony of her expert engineer that "nothing man can do (e.g., constructing houses on pilings in a Zone A or building mounded septic systems) will alter the storm water characteristics of an ocean storm." We need not determine whether this evidence meaningfully implicates the "rational relationship" test, because the validity of the regulations as applied to lot 93 is established easily on other grounds. That the Greniers' proposed residential development would have been in compliance with wetlands and State health regulations also does not lessen the legitimacy of the challenged zoning regulation, which serves different purposes.

b. Takings. While the takings clause is directed primarily at "direct government appropriation or physical invasion of private property," the Supreme Court has "recognized that government regulation of private property may, in some instances, be so onerous that its effect is tantamount to a direct appropriation or ouster — and that such 'regulatory takings' may be compensable under the Fifth Amendment." *Lingle, supra* at 2081. See *Tahoe-Sierra Preservation Council, Inc. v. Tahoe Regional Planning Agency*, 535 U.S. 302, 323-324, 152 L. Ed. 2d 517, 122 S. Ct. 1465 (2002) ("we do not apply our precedent from the physical takings context to regulatory takings claims"). Not every regulation affecting the value of real property constitutes a taking, for "Government hardly could go on if to some extent values incident to property could not be diminished without paying for every such change." *Lingle, supra*, quoting *Pennsylvania Coal Co. v. Mahon*, 260 U.S. 393, 413, 67 L. Ed. 322, 43 S. Ct. 158 (1922). A regulation "goes too far" and becomes a "taking" in any case "where government requires an owner to suffer a permanent physical invasion of her property" or where it "completely deprives an owner of '*all* economically beneficial use' of her property" (emphasis in original). *Lingle, supra*, quoting *Lucas, supra* at 1019. "Outside these two relatively narrow categories...regulatory takings challenges are governed by the standards set forth in [Penn Central]." *Lingle, supra.*

Gove does not claim that the conservancy district regulations effected a physical occupation of her property, so we discuss, first, why Gove has not shown a "total" regulatory taking under *Lucas, supra* at 1026. We then address why she has not shown that she is to compensation under *Penn Central, supra.*

1. "Total" regulatory takings. In *Lucas, supra* at 1019, the Supreme Court concluded that a land use regulation that denies a plaintiff "*all* economically beneficial use of her property," constitutes a taking "except to the extent that 'background principles of nuisance and property law' independently restrict the owner's intended use of the property" (emphasis in original). *Lingle, supra* at 2081, citing *Lucas, supra* at 1026-1032. The

plaintiff in *Lucas* had paid $975,000 for two residential lots at a time when he was "not legally obliged to obtain a permit...in advance of any development activity." *Id.* at 1006, 1008. Two years later, the State enacted laws that, a State court found, rendered the two parcels "valueless," *id.* at 1007, leading the Supreme Court to conclude that a total regulatory taking could be established. *Id.* at 1031-1032.

In *Palazzolo v. Rhode Island*, 533 U.S. 606, 630-631, 150 L. Ed. 2d 592, 121 S. Ct. 2448 (2001) (*Palazzolo*), the Supreme Court further explained that, to prove a total regulatory taking, a plaintiff must demonstrate that the challenged regulation leaves "the property 'economically idle'" and that she retains no more than "a token interest." *Id.* at 631, quoting *Lucas, supra* at 1019. The plaintiff in *Palazzolo* was unable to prove a total taking by showing that an eighteen-acre property appraised for $3,150,000 had been limited, by regulation, to use as a single residence with "$200,000 in development value." *Id.* at 616, 631.

Here, the facts are no more indicative of a total taking than those considered by the Supreme Court in *Palazzolo*. Even if we limit our analysis to lot 93, Gove has failed to prove that the challenged regulation left her property "economically idle." Her own expert testified that the property was worth $23,000, a value that itself suggests more than a "token interest" in the property. See *Rith Energy, Inc. v. United States*, 270 F.3d 1347, 1349 (Fed. Cir. 2001), cert. denied, 536 U.S. 958, 153 L. Ed. 2d 835, 122 S. Ct. 2660 (2002) (discussing "token interest"). Moreover, the expert's $23,000 valuation did not take into account uses allowed in the conservancy district, either as of right or by special permit, which she admitted could make the property "an income producing proposition."...The judge's finding that lot 93 retained significant value despite the challenged regulation invalidates Gove's theory: she cannot prove a total taking by proving only that one potential use of her property — i.e., as the site of a house — is prohibited. *Lucas, supra* at 1019, requires that the challenged regulation "denies *all* economically beneficial use" of land. See *Lingle, supra* at 2082 (in *Lucas* context "the complete elimination of a property's value is the determinative factor").[17] We now turn to the *Penn Central* inquiry.

2. The Penn Central inquiry. Recent Supreme Court opinions have emphasized that almost all regulatory takings cases involve the "essentially ad hoc factual inquiries" described in *Penn Central, supra* at 124. See *Lingle, supra*; *Tahoe-Sierra Preservation Council, Inc. v. Tahoe Regional Planning Agency, supra* at 321-326, 335-336 (2002) (temporary moratoria on development); *Palazzolo, supra.*

The *Penn Central* framework eschews any "set formula" or "mathematically precise variables" for evaluating whether a regulatory taking has occurred, emphasizing instead "important guideposts" and "careful examination . . . of all the relevant circumstances." *Palazzolo, supra* at 633, 634, 636 (O'Connor, J., concurring). The relevant "guideposts" include: the actual "economic impact of the regulation" on the plaintiff; the extent to which the regulation "has interfered with" a landowner's "distinct investment-backed expectations"; and the "character of the governmental action." *Lingle, supra* at 2081-2082, quoting *Penn Central, supra* at 124. See *Tahoe-Sierra Preservation Council, Inc. v. Tahoe Regional Planning Agency, supra* at 320; *Leonard v. Brimfield*, 423 Mass. 152, 154, 666 N.E.2d 1300, cert. denied, 519 U.S. 1028, 136 L. Ed. 2d 513, 117 S. Ct. 582 (1996). In the end, "the *Penn Central* inquiry turns in large part, albeit not exclusively, upon the magnitude of a regulation's economic impact and the degree to which it interferes with legitimate property interests." *Lingle, supra* at 2082.

Considering all of the evidence at trial, we agree with the judge that Gove failed to show that the conservancy district regulations had a substantial "economic impact" on her or deprived her of "distinct investment-backed expectations" in lot 93. *Lingle, supra* at 2082, quoting *Penn Central, supra* at 124. As an initial matter, Gove's failure to introduce a thorough assessment of lot 93's current value left the judge no basis to conclude that she suffered any economic loss at all. But even if we assume that residential development is the most valuable potential use of lot 93, Gove did not prove that the prohibition against a house on lot 93 caused her a loss outside the range of normal fluctuation in the value of coastal property.

See *Tahoe-Sierra Preservation Council, Inc. v. Tahoe Regional Planning Agency, supra* at 332, quoting *Agins v. Tiburon*, 447 U.S. 255, 263 n.9, 65 L. Ed. 2d 106, 100 S. Ct. 2138 (1980) ("fluctuations in value . . . are 'incidents of ownership'").

Lot 93 is a highly marginal parcel of land, exposed to the ravages of nature, that for good reason remained undeveloped for several decades even as more habitable properties in the vicinity were put to various productive uses. Lot 93 is now even more vulnerable than ever to coastal flooding. Nevertheless, recent appreciation in coastal property (belatedly, and for the time being) has given the parcel some development value. Absent the coastal conservancy district regulations, lot 93 might well be worth more. But this is a new — and insofar as it relates to residential development, wholly speculative — value that has arisen after the regulations became effective. Before the enactment of the regulations, Gove had no reasonable expectation of selling the property for residential development, a fact she recognized by removing the property from the market for want of an offer. Nor did Gove have any reasonable expectations of a better outcome as late as the early 1990's, when lot 93 had, by Gove's own estimation, "no value whatsoever." Gove could not have developed reasonable expectations of selling lot 93 for residential development after the early 1990's, by which time the regulations had barred any such development for several years. The takings clause was never intended to compensate property owners for property rights they never had. See *Lucas, supra* at 1030 (outside "total" regulatory takings context, "the Takings Clause does not require compensation when an owner is barred from putting land to a use that is proscribed by . . . 'existing rules or understandings'"); *Boston Chamber of Commerce v. Boston*, 217 U.S. 189, 195, 54 L. Ed. 725, 30 S. Ct. 459 (1910) (Holmes, J.) ("the question is what has the owner lost"); *Leonard v. Brimfield, supra* at 155. See also *Palazzolo, supra* at 641-643 (Stevens, J., concurring in part) (takings claim is "determined by the impact of the event that is alleged to have amounted to a taking"). Gove's argument is not furthered by the Greniers' tentative offer to pay

$192,000 for the parcel contingent on receiving approval to build a single-family house, a proposition that all parties reasonably should have known was highly dubious at best, particularly since the regulations did not permit such variances. It is similarly fallacious for Gove to claim that the regulations diminished the value of her property from $346,000 (the appraiser's estimate of the value of lot 93 at the time of the trial *if* it were suitable for a three-bedroom home) to $23,000 (the appraiser's estimate of the land's "unbuildable" worth).

This is not a case where a bona fide purchaser for value invested reasonably in land fit for development, only to see a novel regulation destroy the value of her investment. Gove did not purchase lot 93; she inherited the property as part of the devise from her mother in which she received other real property of significant value. See note 15, *supra*. By this we do not suggest that Gove's takings claim is defeated simply on account of her lack of a personal financial investment. See *Palazzolo, supra* at 634-635 (O'Connor, J., concurring) ("We . . . have never held that a takings claim is defeated simply on account of the lack of a personal financial investment by a . . . donee, heir, or devisee"). Rather, Gove's failure to show any substantial "personal financial investment" in lot 93 emphasizes her inability to demonstrate that she ever had any *reasonable* expectation of selling that particular lot for residential development, or that she has suffered any substantial loss as a result of the regulations. In these circumstances "justice and fairness" do not require that Gove be compensated. *Penn Central, supra* at 124,

quoting *Goldblatt v. Hempstead*, 369 U.S. 590, 594, 8 L. Ed. 2d 130, 82 S. Ct. 987 (1962). To the contrary, it seems clear that any compensation would constitute a "windfall" for Gove. See *Palazzolo, supra* at 635-636 (O'Connor, J., concurring) (discussing nature of test for "reasonableness" of property owner's expectations).

We add that "the character of the governmental action" here, *Lingle, supra* at 2082, quoting *Penn Central, supra* at 124, is the type of limited protection against harmful private land use that routinely has withstood allegations of regulatory takings. It is not at all clear that Gove has "legitimate property interests" in building a house on lot 93. *Lingle, supra.* The judge found that "it is undisputed that [lot 93] lies in the flood plain and that its potential flooding would adversely affect the surrounding areas" if the property were developed with a house. Reasonable government action mitigating such harm, at the very least when it does not involve a "total" regulatory taking or a physical invasion, typically does not require compensation. See *Agins v. Tiburon*, 447 U.S. 255, 261, 65 L. Ed. 2d 106, 100 S. Ct. 2138 (1980) (regulation reducing "ill effects of urbanization"); *Penn Central, supra* at 138 (regulation restricting alteration of historic landmarks); *Goldblatt v. Hempstead, supra* (regulation restricting extent of excavation below ground water level); *Miller v. Schoene*, 276 U.S. 272, 72 L. Ed. 568, 48 S. Ct. 246 (1928) (statute requiring landowner to destroy disease-harboring trees).

Judgment affirmed.

QUESTIONS

1. In your opinion, what issues are most important to consider when establishing a government limitation on land use such as a Conservation District?

2. Gove inherited Lot 93 when development was permitted on the parcel, 10 years before it was deemed to be part of the flood plain. Should that fact give her the right to use the property in the way permitted when she received it? In other words, should she be "grandfathered" in her use? Why or why not?

3. Under what circumstances should the government have the right to take property by eminent domain? Should such taking be limited to public safety-related issues such as the need to construct a pumping station at a particular location to

prevent flooding for a large populated area? What about uniting properties into a parcel large enough to create economic redevelopment of a blighted area? *See* Kelo v. City of New London, 545 U.S. 469 (2005)

4. The case carefully explains why legally there was no "taking." Do you agree that nothing was taken?

5. Overall, does the outcome in this case strike you as fair from an equity point of view? Why or why not?

For excellent additional treatment of this vital topic, see Thomas, E.A., *Protecting the Property Rights of All: No Adverse Impact Floodplain and Stormwater Management,* Denver, CO: The Rocky Mountain Land Use Institute (2008).

The People's Government. Photo David Barabee.

Chapter 16

POTENTIAL NEGLIGENCE LIABILITY
IN EMERGENCY MANAGEMENT

The following law review article provides an excellent overview of potential negligence liabilities arising from emergency management issues for units of government. As you read the article, consider what parallels may exist for private industry as it performs similar functions. Think on the protections from liability that governments may enjoy, and ways in which the private sector might similarly insulate itself.

GOVERNMENTAL NEGLIGENCE LIABILITY EXPOSURE IN DISASTER MANAGEMENT

Ken Lerner [FNa]

23 Urban Lawyer 333 (1991)

Copyright © 1991 by the American Bar Association; Ken Lerner

I. INTRODUCTION

A DISASTER is, by definition, a situation beyond control. Therefore the term "disaster management" may seem an oxymoron. What can be managed, however, is the efforts of governmental authorities, public service organizations, and volunteers to mitigate the human suffering that results from a disaster. In recent years, response efforts to disasters such as earthquakes, hurricanes, volcanic eruptions, and nuclear power plant accidents have incurred praise, criticism, and a few lawsuits.[1] This article reviews the risk of governmental tort liability associated with disaster management, in light of current statutory and case law regarding governmental tort immunity.

A. Trends in Disaster Management

Disaster management has become a major function of government at all levels. The last two decades have seen considerable growth in the number and scale of governmental programs devoted to preparing for and managing disaster response. At the federal level, programs for disaster response have been established pursuant to the Disaster Relief Act,[2] the Comprehensive Environmental Response, Compensation, and Liability Act,[3] and other statutes. To facilitate administration of these programs, in 1980 the President consolidated administrative responsibilities for disaster preparedness and response into one organization, the Federal Emergency Management Agency (FEMA).[4] In addition to federal relief efforts, FEMA now administers several cooperative federal-state programs, in which FEMA provides assistance, oversight, and partial funding for state and local preparedness efforts.[5]

273

At the state level, every state now has a comprehensive disaster response statute[6] outlining the state's disaster management program. Such statutes generally contain provisions that:

- establish a state emergency or disaster management department;
- designate state and local agency roles in responding to disasters;
- allocate executive power to declare a state of emergency;
- describe special executive powers that attend upon such a declaration; and
- provide for cooperation with other neighboring jurisdictions, as well as addressing many other aspects of disaster preparedness and response.[7] State statutes and regulations define the role of local governments in disaster response. Local and state personnel form an interdependent network of resources for responding to emergency situations.

At all levels of government, the first step in emergency management consists of planning. While it may seem contradictory to "plan" for an emergency – emergencies are by definition unplanned events – in fact there is considerable value in such planning. Appropriate planning means that resources, trained personnel, and procedures are in place when a disaster occurs. Plans usually include an assessment of potential hazards, descriptions of the roles of various agencies and organizations, command structure, protocols for requesting assistance, and treatments of various discrete functions within the total response effort, such as public notification, evacuation, medical assistance, and so on. Emergency plans currently in place range from a few pages to a few pounds.

In addition to planning, most disaster preparedness programs include a regular disaster training regimen for response staff, and operational tests such as drills and exercises. As with planning, the effort devoted to these activities varies; however, holding a major exercise can run into several person-years of effort and hundreds of thousands of dollars.

In an actual disaster, a network of responders is activated, each fulfilling a specific role according to the response plan. The activities of the various parties involved are coordinated through communications and, usually, a military-like command structure. Major decisions, such as evacuations, are made by the legally designated executives. They are implemented through a hierarchy of support agencies and personnel, ranging down to "street-level" operations such as directing traffic.

B. Trends in Governmental Tort Immunity

Concurrent with the growth of disaster management programs has been an erosion of the concept of governmental tort immunity. Governmental tort immunity was once essentially absolute.[8] In recent decades, as a result of legislation and court decisions, tort suits against the government have become permissible at the federal level and in every state.[9] Congress enacted the Federal Tort Claims Act (FTCA)[10] in 1946; since then nearly every state has enacted a similar statute, although the details of these statutes vary considerably (see Section II.B. below[11]). Every jurisdiction has its own peculiar exceptions and limits to tort liability.

The juxtaposition of these two trends raises the question, "What liability risks are assumed by governmental organizations by virtue of their involvement in disaster management?" Handling disasters is tricky, and it is easy to imagine the possibilities for costly error. Failure to order an evacuation could cost lives; on the other hand, an unnecessary evacuation could cause enormous economic costs. Key implementation factors, such as public notification and warning, depend on complex communication systems with many possible points for breakdown. Worries over such possibilities have already prompted some governmental efforts to investigate the question of negligence liability exposure in disaster management.[12] In at least one state, a scarcity of volunteers for local hazardous materials emergency planning committees prompted the legislature to specifically assure them of exemption from liability.[13]

This article considers the range of statutory provisions and cases concerning governmental tort liability, as they apply to all aspects of disaster management, to assess the degree of risk faced by governments engaged in disaster management programs.

II. STATUTORY IMMUNITY PROVISIONS

Most states have extended some form of governmental immunity to disaster management via statute. In general, two layers of statutory immunity protect governmental organizations from having to defend tort suits over their handling of disasters. The first layer, discussed in Section II.A. below,[14] consists of provisions embedded in disaster response statutes limiting liability for actions taken pursuant to those statutes. The second layer, discussed in Section II.B. below,[15] consists of the more general governmental liability statutes.

A. Immunity Provisions in Emergency Services Acts

In most states, a provision regarding immunity is found in the disaster response statute, to the effect that activities undertaken pursuant to the statute are immune from liability. However, the form, scope, and application of these statutes varies from state to state. They are often quite comprehensive in wording. For example, the Alabama statute provides that:

> Neither the state nor any political subdivision thereof nor other agencies of the state or political subdivisions thereof, nor, except in cases of willful misconduct, gross negligence or bad faith, any emergency management worker, individual, partnership, association or corporation complying with or reasonably attempting to comply with this chapter [entitled "emergency management"] or any order, rule or regulation promulgated pursuant to the provisions of this chapter... shall be liable for the death of or injury to persons, or for damage to property, as a result of any such activity:[16]

By including all activities undertaken pursuant to the emergency management statute, this liability provision has a broad sweep. Activities granted immunity include emergency planning,[17] procurement of equipment and supplies for emergency preparedness,[18] training of personnel,[19] and sheltering of evacuees,[20] as well as all manner of response activities such as evacuation, emergency transportation, rescue and medical services, emergency police and fire services, and temporary restoration of public utility service.[21] The Alabama statute also specifically extends immunity to volunteer workers[22] and workers from other jurisdictions who respond pursuant to a "mutual aid" agreement.[23]

Many state disaster response statutes have a provision similar to the one quoted above. However, there are also many variations on this theme, ranging from full protection to none at all. In Maryland and Utah, for example, the emergency services statutes do not contain any immunity provision for disaster management activities.[24] In Kansas, immunity is provided only for response actions taken under a governor's proclamation of a state of emergency; actions taken prior to or in the absence of such a proclamation are not covered, along with preparatory activities such as planning and exercises.[25] In Colorado, immunity is provided for owners of property donated or made available for use in connection with disaster management activities; all other tort liability associated with actions taken pursuant to the Disaster Emergency Act is assumed by the state.[26] In fact, the scope of the immunity granted under the Alabama statute quoted above may be limited by the fact that the statute defines the term "emergency management" in a way that appears not to include hazardous materials spills; it is not clear whether hazardous materials planning and response is an activity undertaken pursuant to the statute.[27]

In view of the many exceptions and gaps in the immunity provisions found in emergency services statutes, it is necessary to consider the consequences of "falling through the cracks" in such immunity provisions. An activity that falls through is not necessarily exposed to tort suit; rather, the activity must be evaluated according to the same standard for immunity that applies to any other government activity.

B. General Governmental Tort Immunity Provisions

The breakdown of traditional sovereign immunity reflects a commonsense concept of fairness –

that when the government runs apartment buildings, hospitals, or automobiles, it should be subject to the same obligations and duties of care as a private party doing the same thing. On the other hand, it is widely recognized that in order to avoid paralysis, the government must be able to perform its essential functions, e.g., enact legislation, maintain public order, and enforce regulations, without fear of reprisal via tort suit. As stated by the U.S. Supreme Court in *U.S. v. Varig Airlines*,[28] it is necessary to "prevent judicial 'second-guessing' of legislative and administrative decisions grounded in social, economic, and political policy through the medium of an action in tort."[29]

Thus, each jurisdiction retains immunity (carves an exception from a general waiver of immunity) for certain types of governmental activity. Two kinds of tests are commonly used for determining what activities are protected: the "governmental function" test and the "discretionary action" test.[30]

The "governmental function" test immunizes activities considered to be traditionally or inherently governmental in nature. The standard has been criticized as vague.[31] However, it is not difficult to characterize in a general fashion: governmental functions are those that are traditionally in the hands of government, or are delegated to the government by constitution or statute, or are essential to the operation of the government, and are exercised or provided for the community as a whole, with no special benefit or profit to the

governmental unit involved.[32] Basic activities such as legislation, collection of taxes, and law enforcement are often given as examples of governmental functions. The converse of governmental functions are proprietary or ministerial functions. Proprietary activities are generally those that make a profit or have private sector counterparts, such as hospitals, parking garages, zoos, recreational facilities, and the like. About fifteen states have adopted a governmental-proprietary function test for tort immunity.[33] In some instances the statute providing for governmental function immunity provides a list of proprietary functions, or of governmental functions, to aid in interpretation.[34]

Discretionary immunity differs conceptually from the governmental function test in that it focuses on the particular act or decision in question, rather than the general type of activity. Under this test an act is discretionary, and therefore immune, if it involves an element of choice, and if the choice involves applying judgment of a sort deemed worthy of protection from suit. Judgments involving matters of policy, balancing competing public interests, or that have been delegated to a given official by statute are generally accorded protection.[35] The discretionary function test was adopted by the federal government in the Federal Tort Claims Act, and some version of discretionary immunity is recognized by nearly every state.[36] Some states retain immunity for both discretionary and governmental functions.[37]

III. APPLICATION OF TRADITIONAL GOVERNMENTAL IMMUNITY TESTS TO EMERGENCY MANAGEMENT IN LITIGATION

Application of the immunity doctrines described above has been litigated in various situations that are relevant for assessing liability exposure associated with disaster management. There are few cases directly involving disaster planning and response situations; however, there are enough cases that are either applicable or analogous to provide for meaningful discussion.

A. Governmental-Proprietary Test

Under the governmental-proprietary test, there is little question that disaster planning and response is immune from tort liability as a uniquely governmental function. It fits all of the usual factors cited in characterizing governmental functions: it is done only by governments, it is conducted for the public benefit, it is not done for profit, nor does it compete with private actions. Aspects of response such as firefighting, public safety, and law enforcement have often been ruled exempt from liability as governmental functions.[38]

B. Discretionary Function Test

Application of the discretionary function test to disaster management is not as straightforward as the governmental function test. The discretionary function test rests on the nature of the particular act alleged to be negligent, rather than the general type of activity in which it is embedded. The challenged act must be evaluated to determine whether it involved the application of protected discretion. Therefore, it is necessary to consider separately the different types of actions that occur within disaster management. The four subsections below evaluate four key components of disaster management, and the things that can go wrong with them in light of current case law defining discretionary functions. They represent, in a general way, the spectrum of disaster management, from the planning stage to street-level operations. Allegations of negligence have been litigated in each area.

1. PLANNING – LACK OF PLAN OR FLAWED PLAN

As described earlier, the general practice is to produce and maintain a plan for emergency response. If having such a plan enhances response, the question naturally arises whether lack of a plan would be actionable. That issue was addressed in *DFDS Seacruises (Bahamas) Ltd. v. United States*,[39] where action was brought against the U.S. Coast Guard under the Suits in Admiralty Act[40] for negligence in fighting a fire aboard a ship in port.[41] The suit alleged, inter alia, that the local Coast Guard authority was negligent in not having a contingency plan for shipboard firefighting.[42] The district court ruled against the plaintiffs, holding that the decision not to develop a plan was within the discretion of the local Coast Guard official, who must allocate the available budgetary resources among the Coast Guard's various local activities.[43] The district court described the decision

whether to establish a contingency plan as a "'textbook' discretionary function,[44] and stated that "however desirable such contingency planning may be, decisions as to whether, where and when to expend time and resources to develop such plans are entrusted to the Coast Guard's judgment and are not reviewable by this court."[45]

The decision whether to expend resources to develop a contingency plan was ruled discretionary in *DFDS Seacruises*; however, it should be noted that a decision not to prepare a plan might not be considered permissible discretion where there is a specific statutory or regulatory requirement to have a plan. Violation of a "specific and mandatory" directive to prepare a plan would not be discretionary under *Berkovitz v. United States*[46] (discussed below[47]). Statutory requirements to prepare and maintain local emergency plans are found in twenty-four states." In addition, SARA Title III requires all states to prepare emergency plans for dealing with hazardous materials releases.[49] It is possible that DFDS Seacruises would be distinguished in such situtations.[50]

What if a plan exists, but it is alleged to be flawed due to negligent drafting? No cases could be found directly concerning errors in disaster planning. However, planning would appear to involve types of decisions and actions that are generally regarded as discretionary: allocation of governmental resources, assignment of responsibilities, determinations of policy, agreements with neighboring jurisdictions, and the like. "Planning" as a generalized activity has been characterized as discretionary in several cases. For example, in *Freeman v. Alaska*,[51] the Alaska Supreme Court stated:

> We have adopted a planning-operational test to determine whether a particular act is immune from liability.... [Under this] "test, decisions that rise to the level of planning or policy formulation will be considered discretionary acts immune from tort liability, whereas decisions that are operational in nature, thereby implementing policy decisions, will not be considered discretionary and therefore will not be shielded from liability."[52]

For these reasons, it appears unlikely that poor drafting of an emergency plan would incur exposure to tort negligence liability under a discretionary function standard. It is difficult to imagine a court placing itself in the position of reviewing an emergency plan to determine its merit.

2. PLAN IMPLEMENTATION – FAILURE TO FOLLOW PLAN

Assuming the existence of a plan governing emergency response, what if, in the actual event, the plan is not followed?

Can response organizations deviate from the plan, and still be protected from liability for their actions? That question apparently has not been directly addressed in any reported cases. However, it can be analyzed based on recent doctrinal developments in federal cases involving situations analagous to disaster response.

In federal court the answer depends on whether, in the circumstances of the case, the plan is interpreted as prescribing a mandatory procedure. This standard derives from the watershed Supreme Court case of *Berkovitz v. United States*.[58] In *Berkovitz*, the petitioners alleged that FDA personnel had deviated from prescribed procedures for regulating production of polio vaccine, thus allowing distribution of vaccine that later proved defective.[54] The FDA claimed immunity on the ground that its regulatory activities were discretionary in nature.[55] The Supreme Court ruled that the presence of protected discretion must be evaluated based on an analysis of the particular tasks involved,[56] did they include actual exercise of policy judgment, or were they simply steps in a prescribed, mandatory procedure? If the latter, then the exception would not apply: "The discretionary function exception will not apply when a federal statute, regulation, or policy specifically prescribes a course of action for an employee to follow. In this event, the employee has no rightful option but to adhere to the directive.[57]

How would this standard be construed in a case involving an emergency plan? Again, there are apparently no cases reported concerning deviance from disaster plans per se. However, some statements can be made based on recent cases. First, a disaster plan could be interpreted as a "specific and mandatory" requirement, even though it is strictly an internal document with no direct legal effect on the public. That can be derived from the language of *Berkovitz*, which refers to "statute, regulation or policy.[58] In fact, the *Berkovitz* plaintiffs had two separate causes of action. One was based on a claim that specific statutory and regulatory mandates had been violated.[59] The other, in the Supreme Court's interpretation, possibly involved action that was within the bounds allowed by regulation, but violated established agency policy.[60] The Court allowed both claims to proceed. Violation of an internal policy, therefore, may provide grounds for suit if the policy "leaves no room for implementing officials to exercise independent policy judgment...."[61] The key factor is not the legal status of the document, but rather how specifically prescriptive it is.

The second statement that can be made is that a careful analysis of the situation and the choices available to the government actors in question is required. Two recent cases, *McMichael v. United States*[62] and *Fortney v. United States*,[63] illustrate the fine line that may separate mandatory from discretionary guidance. *McMichael* and *Fortney* both involved explosions at munitions plants under contract to the Department of Defense (DOD). In *McMichael*, several employees of the plant were killed or injured when an explosion occurred during a thunderstorm.[64] Although the plant was operated by an independent contractor, DOD maintained inspectors there, to enforce both quality and safety standards.[65] The inspectors worked from a DOD safety manual[66] and checklist.[67] However, they had not followed the specific language in the manual, which called for the plant to be evacuated as a precautionary measure during electrical storms.[68] The district court found that the DOD inspectors at the plant were negligent in their enforcement of the safety standards.[69] DOD claimed that the inspectors were performing a discretionary function, since they were entrusted with choosing from a range of options when confronted with a safety violation, up to and including ceasing performance of the contract."[70] The circuit court found, however, that with respect to this particular point (evacuation during electrical storm) the inspectors' guidance was specific and left no room for choice (it was item 16 on the checklist).[71] Thus, the inspectors' actions were found to be outside the discretionary function exception![72]

A contrasting result was reached in *Fortney*. In *Fortney*, as in *McMichael*, suit was brought after a munitions plant explosion, claiming that DOD inspectors were negligent in allowing unsafe practices at the plant.[73] The suit alleged violations of two provisions of an Army safety manual.[74] However, unlike the checklist referred to in *McMichael*, the safety provisions violated in *Fortney* were characterized by the court as being merely advisory in nature.[75] They both used the word "should" (as opposed to "must"), and the manual stated that, "The advisory provisions are those in which 'may' or 'should' are used."[76] The difference between "must" and "should" in this case provided the discretion required to cloak the inspectors' actions with immunity.

These cases indicate the delicacy of the determination as to whether a safety manual or other document constrains an official's choice or merely guides his or her discretionary judgment. Disaster plans generally contain a spectrum of material ranging from general to specific. A general assignment of responsibility for protective action (e.g., to the governor) obviously contemplates the exercise of discretion in carrying it out. On the other hand, where the plan contains specific checklists to follow, there may be no discretion to deviate from the checklist, unless the plan so provides. Some plans contain provisions for taking actions to protect the public – such as recommending evacuation – automatically under some circumstances.[77] A negligent failure to take such steps might incur exposure to tort liability.

3. EXECUTIVE-LEVEL DECISION MAKING – POOR DECISION

Often the most visible aspect of disaster response efforts are the tough decisions that have to be made by those responsible for public safety. The main protective action available in most situations is evacuation. However, evacuation of the public cannot be undertaken lightly; it is very disruptive and costly to local residents and businesses, and carries safety risks of its own.[78]

The official must balance the risks and benefits of taking action. The discretionary nature of such decision making is illustrated by two cases from the state of Washington, both concerning the protective actions taken by the state's governor during the Mt. St. Helen eruption in 1980. Both cases stemmed from the same protective action decision. One set of plaintiffs argued that it was overprotective;[79] the other that it was not protective enough.[80]

The fact situation leading to both cases is as follows. Mt. St. Helen became active in March and April of 1980, producing frequent earthquakes and small-scale eruptions. The Governor of the State of Washington, Dixy Lee Ray, held a series of meetings with representatives of various involved organizations, including the U.S. Geological Survey (USGS), the U.S. Forest Service, the University of Washington, the Washington Department of Emergency Services, the Washington State Patrol, and the Weyerhauser Corporation, to assess the safety hazard posed by the active volcano. On April 30 the Governor, following the recommendations of the USGS, established two restricted zones on the mountain. The inner "red zone" was essentially off limits to everyone. The outer "blue zone" allowed controlled access through a permit system. Thousands of sightseers and journalists came to the area to see the volcano. At the same time, scientists studying the mountain became increasingly concerned about the possibility of a major avalanche or eruption (although none anticipated the magnitude of what in fact occurred). Revisions to the zone system were being considered when the major eruption occurred on May 18, sending 600° gasses, ash, and debris at speeds of 200 to 250 miles per hour over an area 18 miles long and 23 miles wide, extending far beyond the restricted zones. Sixty people were killed or missing, most of them outside the restricted zones. The eruption was limited, however, to the north side of the mountain, leaving the south side virtually unscathed.

In *Karr v. States*,[81] representatives of some of the victims brought suit against the state alleging that the Governor had been negligent in designating the zones.[82] The plaintiffs argued that the discretionary function exception did not apply since the Governor had not "exercised basic policy evaluation, judgment, and expertise."[83] The trial court disagreed, however, and granted summary judgment in favor of the state.[84] The Washington Supreme Court upheld the trial court's decision.[85] The court's opinion reviews the Governor's decision process at some length, recounting the numerous conferences and consultations held with local, state and federal agencies, local private interests, and scientific advisors.[86] The court found that the Governor had exercised reasonable judgment in the face of an uncertain situation. "We hold that reasonable persons could reach but one conclusion: that the Governor made a considered policy decision in closing certain areas around Mt. St. Helen. Consequently, the State was immune from tort liability...."[87]

Whereas the plaintiffs in *Karr* had claimed that the protective measures were inadequate, the plaintiffs in *Cougar Business Owners Association v. State*[88] sued claiming that they were overinclusive.[89] The town of Cougar is about eleven miles southwest of the Mt. St. Helen peak, in an area that was not significantly affected by the eruptions.[90] It was in the "red zone" for about six months.[91] Owners of retail businesses in Cougar sued for the trade they lost during the time access to Cougar was restricted.[92] As in *Karr*, the Washington Supreme Court determined that the Governor's decision had been a reasonable exercise of discretion and was entitled to immunity.[93]

The two cases, taken together, neatly illustrate the essential problem facing the disaster response decision maker: to respond effectively to the disaster, while avoiding unnecessary disruption of normal life. They also illustrate that balancing these factors is exactly the kind of public policy decision that is intended to be protected by the discretionary function exception.

4. STREET-LEVEL OPERATIONS – OPERATIONAL ERROR

Activities such as planning and executive-level decision making are clearly part of the mainstream of what is usually characterized as a discretionary function. However, they are only part of the total disaster management effort. The best planning and leadership is of little value without the apparatus to implement protective measures. A disaster response effort consists in large part of specific, street-level efforts to implement established procedures and decisions. These are activities such as firefighting, security and traffic control, door-to-door warning, search and rescue, sandbagging, collecting environmental samples, and so on. To what extent are these activities also covered by the discretionary function exception?

First, it is well-settled that faithful execution of specific, discretionary decisions or policies will not result in liability. Obviously, the protection afforded a discretionary decision would be of little practical value if the decision could be disputed through suits against those carrying it out. That determination was made in the first Supreme Court case interpreting the FTCA, *Dalehite v. United States*.[94] *Dalehite* concerned a postwar government program to produce and ship fertilizer to Europe.[95] The fertilizer was produced in converted munitions factories and was highly flammable; it had a composition similar to some explosives.[96] A cargo ship loaded with the fertilizer caught fire and exploded in port causing extensive property damage and considerable loss of life.[97] The explosion was alleged to have resulted from the improper way in which the fertilizer was loaded, which allowed it to become overheated and ignite.[98] The plaintiffs argued that even if establishment of the fertilizer program was a protected discretionary decision, its negligent implementation was not, and the government should be liable for the operational level mistakes that resulted in the fertilizer being loaded improperly.[99] The Supreme Court noted, however, that the particular schedules and manner of loading were specified "from the top" by the program manager."[100] Since the program manager's decision was within the ambit of the discretionary function exception, the government would not be liable for the actions of its personnel in carrying it out.[101] As the Court put it, the concept of "discretionary function" includes:

more than the initiation of programs and activities. It also includes determinations made by executives or administrators in establishing plans, specifications, or schedules of operations. Where there is room for policy judgment and decision there is discretion. It necessarily follows that acts of subordinates in carrying out the operations of government in accordance with official directions cannot be actionable.[102]

Thus, the faithful implementation of a protected decision will enjoy the same immunity as the original decision itself.

It is likewise well-settled that discretionary immunity will not apply when specific instructions are not faithfully carried out; that was the lesson from *Berkovitz v. United States.*[103] The *McMichael v. United States* case,[104] discussed in conjunction with *Berkovitz* above,[105] is arguably an example of an operational level error that was found to be outside the immunity protection since a specific checklist procedure was not followed.[106]

The remaining, and most common case is one in which personnel at the operational level are acting without specific, mandatory instructions covering the action in question. Such situations often involve making decisions, sometimes requiring the application of judgment in an uncertain situation, much like the executive-level decisions discussed in Section III.B.3. above.'" Discretionary immunity has sometimes been applied to these decisions, and sometimes not. Several federal cases have considered and rejected the argument that operational-level decisions are nondiscretionary per se. One such case involved emergency cleanup operations at a hazardous chemical dump site. *United States v. Fidelity & Guaranty Co.*[108] concerned an EPA-supervised effort to neutralize hazardous chemicals found at an abandoned factory site. The operation went awry, producing a cloud of toxic acid vapor that drifted into a nearby town, causing damage to property.[109] The suit charged that the EPA On-Scene Coordinator (OSC)[110] had directed the operation to continue despite a contractor's recommendation that it be performed only on a day with more favorable wind conditions (i.e., blowing away from the town).[111] The circuit court found that although the decision could be

characterized as "operational" in nature, it was nonetheless discretionary, since the OSC had to weigh the comparative risks, costs, and advantages of delaying the operation or using another method to deal with the hazardous chemicals.[112]

Since a per se rule has been rejected, what guidelines are available to distinguish operational level actions that are discretionary? The best development of discretionary function doctrine in this context is found in the area of suits against local fire departments. Courts are generally reluctant to superimpose their judgment of proper fire response over that of the firefighters. In an oft-cited case, *City of Daytona Beach v. Palmer*,[113] the plaintiff described various alleged faults in the city's firefighting efforts, including improper training and supervision of the firefighters, and negligent firefighting methods, resulting in property loss when his office was destroyed by fire.[114] In considering the question of discretion, the Florida Supreme Court summarized the law in the following fashion:

> The decisions of how to properly fight a particular fire, how to rescue victims in a fire, or what and how much equipment to send to a fire, are discretionary judgmental decisions which are inherent in this public safety function of fire protection. A substantial majority of jurisdictions that have addressed the issue of governmental liability for asserted negligent conduct in responding to and fighting fires have reached this same conclusion.[115]

Daytona was followed in an Indiana case involving some very specific complaints of misconduct, *Ayres v. Indian Heights Volunteer Fire Department*.[116] In *Ayres*, materials stored in the plaintiffs' truck caught fire while parked in their driveway.[117] The actions of the fire department[118] are described in the case as follows:

> The Defendant, Indian Heights Volunteer Fire Department, was called by a neighbor of Plaintiffs and upon arrival at the scene told a neighbor who was extinguishing the fire with his hand extinguisher to get out of the way; whereupon the firemen sprayed a large fire extinguisher into the rear of the truck with such force that it blew the burning materials out of the truck and against the fiberglass door of Plaintiffs' garage causing it to burn.... [The] firemen had a large fire hose, but were unable

to get it to work until after setting the garage door afire; then, when they got the hose working, they ignored the request of Plaintiff Helen Ayres to enter the service entrance and spray from the inside so as to keep the fire from entering the garage where Plaintiffs had stored valuable merchandise; instead, they sprayed from the outside, blowing the fire from the burning door into the garage and totally destroying the garage and its contents.[119]

The court tersely held that the manner in which a particular fire is fought is a discretionary function since "all fires are different and require separate and distinct judgments as to the proper manner of combatting."[120]

Notwithstanding *Daytona* and *Ayres*, other jurisdictions have allowed suits against fire departments to proceed. For example, *Gordon v. City of Henderson*[121] reversed dismissal of a complaint against a local fire department where it was alleged, inter alia, that the firemen "were absent from their regular duty station, and had to be located by the Henderson Police Department"; that their response time was considerably longer as a result, and that some of the firemen "had the smell of liquor on their breath and were unable to respond as trained and professional firemen."[122] The Tennessee Supreme Court opined:

> It may be on a full development of facts that some of the acts of the firemen logically will be classified as "discretionary functions," but we find it difficult to categorize the apparent intoxication of firemen as a "discretionary function," nor, without an expla-

nation by defendants, the absence of firemen from their duty station and the resultant undue delay in response time.[123]

Similarly, the Alabama Supreme Court allowed a complaint to proceed against a fire department in *Williams v. Tuscumbia,*[124] saying:

> We recognize the fact that firemen may act with extreme skillfulness and yet be unable to get to a fire to prevent a building from burning to the ground. But, here the complaint alleges that the reason the fire department did not immediately respond was that the driver of the truck had gone home sick and had not been replaced. We opine that the fire department acted unskillfully by not having a back-up driver who could have immediately taken the place of the sick driver.... In other words, the fire department lacked proficiency.[125]

These cases appear to illustrate a split among states as to the discretionary nature of firefighting. There may be a common thread among them, however. The allegations in *Tuscumbia* and *Henderson* involved issues, such as absence or intoxication of personnel, that are easily accessible to the layperson. The allegations in *Daytona* and *Ayres*, however, went to the heart of the firefighters' skill and technique, requiring analysis of their technical firefighting competence. The courts' natural reluctance to evaluate technical issues may translate into a rule that fine points of technique and judgment are discretionary, whereas errors recognizable by any layperson, such as absence or intoxication, are not.[126]

IV. CONCLUSION

In general, those involved in disaster management should take comfort from the findings of this paper. Tort suits over alleged flaws in disaster response have almost always been dismissed. In many jurisdictions, statutory protection is comprehensive. Where that is not the case, a given aspect of response is still likely to be immune from suit, either as a "governmental function," or as a "discretionary action." At the level of planning and executive decision making, legislatures and the courts have been cognizant that fear of

suit could create paralysis. Considerable protection has been accorded to any well-intended effort. At the operational level, disaster management often involves making life and death decisions, under emergency conditions, and in the face of uncertainty, personal risk, and lack of time. Courts are reluctant to question the judgment of individuals who are called upon to act under such circumstances. This reluctance translates into a proclivity to find that the acts in question were discretionary. Only the most flagrant or obvious deviations from procedure or good practice have been exposed to actions for negligence.[127]

Two useful lessons can be learned from reviewing the law in this area. First, it may be inadvisable from a liability standpoint to avoid or refuse to draft a disaster plan, or otherwise prepare for a disaster, where there is a statutory requirement to do so. A "specific and mandatory" requirement must be obeyed to preserve immunity, per *Berkovitz v. United States*.[128] Local jurisdictions have sometimes delayed or refused to participate in disaster preparedness associated with projects they opposed, such as nuclear power plants.[129] Second, disaster planners should try to write plans and procedures in such a way as to make clear the level of discretion afforded to those who will implement them. One can imagine a variety of scenarios leading to deviation from the plan, ranging from simple errors or ignorance of the plan, up to a considered decision on the part of an experienced public safety official that the plan's prescriptions are inappropriate under the circumstances at hand. Therefore, planners working under a "discretionary function" rule may wish to make the plan explicit as to the scope and range of judgment expected to be exercised by those implementing the plan. A broad disclaimer, stating that the plan is merely advisory in nature, might have some effect. However, such a disclaimer might not receive much weight if the circumstances of the case indicate that the actor in question was not expected to apply judgment, but rather to simply follow instructions. Where a "specific, mandatory" procedure is desired, it should be clearly labeled as such. On the other hand, planners should be careful to explicitly preserve discretion, at any level of implementation, where discretion is desired. If it is desired that the county sheriff, or the school district superintendent, or the sheriff's deputy directing traffic use their best judgment under the circumstances, the plan should provide them with guidelines, but not box them in with mandatory procedures.

NOTES

a. Assistant Energy and Environmental Program Attorney, Argonne National Laboratory; J.D., University of Michigan Law School, 1982; B.A., Philosophy and Economics, University of Illinois, Urbana, 1977.

1. See, e.g., CRISIS MANAGEMENT: A CASEBOOK, (M. Charles & J. Kim, eds. 1988).
2. 42 U.S.C. § 5121 (1988).
3. 42 U.S.C. § 9601 (1988).
4. Exec. Order No. 12,148, 3 C.F.R. 412 (1979).
5. Examples of such programs include: the Flood Insurance program for communities located in floodplains; the Radiological Emergency Preparedness program (administered cooperatively by FEMA and the Nuclear Regulatory Commission, primarily for communities near nuclear power plants), the Hazardous Materials Emergency Preparedness program, established by the Emergency Planning and Community Right-to-Know Act (Title III of the Superfund Amendment and Reauthorization Act of 1986, Pub. L. No. 99-499, § 300, 100 Stat. 1613, 1729- 60 (1986)); and most recently the Chemical Stockpile Emergency Preparedness program, administered cooperatively by FEMA and U.S. Army to improve disaster preparedness in communities near chemical weapon storage sites.

6. The titles of these statutes vary, but their contents are generally similar, as outlined immediately below. Examples of titles are the "Colorado Disaster Emergency Act of 1973" (COLO. REV. STAT. § 24-33.5-701 (1988)), and "Arkansas Emergency Services Act of 1973" (ARK. STAT. ANN. § § 12-75-102 to 12-75-130 (1987)).

7. A summary table of these statutes can be found in J. PINE, TORT LIABILITY OF GOVERNMENTAL UNITS IN EMERGENCY ACTIONS AND ACTIVITIES (1988) (funded under a grant from FEMA). Other indications of the growth of disaster preparedness include initiation of scholastic disaster management programs (see Popkin, The Great Degree Debate: College Level Programs for Emergency Management, HAZARD MONTHLY, Sept. 1986) and the initiation of several trade journals in the area, e.g., DISASTER MANAGEMENT, HAZARD MONTHLY, INDUSTRIAL CRISIS QUARTERLY, and EMERGENCY PREPAREDNESS NEWS.

8. A history of federal governmental immunity is provided in Shimomura, The History of Claims Against the United States: The Evolution from a Legislative Toward a Judicial Model of Payment, 45 LA.L.REV. 625 (1985).

9. See generally 57 AM. JUR. 2D Municipal, County, School, and State Tort Liability (1988).

10. Part of the Legislative Reorganization Act, 60 Stat. 142 (1946).

11. See infra text accompanying notes 28-37.

12. See PINE, supra note 7; R. DONOVAN, LIABILITY AND RISK MANAGEMENT: EMERGENCY MANAGEMENT ISSUES (1987).

13. See COLO. REV. STAT. § 24-33.5-1505(2) (1988).

14. See infra text accompanying notes 16-27.

15. See infra text accompanying notes 28-37.

16. ALA. CODE § 31-9-16(b) (1989).

17. Id. at § 31-9-6.

18. Id.

19. Id.

20. Id. at § 31-9-17.

21. Id. at § 31-9-3(1). It should be noted, however, that the inclusion of fire services in this provision did not prevent the Alabama Supreme Court from affixing liability to the Tuscumbia Fire Department for a mishandled response effort. See infra text accompanying notes 124-25 (discussion of *Williams v. Tuscumbia*, 426 So. 2d. 824 (Ala. 1983)).

22. ALA. CODE § 31-9-16(d) (1989).

23. Id. at § 31-9-16(e). Mutual aid refers to a reciprocal agreement with a neighboring jurisdiction to provide help with disaster response services when needed. Most local jurisdictions are party to such agreements.

24. The Maryland disaster management statute does provide that disaster response personnel are entitled to various special legal privileges when engaged in response to a declared emergency; for example, they may obtain a stay in civil actions to which they are a party (MD. ANN. CODE art. 16A, § 17 (1990)), and are temporarily immune from actions such as mortgage foreclosures and repossession of personal property bought on an installment contract. Id. at § 23. However, the statute does not provide immunity from liability for actions taken in the course of the emergency response work itself.

25. KAN. STAT. ANN. § 48-915 (1983). Similar or related provisions, limiting immunity to situations where there is a declared state of emergency, are also found, at least in Idaho (IDAHO CODE § 46-1017 (1977)) and Tennessee (TENN. CODE ANN. § 58-2-129 (1989)), and per PINE, supra note 7, also in Massachusetts, Montana, and Pennsylvania.

26. COLO. REV. STAT. § § 24-33.5-902, 903 (1988).

27. ALA. CODE § 31-9-3 (1989).

28. 467 U.S. 797 (1984).

29. Id. at 814.

30. The two functional tests, described below generally, define the scope of tort immunity accorded to governmental bodies. Two other common statutory provisions also bear mention. First, in some states the purchase of insurance by a governmental organization constitutes a waiver of immunity to the extent of the insurance coverage. See, e.g., *Avallone v. Board of County Comm'rs*, 493 So. 2d 1002 (Fla. 1986). Second, most states limit the amount of compensatory damages that can be recovered in actions against the government. E.g., IND. CODE ANN. § 34-4-16.5-4 (West Supp.1989) sets recovery limits for injury or death to $300,000 per person and $5 million total per occurrence. Punitive damages are typically disallowed.

31. *Indian Towing Co. v. United States*, 350 U.S. 61 (1955) (Frankfurter, J.). This was an FICA case that rejected use of the governmental/proprietary distinction; Frankfurter characterized the distinction as a "quagmire" and "inherently unsound." Id. at 65.

32. See generally 57 AM. JUR. 2D Municipal, County, School, and State Tort Liability, § § 87-110 (1988).

33. See PINE, supra note 7.

34. E.g., OHIO REV. CODE ANN. § 2743.03 (Anderson 1989). The Ohio statute lists proprietary functions, including operation of a hospital, cemetery, playground, swimming pool, utility, stadium, golf course, or auditorium. On the other hand, Utah provides a list of activities for which immunity is not waived, including tax collection, building inspections, incarceration of prisoners, firefighting, hazardous waste handling, and emergency evacuations. UTAH CODE ANN. § 63-30-10 (1989).

35. As stated in *Berkovitz v. United States*, 486 U.S. 531, 536 (1988) (discussed below, see infra text accompanying notes 53-61), "The [discretionary function] exception, properly construed, protects only governmental actions and decisions based on considerations of public policy."

36. See PINE, supra note 7, for a list of states with discretionary immunity. For an excellent discussion

of the discretionary function exception in federal cases, see Fishback & Killefer, The Discretionary Function Exception to the Federal Tort Claims Act: Dalehite to yang to Berkovitz, 25 IDAHO L.REV. 291 (1988-89).

37. E.g., KY. REV. STAT. ANN. § 44.073(13) (Michie 1986) provides, "The preservation of sovereign immunity...includes...(a) Discretionary acts or decisions; (b) Executive decisions; (c) Ministerial acts...."

38. See, e.g., *Zavala v. Zinser*, 123 Mich. App. 352, 333 N.W.2d 278 (1983); *Ozark Silver Exch., Inc. v. City of Rolla*, 664 S.W.2d 50 (Mo. Ct. App. 1984); *Valevais v. City of New Bern*, 10 N.C. App. 215, 178 S.E.2d 109 (1970); *Davis v. City of Lexington*, 509 So.2d 1049 (Miss. 1987).

39. 676 F.Supp. 1193 (S.D. Fla. 1987).

40. 46 U.S.C. § § 741-52 (1988).

41. 676 F.Supp. at 1195.

42. Id.

43. Id. at 1205.

44. Id.

45. Id.

46. 486 U.S. 531 (1988).

47. See infra text accompanying notes 53-61.

48. See PINE, supra note 7, § VIII.E, App. A.

49. 42 U.S.C. § 11003 (1988).

50. In fact, the plaintiffs in DFDS Seacruises argued that the Coast Guard's "Marine Safety Manual" and other Coast Guard policy pronouncements represented a mandatory requirement for planning in that case. 676 F.Supp. at 1205. That argument was dispatched by the district court on the ground that the Coast Guard could not create a duty to private citizens on the part of the U.S. government through its own internal guidance or regulations. Id. However, that reasoning is questionable in light of subsequent cases where regulations and internal manuals have been found to be "specific and mandatory directives," thus removing discretion on the part of agency personnel. See discussion of *Berkovitz v. United States, McMichael v. United States* and *Fortney v. United States*, infra text accompanying notes 53-77.

51. 705 P.2d 918 (Alaska 1985).

52. Id. at 920 (quoting *Johnson v. State*, 636 P.2d 47, 64 (Alaska 1981), and *State v. Abbott*, 498 P.2d 712, 721, (Alaska 1972)), see also *Grossman v. School Bd. of I.S.D.* #640, 389 N.W.2d 532 (Minn. 1986); *Nunn v. California*, 35 Cal. 3d 616, 200 Cal. Rptr. 440, 677 P.2d846 (Cal. 1984).

53. 486 U.S. 531 (1988).

54. Id.

55. Id. at 533-34.

56. Id. at 544-55.

57. Id. at 536.

58. Id. (emphasis added).

59. Id.

60. See id. at 533.

61. Id. at 547.

62. 856 F.2d 1026 (8th Cir.1988).

63. 714 F.Supp. 207 (W.D. Va. 1989).

64. 856 F.2d at 1028.

65. Id. at 1029.

66. See id. (DEFENSE CONTRACTOR'S SAFETY MANUAL FOR AMMUNITION, EXPLOSIVES AND RELATED DANGEROUS MATERIALS (DOD Manual 4145.26M)).

67. Id.

68. 856 F.2d at 1030.

69. Id. at 1028.

70. Id. at 1030.

71. Id. at 1033.

72. Id.

73. See 714 F.Supp. 207 (W.D. Va. 1989).

74. See id. at 207-08.

75. Id. at 208.

76. Id. (quoting AMCR 385-100, 1-3b) (emphasis added).

77. For example, nuclear power plants use a standard four-level system for classifying plant emergencies. Nearby communities sometimes incorporate clauses in their emergency response plans to the effect that specific actions-such as sounding warning sirens or evacuating a two-mile radius around the plant-are to be taken automatically when the plant reaches the third or fourth level, irrespective of whether there is any actual release of radioactivity.

78. For example, traffic accidents may occur during the evacuation. In addition it should be noted that, in some situations, other protective actions may be preferable. For example, exposure to a passing airborne plume of harmful material will be reduced by in-place "sheltering" (i.e., staying indoors with windows and doors shut). If the plume is a short duration "puff" then this method may be more effective than evacuation, where evacuation would mean getting out into the plume.

79. *Cougar Business Owners Ass'n v. State*, 97 Wash. 2d 466, 647 P.2d 481, cert. denied, 459 U.S. 971 (1982).

80. *Karr v. State*, 53 Wash. App. 1, 765 P.2d 316 (1988).
81. 53 Wash. App. 1, 765 P.2d 316.
82. Id. at –, 765 P.2d at 320.
83. Id.
84. Id. at –, 765 P.2d at 321.
85. Id.
86. Id. at –, 765 P.2d at 317-19.
87. Id. at –, 765 P.2d at 319.
88. 97 Wash. 2d 466, 647 P.2d 481, cert. denied, 459 U.S. 971 (1982).
89. Id. at –, 647 P.2d at 482.
90. Id.
91. Id.
92. Id. at –, 647 P.2d at 484.
93. Id. at –, 647 P.2d at 488.
94. 346 U.S. 15 (1953).
95. Id. at 19.
96. Id.
97. Id. at 23.
98. Id. at 46-47.
99. Id. at 36.
100. Id. at 40.
101. Id. at 42.
102. Id. at 35-36.
103. 486 U.S. 531 (1988).
104. 856 F.2d 1026 (8th Cir.1988).
105. See supra text accompany notes 62-72.
106. See 856 F.2d at 1033-34.
107. See supra text accompanying notes 78-93.
108. 837 F.2d 116 (3d Cir.1988), cert. denied, 437 U.S. 1235 (1988).
109. Id. at 119.
110. An EPA On-Scene Coordinator was appointed to direct cleanup of the site pursuant to the Comprehensive Environmental Response, Compensation, and Liability Act of 1980 (CERCLA), 42 U.S.C. § § 9601-9657, and the National Oil and Hazardous Substance Pollution Contingency Plan, 40 C.F.R. § 300.1- .920. 837 F.2d at 118.
111. Id.
112. Id. at 122.
113. 469 So. 2d 121 (Fla. 1985).
114. Id. at 122.
115. Id. at 123 (citations to eleven states omitted).
116. 493 N.E.2d 1229 (Ind. 1986).
117. Id. at 1231.
118. These actions are reminiscent of the fire department in the movie *Roxanne* (Columbia Pictures 1987).
119. 493 N.E.2d at 1231.
120. Id. at 1232.
121. 766 S.W.2d 784 (Tenn. 1989).

122. Id. at 785.
123. Id. at 786.
124. 426 So.2d 824 (Ala. 1983).
125. Id. at 826; see also *Davis v. Lexington*, 509 So.2d 1049 (Miss. 1987).
126. The cited cases speak in terms of the application of judgment on the part of the firefighters, but do not address the question of whether it is the type of judgment that the discretionary function exception was intended to protect. It should be noted that many formulations of the discretionary function rule include reference to considerations of public policy; in other words, only judgments involving considerations of public policy are protected. Is a firefighter's judgment on firefighting method one of public policy, or is it merely a technical judgment? Application of the discretionary function exception to matters of purely technical or scientific judgment has been litigated in cases such as *Arizona Maintenance Co. v. United States*, 864 F.2d 1497 (9th Cir.1989), with the result that such matters were not found to involve the sort of "policy choice" that is protected under a discretionary function test.
127. It should be noted that even if immunity is not found to apply, two considerable hurdles may still stand in the way of a recovery for negligence. One, in many jurisdictions a plaintiff pursuing a negligence action against a governmental employee must show that a "special duty" of care existed based on a specific relationship between the plaintiff and defendant employee, as opposed to a "general duty" owing to the public at large. The distinction has been criticized, however, e.g., in *Williams v. Tuscumbia*, 426 So.2d 824 (Ala. 1983), where the Court stated, "Tuscumbia contends that a duty imposed upon a municipal fire department is owed to the general public-not to an individual. Does this mean that the whole town has to be on fire before the fire department responds to a call?" Id. at 826. Two, in determining negligence, a greater degree of latitude is accorded those who must act in an emergency situation. This rule is discussed in 57A AM. JUR. 2D Negligence, § § 213-232 (1988).
128. 486 U.S. 531, 544 (1988).
129. See, e.g., *Commonwealth of Massachusetts v. United States*, 856 F.2d 378 (1st Cir.1988) (regarding licensing of the Seabrook nuclear power plant in the absence of certain state and local disaster response plans).

QUESTIONS

1. Why do statutes and the courts protect emergency management from liability for negligent acts?

2. Why do they not excuse units of government from grossly negligent or intentional acts?

3. Describe the kind of acts that would make the immunities discussed above inapplicable.

4. Discuss why failure to perform acts mandated by statute (such a creating a plan) might not be immune, while failing to create a plan in the absence of specific direction to do so probably will be immune to legal attack.

5. Think back to the discussion of special duty vs. general duty in Chapter 1. Consider the arguments on both sides of the issue regarding the advisability of such an exception to immunities that might otherwise apply to governmental acts.

6. The article talks about the advisability of drafting a plan with mandated levels of discretion for those who must carry out the plan. What are the pros and cons of such an approach? How much discretion would be advisable? What might the legal result be of a plan that simply states that the emergency manager or on scene commander is vested with full discretion to respond in a prudent manner?

7. Recall the Victor Microwave case in Chapter 7. Consider the responsibilities of a private entity tasked with performing duties similar to those performed by government, such as preparing an emergency response plan under OSHA and responding in a proper manner in that case. How and why might the outcome of Victor Microwave have differed if the entity involved had been a unit of government? How and why might the results have been similar?

8. For private industry, what avenues are available to provide protection like that units of government receive through the means described in the article above? Consider both formal and informal means of protection.

Biloxi, MS, September 27, 2005 – FEMA Community Relations (CR) interpreter, Phuong Huynh, speaks to Vietnamese residents of Biloxi in line to get donated clothes. CR representatives take the FEMA message directly to the affected residents in a disaster. FEMA/Mark Wolfe. As Chapter 17 makes clear, state and local emergency managers cannot depend on FEMA to provide interpreters for local needs.

Chapter 17

LIMITED ENGLISH PROFICIENT (LEP) POPULATIONS AND EMERGENCY MANAGEMENT: LEGAL REQUIREMENTS AND INTERPRETING/TRANSLATING ASSISTANCE

Nancy Schweda Nicholson

I. INTRODUCTION

The Time: 2005 Shortly After Hurricane Katrina

The Place: Gulfport, Mississippi

The Issue: Inadequate Language Services

Experience 1: When finally rescued after drifting for five days in a heavily damaged fishing vessel, a Vietnamese man indicated that he had not understood the announcements to evacuate.

Experience 2: During the recovery stage, a 13-year-old seventh grade youngster grappled with the daunting and difficult task of translating information for adults (including his grandparents) staying in a Buddhist Temple in East Biloxi (Musgrave, 2005: A11).

These chilling examples are only several of an uncounted number which illustrate the chaos and despair that could have been avoided with adequate language assistance.

> It's disconcerting that there isn't any infrastructure to offer Vietnamese/English translation to reassure these Vietnamese people....Someone should pay attention to these people because they have no advocates....Please send out information in Vietnamese. Send it really soon. (Musgrave, 2005: A11)

Now, more than ever, as our linguistic and cultural diversity continues to grow, the provision of professional language services must be an important component of all phases of emergency management (EM): mitigation, preparedness, response and recovery. Specific legal enactments require that language services be furnished to members of the general public. People must be able to understand all relevant aspects of EM. They must be given the linguistic tools to ensure their safety. Working with language services professionals on an ongoing basis will result in a solid relationship that fosters full communication with all affected persons.

As with any other human resource, interpreters and translators vary greatly in their background, training, and skills. The prudent emergency manager is therefore well advised to address this matter during the planning process, so that those who will be utilized are competent and aware of the jurisdiction's specific requirements.

More specifically, this chapter: (1) sets out legal requirements and standards for the provision of language assistance; (2) offers an overview of the language services profession, highlighting various categories of interpreters and translators as well as commonly-held misconceptions about the field; (3) highlights the need for greater awareness of LEP issues in the EM domain; (4) includes information on ethical considerations, such as confidentiality, impartiality and scope of practice (among others); (5) describes how to build a network of language services providers; (6) discusses the role of professional organizations/associations in training, testing and certifying interpreters and translators in the medical, legal and community services fields; (7) examines federal and state court interpreter certification programs; (8) presents an overview of language services initiatives in both the public and private sectors; (9) directs the reader to relevant websites and information in languages other than English; (10) addresses the need for targeted recruitment of fluent bilingual EM personnel; and (11) critically assesses the value of social networking sites across all four phases of EM.

This chapter will explore how to incorporate language services so that all persons affected by an emergency or disaster will have the most basic of their needs fulfilled – information that allows them to act to ensure their safety.

Census 2010 data show that both the Hispanic/Latino and Asian populations in the US increased by 43% in the period 2000–2010 (http://2010.census.gov). It is a remarkable coincidence that the growth pattern should be virtually identical for each group. More specifically, Hispanics/Latinos account for "more than half of the growth in the U.S. population between 2000 and 2010...." (2010 Census Shows America's Diversity, 2011:1). The most recent Census information indicates that they currently make up 16% of America's total population. Moreover, Asians grew more rapidly than any other major racial group during the period 2000–2010, and increased from 4% (2000) to 5% (2010) in terms of the total U.S. population (2010 Census Shows....2011:2). Comparatively speaking, those individuals who identify themselves as "white alone" within the non-Hispanic population grew at a very slow rate of only 1% (2010 Census Shows....2011:1). With respect to geographic distribution, the South and the West continued to grow the most, following a pattern that was also present in the 1990s. In terms of metropolitan areas, Palm Coast, Florida led with a growth rate of 92%, and Raleigh, North Carolina tied with Las Vegas, Nevada at 41.8% (Mackun & Wilson 2011: 4–5).

Although the U.S. populace has become increasingly diverse in terms of languages and cultures in the past 20 years, Census 2010 statistics show that those who checked "non-Hispanic white alone" still comprise 64% of the total. This number, however, constitutes a decrease of 5% from 69% in Census 2000 (2010 Census Shows.... 2011:1). One cannot, of course, overgeneralize that all non-Hispanic white alone individuals are speakers of English only; however, given a general American lack of interest in studying languages other than English (not to mention the very small numbers among this latter group that actually *learn* another language and are able to function within a society that speaks the language natively), one can safely assume that *many* in this racial category speak only English.

The American Community Survey (ACS) gathers demographic information on an ongoing basis which is meant to supplement the broader once-a-decade Census data. The ACS examines many characteristics of the U.S. population, such as age, race, family and relationships and disabilities. Another focus is "linguistic isolation":

> A linguistically isolated household is one in which no member 14 years and over (1) speaks only English or (2) speaks a non-English language and speaks English 'very well.' In other words, all members of the household 14 years and over have at least some difficulty with English (American Community Survey, S1602, 2009:1).

With this definition in mind, the most recent ACS data (a 1-year estimate) available on the U.S. Census website show that 25.9% of Spanish-speaking households and 27.5% of those who speak Asian and Pacific Island languages are considered to be "linguistically isolated." Moreover, linguistic isolation is also a factor within two additional household categories: Other Indo-European Languages: 16.6% and Other languages: 17.2% (American Community Survey, S1602, 2009:1).

All of these U.S. Census and ACS data demonstrate that there are many millions of people who are linguistically isolated and fall into the "limited English proficient" (LEP) category. The challenge for emergency managers is to reach these populations before, during and after disaster strikes so that they may not only contribute their ideas and cooperate with EM officials, but also so that they receive the information and services to which the law entitles them.

II. LEGAL REQUIREMENTS AND STANDARDS FOR THE PROVISION OF LANGUAGE SERVICES

A. Federal Statutory Enactments

1. The Civil Rights Act of 1964
   A seminal and ground-breaking document, Section 601 of Title VI of the Civil Rights Act of 1964 (42 U.S.C.2000d) states that no person shall "on the ground of race, color, or national origin, be excluded from participation in, be denied the benefits of, or be subjected to discrimination under any program or activity receiving Federal financial assistance." A famous 1974 U.S. Supreme Court case, *Lau v. Nichols* (414 U.S.563), affirmed the provisions of Title VI regarding equal access for LEP Chinese students in San Francisco. The school district received Federal funding, so it was obligated to implement measures to permit these students to take part. Although *Lau* deals specifically with an educational environment, Title VI "include[s] all of the programs and activities of all recipients of federal financial assistance" (Language Assistance Self-Assessment and Planning Tool for Recipients of Federal Financial Assistance www.lep.gov/ selfassesstool.htm). This 14-page document includes an assessment instrument composed of questions to assist the parties in identifying their needs as well as guidelines regarding the creation of a Language Assistance Plan (LAP) and the effective use of language services. Although designed to apply to all federal programs, it is particularly relevant to the challenges that emergency managers face with respect to LEP individuals.
2. The Court Interpreters Act of 1978 (Public Law 95-539)
   The Court Interpreters Act of 1978 (Public Law 95-539) mandated the development of a certification instrument for interpreters who work in the Federal Courts. The certification program began with Spanish/English written and oral testing in 1980. It added Haitian Creole/English and Navajo/ English exams in the 1980s; however, these language combinations are no longer being tested. The Federal Court Interpreter Certification Examination in Spanish/English continues and thrives. It is currently administered by the National Center for State Courts. (See www.ncsc.org, Schweda Nicholson 1986 and the discussion of "Federal and State Court Interpreter Certification Programs" in Section VI. for further information.)

B. Executive Order 13166

Issued by President William Clinton on August 11, 2000, Executive Order (EO) 13166, "Improving Access to Services for Persons with Limited English Proficiency," reinforces the provisions of Title VI of the Civil Rights Act of 1964 and specifically addresses the needs of LEP individuals. (See Appendix A for not only a complete text of EO 13166, but also enforcement guidelines for Title VI of The Civil Rights Act of 1964, as published in *The Federal Register* (2000).) Very importantly, EO 13166 does not create new rights or obligations. It simply reiterates existing rights and the requirement that recipients of Federal financial assistance provide "meaningful access" to their services and programs.

Relevant to the Civil Rights Act of 1964 and EO 13166 is an excellent website maintained by the Federal government on LEP issues (www.lep .gov). This site acts as a clearinghouse, providing and linking to a wide variety of information, tools, and technical assistance regarding LEP and language services for (a) Federal agencies; (b) recipients of Federal funds; (c) users of Federal programs and Federally-assisted programs; and (d) other stakeholders. More specifically, there is a webpage dedicated to EO 13166, which includes many additional links (http://www.lep .gov/13166/eo13166.html). For example, the link to "Interpretation and Translation" provides a plethora of resources, including information on medical interpreters, certification and training opportunities as well as a listing of interpreter and translator organizations on a region-by-region and state-by-state basis.

EO 13166 also directed every Federal agency to develop a plan to provide better access to LEP individuals. "Commonly Asked Questions and Answers Regarding Executive Order 13166" is an excellent document that addresses many situations with respect to the implementation of EO 13166 (http://www.lep.gov/faqs/faqs.html#Two_EO13166_FAQ).

For example, it treats: (a) agencies' and programs' responsibilities if an Official English (or "English-Only") law has been enacted in a particular state; (b) how to identify "vital" documents for translation; and (3) the standards for oral interpreting. Very important to the emergency manager is the section that reads:

> In rare emergency situations, the agency or recipient may have to rely on an LEP person's family members or other persons whose language skills and competency in interpreting have not been established. Proper agency or recipient planning and implementation is important in order to ensure that those situations rarely occur (Commonly Asked Questions...#11).

A page entitled "Federal Agency LEP Guidance & Language Access Plans" (http://www.lep .gov/guidance/guidance_index.html) indicates that the Department of Homeland Security's LEP guidance and assistance materials are "pending."

On February 17, 2011, U.S. Attorney General Eric Holder issued a memorandum entitled: "Federal Government's Renewed Commitment to Language Access Obligations Under Executive Order 13166" (Appendix B). In this document, he primarily addresses the requirement for all Federal agencies to (a) "...develop and implement a system by which limited English proficiency (LEP) persons can meaningfully access the agency's services" and (b) issue guidelines to recipients of federal financial aid regarding their "legal obligations" to provide "meaningful access" under Title VI of the Civil Rights Act of 1964..." (Holder, 2011:1). The Attorney General lists eight "action items" that address ways to comply with this law and the Executive Order. All of these points are critical and relevant to the language challenges regularly encountered. Of great interest in the context of this chapter are: Number 2 ("Evaluate and/or update your current response to LEP needs...."); Number 6 ("When considering hiring criteria, assess the extent to which non-English language proficiency would be necessary for certain positions or to fulfill your agency's mission"); and Number 7 ("For written translations, collaborate with other agencies to share resources" and gather community input on the quality and accuracy of "professional translations intended for mass distribution"). It is important to note that, although this memo is directed to Federal agencies, local emergency managers are bound by its dictates because they receive Federal grant funds, the states as grantees and locals as sub-grantees.

The Federal Coordination and Compliance Section of the Department of Justice maintains an extremely useful website: www.justice.gov/crt/about/cor. Unlike the case with many other Federal and state resources, some of the documents found there appear in various language versions. In fact, Holder's Memorandum has been translated into Spanish, Russian, Chinese (simplified characters), Korean and Vietnamese. Some new FAQs regarding LEP issues also appear on this site, such as "Frequently Asked Questions about the Protection of Limited English Proficient (LEP) Individuals under Title VI of the Civil Rights Act of 1964 and Title VI Regulations."

C. National Incident Management System (NIMS) and the National Response Plan (NRP) – 2004

After much initial work, consideration of comments, and revisions, both the National Incident

Management System (NIMS) and the National Response Plan (NRP) were adopted in 2004 (NIMS in March and the NRP in November). (For a detailed discussion of the origins of NIMS and the NRP and their implementation as well as relevant sections on preparedness challenges and training issues, see Nicholson 2006a as well as Chapter 6 of this book). Since 2004, there have been various incarnations of NIMS and the NRP. In fact, the name of the NRP was changed to National Response Framework (NRF) in 2008. (http://www.dhs.gov/files/programs/editorial_0566.shtm).

"NIMS works hand in hand with the *National Response Framework (NRF)*. NIMS provides the template for the management of incidents, while the NRF provides the structure and mechanisms for national-level policy for incident management." http://www.fema.gov/emergency/nims/About NIMS.shtm

NIMS includes specific language-related information in Emergency Support Function (ESF) #8: Public Affairs Information Requests and #15: External Affairs for Action and Response. ESF #8 makes available language-assistance services (such as interpreters for different languages, telecommunications devices for the deaf, and accessible print media) to facilitate communication with all members of the public. Requests for information may be received from various sources, such as the media and the general public, and are referred to ESF #15. On page 3 of the External Affairs Annex for ESF #15, the Public Affairs section stresses the need to provide "...incident-related information through the media and other sources in accessible formats and multiple languages to individuals, households, businesses, and industries directly or indirectly affected by the incident." http://www.fema.gov/pdf/emergency/nrf/nrf-esf-15.pdf. Within more than 180 pages of Standard Operating Procedures for ESF #15, only one short paragraph deals with language services issues:

> **4.4** Individuals who do not speak English or have limited English proficiency may need information in a language other than English or an interpreter who can relay information to them. It is important to identify groups and organizations that can

provide interpreters for local populations with limited English proficiency. (http://www.fema.gov/pdf/emergency/nrf/nrf_jfo_sop.pdf, ANNEX E Community Relations: 5, 2009)

D. CPG 301: Interim Emergency Management Planning Guide for Special Needs Populations – Federal Emergency Management Agency and DHS Office for Civil Rights and Civil Liberties. *Version 1.0 (August 15, 2008)* (http://serve.mt.gov/wp-content/uploads/2010/10/CPG-301.pdf)

Although CPG 101, Version 2.0 (discussed in II. D.) supercedes CPG 301, Version 1.0, it is worthwhile to explore pertinent sections of 301 in terms of the treatment of language services issues. In some ways, it appears to this researcher that 301 offers better and more detailed direction to emergency managers in this domain.

Unlike CPG 101, Version 1.0 and CPG 101, Version 2.0 (which address all populations as well as general settings and situations), Comprehensive Preparedness Guide (CPG) 301,'"Emergency Management Planning Guide for Special Needs Populations,' identifies and provides guidance on considerations that should be made during the plan development process to support citizens with special needs. This guide was developed jointly between the Federal Emergency Management Agency and the DHS Office of Civil Rights and Civil Liberties, with significant input from key Federal partners, non-profit organizations and other groups active in the special needs community." http://www.fema.gov/pdf/media/factsheets/2009/npd_comp_guide_301.pdf

Within CPG 301's Appendix F, "Civil Rights Considerations Related to Special Needs Planning," there is a complete section devoted to a discussion of "limited English proficient/proficiency" that begins on page 67. For more than 30 years, the US English Organization has been devoted to making English the official language of the United States (www.us-english.org). In 1981, Senator H. I. Hayakawa introduced Senate Joint Resolution 72 (S.J.RES.72), more commonly know as the "English Language Amendment." This proposed legislation called for a Constitutional

Amendment to declare English as the official language of the United States. If enacted, it would have prohibited the United States or any state from passing a law that requires "the use of any language other than English." More specifically, it would have banned the use of any language other than English in all federal and state courts (http://www.languagepolicy.net/archives/ela97.htm). The most recent incarnation in a long line of official English bills is H.R. 997, the English Language Unity Act, which was introduced in the 112th Congress in March 2011. H.R. 997 currently has the support of more than 100 members of Congress. It is important to note, however, that the great majority of the proposed legislation over the past 3 decades lists certain exclusions regarding the enforcement of an official English policy. These include "common sense exceptions to protect public health and safety, national security, and to provide for the needs of commerce and the criminal justice system" (http://www.us-english.org/view/820). On its website, US English lists thirty-one states that have passed "Official English" laws to date (www.us-english.org). Inasmuch as EMs operate within the constraints of Federal as well as local and state laws, it is important for them to note that "[r]ecipients of Federal financial assistance operating in jurisdictions in which English has been declared the official language are subject to Federal nondiscrimination requirements, including those applicable to the provision of federally assisted services to persons who are LEP" (CPG 301 2008: 67). In other words, if their program is the beneficiary of Federal monies, they must provide LEP individuals with "meaningful access" to all services in spite of the fact that their state may have enacted an Official English law.

EO 13166 was incorporated into CPG 301 because of its relevance to the provision of language services to a variety of "special needs" and/or "disadvantaged" populations. Specific references to language assistance appear in the "Triage" and "Recovery" sections as well. CPG 301 provides additional language services guidance to emergency managers in terms of "meaningful access" to LEP speakers:

Examples of Language Assistance Services

- Direct foreign language communication by **fluent bilingual staff**
- Interpretation (oral), conducted in-person or via telephone by **qualified** interpreters
- Translation (written) by **qualified** translators" (author bold) (CPG 301 2008:69)

It is heartwarming to see the use of the words "fluent" and "qualified" in this text. It sends the message that there is more to providing competent language services than just basic knowledge of a particular language. Moreover, the indication that some of the EM staff might themselves be bilingual is truly a step in the right direction. This would obviate the need for outside assistance in certain languages, and the EM would be assured that the individual was knowledgeable about EM procedures and knew all of the pertinent EM vocabulary. It also follows naturally that a bilingual staff member would work to create solid ties to the minority community, thereby resulting in an all-important relationship of trust and mutual respect. Many minority groups are distrustful of government and/or law enforcement groups because of bad experiences in their home countries. Seeing a fellow countryman/woman in an official position can serve to strengthen a connection and, perhaps, encourage other bilingual minority community members to seek employment with the agency.

E. Comprehensive Preparedness Guide (CPG) 101, Version 1.0: Developing and Maintaining State, Territorial, Tribal and Local Government Emergency Plans

Comprehensive Preparedness Guide (CPG) 101, Version 1.0 appeared in March 2009. This 172-page document contains minimal information relevant to language services. For example, a search of the terms "bilingual" and "translator" returned zero results. Moreover, the specific expression "limited English proficient" was not found. However, people who have "limited proficiency in English or are non-English-speaking" are included in a definition of "Special-Needs

Population" in the Glossary. (See Section II. F. for further discussion.) "Special Needs" does occur several times in the body of CPG 101 Version 1.0 as well. Worthy of mention is the inclusion of difficulty with "language comprehension" as an example of a special need (CPG 101 Version 1.0 2009: 6-2). "English" is included only twice – once in the context of "Keep the language simple and clear by writing in Plain English (CPG 101 Version 1.0 2009: 3-18) and in the "Special Needs" Glossary entry referred to earlier in this paragraph. The term "interpreter" appears only once under the section on Mass Care/Emergency Assistance:

> Describe the agencies and methods used to provide care and support for institutionalized or special-needs individuals (e.g., medical and prescription support, durable medical equipment, child care, transportation, **foreign language interpreters**) and their caregivers (author bold) (2009: C-18).

F. Developing and Maintaining Emergency Operations Plans, Comprehensive Preparedness Guide (CPG) 101, Version 2.0, November 2010

Comprehensive Preparedness Guide (CPG) 101, Version 2.0 (November 2010) is the most recent document available. It replaces and supercedes (a) CPG 301 Version 1.0, which appeared on August 15, 2008; and (b) CPG 101 Version 1.0, issued in March of 2009.

CPG 101 Version 2.0 stresses active participation of the entire community in the planning process, as evidenced by the following quotations:

> CPG 101 Version 2.0 encourages emergency and homeland security managers to engage the *whole community* in addressing all of the risks that might impact their jurisdictions. (author italics) (http://www.fema.gov/prepared/plan.shtm).

> Planning that engages and includes the whole community serves as the focal point for building a collaborative and resilient community (Introductory Letter to CPG 101-2.0 Craig Fugate 2010 http://www.fema.gov/pdf/about/divisions/npd/CPG_101_V2.pdf).

> Engaging in community-based planning – planning that is *for* the whole community and *involves*

the whole community – is crucial to the success of any plan (CPG 101-2.0: 4-4).

Whereas Executive Order 13166 was fully incorporated into CPG 301 Version 1.0, its full text does not appear in CPG 101 Version 2.0. It should be noted, however, that on page 1-1 "Planning Fundamentals," Footnote 4 states that "planners should ensure compliance with the requirements of Title VI of the Civil Rights Act of 1964, Executive Order 13166, the Americans With Disabilities Act, Section 504 of the Rehabilitation Act, and other Federal, state or local laws and anti-discrimination laws." Moreover, Appendix A: "Authorities and References" includes citations to EO 13166, and Homeland Security Presidential Directive 5, under whose authority NIMS and the NRF were promulgated (2011: A1). This Appendix includes additional authorities as well.

In terms of relevant Glossary definitions (Appendix B), the "disability" entry is more fleshed out than those found in CPG 301 and CPG 101 Version 1.0. The Version 2.0 information includes reference to the Americans With Disabilities Act. Note that LEP is not included, as it is not considered a "disability." (http://www.fema.gov/pdf/about/divisions/npd/CPG_101_V2.pdf)

Unlike CPG 301, CPG 101 Version 2.0 does not contain a definition for "sign language interpreter." (Glossary information for "spoken language interpreter" is not incorporated into either version of CPG 101 or in CPG 301). Notably as well, there is no entry for "special needs population" in CPG 101 Version 2.0 (which appeared in both CPG 101 Version 1.0 and CPG 301). However, in CPG Version 1.0's Appendix B: Glossary and List of Acronyms, the broad description of "Special Needs Population" incorporates LEP persons:

Special-Needs Population

A population whose members may have additional needs before, during, or after an incident in one or more of the following functional areas: maintaining independence, communication, transportation, supervision, and medical care. Individuals in need of additional response assistance may include those who have disabilities; live in institutionalized settings; are elderly; are children; are from diverse cultures, **have limited proficiency in English or** are **non-English-speaking**; or are transportation disadvantaged (author bold) (CPG 1.0 2009: B-7).

Inasmuch as there is no "special needs population" entry in CPG 101 Version 2.0, "limited English proficiency" now stands alone in the revised version's Glossary:

Limited English Proficiency

Persons who do not speak English as their primary language and who have a limited ability to read, speak, write, or understand English (CPG 101 Version 2.0: B-7).

Perhaps this is also a way to ensure that LEP individuals are mainstreamed into the EM process rather than including their numbers in a very broad definition that comprises many groups.

As a result of these wording and classification changes, the typical text employed when referring to particular at-risk groups within CPG 101 Version 2.0 reads:

Integrates the needs of the general population, children of all ages, individuals with disabilities and others with access and functional needs, *immigrants*, **individuals with limited English proficiency**, *and diverse racial and ethnic populations* (author italics and bold) (CPG 101-2.0: 4-18).

Full incorporation of LEP persons within Emergency Operations Plans (EOPs) is paramount.

In searching the 124-page Guide, the term "limited English proficient/proficiency"(LEP) occurs twenty (20) times. For example, in the all-important initial stage of EM,

... it is essential to incorporate individuals with disabilities or specific access and functional needs and **individuals with limited English proficiency**, as well as the groups and organizations that support these individuals, in all aspects of the planning process. When the plan considers and incorporates

the views of the individuals and organizations assigned tasks within it, they are more likely to accept and use the plan (author bold) (CPG 101, Version 2.0: 1-2).

Also noteworthy is the reference to the use of oral and written language assistance:

[c]ommunity-based planning should also include notifying affected, protected groups of opportunities to participate in planning activities and making such activities accessible to the entire community (e.g., **use of interpreters and translated announcements**) (author bold) (CPG 101, Version 2.0: 4-4).

The preceding quotations are fully consistent with CPG 101 Version 2.0's goal of working with and engaging the entire community in all phases of emergency management, especially in the preparedness stages.

CPG 101, Version 2 includes only one reference to "bilingual":

The information gathered during the jurisdictional assessment of individuals with disabilities and others with access and functional needs requires a detailed analysis. Emergency planners need to review the assessment findings and analyze the quantity and types of resources (including personnel) needed during different types of incidents. For example, a jurisdiction with a large number of limited English proficiency residents might need to identify methods by which language assistance will be provided (e.g., **bilingual personnel**, interpreters, translated documents) to support operations, such as evacuation, sheltering, and recovery (author bold) (CPG 101, Version 2.0: 4-9).

Noticeably lacking when terms like "bilingual personnel" and "interpreters" appear are all-important adjectives like "qualified" and "fluent," which were incorporated into CPG 301 in its citation of EO 13166. (See Section II.D.)

III. OVERVIEW OF THE LANGUAGE SERVICES PROFESSION

A. Modes of Interpreting

There are three basic modes of interpreting: simultaneous, consecutive and sight translation.
Simultaneous: "In simultaneous interpretation (SI), interpreters listen, analyze, translate

and speak with a lag of only seconds/words separating the original source-language (SL) version and the target-language (TL) rendition. The goal of this complicated process is to communicate a message from the SL to the TL with minimal analysis time" (Schweda-Nicholson, 1987:194).

SI is that type of interpreting most associated with the United Nations and other international organizations. Essentially, simultaneous interpreters listen and speak at the same time. More specifically, they are attending to the incoming SL message while at the same time uttering material in the TL that has already been analyzed. Normally, sophisticated equipment is required in order to perform SI. For this reason, it is not the mode which is most commonly employed in emergency management situations/settings.

Consecutive: Consecutive interpreting (CI) is most often associated with the back-and-forth structure of a conversation. For example, Speaker One utters a message in a particular language (L1) and then pauses. The interpreter, in turn, communicates the message to Speaker 2 in another language (L2). Speaker 2 then responds in L2, and the conversation continues. Unlike SI, the interpreter in not required to speak and listen at the same time. Normal conversational pauses allow for the interpreter to offer his/her rendition of what has just been said in L1 into L2. Because there is no overlap of listening and speaking, the interpreter can place his/her full attention on listening and analyzing. Often, a consecutive interpreter takes notes as a short-term memory aid which, among other things, permits him/her to maintain the speaker's order of events and relay dates, numbers and statistics accurately. Based on the needs of emergency managers, CI is the mode that is most often used to communicate with LEP community members.

Sight Translation: Sight translation (ST) is a hybrid form of interlingual communication. The language professional uses a written document for the SL, and translates the message into the TL out loud in an oral form. This mode is most frequently used when an individual needs to quickly learn/communicate the contents of a document. Perhaps an emergency manager has prepared an evacuation order, but does not have time to have it professionally translated. An interpreter could be riding in a car sight-translating the document and speaking into a microphone, which broadcasts the order to the target populations in their neighborhoods. This mode is generally not used for lengthy written material, as a real translation is preferable in this situation. However, when time is of the essence and the document is not too long, sight translation is the preferred mode for quick communication. (For additional information on the three modes of interpreting, go to the National Association of Judiciary Interpreters and Translators (NAJIT) website at www.najit.org. On the homepage, click on "Position Papers," and then click on "Modes of Interpreting.")

B. The Role of *ad hoc* Interpreters

The definition of *ad hoc* varies from source to source. In essence, it is "a Latin phrase which is used to mean that something has been made up as we went along; it was not planned; there is no particular reason why it should be the way it is" (http://thesaurus.maths.org/mmkb/entry.ht). Located through http://www.encyclo.co.uk/define/ad-hoc). Interpreters are often selected on an *ad hoc* basis in numerous settings (Schweda Nicholson, 1989). For example, a need for language services may be identified in an emergency situation, such as providing a warning to potential tornado victims as the storm bears down. People scramble to find someone whom they believe will be able to communicate the information, but they have no assurance or external validation that the individual possesses the requisite skills and abilities. Grab Ms. Jimenez, the sheriff's secretary or Mr. Chin, a worker on the custodial staff. After all, their names identify them with a minority group and, as a result, they *must* speak their language well. What people often fail to realize is that English skills may be lacking. If so, how can these individuals possibly understand what is communicated to them in English so that they may convey the message to their fellow countrymen and women? And how can the emergency manager be assured that the English translation that he/she hears in response is fully and accurately expressed? However, in the heat of the moment, rational thinking may go out the window!

Let's now examine why *ad hoc* "interpreters" are so common in the United States. Trabing (2009) delineates a number of oft-repeated "cop-outs":

- "Immigrants should learn English!" After all, it is the dominant language of the United States, and newcomers will be able to more fully participate in American life if they can communicate in English.
- "We have a "bilingual" person here who can help." Does this person have the linguistic and cultural knowledge necessary to competently act as an interpreter? Volunteers may mean well, but their abilities must be questioned. Remember the old adage: "You get what you pay for."
- "Who will fund this service? We don't have any money in our budget for language assistance." With cutbacks in virtually every sector of American life due to a languishing economy and a high unemployment rate, many view remuneration for language assistance as an unnecessary expense. Clearly related to this mindset is one that is typified by the following attitude: "What's the big deal? Everybody speaks a language!!" It is sad but true that, unlike in other countries, high-level knowledge of languages is not considered to be specialized, arcane, technical expertise which merits recognition (such as one would consider skills in business, computer technology, law and/or medicine, to name only a few).
- "Interpreters get in the way of direct communication." This is truly a ludicrous statement. Interpreters facilitate communication between those who do not share a common language. Without language professionals, attempts at communicating are reduced to pointing and hand signals, which may be imprecise and potentially confusing, thereby detracting from mutual understanding. It is true that using interpreters during a conversation takes more time because of the necessary pauses in speech during which the interpreter offers his/her rendition of what has just been uttered in the other language. However, if the goal is complete and accurate communication, working with a competent interpreter is really the only choice.
- "It's not my/our problem!" It IS your problem if you are an emergency manager or a provider who is working with the public in any/all of the four stages of EM. Burgener (2009) conducted a survey within the field and determined from his research that there is definitely a problem with reaching linguistic and cultural minorities. As discussed in Section II of this chapter, there are various statutes and rules that require, for example, that organizations which receive federal funding offer non-English-speakers or LEP speakers "meaningful access" to services and information (Executive Order 13166 2000:1).

On the positive side, there may be less need for *ad hoc* interpreters in the future. The U.S. Bureau of Labor Statistics (BLS) predicts a 22% growth rate among "Interpreters and Translators" during the period 2008–2018. These projections have been abstracted from the National Employment Matrix and are part of the Occupational Outlook Handbook 2008–2018 (http://www.bls.gov/oco/). The rate of 22% situates interpreters and translators in a category that will experience "much faster than average employment growth" over the next decade (http://www.bls.gov/oco/pdf/ocos175.pdf). Expansion is anticipated especially in both the legal and healthcare fields. In terms of demand for specific languages, Spanish is expected to remain very strong. Moreover, the widely-spoken languages of East Asia – Japanese, Korean, and Mandarin Chinese will also continue to be required. Arabic and additional Middle Eastern languages play major roles in the interpreting and translation marketplace. Finally, there is a burgeoning need for American Sign Language interpreters, especially now with broader use of video relay, a technology that enables client and interpreter who are physically far apart to be linked via the Internet.

C. General Misconceptions Regarding Language Services

At this juncture, it is helpful to examine some commonly held misconceptions with respect to language services:

- Many people who speak only one language believe that *anyone* can function as an interpreter (oral communication) or translator (written communication), even those individuals who possess only the most basic knowledge of another language (such as one or two years of high school Spanish). This is simply not true. Within the profession, we view excellent abilities in at least two languages as merely a starting point for interpreter/translator training. And, even those with superior linguistic knowledge do not always make good language professionals. Interpreting and translating skills (such as cognitive flexibility, speed, outstanding short-and long-term memory, note-taking and the ability to speak and listen at the same time, among many others) are viewed separately from language skills.

- There is also the mistaken belief that interpreting and translating are simple word-for-word substitution processes. Equivalents for all words exist in both languages, so what's the big deal (what is the issue?) Additionally, because of the "equivalency" misconception, the task of converting a message from one language to another language is often viewed as robotic and mechanical, one that requires no real thinking at all. It is actually the case that *true* equivalents are rare and constitute the exception to the rule. And the word "meaning" is the key. It is the meaning of the original ("source language") message that interpreters and translators strive to communicate into the other ("target language"). It is their responsibility to select the words and phrases from a vast target language inventory that will accurately convey the meaning of the source language utterance. And languages do not exist in a vacuum. Language and culture are inextricably tied. As a result, interpreters and translators must possess firsthand knowledge of the cultures that are associated with the languages involved. (For an overview of the importance of intercultural understanding, see Hudson, 2008.)

- To summarize, excellent linguistic skills, interpreting/translating abilities and multicultural understanding all contribute to the total package that is the language professional. Moreover, with the exception of a very few individuals who are "born interpreters", attainment of this high knowledge level requires specialized training. In addition to mastering the nuts and bolts of the interpreting process, interpreters learn to deal with internal and external distractions as well as stress. No matter how qualified you are in the aforementioned areas, if you cannot perform under stress, you will not be successful. (See Gard, 2009 for a detailed discussion of the psychological side of interpreting during disasters and emergencies.)

D. Additional Discussion of EM Issues in the Interpreting/Translation Literature

Although far less numerous than publications on legal and conference interpreting, a limited number of articles that treat language services in emergency and disaster situations as well as public health crises do appear in the interpreting and translating literature. For example, Russell-Bitting (2002) describes the need for linguists after the tragic events of 9/11. Moreover, Thickstun (2009) writes about how medical interpreters and translators may be of service during an influenza pandemic, such as the H1N1 outbreak in 2009. She discusses the need for language services providers to research pertinent medical terminology as well as to read Centers for Disease Control (CDC) publications and updates for the latest news on the spread of the virus. She also suggests that language professionals be aware of emergency preparedness plans in place at facilities that may become involved in treating infected persons. LEP individuals are often unaware of impending threats because they have limited access to information in their native languages. As mentioned earlier in this chapter, the poor warning system for notifying the Vietnamese population in Gulfport, Mississippi prior to Hurricane Katrina resulted in many deaths and injuries (Musgrave, 2005). Thickstun also cites the usefulness of message maps, in which critical, complex information is broken down into smaller, more easily comprehensible sentences, thereby offering key concepts and guidelines in plain language and in a very simple format. Employing message maps is not only helpful to emergency managers in getting their message out in palatable form to native English-speakers, but also to LEP community members. Whereas she writes specifically about the flu, Thickstun's advice about risk communication and crisis management is valuable for use during any such emergency situation. Farkas (2010) describes the monumental task of mobilizing language services professionals to assist after the 2010 earthquake in Haiti. She recounts the story of the creation of the extremely successful "Together We Can Find 100,000 Translators and Linguists" initiative, launched by Nicholas

Ferreira on Facebook. Farkas also writes of the involvement of the International Medical Interpreters Association (IMIA) as well as Translators Without Borders (http://tsf.eurotexte.fr/?lang=en) in the relief effort. From a language perspective, Haitian Creole presents a challenge because it is not widely spoken by those who are not Haitian nationals themselves. Although telephone interpreting is not completely well-suited to the types of language assistance required after the earthquake, both Language Line Services and Pacific Interpreters donated free Haitian Creole interpreter services to people who were adversely affected by the strong tremor. Farkas sums up the wide-ranging effects of the coordinated venture: "The large-scale cooperative effort that the industry undertook has raised public awareness of our value and made us better prepared to respond to future disasters" (2010:28).

IV. THE NEED FOR GREATER AWARENESS OF LANGUAGE ISSUES IN THE EMERGENCY MANAGEMENT FIELD

As someone who has been working for more than 30 years in the area of language planning and policy development for interpreter services (primarily in the American courts and other legal settings), Dr. Nicholson became interested in language services in EM situations after reading the Musgrave (2005) article about Hurricane Katrina cited in the Introduction to this chapter. It appears that evacuation warnings had not been provided in Vietnamese to residents of Gulfport, Mississippi, which resulted in greater danger to this largely LEP population. Such individuals were unaware of the impending storm. Moreover, information in Vietnamese after the hurricane was offered only sporadically. Apparently, there was no infrastructure in place to provide the necessary language services to this linguistically isolated population. This story is not unique. The use of *ad hoc* interpreters in legal, medical and emergency situations is widespread in the United States, although things have been changing for the better in the past 30 years (Schweda Nicholson, 2005a; 2005b; 1989).

A. Treatment of Language Issues in Selected EM Publications

Dr. Nicholson conducted a brief search of the indices of a variety of emergency management books, some of which are intended as textbooks in introductory courses. She looked for the terms "bilingual," "communication/s." "culture," "disadvantaged," "interpreter," "language," "limited English proficient/proficiency," "minority," "Spanish," and "translation/translator." Only one of these terms appeared in the Haddow and Bullock (2006) index: "communications." Although there were many sub-entries under this topic, the only one relevant to language services was "identifying audiences and customers" (199). Haddow and Bullock devote about one-half page to this subject, and simply mention "minority" as a group within the "general public" that needs to be notified. There is neither elucidation nor are there guidelines provided regarding the need to establish close ties with minority communities and/or the importance of disseminating information in these groups' languages. Perry and Lindell (2007) devote about one-half page to the issue of "language barriers" (361). To their credit, they stress that it is critical for emergency managers to know their community (as well as available minority group resources, such as language-specific radio and television stations). Perry and Lindell raise the diversity issue and emphasize that Spanish is not the only language for which interpreter/translator services are required in the United States. They also state: "All written information disseminated should be multilingual. The translations need to be professionally done" (361). In Waugh and Tierney (2007), one finds the sub-topic "socially vulnerable populations" under "communication." Enarson (2007) mentions important issues, including identification of required languages and the necessity for translators. She also stresses the importance of knowing one's community and partnering with local organizations, such as those that offer English as a Second Language (ESL) programs. "Language, communication obstacles" and "cultural understanding" both appear in the index of Coppola (2007). He writes: "It would seem that simply learning the language of the target population would mitigate this issue, but the answer is not always that straightforward" 2007: 234). This statement strikes the author as ingenuous, as it implies, to a great extent, that learning another language in order to communicate effectively with native speakers is a mere trifle and requires very little effort. His use of the word "simply" adds fuel to the fire. The skill of knowing another language well is trivialized, and this attitude is in line with attaching little value to linguistic abilities. In this context, there is no mention of the importance of recruiting competent bilinguals to work in the EM field, something that seems to be lacking in many jurisdictions. To the author, this approach would follow naturally and should be a priority. All in all, it is truly surprising that a book specifically written about "'international' disaster management" would have so little to contribute on the need for effective multilingual communication between EM officials and the diverse populations they are legally bound to serve.

Although Erickson (2006) spends very little time talking about language and cultural issues, his less-than-one-page treatment is the most realistic and honest evaluation of the American ethnic and linguistic minority situation. Whereas he primarily focuses on corporate training for emergency response in the section where these paragraphs appear, his comments are generally applicable to broader language and cultural contexts in the United States as well as all phases of emergency management:

> That there can be no effective training without effective communication is a bromide so logically soporific it is usually ignored in practice, especially in the United States where the Americanized English language is consider [sic] the *lingua franca* that not only overcomes all linguistic and cultural barriers but also obviates any and all distinctions imposed by diverse personal experience and values. The perception is, of course, quite wrong. . . . " (348).

Moreover, in paging through Coppola and Maloney (2009), one notes that they include people "who speak different languages" (20) on a list of nine groups to consider in the warning planning stages. They also provide elaborate detail on a "Public Information Campaign" and the importance of defining the target populations (35, 78). Coppola and Maloney recognize the importance of (1) involving special groups in the planning process; (2) making an effort to understand cultural differences and consider them when devising warning strategies; and (3) building trust between government entities and the constituents they serve. A number of "Emergency Management Public Education Case Studies" are included at the end of the text. Several of these address the involvement of local linguistic minority groups in outreach and educational programs (235).

The information on the provision of language services gathered from the aforementioned sources demonstrates that this critical issue is often treated only very briefly in the context of general populations that must be contacted. Specific guidelines regarding how to locate competent, qualified interpreters and translators as well as the pitfalls to be avoided are generally not included. Later in this chapter, we will see that a 2009 webinar demonstrated that, at least in the jurisdictions that participated, there is a dearth of linguistic and ethnic minorities employed in the emergency management field. As a result, it seems logical that more detailed guidance should be offered to monolingual English-speaking emergency managers and their staffs.

Over the years, researchers and activists have called for expanded contact efforts and creative solutions to the problem of reaching disadvantaged populations. (See, for example, Bates & Swan, 2007; Drabek, 2010; Enarson, 2007; Lachlan, Spence & Eith, 2007; Leong *et al.*, 2007; Mileti, Drabek & Haas, 1975; Mileti and Fitzpatrick, 1991; Phillips, 2009; and Trujillo-Pagán, 2007). More specifically, Leong *et al.* (2007) discuss the importance of going beyond a Black/White investigation of Hurricane Katrina and focusing on the problems of the Vietnamese population. Trujillo-Pagán (2007) examines the Latino community in New Orleans in terms of disaster response and recovery issues. Moreover, people who are Deaf or hard of hearing have special needs. Just as rudimentary knowledge of English or another language does not an interpreter make, the same holds true for American Sign Language (ASL). In her presentation entitled, "Understanding, Motivating and Supporting Evacuation with Special Needs Populations," which appears on FEMA's website, Phillips cites a graduate student from Gallaudet University who was helping FEMA in a post-impact situation: "It is not just a matter of knowing sign language. It takes years of training to become an interpreter who can work with a group like this, and it takes an understanding of the deaf culture to be effective." http://www.fema.gov/pdf/about/ regions/regionii/phillips.pdf (2009). Always remember that individuals' language skill levels can vary greatly. Just because people say they speak Tagalog, Vietnamese or English doesn't mean that they speak it fluently. Nor does it mean that they have the requisite breadth and depth of vocabulary to assist effectively.

A good emergency manager knows his/her community well in terms of socioeconomic levels/status, specific ethnic and cultural groups and the language challenges they may present. It is critically important that emergency managers reach out to local ethnic and linguistic communities in order to work hand in hand to foster a trusting, cordial and collaborative relationship. Identify trusted, respected minority community leaders as spokespeople whose instructions to evacuate and other mandates, for example, will be taken seriously and followed at the time of a disaster. Get together and open a dialogue before a problem arises so you can work together after the fact. For example, the emergency manager must know the languages other than English that are spoken locally. Moreover, if a particular variety predominates, he/she must be aware of this in advance so that the most effective language assistance will be provided when necessary. For example, there are many dialects of "Chinese": Mandarin (also known as Putongha), Cantonese, Foochow, Sichuan, and others. Unknown to most is that there are 293 languages in China (http://www.ethnologue.com/show_country.asp?name=cn). Learn if the Chinese speakers are from the Beijing area or from Taiwan, for example. The writing systems are different: the People's Republic of China (PRC) uses simplified characters, whereas Taiwan employs traditional characters. Translations must be specifically tailored to each group in order for them to be comprehensible. In this same vein, just because a person is from Mexico does not automatically mean that he/she is a Spanish speaker. Many indigenous Indian languages are spoken there, such as Mixtec and Nahuatl. Additionally, there are numerous dialects of Mixtec and Nahuatl which are found in different regions of the country (www.ethnologue.com).

It is essential in EM for all parties to work together in mitigation and preparedness as well as response and recovery. In terms of mitigation and preparedness, have basic documents, such as warnings, translated into target languages as standard operating procedure. Don't wait until a disaster strikes and find yourself scurrying around to locate a competent language professional. A difficult and challenging time becomes even more stressful if these important matters are left until the last minute. Moreover, working in haste and under extreme pressure, you may make the wrong choice of a translator and be left with materials of poor quality. If there are "boilerplate" forms that remain the same except for specific details, have these translated so that only the blanks need to be filled in at the time of an emergency.

A good rule of thumb is to extend everything you do to accommodate LEP speakers. Potential legal liability for "failure to act" is an important consideration. Clearly, there is a regulatory duty to act under CPG 101 Version 2.0. EO 13166 reiterates the statutory duty required by the Civil Rights Act of 1964. In this case, emergency managers have a responsibility to notify LEP community members in their own language(s). As discussed at greater length elsewhere in this book, failure to perform legally-mandated duties is not protected by immunities that might otherwise apply to governmental actions. Note that failure to warn or provision of an incomplete warning can be the basis for negligence liability. This applies not only in one's own jurisdiction, but may also be a cause of action under international law (Nicholson, 2006b).

B. Problems and Pitfalls in Locating Competent Language Services Providers

Emergency managers' first instinct may be to contact ethnic and linguistic minority community leaders within their jurisdiction in an attempt to secure volunteer interpreting and translation assistance. While this approach seems logical, it may not be prudent for a number of reasons. First, there are many self-styled "interpreters" whose skills may be lacking. Native English-speakers may be quick to say that Ms. Morales is a native Spanish speaker and that Mr. Wang is a native Mandarin speaker, as if these are the only qualifications necessary. Emergency managers need to know how these individuals' English is, for they will be communicating back and forth into and from English. If English skills are sufficient, is the person's accent a problem? Many native English-speakers who do not have exposure to non-native speakers have great difficulty understanding someone who speaks English with a heavy accent. In the same way that an untrained member of a minority community may not be the best choice as an interpreter, the same goes for native English-speakers whose foreign language skills may not be of the necessary high caliber.

Second, if an untrained community member is selected to act as an interpreter, he/she is probably not aware of the codes of ethics that professional interpreters follow (NAJIT Code of Ethics and Professional Responsibilities (no date) http://www.najit.org/about/NAJITCodeofEthics FINAL.pdf; Registry of Interpreters for the Deaf Code of Professional Conduct (no date) http://www.rid.org/ethics/code/index.cfm; Schweda Nicholson, 1994a, 1994b; Trabing, 2004). For example, confidentiality can be a big problem in ethnic and linguistic minority communities. Certain information about individuals' personal situations, for example, that the "interpreter" may learn during the course of his/her duties must be kept to oneself. Often, if other community members know that a particular person has acted as an interpreter in a highly personal and/or potentially embarrassing situation, they may press him/her for details. A professionally trained interpreter knows that confidentiality must be respected. Moreover, accuracy and completeness are critical. Interpreters must refrain from: (a) adding or deleting information; (b) altering the emotional tone; (c) offering explanations that were not part of the original statement; and (d) engaging any of the participants in a conversation that will be inaccessible to the other parties. Maintaining impartiality is also an important ethical requirement. An interpreter cannot take sides in a matter with the intention of influencing the outcome. His/her role is to be a conduit of information and to represent the parties as they are. Finally, an interpreter's scope of practice must be limited to his/her abilities/qualifications. If an interpreter is unfamiliar with a particular subject matter and/or lacking in the requisite technical vocabulary to perform in a specific setting, he/she must refuse the assignment. In the context of EM, pertinent technical terms and procedures must be within the interpreter's knowledge base.

In sum, although a member of a particular linguistic minority community may seem like the natural choice, he/she may not be the best person for the job. Moreover, a native-born American English-speaker with limited foreign language skills who volunteers to assist may be problematic as well.

Volunteer interpreters generally have good intentions, but their knowledge often does not go beyond ordering in an ethnic restaurant or conversing socially about family matters. Their skill levels in both English and the foreign language are not known and have not been verified or tested. Poor volunteer language services are worth just what you paid for them. And, unless you speak the foreign language in question, you have no way of knowing if what the person is saying in the other language is correct. Just because someone is speaking fluently and does not hesitate is no guarantee that they are interpreting completely and accurately. People selected to work as interpreters should have *documented* language skills, demonstrated through formal testing and certification procedures. "Assume that you need the best in order to do your best work" (Croon, 2009:42).

The use of untrained and poorly skilled volunteers could actually end up costing more money in the long run and not only on the language services side. There may be additional consequences as well, resulting in the loss of lives/property because people acted on misinformation taken from poorly translated documents or poorly interpreted conversations. The potential for lawsuits is great. From the legal perspective, one must ask: "Would a reasonable person have ensured that there was proper communication?" There are legal liability issues here!

Emergency managers may also contact local interpreting and translation businesses in order to locate language services. In fact, CPG 301 includes "translation and interpretation service agencies" as suggested possible resources under the heading "Planning Networks" (CPG 301, 2008: 9-10). However, it is important to do one's homework before engaging an interpreter using such a resource. There are many unscrupulous agencies that misrepresent the skills of their employees and charge high fees for their services to boot.

Beware of the word "certified." It is often misused and abused within the language services field. An agency representative may have a five-minute conversation with a potential interpreter in his/her native language and neglect to check English skills. Moreover, someone may claim to be "certified" as a translator or interpreter simply because they are registered with an agency and they have been told that they are now "certified." Unlike may other professions, there is no single certifying body for interpreters and translators. As will be discussed in greater detail later in this chapter, language professionals in the United States may be certified by professional organizations as well as Federal and state court entities. And also beware of using someone who is certified as a translator in an interpreter role. While there is some overlap of skills set, most translators don't make good interpreters, primarily because of the high pressure, time-sensitive nature of the work and the immediate reaction required.

V. THE ROLE OF PROFESSIONAL INTERPRETER AND TRANSLATOR ASSOCIATIONS IN LEP ACCESS

The **American Translators Association (ATA)** (www.atanet.org) is a longstanding professional organization founded in 1959. Membership includes primarily translators, but many interpreters also belong. The ATA has certification tests for written translation *only* from 10 languages into English and from English into 14 languages. It publishes a client education brochure, *Translation: Getting It Right*, available for free download at http://www.atanet.org/publications/getting_it_right.php).

The **National Association of Judiciary Interpreters and Translators (NAJIT)**: (www.najit.org) is primarily composed of legal interpreters, but many members work in other domains, such as social services and healthcare. NAJIT has a criterion-referenced, professionally-developed certification exam that includes interpreting and translating sections in Spanish and English. In April, 2009, NAJIT signed an MOU with the American Red Cross to provide volunteer interpreters during disasters. At present, more than

1,300 interpreters in more than 100 languages are part of the on-call network (Bridging the Gap, 2009; Memorandum of Understanding, 2009; Red Cross and NAJIT Collaborate, 2009).

Registry of Interpreters for the Deaf (RID): (www.rid.org) is a national organization dedicated to excellence in the fields of education and certification of interpreters and transliterators for the deaf and hard-of-hearing. Founded in 1964, it has an impressive reputation in training and testing. RID certifications fall into the "generalist" and "specialist" categories, and there are stringent continuing education requirements for all (certified) members. This professional organization has state chapters as well for local contacts.

International Medical Interpreters Association (IMIA): (www.imiaweb.org) is the largest and oldest medical interpreter association in the United States. Founded in 1986, it boasts over 2,000 members who work in more than 70 languages. IMIA's Disaster Relief Language Services Database is comprised of professional medical translators and interpreters who volunteer their services in times of emergency on a worldwide basis. The IMIA is dedicated to furthering all aspects of professionalism among medical interpreters, including certification and continuing education. In 2009, the National Board of Certification for Medical Interpreters (www.certifiedmedicalinterpreters.org) announced the beginning of the certification process. Since that time, a Written Exam (in English) and an Oral Exam (in Spanish) have been developed and implemented. As of this writing, it is anticipated that additional Oral Exams in Cantonese, Korean, Mandarin, Russian and Vietnamese will be available in late 2011.

VI. FEDERAL AND STATE COURT INTERPRETER CERTIFICATION PROGRAMS

Although the following entities test and certify interpreters for the judicial system, many of those who hold certification work in areas other than the courts. The great advantage to having the services of certified interpreters is that they have demonstrated a high level of language skill as well as interpreting ability, which are the basis for working in any field. Certified interpreters can become familiar with pertinent vocabulary and procedure in a wide variety of disciplines, professions, settings and situations. Of importance as well is that certified interpreters are aware of and abide by codes of ethics, which is rarely the case with untrained volunteers. The organizations that certify interpreters clearly indicate the languages in which an individual is qualified/certified and, if relevant, into and from which he/she is certified to work. For example, in simultaneous interpretation, a person may be certified to work *into* English from another language, but not into the other language. State and Federal certifying bodies also provide continuing education seminars in order to permit interpreters to maintain and upgrade their skills.

The National Center for State Courts (NCSC): www.ncsc.org currently houses both the Federal Court Interpreter Certification Examination (FCICE) and the Consortium for Language Access in the Courts. The FCICE began in 1980 and has certified hundreds of Spanish/ English interpreters during this 30-year period. The latter is a group of 41 U.S. states that share certification tests in 19 languages. The NCSC has a very detailed website which describes not only the Consortium's work but includes numerous external informational links as well. Names and contact information for each of the 41 states' administrators are listed, so emergency managers from most states are provided with a local resource. http://www.ncsc.org/Web%20Document%20Library/EC_StateInterpCert.aspx.

Member States: http://www.ncsconline.org/D_Research/CIConsortContactspage.html

State Links: http://devlegacy.ncsc.org/WC/CourTopics/statelinks.asp?id=16&topic=CtInte

Languages: http://www.ncsconline.org/D_Re
search/CourtInterp/OralExamReadyforAdmin
istration.pdf

General Language Services Resources: http://
devlegacy.ncsc.org/WC/CourTopics/Resource
Guide.asp?topic=CtInte.)

VII. LANGUAGE SERVICES INITIATIVES IN THE PUBLIC AND PRIVATE SECTORS

Language Line is a major player in the area of "over the phone interpreting" (OPI) services. It works with a large variety of governmental and non-governmental organizations such as courts, law enforcement, healthcare entities, private sector corporations and many others (http://www.languageline.com). Although some people within the profession eschew telephone interpreting, it is generally agreed that it is better to have a skilled interpreter on the phone than a poor one in person. Language Line University offers not only skills assessment for companies' bilingual employees, but also medical and legal consecutive interpreting certification tests for individuals in a variety of languages over the phone. http://www.languageline.com/page/cert_prod/.) Geersten and Romero (2000) discuss Language Line's training and certification initiatives. Bostrom (2006) describes the use of Language Line services in several rural North Carolina counties.

The **American Translators Association (ATA)** has joined forces with the American Red Cross (ARC) in order to recruit volunteer interpreters and translators to assist with language services during relief operations after disasters. See www.atanet.org/red_cross for further information.

The **National Association of Judiciary Interpreters and Translators** has also partnered with the **American Red Cross**. Along with eleven NGOs and the FEMA Voluntary Agency Liaisons, NAJIT and the ARC formed the American Red Cross Partners, which held its inaugural meeting in Washington, D.C. on April 14, 2009. The majority of the interpreting services will be furnished via telephone, thereby obviating the need for personnel to travel to disaster sites (Red Cross & NAJIT Collaborate, 2009). NAJIT and the ARC also signed a Memorandum of Understanding in 2009, which not only addresses volunteers' provision of language services to non-English-speaking and LEP populations but also to the Deaf and hard-of-hearing. Although not required to do so, participants will be encouraged to complete a course on disaster response. A basic introductory training class is available online at www.redcross.org/flash/course 01_v01/ (Memorandum of Understanding...2009).

The National Virtual Translation Center (NVTC) was created by H.R. 3162: The USA Patriot Act of 2001. Section 907 describes this government entity which strives to provide timely translation services within the intelligence community. The NVTC is also interested in recruiting volunteer and contract language services providers to assist in its mission. Although the emphasis is placed on foreign intelligence, the NVTC maintains a roster of qualified linguists who are willing to help in times of natural and national emergencies/disasters. For further information, see www.nvtc.gov.

Created in 2006 by the Defense Authorization Act (and originally known as the Civilian Linguist Reserve Corps), **The National Language Service Corps (NLSC)** (www.nlscorps.org) is a group of more than 2,000 volunteer language services personnel who assist with government activities to reach LEP individuals in times of crisis and disaster. For example, a Vietnamese NLSC interpreter assisted Secretary of the Navy, Ray Mabus, when he traveled to the Gulf for Town Hall meetings in Louisiana, Alabama and Mississippi after the oil leak in the summer of 2010. As of June 2011, the NLSC is represented in all 50 states as well as in the District of Columbia and Puerto Rico. The selection process includes two stages. First, a prospective member must offer a self-assessment of his/her language skills. The NLSC generally requires a minimum of a Level Three in speaking, reading and writing on the Interagency Language Roundtable scale in the target language. If an individual is a non-native speaker of English, he/she must self-rate at Level Three in English as well. Prior to being sent into the field, some members are also required to go through a battery of formal tests in order to verify the self-assessments (http://www.nlscorps.org/Forms_Kernel/FAQCS.aspx).

Iowa's **Multilingual Emergency Response Network (MLERN)** grew out of a special needs committee formed by the Iowa Disaster Human Resource Council (IDHRC). The MLERN task force was established by a number of entities, including Iowa Homeland Security and Emergency Management and the State of Iowa Bureau of Refugee Services (Bawn & Peterson, 2008; De Rouchey, 2011). According to Blake DeRouchey (Volunteer and Donations Management Coordinator, Iowa Homeland Security & Emergency Management), "extensive training" was provided to the MLERN volunteers several years ago, but his agency has never tapped into this resource during a disaster. De Rouchey added that they are trying to revive the program and conduct outreach to the volunteers (Personal communication. July 6, 2010).

In an April 2011 e-mail update, DeRouchey reiterated that the MLERN volunteers have not been used, but that his agency works with the Iowa Council for International Understanding (ICIU), a for-profit organization, which also offers free translator services during times of emergency. He indicated that Iowa Homeland Security & Emergency Management would most likely contact the ICIU if language services were needed. DeRouchey added that the MLERN volunteers would probably be called on if language assistance were to be required for an extended period (Personal communication: April 26, 2011). However, to date, the State of Iowa has not used either of these translator services in a disaster.

In **Nebraska**, efforts are underway to reach speakers of "languages other than Spanish" (LOTS), not only to provide community members with necessary services in a variety of languages, but also to recruit fluent speakers of these languages and English who can act as interpreters and translators (Conroy, 2008). As has been the case throughout the U. S. over the past decade or so, many states that have traditionally been characterized by small linguistic and ethnic minority populations have witnessed unprecedented growth in diversity. Nebraska is one such state, as Lancaster County (in which Lincoln, the capital, is found) ranks 14th in the nation in terms of refugee resettlement from a per capita perspective (Conroy, 2008:14). It has been estimated that more than 50 languages are spoken in Lincoln alone. Moreover, Nebraska's Sudanese population is the largest of any state. Public and private sector entities were asked about the languages they need most after Spanish. The top three listed were: Nuer, Somali and Arabic. This response is quite unanticipated, and steps must be taken to reach community members on a number of levels.

In 2008, the **Federal Emergency Management Agency** (FEMA) began employing a new communications system called **Deaf Link** to assist Deaf or hard of hearing individuals file applications for disaster assistance at FEMA/State Disaster Recovery Centers. FEMA is utilizing Deaf Link's program, Video Remote Interpreting, which forms a communication triad: a hard-of-hearing/Deaf individual and a hearing person (who are both in the same place) are linked to a competent American Sign Language (ASL) interpreter at Deaf Link's secure offsite center via an internet communication system which offers both video and audio modes. The Deaf Link interpreter views the applicant's ASL on the video monitor and translates it into English audio. This information is then processed by the FEMA representative, who is then able to input the data into the FEMA system (Technology Assists Hearing-Impaired Register with FEMA, 2008) http://www.fema.gov/pdf/hazard/hurricane/2008/ike/enews_10-20-08.pdf. The term "hearing-impaired" (which is unacceptable to this community) is sometimes employed in FEMA materials. However, the preferred expression, "hard of hearing," is also found in FEMA publications. To FEMA's credit, a very recent brochure, *Planning for the Whole Community* includes the following: "Over 30 million people have a hearing disability – they may be deaf, hard of hearing, or deaf/blind" (April 2011:8).

http://www.fema.gov/pdf/about/odic/all_hands_0411.pdf

Also available through FEMA, the **National Disaster Housing Strategy** (NDHS) site offers information on a toll-free language bank which provides interpreter services. Language assistance can be accessed at a Disaster Recovery Center (DRC) or when applying via telephone for FEMA's help. http://www.fema.gov/pdf/emergency/disaster housing/NDHSAnnex3.pdf

VIII. FEMA AND OTHER WEBSITE MATERIALS IN LANGUAGES OTHER THAN ENGLISH (LOES)

The majority of this chapter has dealt with the provision of language services to non-English-speakers and LEP speakers. An ideal solution to the problem/challenge of reaching speakers of multiple languages would be to prepare materials related to the four phases of emergency management in LOEs so as to reduce the need for interpreters. To a certain extent, FEMA has done this. For example, if Vietnamese speakers can read documents in their language about preparedness and the importance of heeding warnings to evacuate, they will be better informed and able to proceed in a manner consistent with the required procedures. This approach not only contributes to a potentially increased level of safety for them, but also for the community at large by reducing strain on the system at the time of a disaster.

On the FEMA homepage, there is a separate link for "Spanish" and another one entitled "Additional languages" in the upper right corner. A click on this heading takes the reader to "Resources for Other Languages." On this "multilingual webpage", one finds prose indicating that information on the four phases of emergency management is available through brochures, flyers, public service announcements and press releases, for example. There are links to no fewer than 20 languages in addition to Spanish, making the grand total 21. At first glance, this is a very encouraging result to a researcher investigating the dissemination of news, notifications and warnings, for example, among ethnic and linguistic minorities. However, there is a great discrepancy from one language to another in the resources available. For example, the Spanish link connects to a homepage with numerous additional links, including photos, a "What are you looking for?" section (with links to governmental and private sector resources, brochures and forms, information for those with disabilities, children in a disaster, and a listing for FEMA regional offices, among others). There are also well-developed sections on preparedness and what to do if you are a disaster survivor. Moreover, there are a large map and additional links to information on recent/current disasters, such as the floods and tornados in the Northeast and Midwest. On the other hand, if one clicks on "Thai", there is only one PDF file about getting help after a disaster. Information in Russian, for example, features just three translated documents, including one about results from formaldehyde testing.

FEMA has a brochure called *Ready* under "Planning and Preparation" which is translated into only eight of the 21 languages: Chinese, French, Haitian-Creole, Korean, Russian, Spanish, Tagalog, and Vietnamese. It seems logical that such important information would be available in all 21 languages, especially with FEMA's continued emphasis on community involvement in the planning stage with the goal of reducing the number of problems when a disaster strikes. In this connection, the most common brochure is entitled: *Help After a Disaster: Applicant's Guide to the Individuals & Households Program*, which appears for 18 languages. The Croatian, Khmer and Hmong sites are lacking in this regard. However, the Hmong page includes *Planning Ahead is Essential for Flood Evacuations* and the information for Khmer offers a *Generic Guide to Assistance Programs*. The only brochure on the Croatian page discusses formaldehyde issues in temporary housing. The Vietnamese and Spanish sites are the only ones that feature information on hurricanes, whereas several discuss floods and earthquakes.

Clearly, the best-developed page after Spanish is Vietnamese, with a total of 21 links to additional materials. (Links for all of the aforementioned languages (except Spanish) may be found at http://www.fema.gov/media/resources/languages .shtm.)

In sum, it is suggested that FEMA strive to offer translations of all key informational documents to numerous linguistic minority populations. Moreover, FEMA should work to expand its list of languages. Perhaps it could partner with local emergency managers in specific jurisdictions where certain languages are represented to provide translations for use on a regional basis. Looking down the road, if a particular group were to appear in another area, translations in their language would already be available and could easily be shared with them.

At this point, it is also relevant to direct the reader to the National Resource Center on Advancing Emergency Preparedness for Culturally Diverse Populations (http://www.diversitypreparedness.org/). The Resource Center was created by the Drexel University School of Public Health's Center for Public Health Readiness & Communication. It is an excellent resource for emergency managers and planners. Its well-developed site includes numerous links on public health and related domains, all of great relevance to the efforts of emergency operations. One can browse by language, for example. Clicking this link brings up no fewer than 59 languages in which publications are available. Languages include those that are often needed, such as Hmong, Haitian Creole and Vietnamese as well as the less common Sinhalese, Mien and Marshallese. Like the FEMA site, however, there is no consistency across languages in terms of the publications that have been translated. For example, there is only one document for Chuukese (spoken in Micronesia), yet there are 174 available in Arabic.

IX. THE NEED FOR RECRUITMENT OF FLUENT BILINGUAL EM STAFF PERSONNEL

Relevant to the discussion of CPG 301 language assistance issues (treated in Section II. D.), Dr. Nicholson participated in a 2009 webinar entitled, "Preparing a Diverse Community: Emergency Response Plans for Limited English Proficient (LEP) Communities" (Language Access Webinar V, September 9). It was sponsored by the Migration Policy Institute, National Center on Immigrant Integration Policy. EM officials from a variety of jurisdictions offered their input on the most pressing LEP issues. At the end of the presentations, they asked for questions from the web audience.

Dr. Nicholson's question: "You have all spoken about the use of volunteer and paid translators and interpreters as well as the importance of partnering with community groups and buy-in on their part. With ever-increasing numbers of LEP speakers and immigrant communities, have you seen a corresponding increase in the number of minority group members employed full time in the EM field?"

People responded in a variety of ways. A representative from New York City stated that there had been an increase in diversity among employees there and that some outreach had been successful. However, he also indicated that a NYC preparedness poll was only conducted in English! Moreover, he addressed cultural differences in terms of the concept of 'emergency preparedness', stating that the Chinese population in NYC is interested whereas the Russians are not. In Hawaii, EM personnel reflect a very diverse population. In fact there are Language Access Officers in every state agency. However, she also stated in the next breath that bilingual workers are not really engaged in preparedness, but this is "starting," and we are "getting there."

In Montgomery County, Maryland (a large community just outside of Washington, D.C.), there are more than four hundred county workers with certified language abilities, but many are not working in EM. There is ongoing active recruitment of minorities, which has resulted in success among the Latino population, but not among Asian communities. They are hopeful with respect to future additional recruitment of linguistic and ethnic minorities.

The overriding message of this webinar is that there is still much work to be done, not only in the recruitment of minorities, but in building interest and trust among particular groups. It also appears that, perhaps, EM personnel could be more aggressive in working to attract current state/county employees with strong bilingual skills to the EM field and/or to at least create a register of such language-skilled employees who could be called on to assist in times of emergency. In this connection, the Migration Policy Institute has a website, the Language Portal, which is an excellent source of information regarding LEP access: http://www.migrationinformation.org/integration/language_portal/index.cfm. An article on this homepage (Reed 2011) discusses the importance of identifying, testing and certifying current bi/multilingual employees. Moreover, one possibility might be to offer incentive pay to language-and-interpreting-skilled individuals through the recruitment process. In this way, they would be paid more than someone who does not possess these special abilities. Increased pay is certainly one way to attract potential employees (and retain them as well) (Burgener, 2009:18).

Relevant as well to the current discussion of the overwhelming need for competent language assistance within the ranks is the issue of language training programs for emergency managers and Emergency Medical Technicians (EMTs). From time to time, one reads about emergency workers and/or police officers taking basic courses in a variety of languages with the goal of communicating better with LEP speakers in their communities (Boyle, 2001; Evans, 2000). While a noble prospect, there are serious potential risks to embarking on such a program. These courses generally stress acquiring fundamental conversational skills as well as learning pertinent terminology in fields like medicine, emergency management, firefighting and law enforcement, for example. It goes without saying that firefighters and paramedics can pro-

vide better assistance to those who are hurt if they understand what their condition is. Written reference lists of pertinent terms in English and other languages can be very helpful, but only if the injured know how to read. Of course, emergency personnel can also read the terms out loud, as long as their pronunciation skills in the target language are at a level that is comprehensible to a native speaker. As another example, advanced high school foreign language students who are selected as instructors for emergency personnel must be carefully screened before they are placed in such a role and closely monitored during the process.

Traditionally, classroom foreign language learning in the United States has tended to be conducted in a relatively closed atmosphere, isolated from native speakers who may actually reside in the region. For this reason, such students may have had limited or no opportunities to interact with a native Spanish-speaker. Moreover, as an Advanced Placement (AP) Spanish student herself as a high school senior, the author's vocabulary revolved around that required to read and discuss Spanish and Latin American literature. She never learned to say: "Do you have any weapons/ammunition in the house?" or "Are there any hazardous materials in your home that could explode?" Moreover, the emergency worker's ability to ask basic questions does not equate with his/her ability to comprehend the answers that will be proffered. As individuals with very rudimentary knowledge of the language in question, it will be very difficult for them to follow a constant stream of speech, especially if the person involved is injured, in pain, or worried about the house burning down. In such a situation, the native speaker will probably speak much more rapidly than usual. Moreover, speech may be slurred or garbled due to any number of factors. In sum, monolingual English-speaking EM supervisors must understand the limitations of such courses at the outset and not naively assume that someone with several hours of extremely basic instruction in another language will be able to competently communicate (a) *at the level of* or (b) *with* a native speaker.

X. USE OF THE INTERNET AND SOCIAL NETWORKING SITES LIKE FACEBOOK IN THE EMERGENCY MANAGEMENT FIELD

The internet and web resources have expanded exponentially in the past 10 years. The available information seems limitless, and search engines like Google can locate multiple answers to questions in less than a second. It seems only natural that emergency managers should take advantage of these vast resources, not only for gathering information, but also for disseminating it. In NIMS, Annex R to Emergency Support Function #15 – External Affairs: "Social Media" reviews a variety of internet options available for reaching communities in times of disaster, such as text messaging, social networks (Facebook, MySpace), Wikipedia and Twitter as well as photo- and video-sharing (http://www.fema.gov/pdf/emergency/nrf/nrf_jfo_sop.pdf, R-1 to R-3). As we will see below, however, social media sites may not be the best way to reach LEP populations and cultural minorities, largely because of limited access.

For over 50 years, the Current Population Survey (CPS) (an arm of the U.S. Census Bureau) has been compiling and analyzing data taken from monthly surveys of approximately 50,000 households. "CPS data are used by government policymakers and legislators as important indicators of our nation's economic situation and for planning and evaluating many government programs" (http://www.census.gov/cps/: April 2011). The CPS examines information in the areas of age, race, employment, income, and many others.

The most recent statistics regarding American households' internet availability and use offer interesting insights into the web's potential for communicating warnings and evacuation orders, for example, as well as recovery and FEMA/insurance claims information. Of particular interest is a table entitled: "Reported Internet Usage for Households, by Selected Householder Characteristics (2009) U.S. Census Bureau, Oct. 2009. Internet Release Date: Feb 2010" http://www.census.gov/population/socdemo/computer/2009/tab01.csv.

For those individuals who fall into the category "Households with Internet Use at Home,"

access numbers increase with age and educational level. People aged 55 and over and those with a Bachelor's degree or higher represented the two largest groups. U.S. Census data demonstrate that many LEP speakers in the United States fall in the "Less than a high school education" group (www.census.gov).

In terms of "Race and Hispanic Origin" within households who have internet access at home, whites far outnumber all other groups. In second place are Blacks, followed by Hispanics in third place. Noteworthy is that Asians are in last place, with twice as many Hispanics reporting internet access than the Asian population. Statistics such as these, combined with Linguistic Isolation data, suggest that Asians may more marginalized than other minority populations. Such information may be extremely valuable to emergency managers as they consider their notification options. And in terms of posting information in languages other than English, such a strategy may prove helpful; however, the EM still needs a skilled professional to create the target language messages in the first place!

All in all, the internet may only be of limited assistance in reaching linguistically disadvantaged populations. While these statistics may be interesting and useful, it is also important to remember that, sometimes in the aftermath of a major emergent event, the web (and the electricity needed to access it) may be unavailable for lengthy periods of time. For this reason, perhaps the internet should not be counted on for use in the "response phase."

Let's now examine the demonstrated and potential value of Facebook, a social networking system. Founded in 2004, its "mission is to give people the power to share and make the world more open and connected" (www.facebook.com). Millions of people use Facebook every day. In terms of its usefulness to emergency managers, pre-event warnings, for example, can be posted on Facebook. However, the primary use for Facebook over the past several years has been to issue immediate post-event warnings and ascertain damage assessment (Social media a possible lifesaver during disasters, 2009).

In conjunction with the CPS data included above, it goes without saying that one must have a computer in order to have access to Facebook. Web resources can be phenomenal in terms of breadth and depth (and virtually instantaneous), but those individuals/households without computers must be contacted as well through use of other means. Although no longer available as of April 21, 2011, the website www.checkfacebook .com used to be updated on a daily basis with Facebook statistics and other information. As of July, 2010, the U.S. age distribution for Facebook customers was the following: Most of its users fell into the 25–34-year group. In second place was the age range 18–24, and in third were those individuals from 35–44 years of age. People over 44 years of age constituted that smallest group of Facebook aficionados. The number of females was slightly higher than males, but the distribution was very close to 50/50 in terms of gender. Given the CPS age and racial data cited above, it is interesting to note that, whereas older (white) people have the highest access to the internet at home, they are also the smallest group of

Facebook users. As a result, the internet (in general) may be a desirable method through which to disseminate information, but it appears that Facebook will reach very few individuals over the age of 44. (Note: The "checkfacebook" site did not contain any information about the racial breakdowns of its users.) With respect to posting messages on Facebook in other languages, one notes that there are many foreign language Facebook sites. However, the emergency manager still needs a competent individual to prepare the translated information accurately before it can be disseminated.

Given the possibility that the internet may not be accessible after a disaster, but also given the fact that Facebook has been most used to disseminate information and assist people in locating resources after the fact, it seems reasonable to suggest that emergency managers have many arrows in their quivers. In other words, ensure that multiple avenues are available and ready to be activated. If one is not accessible, then there is a backup communication network at the ready. After all, taking an "all hazards" approach can easily translate to building an "all modes" system for reaching populations in harm's way.

XI. CONCLUSION

The current chapter examined many aspects of LEP issues with a focus on the legal side. In addition to citing laws, an executive order, regulations and standards, it provided resources on language assistance and the interpreting/translation profession. Moreover, it included very timely issues (such as the value of social networking in all four phases of EM), and called for the recruitment of fluent bilingual EM personnel to further advance the goals of FEMA. This contribution also offered practical suggestions to emergency managers regarding the nuts and bolts of language services as well as situations that should send up red flags. The following quote suggests that LEP individuals (as well as others with special needs) should not be considered as a separate group, one that is outside the mainstream:

We hear it all the time – "special needs" and "vulnerable." Both terms do damage. When people with disabilities are thought of as "special," they are often thought of as marginal individuals who have needs, not rights. The word "vulnerable" has a similarly unfortunate effect. Vulnerable people must have things done for them; they're recipients, not participants.

Don't think 'special' or 'vulnerable;' think 'universal access.' Integrate access into all aspects of emergency services: transportation, sheltering, education, evacuation, etc. And remember that access is a civil right, not a favor or an amenity.

-CT P&A

(Slide 26 of 42) http://www.fema.gov/pdf/about/odic/all_hands_0411.pdf

It is true that the attitude/approach cited in the preceding quotation should be the dominant one. It certainly represents the spirit of CPG 101 Version 2. However, it is clear from the information provided in this chapter that this is not universally the case. We must always ask two questions:

(1) Who are the people who require services?
(2) What do they need?

After identifying the target populations, we must *communicate* with them not only to ascertain their needs, but to foster an ongoing dialog that will be mutually beneficial to all involved. LEP speakers will not be part of the mainstream until we have accomplished a great deal more in the area of language services. "It is time children, people with disabilities or any other segment of our communities who have traditionally been underserved, to be more fully and consistently integrated into preparedness and planning efforts at every level of government."

Craig Fugate, FEMA Administrator (Slide 37 of 42) http://www.fema.gov/pdf/about/odic/all_hands_0411.pdf

FEMA's key goal is to better serve *all* communities and *all* community members. Innovative thinking allows for partnering with business entities like "big box" stores to provide efficient services in times of crisis and afterwards. "Outside the box" thinking typifies Craig Fugate's approach to EM. As the former EM Director for the multilingual State of Florida, Mr. Fugate knows that full communication with all stakeholders is an essential requirement if all communities and community members are to receive prompt, efficient and even-handed service. Mr. Fugate's experience at the state level will encourage and support "outside the box" thinking and solutions that will also be part of EM preparedness, mitigation, response and recovery in terms of LEP issues on the national stage. The challenge will be to spread Mr. Fugate's inclusive approach regarding LEP issues to every state and local government. As this chapter illustrates, FEMA's views reflect and reinforce Federal legal requirements to ensure that all people, regardless of language barriers, be fully included in all phases of emergency management.

QUESTIONS

1. As you reflect on this chapter, what governmental decisions and ethnic minority social group pressures may influence future needs for LEP services?
2. Currently, many politicians and other officials are proposing that services which have traditionally been provided by government entities be contracted to private companies. Describe how LEP services might be furnished in the future and who might pay for them.
3. What suggestions do you have for strategies to recruit members of linguistic and ethnic minority groups to the EM field? Divide the class into small groups and create an advertising campaign/public service announcement that showcases your ideas.
4. Are there LEP groups living in your local community? If so, which ones? Do some research on the procedures in place to work in your area with these special needs populations in all four phases of EM.
5. Do you agree or disagree with the quote in the Conclusion about not referring to LEP populations as "special needs" groups? Defend your position.
6. In these difficult times, emergency managers are asked to do more with less. What suggestions do you have for your financially-strapped local emergency manager to best deal with the challenges of providing the legally-required services to LEP populations?
7. What are your thoughts on the usefulness of social networking sites to reach LEP groups in an emergency/disaster? Use the Virginia Tech shootings as a jumping-off point for your discussion.

8. What legal advantages result from complying with CPG 101's requirement to incorporate LEP populations in all aspects of planning?

9. What are the potential legal ramifications of not providing access to language services for LEP groups?

Persons interested in this material may wish to access the following recent release:

Federal Coordination and Compliance Section
Civil Rights Division
U.S. Department of Justice

September 2011

Consideration for Providing Language Access in a Prosecutorial Agency found on line at http://www.lep.gov/resources/092111_Prosecutors_Planning_Tool.pdf

Appendix A

PDF TEXT OF EXECUTIVE ORDER 13166 AND ACCOMPANYING *FEDERAL REGISTER* COMMENTARY (AUGUST 16, 2000)

Federal Register

Wednesday,
August 16, 2000

Part V

The President

Executive Order 13166—Improving Access to Services for Persons With Limited English Proficiency

---

Department of Justice

Enforcement of Title VI of the Civil Rights Act of 1964—National Origin Discrimination Against Persons With Limited English Proficiency; Notice

50121

Federal Register

Vol. 65, No. 159

Wednesday, August 16, 2000

Presidential Documents

Title 3—

The President

Executive Order 13166 of August 11, 2000

Improving Access to Services for Persons With Limited English Proficiency

By the authority vested in me as President by the Constitution and the laws of the United States of America, and to improve access to federally conducted and federally assisted programs and activities for persons who, as a result of national origin, are limited in their English proficiency (LEP), it is hereby ordered as follows:

Section 1. *Goals.*

The Federal Government provides and funds an array of services that can be made accessible to otherwise eligible persons who are not proficient in the English language. The Federal Government is committed to improving the accessibility of these services to eligible LEP persons, a goal that reinforces its equally important commitment to promoting programs and activities designed to help individuals learn English. To this end, each Federal agency shall examine the services it provides and develop and implement a system by which LEP persons can meaningfully access those services consistent with, and without unduly burdening, the fundamental mission of the agency. Each Federal agency shall also work to ensure that recipients of Federal financial assistance (recipients) provide meaningful access to their LEP applicants and beneficiaries. To assist the agencies with this endeavor, the Department of Justice has today issued a general guidance document (LEP Guidance), which sets forth the compliance standards that recipients must follow to ensure that the programs and activities they normally provide in English are accessible to LEP persons and thus do not discriminate on the basis of national origin in violation of title VI of the Civil Rights Act of 1964, as amended, and its implementing regulations. As described in the LEP Guidance, recipients must take reasonable steps to ensure meaningful access to their programs and activities by LEP persons.

Sec. 2. *Federally Conducted Programs and Activities.*

Each Federal agency shall prepare a plan to improve access to its federally conducted programs and activities by eligible LEP persons. Each plan shall be consistent with the standards set forth in the LEP Guidance, and shall include the steps the agency will take to ensure that eligible LEP persons can meaningfully access the agency's programs and activities. Agencies shall develop and begin to implement these plans within 120 days of the date of this order, and shall send copies of their plans to the Department of Justice, which shall serve as the central repository of the agencies' plans.

Sec. 3. *Federally Assisted Programs and Activities.*

Each agency providing Federal financial assistance shall draft title VI guidance specifically tailored to its recipients that is consistent with the LEP Guidance issued by the Department of Justice. This agency-specific guidance shall detail how the general standards established in the LEP Guidance will be applied to the agency's recipients. The agency-specific guidance shall take into account the types of services provided by the recipients, the individuals served by the recipients, and other factors set out in the LEP Guidance. Agencies that already have developed title VI guidance that the Department of Justice determines is consistent with the LEP Guidance shall examine their existing guidance, as well as their programs and activities, to determine if additional guidance is necessary to comply with this order. The Department of Justice shall consult with the agencies in creating their guidance and, within 120 days of the date of this order,

50122   Federal Register / Vol. 65, No. 159 / Wednesday, August 16, 2000 / Presidential Documents

each agency shall submit its specific guidance to the Department of Justice for review and approval. Following approval by the Department of Justice, each agency shall publish its guidance document in the **Federal Register** for public comment.

Sec. 4. *Consultations.*

In carrying out this order, agencies shall ensure that stakeholders, such as LEP persons and their representative organizations, recipients, and other appropriate individuals or entities, have an adequate opportunity to provide input. Agencies will evaluate the particular needs of the LEP persons they and their recipients serve and the burdens of compliance on the agency and its recipients. This input from stakeholders will assist the agencies in developing an approach to ensuring meaningful access by LEP persons that is practical and effective, fiscally responsible, responsive to the particular circumstances of each agency, and can be readily implemented.

Sec. 5. *Judicial Review.*

This order is intended only to improve the internal management of the executive branch and does not create any right or benefit, substantive or procedural, enforceable at law or equity by a party against the United States, its agencies, its officers or employees, or any person.

William J. Clinton

THE WHITE HOUSE,
August 11, 2000.

[FR Doc. 00–20938
Filed 8–15–00; 8:45 am]
Billing code 3195–01–P

DEPARTMENT OF JUSTICE

Enforcement of Title VI of the Civil Rights Act of 1964—National Origin Discrimination Against Persons With Limited English Proficiency; Policy Guidance

AGENCY: Civil Rights Division, Department of Justice.

ACTION: Policy guidance document.

SUMMARY: This Policy Guidance Document entitled "Enforcement of Title VI of the Civil Rights Act of 1964 " National Origin Discrimination Against Persons with Limited English Proficiency (LEP Guidance)" is being issued pursuant to authority granted by Executive Order 12250 and Department of Justice Regulations. It addresses the application of Title VI's prohibition on national origin discrimination when information is provided only in English to persons with limited English proficiency. This policy guidance does not create new obligations, but rather, clarifies existing Title VI responsibilities. The purpose of this document is to set forth general principles for agencies to apply in developing guidelines for services to individuals with limited English proficiency. The Policy Guidance Document appears below.

DATES: Effective August 11, 2000.

ADDRESSES: Coordination and Review Section, Civil Rights Division, P.O. Box 66560, Washington, D.C. 20035–6560.

FOR FURTHER INFORMATION CONTACT: Merrily Friedlander, Chief, Coordination and Review Section, Civil Rights Division, (202) 307–2222.

Helen L. Norton,

Counsel to the Assistant Attorney General, Civil Rights Division.

Office of the Assistant Attorney General

Washington, D.C. 20530

August 11, 2000.

TO: Executive Agency Civil Rights Officers

FROM: Bill Lann Lee, Assistant Attorney General, Civil Rights Division

SUBJECT: Policy Guidance Document: *Enforcement of Title VI of the Civil Rights Act of 1964—National Origin Discrimination Against Persons With Limited English Proficiency* ("LEP Guidance")

This policy directive concerning the enforcement of Title VI of the Civil Rights Act of 1964, 42 U.S.C. §§ 2000d *et seq., as amended,* is being issued pursuant to the authority granted by

Executive Order No. 12250 [1] and Department of Justice regulations.[2] It addresses the application to recipients of federal financial assistance of Title VI's prohibition on national origin discrimination when information is provided only in English to persons who do not understand English. This policy guidance does not create new obligations but, rather, clarifies existing Title VI responsibilities.

Department of Justice Regulations for the Coordination of Enforcement of Non-discrimination in Federally Assisted Programs (Coordination Regulations), 28 C.F.R. 42.401 *et seq.,* direct agencies to "publish title VI guidelines for each type of program to which they extend financial assistance, where such guidelines would be appropriate to provide detailed information on the requirements of Title VI." 28 CFR § 42.404(a). The purpose of this document is to set forth general principles for agencies to apply in developing such guidelines for services to individuals with limited English proficiency (LEP). It is expected that, in developing this guidance for their federally assisted programs, agencies will apply these general principles, taking into account the unique nature of the programs to which they provide federal financial assistance.

A federal aid recipient's failure to assure that people who are not proficient in English can effectively participate in and benefit from programs and activities may constitute national origin discrimination prohibited by Title VI. In order to assist agencies that grant federal financial assistance in ensuring that recipients of federal financial assistance are complying with their responsibilities, this policy directive addresses the appropriate compliance standards. Agencies should utilize the standards set forth in this Policy Guidance Document to develop specific criteria applicable to review the programs and activities for which they offer financial assistance. The Department of Education [3] already has

---

[1] 42 U.S.C. § 2000d–1 note.

[2] 28 C.F.R. § 0.51.

[3] Department of Education policies regarding the Title VI responsibilities of public school districts with respect to LEP children and their parents are reflected in three Office for Civil Rights policy documents: (1) the May 1970 memorandum to school districts, "Identification of Discrimination and Denial of Services on the Basis of National Origin," (2) the December 3, 1985, guidance document, "The Office for Civil Rights' Title VI Language Minority Compliance Procedures," and (3) the September 1991 memorandum, "Policy Update on Schools Obligations Toward National Origin Minority Students with Limited English Proficiency." These documents can be found at the Department of Education website at www.ed.gov/office/OCR.

established policies, and the Department of Health and Human Services (HHS) [4] has been developing guidance in a manner consistent with Title VI and this Document, that applies to their specific programs receiving federal financial assistance.

Background

Title VI of the Civil Rights Act of 1964 prohibits recipients of federal financial assistance from discriminating against or otherwise excluding individuals on the basis of race, color, or national origin in any of their activities. Section 601 of Title VI, 42 U.S.C. § 2000d, provides:

No person in the United States shall, on the ground of race, color, or national origin, be excluded from participation in, be denied the benefits of, or be subjected to discrimination under any program or activity receiving Federal financial assistance.

The term "program or activity" is broadly defined. 42 U.S.C. § 2000d–4a.

Consistent with the model Title VI regulations drafted by a Presidential task force in 1964, virtually every executive agency that grants federal financial assistance has promulgated regulations to implement Title VI. These regulations prohibit recipients from "restrict[ing] an individual in any way in the enjoyment of any advantage or privilege enjoyed by others receiving any service, financial aid, or other benefit under the program" and "utiliz[ing] criteria or methods of administration which have the effect of subjecting individuals to discrimination" or have "the effect of defeating or substantially impairing accomplishment of the objectives of the program as respects individuals of a particular race, color, or national origin."

In *Lau* v. *Nichols,* 414 U.S. 563 (1974), the Supreme Court interpreted these provisions as requiring that a federal financial recipient take steps to ensure that language barriers did not exclude LEP persons from effective participation in its benefits and services. *Lau* involved a group of students of Chinese origin who did not speak English to whom the recipient provided the same services—an education provided solely in English—that it provided students who did speak English. The Court held that, under these circumstances, the school's practice violated the Title VI prohibition against discrimination on

---

[4] The Department of Health and Human Services is issuing policy guidance titled: "Title VI Prohibition Against National Origin Discrimination As It Affects Persons With Limited English Proficiency." This policy addresses the Title VI responsibilities of HHS recipients to individuals with limited English proficiency.

50124   **Federal Register**/ Vol. 65, No. 159/ Wednesday, August 16, 2000/ Notices

the basis of national origin. The Court observed that "[i]t seems obvious that the Chinese-speaking minority receive fewer benefits than the English-speaking majority from respondents' school system which denies them a meaningful opportunity to participate in the educational program—all earmarks of the discrimination banned by" the Title VI regulations.[5] Courts have applied the doctrine enunciated in *Lau* both inside and outside the education context. It has been considered in contexts as varied as what languages drivers' license tests must be given in or whether material relating to unemployment benefits must be given in a language other than English.[6]

Link Between National Origin And Language

For the majority of people living in the United States, English is their native language or they have acquired proficiency in English. They are able to participate fully in federally assisted programs and activities even if written and oral communications are exclusively in the English language.

The same cannot be said for the remaining minority who have limited English proficiency. This group includes persons born in other countries, some children of immigrants born in the United States, and other non-English or limited English proficient persons born in the United States, including some Native Americans. Despite efforts to learn and master English, their English language proficiency may be limited for some time.[7] Unless grant recipients take steps to respond to this difficulty, recipients effectively may deny those who do not

speak, read, or understand English access to the benefits and services for which they qualify.

Many recipients of federal financial assistance recognize that the failure to provide language assistance to such persons may deny them vital access to services and benefits. In some instances, a recipient's failure to remove language barriers is attributable to ignorance of the fact that some members of the community are unable to communicate in English, to a general resistance to change, or to a lack of awareness of the obligation to address this obstacle.

In some cases, however, the failure to address language barriers may not be simply an oversight, but rather may be attributable, at least in part, to invidious discrimination on the basis of national origin and race. While there is not always a direct relationship between an individual's language and national origin, often language does serve as an identifier of national origin.[8] The same sort of prejudice and xenophobia that may be at the root of discrimination against persons from other nations may be triggered when a person speaks a language other than English.

Language elicits a response from others, ranging from admiration and respect, to distance and alienation, to ridicule and scorn. Reactions of the latter type all too often result from or initiate racial hostility * * *. It may well be, for certain ethnic groups and in some communities, that proficiency in a particular language, like skin color, should be treated as a surrogate for race under an equal protection analysis.[9]

While Title VI itself prohibits only intentional discrimination on the basis of national origin,[10] the Supreme Court has consistently upheld agency regulations prohibiting unjustified discriminatory effects.[11] The Department of Justice has consistently adhered to the view that the significant

discriminatory effects that the failure to provide language assistance has on the basis of national origin, places the treatment of LEP individuals comfortably within the ambit of Title VI and agencies' implementing regulations.[12] Also, existing language barriers potentially may be rooted in invidious discrimination. The Supreme Court in *Lau* concluded that a recipient's failure to take affirmative steps to provide "meaningful opportunity" for LEP individuals to participate in its programs and activities violates the recipient's obligations under Title VI and its regulations.

All Recipients Must Take Reasonable Steps To Provide Meaningful Access

Recipients who fail to provide services to LEP applicants and beneficiaries in their federally assisted programs and activities may be discriminating on the basis of national origin in violation of Title VI and its implementing regulations. Title VI and its regulations require recipients to take reasonable steps to ensure "meaningful" access to the information and services they provide. What constitutes reasonable steps to ensure meaningful access will be contingent on a number of factors. Among the factors to be considered are the number or proportion of LEP persons in the eligible service population, the frequency with which LEP individuals come in contact with the program, the importance of the service provided by the program, and the resources available to the recipient.

(1) Number or Proportion of LEP Individuals

Programs that serve a few or even one LEP person are still subject to the Title VI obligation to take reasonable steps to provide meaningful opportunities for access. However, a factor in determining the reasonableness of a recipient's efforts is the number or proportion of people who will be excluded from the benefits or services absent efforts to remove language barriers. The steps that are reasonable for a recipient who serves one LEP person a year may be different than those expected from a recipient that serves several LEP persons each day. But even those who serve very few LEP persons on an infrequent basis should utilize this balancing analysis to determine whether reasonable steps are

---

[5] 414 U.S. at 568. Congress manifested its approval of the *Lau* decision requirements concerning the provision of meaningful education services by enacting provisions in the Education Amendments of 1974, Pub. L. No. 93–380, §§ 105, 204, 88 Stat. 503–512, 515 codified at 20 U.S.C. 1703(f), and the Bilingual Education Act, 20 U.S.C. 7401 *et seq.*, which provided federal financial assistance to school districts in providing language services.

[6] For cases outside the educational context, *see, e.g., Sandoval v. Hagan,* 7 F. Supp. 2d 1234 (M.D. Ala. 1998), *affirmed,* 197 F.3d 484, (11th Cir. 1999), *rehearing and suggestion for rehearing en banc denied,* 211 F.3d 133 (11th Cir. Feb. 29, 2000) (Table, No. 98–6598–II), *petition for certiorari filed* May 30, 2000 (No. 99–1908) (giving drivers' license tests only in English violates Title VI); and *Pabon v. Levine,* 70 F.R.D. 674 (S.D.N.Y. 1976) (summary judgment for defendants denied in case alleging failure to provide unemployment insurance information in Spanish violated Title VI).

[7] Certainly it is important to achieve English language proficiency in order to fully participate at every level in American society. As we understand the Supreme Court's interpretation of Title VI's prohibition of national origin discrimination, it does not in any way disparage use of the English language.

[8] As the Supreme Court observed, "[l]anguage permits an individual to express both a personal identity and membership in a community, and those who share a common language may interact in ways more intimate than those without this bond." *Hernandez v. New York,* 500 U.S. 352, 370 (1991) (plurality opinion).

[9] *Id.* at 371 (plurality opinion).

[10] *Alexander v. Choate,* 469 U.S. 287, 293 (1985).

[11] At 293–294; *Guardians Ass'n v. Civil Serv. Comm'n,* 463 U.S. 582, 584 n.2 (1983) (White, J.), 623 n.15 (Marshall, J.), 642–645 (Stevens, Brennan, Blackmun, JJ.); *Lau v. Nichols,* 414 U.S. at 568; *id.* at 571 (Stewart, J., concurring in result). In a July 24, 1994, memorandum to Heads of Departments and Agencies that Provide Federal Financial Assistance concerning "Use of the Disparate Impact Standard in Administrative Regulations Under Title VI of the Civil Rights Act of 1964," the Attorney General stated that each agency "should ensure that the disparate impact provisions of your regulations are fully utilized so that all persons may enjoy equally the benefits of federally financed programs."

[12] The Department's position with regard to written language assistance is articulated in 28 CFR § 42.405(d)(1), which is contained in the Coordination Regulations, 28 CFR Subpt. F, issued in 1976. These Regulations "govern the respective obligations of Federal agencies regarding enforcement of title VI." 28 CFR § 42.405. Section 42.405(d)(1) addresses the prohibitions cited by the Supreme Court in *Lau.*

possible and if so, have a plan of what to do if a LEP individual seeks service under the program in question. This plan need not be intricate; it may be as simple as being prepared to use one of the commercially available language lines to obtain immediate interpreter services.

(2) Frequency of Contact with the Program

Frequency of contacts between the program or activity and LEP individuals is another factor to be weighed. For example, if LEP individuals must access the recipient's program or activity on a daily basis, *e.g.*, as they must in attending elementary or secondary school, a recipient has greater duties than if such contact is unpredictable or infrequent. Recipients should take into account local or regional conditions when determining frequency of contact with the program, and should have the flexibility to tailor their services to those needs.

(3) Nature and Importance of the Program

The importance of the recipient's program to beneficiaries will affect the determination of what reasonable steps are required. More affirmative steps must be taken in programs where the denial or delay of access may have life or death implications than in programs that are not as crucial to one's day-to-day existence. For example, the obligations of a federally assisted school or hospital differ from those of a federally assisted zoo or theater. In assessing the effect on individuals of failure to provide language services, recipients must consider the importance of the benefit to individuals both immediately and in the long-term. A decision by a federal, state, or local entity to make an activity compulsory, such as elementary and secondary school attendance or medical inoculations, serves as strong evidence of the program's importance.

(4) Resources Available

The resources available to a recipient of federal assistance may have an impact on the nature of the steps that recipients must take. For example, a small recipient with limited resources may not have to take the same steps as a larger recipient to provide LEP

assistance in programs that have a limited number of eligible LEP individuals, where contact is infrequent, where the total cost of providing language services is relatively high, and/or where the program is not crucial to an individual's day-to-day existence. Claims of limited resources from large entities will need to be well-substantiated.[13]

Written vs. Oral Language Services

In balancing the factors discussed above to determine what reasonable steps must be taken by recipients to provide meaningful access to each LEP individual, agencies should particularly address the appropriate mix of written and oral language assistance. Which documents must be translated, when oral translation is necessary, and whether such services must be immediately available will depend upon the factors previously mentioned.[14] Recipients often communicate with the public in writing, either on paper or over the Internet, and written translations are a highly effective way of communicating with large numbers of

---

[13] Title VI does not require recipients to remove language barriers when English is an essential aspect of the program (such as providing civil service examinations in English when the job requires person to communicate in English, *see Frontera* v. *Sindell*, 522 F.2d 1215 (6th Cir. 1975)), or there is another "substantial legitimate justification for the challenged practice." *Elston* v. *Talladega County Bd. of Educ.*, 997 F.2d 1394, 1407 (11th Cir. 1993). Similar balancing tests are used in other nondiscrimination provisions that are concerned with effects of an entity's actions. For example, under Title VII of the Civil Rights Act of 1964, employers need not cease practices that have a discriminatory effect if they are "consistent with business necessity" and there is no "alternative employment practice" that is equally effective. 42 U.S.C. § 2000e–2(k). Under Section 504 of the Rehabilitation Act, 29 U.S.C. § 794, recipients do not need to provide access to persons with disabilities if such steps impose an undue burden on the recipient. *Alexander* v. *Choate*, 469 U.S. at 300. Thus, in situations where all of the factors identified in the text are at their nadir, it may be "reasonable" to take no affirmative steps to provide further access.

[14] Under the four-part analysis, for instance, Title VI would not require recipients to translate documents requested under a state equivalent of the Freedom of Information Act or Privacy Act, or to translate all state statutes or notices of rulemaking made generally available to the public. The focus of the analysis is the nature of the information being communicated, the intended or expected audience, and the cost of providing translations. In virtually all instances, one or more of these criteria would lead to the conclusion that recipients need not translate these types of documents.

people who do not speak, read or understand English. While the Department of Justice's Coordination Regulation, 28 CFR § 42.405(d)(1), expressly addresses requirements for provision of written language assistance, a recipient's obligation to provide meaningful opportunity is not limited to written translations. Oral communication between recipients and beneficiaries often is a necessary part of the exchange of information. Thus, a recipient that limits its language assistance to the provision of written materials may not be allowing LEP persons "effectively to be informed of or to participate in the program" in the same manner as persons who speak English.

In some cases, "meaningful opportunity" to benefit from the program requires the recipient to take steps to assure that translation services are promptly available. In some circumstances, instead of translating all of its written materials, a recipient may meet its obligation by making available oral assistance, or by commissioning written translations on reasonable request. It is the responsibility of federal assistance-granting agencies, in conducting their Title VI compliance activities, to make more specific judgments by applying their program expertise to concrete cases.

Conclusion

This document provides a general framework by which agencies can determine when LEP assistance is required in their federally assisted programs and activities and what the nature of that assistance should be. We expect agencies to implement this document by issuing guidance documents specific to their own recipients as contemplated by the Department of Justice Coordination Regulations and as HHS and the Department of Education already have done. The Coordination and Review Section is available to assist you in preparing your agency-specific guidance. In addition, agencies should provide technical assistance to their recipients concerning the provision of appropriate LEP services.

[FR Doc. 00–20867 Filed 8–15–00; 8:45 am]

BILLING CODE 4410–13–P

Appendix B

MEMORANDUM FROM ATTORNEY GENERAL ERIC HOLDER REGARDING RENEWED ENFORCEMENT OF THE CIVIL RIGHTS ACT OF 1965 AND EO 13166 (FEBRUARY 17, 2011)

MEMORANDUM FOR: HEADS OF FEDERAL AGENCIES, GENERAL COUNSELS, AND CIVIL RIGHTS HEADS

FROM: THE ATTORNEY GENERAL

February 17, 2011

SUBJECT: Federal Government's Renewed Commitment to Language Access Obligations Under Executive Order 13166

Executive Order 13166[1] was issued in August of 2000 and this memorandum reaffirms its mandate. The Executive Order has two primary parts. First, it directs each federal agency to develop and implement a system by which limited English proficient (LEP) persons can meaningfully access the agency's services. Second, it directs each agency providing federal financial assistance to issue guidance to recipients of such assistance on their legal obligations to take reasonable steps to ensure meaningful access for LEP persons under the national origin nondiscrimination provisions of Title VI of the Civil Rights Act of 1964, and implementing regulations.

Whether in an emergency or in the course of routine business matters, the success of government efforts to effectively communicate with members of the public depends on the widespread and nondiscriminatory availability of accurate, timely, and vital information. Events such as the HINI influenza pandemic, Hurricanes Katrina and Rita, the Gulf Oil Spill, and the 2010 Decennial Census highlight the need for federal agencies to ensure language access both in their own activities, as well as in those of the recipients of federal financial assistance.

Despite the legal and public service obligations that compel federal agencies and recipients to ensure language access, a 2006 language access survey of the federal government revealed significant variations in the extent to which federal agencies are aware of, and in compliance with, principles of language access. This conclusion is buttressed by an April 2010 Government Accountability Office (GAO) report on language access at federal agencies. That report offers concrete suggestions, some of which are incorporated in this memorandum, for improving our efforts to comply with Executive Order 13166. Further, federal interagency language access conferences held over the last few years reveal that, while the federal government as a whole has taken commendable strides toward providing language access in certain areas, the implementation of comprehensive language access programs remains uneven throughout the federal government and among recipients of federal financial assistance, especially in the face of limited resources and personnel.

In an effort to secure the federal government's full compliance with Executive Order 13166, and under the Department of Justice's (DOJ's) coordination authority conferred by Executive Order 12250, I request that your agency join DOJ in recommitting to the implementation of Executive Order 13166 by undertaking the following action items:

(1) Establish a Language Access Working Group that reflects your agency's organizational structure and is responsible for implementing the federally conducted and

---

[1]65 Fed. Reg. 50,121 (Aug. 16, 2000).

federally assisted provisions of the Executive Order.

(2) Evaluate and/or update your current response to LEP needs by, among other things, conducting an inventory of languages most frequently encountered, identifying the primary channels of contact with LEP community members (whether telephonic, in person, correspondence, web-based, etc.), and reviewing agency programs and activities for language accessibility.

(3) Establish a schedule to periodically evaluate and update federal agency LEP services and LEP policies, plans, and protocols. As an initial step, within six months after the date of this memorandum, submit updated LEP plans and an anticipated time frame for periodic reevaluation of LEP plans and related documents to the Federal Coordination and Compliance Section (previously named the Coordination and Review Section) of DOJ's Civil Rights Division.

(4) Ensure that agency staff can competently identify LEP contact situations and take the necessary steps to provide meaningful access.

(5) Notify the public, through mechanisms that will reach the LEP communities you serve, of your LEP policies, plans, and procedures, and LEP access-related developments. Provide a link to materials posted on your website to the Federal Coordination and Compliance Section so that it can be posted on LEP.gov.

(6) When considering hiring criteria, assess the extent to which non-English language proficiency would be necessary for particular positions or to fulfill your agency's mission.

(7) For written translations, collaborate with other agencies to share resources, improve efficiency, standardize federal terminology, and streamline processes for obtaining community feedback on the accuracy and quality of professional translations intended for mass distribution.

(8) For agencies providing federal financial assistance, draft recipient guidance. Note that such assistance is broadly defined to include not only financial grants, but also equipment, property, rental below fair market value, training, and other forms of assistance. Agencies that have not already done so should issue recipient guidance on compliance with language access obligations, and submit that guidance to the Federal Coordination and Compliance Section of DOJ's Civil Rights Division within six months after the date of this memorandum. Agencies that have determined that they do not provide federal financial assistance and, therefore, do not need to issue recipient guidance, should include a statement of this determination when transmitting the federally conducted language access plan.[2] Federal funding agencies should also regularly review recipient compliance, and provide vigorous technical assistance and enforcement action in appropriate cases.

DOJ's Civil Rights Division, in cooperation with the Federally Conducted Committee of the Interagency Working Group on Limited English Proficiency, will undertake periodic monitoring of these action items through follow-up language access surveys of the type distributed in 2006. Agencies should expect the first of these follow-up surveys in 2011.

For your convenience, the addendum to this memorandum contains a variety of useful information, including links to resources and further guidance on some of the action items outlined above. Should you require further technical assistance or support in implementing the goals of Executive Order 13166, please do not hesitate to

---

[2] Agencies disputing coverage under the Executive Order's provision relating to federally conducted programs and activities should file with the Department a report indicating the basis for disputing coverage, the number of contacts they have had with LEP individuals, the frequency of such contacts, and the nature and importance of such contacts. The report should capture phone contacts, in person contacts, correspondence, and all other interactions with LEP individuals (including via agency websites). Finally. the report should describe the standards such agencies are using to determine LEP status.

contact Christine Stoneman, Special Legal Counsel, or Bharathi Venkatraman, Attorney, at the Federal Coordination and Compliance Section, at (202) 307-2222. Thank you for your continued commitment to ensuring that federal resources and services are available and accessible to the LEP community and the public as a whole.

SUPPLEMENT TO THE ATTORNEY GENERAL'S MEMORANDUM TO FEDERAL AGENCIES ON EXECUTIVE ORDER 13166 COMPLIANCE

SPECIFICS OF IMPLEMENTATION FOR THE ACTION ITEMS

1. *Action Item:* Each agency should establish a Language Access Working Group that reflects its organizational structure and is responsible for implementing the federally conducted and federally assisted provisions of the Executive Order.

   Specifics: The Working Group should be chaired by an LEP Coordinator who reports to a designee of the Secretary (or to a designee of a Secretary-level official in charge of the agency). The Working Group should be comprised of individuals from multiple components or operational subdivisions of the agency, and should include members from field offices, as appropriate. Members of the Working Group should be responsible for identifying barriers to language access, consulting with stakeholders, formulating strategies and responses to overcome the barriers to meaningful language access, ensuring consistency within the agency on its federally assisted enforcement activities. They also should be accountable for implementation. Staff should also be apprised of the agency's Language Access Working Group and its mission.

2. *Action Item:* Each agency should evaluate and/or update its current response to LEP needs by, among other things, conducting an inventory of languages most frequently encountered, identifying the primary channels of contact with LEP community members (whether telephonic, in person, correspondence, web-based, etc.), and reviewing agency programs and activities for language accessibility.

   Specifics: Agencies may need to update program operations, services provided, outreach activities, and other mission-specific activities to reflect current language needs. Further, each agency should ensure that its in-house and contract language services, directory of translated documents, signs, and web-based services meet current language needs.

3. *Action Item:* Each agency should establish a schedule to periodically evaluate and update agency LEP services and LEP policies, plans, and protocols. As an initial step, updated LEP plans and an anticipated timeframe for periodic reevaluation of LEP plans and related documents should be submitted within six months after the date of this memorandum to the Federal Coordination and Compliance Section of the Department of Justice's (DOJ's) Civil Rights Division.

   Specifics: Requested information can be sent to the Federal Coordination and Compliance Section at 950 Pennsylvania Avenue, NW (NW Bldg), Washington, D.C. 20530, Attention: Christine Stoneman and Bharathi Venkatraman. You may also email information to christine.stoneman@usdoj.gov or bharathi.a.venkatraman @usdoj.gov. Note that an agency's contemplated schedule should not serve to bar the agency from conducting more frequent inventories/reinventories of languages encountered to ensure that agency services are meeting current language needs and demands.

4. *Action Item:* Agencies should ensure that staff can competently identify LEP contact situations and take the necessary steps to provide meaningful access.

Specifics: Agency staff should be able to, among other tasks, identify LEP contact situations, determine primary language of LEP individuals, and effectively utilize available options to assist Iin interpersonal, electronic, print, and other methods of communication between the agency and LEP individuals.

5. *Action Item:* Agencies should notify the public, through mechanisms that will reach the LEP communities it serves, of its LEP policies and LEP access-related developments.

Specifics: Examples of methods for publicizing LEP access information include, but are not limited to, posting on agency websites, issuing print and broadcast notifications, providing relevant information at "town hall" style meetings, and issuing press releases. Agencies should consult with their information technology specialists, civil rights personnel, and public affairs personnel to develop a multipronged strategy to achieve maximum and effective notification to LEP communities.

6. *Action Item:* When considering hiring criteria, agencies should assess the extent to which non-English language proficiency would be necessary for particular positions or to fulfill an agency's mission.

Specifics: Determine whether the agency would benefit from including non-English language skills and competence thresholds in certain job vacancy announcements and position descriptions.

7. *Action Item:* For written translations, collaborate with other agencies to share resources, improve efficiency, standardize federal terminology, and streamline processes for obtaining community feedback on the accuracy and quality of professional translations intended for mass distribution.

Specifics: Agencies should actively participate in the Interagency Working Group's efforts to develop collaborations and clearinghouse options to produce high quality and effective translations. While improving efficiency is a priority, ensuring the quality of translations is equally, if not more, important. As such, agencies should avoid pursuing free translations from community groups. Rather, community input can serve to ensure that professional translations meet community needs and are appropriate to the audience.

8. *Action Item:* For agencies providing federal financial assistance, draft recipient guidance.

Specifics: Agencies should refer to the DOJ Recipient Guidance document and LEP.gov, both of which are referenced in the Resources section below, for templates. Agencies should submit their recipient guidance documents for review and approval to the Federal Coordination and Compliance Section of *DOJ's* Civil Rights Division, at 950 Pennsylvania Avenue, NW (NW Bldg), Washington, D.C. 20530, Attention: Christine Stoneman and Bharathi Venkatraman. You may also email agency recipient guidance to christine.stoneman@usdoj.gov or bharathi.a. venkatraman@usdoj.gov.

Resources

Executive Order 13166: http://www.justice.gov/crt/cor/Pubs/eolep.pdf

DOJ LEP Guidance: http://www.justice.gov/crt/cor/lep/DOJFinLEPFRJunI82002.php

Website of the Federal Interagency Working Group on LEP: http://www.lep.gov

Top Tips from responses to the 2006 language access survey of federal agencies: http://www.lep.gov/resources/2008 _Conference _ Materials/TopTips.pdf

The 2006 Language Access Survey: http://www.lep.gov/resources/2008_Conference_Materials/FedLangAccessSurvey.pdf

GSA Language Services Schedule: http://www.gsa.gov/portal/contentlI04610

I Speak Language Identification flashcards: http://www. lep.gov/ISpeakCards2004.pdf

LEP rights brochure: http://www.lep.gov/resources/lep_aug2005.pdf

References

2010 Census Shows America's Diversity (U.S. Census Bureau). (2011). http://2010census.gov/news/releases/operations/cb11-cn125.html

American Community Survey. (2009). S1602: Linguistic Isolation. Data Set: 2009 American Community Survey 1-Year Estimates. http://factfinder.census.gov/servlet/STTable?_bm=y&-geo_id=01000US&-qr_name=ACS_2009_1YR_G00_S1602&-ds_name=ACS_2009_1YR_G00_

Bates, K. A. & Swan, R. S. (Eds.). (2007). *Through the eyes of Katrina: Social justice in the United States.* Durham, NC: Carolina Academic Press.

Bawn, L. & Peterson, C. (2008). Iowa's multilingual emergency response network. *American Translators Association (ATA) Chronicle* (April, pp. 20–22).

Bostrum, M. (2006). Language is barrier to 911 help. *Durham News & Observer* (NC), 7B.

Boyle, P. (2001). Bethlehem police learning to hable español. *The Morning Call* (Allentown, PA). B3, May 1.

Bridging the gap. (2009). *Proteus 18/2:* 23 (Available at www.najit.org).

Burgener, R. (Ed.). 2009. *Communication challenges during delivery of emergency services: Is there a problem?* Research Report: International. Connections (INTERNECT) www.internect.org

Comprehensive Preparedness Guide (CPG) 101, Version 2.0. 2010. Developing and Maintaining Emergency Operations Plans. http://www.fema.gov/pdf/about/divisions/npd/CPG_101_V2.pdf

Comprehensive Preparedness Guide (CPG) 101, Version 1.0. 2009. Developing and Maintaining State, Territorial, Tribal and Local Government Emergency Plans.

Comprehensive Preparedness Guide (CPG) 301, Version 1.0. 2008. Interim Emergency Management Planning Guide for Special Needs Populations.

Conroy, M. (2008). Reaching out to LOTS in Nebraska. *The ATA Chronicle* (November/December) 14–16.

Coppola, D. P. (2007). *Introduction to international disaster management.* Burlington, MA: Butterworth-Heinemann (Elsevier).

Coppola, D. P. & Maloney, E. K. (2009). *Communicating emergency preparedness.* Boca Raton, FL: CRC Press.

Croon, C. (2009). Our assumptions about language abilities can lead us astray. In R. Burgener, (Ed.). *Communication challenges during delivery of emergency services: Is there a problem?* 39–42. Available online at: www.internect.com.

DeRouchey, B. (2011). Personal communication (April 26).

DeRouchey, B. (2010). Personal communication (July 6).

Drabek, T. E. (2010). *The human side of disaster.* Boca Raton, FL: CRC Press.

Enarson, E. (2007). Identifying and addressing social vulnerabilities. In W. Waugh & K. Tierney, (Eds.). *Emergency management: Principles and practice for local government* (2nd Ed.). pp. 257–278. Washington, D.C.: International County/City Management Association.

Erickson, P. A. (2006). *Emergency response planning for corporate and municipal managers* (2nd Ed.). Burlington, MA: Elsevier/Butterworth Heinemann.

Evans, T. (2000). Emergency workers learning Spanish. *Indianapolis Star* (August 7, A1-2).

Farkas, A. (2010). Language industry comes together for Haiti and the future. *ATA Chronicle 8:*26–29.

Frequently asked questions about the protection of limited English proficient (LEP) individuals under Title VI of the Civil Rights Act of 1964 and Title VI Regulations. (2011). (March). http://www.justice.gov/crt/about/cor/FAQ_About_LEP_Title_VI_and_Title_VI_Regs.pdf

Gard, B. (2009). The role of interpreters during disasters. *ATA Chronicle 7:*19–23).

Geersten, D. & Romero, N. (2000). The development of a comprehensive interpreter certification program. *ATA Chronicle 6:*13–18).

Haddow, G. D. & Bullock, J. A. (2006). *Introduction to emergency management.* (2nd Ed.). Burlington, MA: Elsevier Butterworth-Heinemann.

Holder, E. (2011). Federal Government's Renewed Commitment to Language Access Obligations Under Executive Order 13166. Memorandum to HEADS OF FEDERAL AGENCIES, GENERAL COUNSELS, AND CIVIL RIGHTS HEADS (February 17).

Hudson, V. F. (2008). Five keys to successful intercultural communication. *ATA Chronicle 11/12:*18–22.

Lachlan, K., Spence, P. & Eith, C. (2007). Access to mediated emergency messages: Difference in crisis knowledge across age, race and socioeconomic status. In Bates and Swan (Eds.). *Through the eyes of Katrina: Social justice in the United States.* Durham, NC: Carolina Academic Press.

Leong, K. J., Airriess, C. Chia-Chen Chen, A., Keith, V., Li, Wei, Wang, Ying & Adams, K. From invisibility to hypervisibility: The complexity of race, survival, and resiliency for the Vietnamese-American community in eastern New Orleans. 2007. In Bates and Swan (Eds.). *Through the eyes of Katrina: Soical justice in the United States.* Durham, NC: Carolina Academic Press.

Mackun, P. & Wilson, S. (2011). Population Distribution and Change 2000–2010 – 2010 Census Briefs. http://www.census.gov/prod/cen2010/briefs/c2010br-01.pdf

Memorandum of Understanding between the American National Red Cross and the National Association of Judiciary Interpreters and Translators. (2009). *Proteus 18/2*: 19, 21. (Available at www.najit.org).

Mileti, D. & Fitzpatrick, C. (1991). *The great earthquake experiment: Risk communication and public action.* Boulder, CO: Westview Press.

Mileti, D., Drabek, T. & Haas, J. E. (1975). *Human systems in extreme environments: A sociological perspective.* Boulder: Institute of Behavioral Science, University of Colorado.

Musgrave, B. (2005). Language barrier nearly fatal. *The News & Observer* (Raleigh, NC). September 12, p. 11A.

Nicholson, W. C. (2006a). Seeking consensus on homeland security standards: Adopting the National Response Plan and the National Incident Management System. *Widener Law Review* XII/2: 491–559.

Nicholson, W. C. (2006b). Legal issues arising from inadequate warning systems prior to the Indian Ocean tsunami. *Indian Ocean Survey* 2/2 (July–December 2006).

Perry, R. W. & Lindell, M. K. (2007). *Emergency planning.* Hoboken, NJ: John Wiley & Sons, Inc.

Phillips, B. (2009). *Disaster recovery.* Boca Raton, FL: CRC Press.

Red Cross and NAJIT collaborate. (2009). *Proteus 18/2*:19. (Available at www.najit.org).

Reed, J. (2011). Tips for testing and certifying multilingual employees. Found at: http://www.migrationinformation.org/integration/language_portal/index.cfm

Schweda Nicholson, N. (2010). The Tubes Are Clogged: Disaster Messages via Social Media. Panel presentation at the 35th Annual Natural Hazards Research and Applications Workshop (University of Colorado) in Broomfield, CO.

Schweda Nicholson, N. (2005a). The Court Interpreters Act of 1978: A 25-year retrospective: Part I. *ATA Chronicle 9*:36–41.

Schweda Nicholson, N. (2005b). The Court Interpreters Act of 1978: A 25-year retrospective: Part II. 2005. *ATA Chronicle 9*:32–37, 39.

Schweda Nicholson, N. (1994a). Professional ethics for court and community interpreters. In Deanna L. Hammond, Editor. *American Translators Association Scholarly Monograph Series, Volume VII: Professional Issues in Translation and Interpretation.* 79–97. Amsterdam and Philadelphia: John Benjamins Publishing Company.

Schweda Nicholson, N. (1994b). Community interpreter training in the United States and the United Kingdom: An overview of selected initiatives. *Hermes, Journal of Linguistics 12*:127–139.

Schweda Nicholson, N. (1989). *Ad hoc* court interpreters in the United States: Equality, inequality, quality? *Meta 34*/4:711–723.

Schweda Nicholson, N. (1986). Language planning and policy development for court interpretation services in the United States. *Language Problems and Language Planning 10*/2:140–157.

Social media a possible lifesaver during disasters. (No author). (2009). *Coloradan* (December, p. 20).

Thickstun, P. (2009). Role of medical linguists in an influenza pandemic. *The ATA Chronicle 11/12*:12–18).

Trabing, E. (2009). Designing programs and materials for OMH (Office of Mental Health) and DHHS (Department of Health and Human Services). Presentation at the Annual Conference of the Tennessee Association of Professional Interpreters and Translators (TAPIT). Nashville, September 11.

Trabing, E. (2004). Ethics for community interpreters. *The ATA Chronicle 4*:15–16, 22)

Trujillo-Pagán, N. (2007). Katrina's Latinos: What natural events tell us about social disasters among New Orleans' Latinos. In Bates & Swan (Eds.). *Through the eyes of Katrina: Social justice in the United States.* Durham, NC: Carolina Academic Press.

Waugh, W. & Tierney, K. (Eds.). (2007). *Emergency management: Principles and practice for local government* (2nd Ed.). Washington, D.C.: International County/City Management Association.

Cordova, Alabama, August 6, 2011 – Cordova resident Ester Lee Russell (right) enjoys a moment of conversation wirh Auburn University Professor Cheryl Morgan at a community meeting for Long-Term Coummnity Recovery (LTCR) at Cordova High School and the tornado that struck Cordonva on April 27, 2011. Photo by Christopher Mardorf/FEMA.

Chapter 18

PREPAREDNESS CASES

PREPAREDNESS INCLUDES PLANNING, TRAINING, AND EXERCISING

Coates v. U.S.

United States District Court for the Central District of Illinois

612 F. Supp. 592 (1985)

FINDINGS OF FACT

On the day Terry Coates drowned, he was camping with his wife and two children in an area of Rocky Mountain National Park known as the Aspenglen Campground. He was, on that day, 36 years of age, healthy, and employed by the Peoria Public Schools. He had begun as a teacher with the school district and had, by virtue of additional schooling and effective performance, moved up into an administrative position. The evidence showed that Terry Coates was an accomplished handyman, performing many of the routine duties of maintenance and repair around his home. In addition, he completely remodeled the family basement, doing all but the most basic plumbing in order to convert it to a family room. He also made furniture for the home. He built the patio, repaired the family cars, and did everyday yard work and housework. He also was actively involved with his children, helping them with their school work and taking opportunities to teach them through the family's recreational activities. This latter activity included teaching the children biology while they were on camping trips.

Theirs was a family of avid campers, and in July of 1982 they planned the longest camping

trip they had ever taken, going to Rocky Mountain National Park in Colorado. Upon their arrival at the camp they paid the fee, which was assessed by the United States for camping, and were assigned a location in the Aspenglen Campground.

At this point, the Court will attempt to create a word picture of that area in Rocky Mountain National Park relevant to an understanding of the events which occurred on the day of Terry Coates' death.

At the top of a steep mountainous incline is the Lawn Lake Dam (the dam), which is situated within the park but owned by the Farmers Irrigation Ditch and Reservoir Company. Near the dam and proceeding down the side of the mountain is a water channel known as the Roaring River Drainage. At the base of the Drainage there is a large, relatively flat meadow area which is known as Horseshoe Park. Although the terrain of Horseshoe Park itself is substantially flat, the elevation of the west end is higher than that of the east end so that the park is tilted downward in the direction of Aspenglen Campground. At the lower east end of Horseshoe Park is a second dam, known as Cascade Dam, which is approximately 40 feet wide and located about one-half mile from the Aspenglen Campground. The campground itself is downstream from the Cascade Dam, sitting at an elevation lower than that of Horseshoe Park. It consists of two separate areas; the first a "mainland" area consisting of a parking lot and walk-in campsites, and the second being an island of still lower elevation, connected to the mainland by a footbridge.

On the morning of July 15, 1982, the tent of Terry Coates and his family was pitched in a mainland campsite at the Aspenglen Campground. Sometime before 6:30 a.m. that morning, the Lawn Lake Dam had failed and a garbage man, observing the rushing [**4] water, notified the ranger station which served that portion of the park. The following events all occurred within a period of slightly more than one hour. Park rangers were alerted to the potentially dangerous situation and, within approximately 20 minutes of the report of the dam's failure, Ranger Schultz was dispatched to warn the "walk-in" (mainland) campsites. At 6:50, he entered the campsite and, within 15 minutes, had warned several of the campers, but not all of them assigned to that area. He did not, however, tell them that a dam had failed or that a flood was approaching; rather, he suggested, apparently without urgency, that it might be wise for campers to evacuate the area. Neither Terry nor Rosemary Coates received any warning from Ranger Schultz, although they did have second-hand information that campers were being warned to leave the area. At that time, they could see what Rosemary Coates described as a "little water," but nothing to warn them of the actual situation. This was approximately 7:15. The Coates had planned to leave the Aspenglen Campground that day anyway. While Rosemary Coates woke the children and prepared to leave, Terry Coates went to the parking area, got his camera from the car, and began taking pictures. After she had packed up, Mrs. Coates heard a loud noise, saw a rush of water, and began to move to high ground with the children. She had no knowledge at that point of where her husband was. It is, however, apparent that he had

returned to the car because his camera was later found there. The testimony shows that witnesses saw Terry Coates and Bridget Dorris, who also died in the flood, crossing the footbridge to the island portion of the campground.

While the decision to warn was being made and the subsequent events were unfolding, the water, freed from Lawn Lake Dam, was moving down the Roaring River Drainage and entering the west end of Horseshoe Park. This situation was known to the park rangers by approximately 6:45. As more and more water entered the park, it inevitably flowed to the lower end, putting increasing pressure on Cascade Dam. At about 7:40 a.m., Cascade Dam failed, sending a flood of water into the Aspenglen Campground. Terry Coates was drowned by those flood waters as they passed through the campground.

Although the testimony indicates a good deal of ranger activity during the course of these [**6] events, the Court finds by a preponderance of the evidence that the conduct of the Defendant was negligent in the following respects: (1) The Defendant failed to post an observer at the Cascade Dam to monitor the flood waters and their impact on the dam; (2) they failed to have a ranger in Aspenglen Campground at all times during the flood; (3) they failed to give adequate warning of the danger resulting from the failure of the dam to Terry Coates; and (4) they failed to maintain or implement a plan for dealing with emergencies, such as the one involved here.

On the issue of damages, the Court finds that Terry Coates' widow and two minor children sustained a pecuniary loss, as a result of his death, in the amount of Eight Hundred Thousand Dollars ($800,000). Since the time of his death, Rosemary Coates and her children have received Social Security survivor benefits from the Social Security Administration. At the time of trial, the amount paid to them, collectively, was $10,259. Because of her remarriage in January of 1985, Rosemary Coates is no longer eligible to receive benefits; Marcian and Adam, however, will continue to receive payment until each has attained the age of 18. Extrapolating from the current rate of payment and making reasonable projections, it appears that Marcian is entitled to an additional $9,036 and Adam to an additional $13,806.

A. Failure to Inspect

Turning to the failure of the Lawn Lake Dam, which was the precipitating event, the Court finds that, although the dam is located within the boundaries of the Rocky Mountain National Park, it was, in fact, built in 1902 by the Farmers Irrigation Ditch and Reservoir Company and is still owned by them pursuant to an easement granted them by the Government. Although the federal Government had never formally inspected the dam, it was inspected in 1975, 1977, and 1978 by a Colorado State Engineer. It was scheduled for federal inspection in 1982 on a date later than July 15. The most probable cause of the dam's collapse was rapid erosion caused by water seeping through a breach in the junction of an inlet pipe and a valve. Since it appears that such erosion would be extremely rapid, it is unlikely that a normal inspection of the dam would have disclosed the defect unless that inspection took place within a few hours prior to its failure.

It may well be that, as Plaintiffs allege, Lawn Lake Dam was a ticking bomb. Inspection by the federal Government may or may not have revealed critical structural defects, including the one which appears to have been the most probable cause of the dam's failure on July 15, 1982. The record in this case is such that the Court finds Plaintiffs have not shown by a preponderance of the evidence that the federal Government's failure to inspect was a proximate cause of the death of Terry Coates.

B. Failure to Have a Plan in Place

Rocky Mountain National Park has been preserved by the federal Government, as have all of the national parks, from the incursions of a technological society. It is made available to lovers of parks and those who enjoy ranging freely through what nature has offered in that section of the country. For those who wish to utilize these preserved enclaves, the Government exacts a monetary fee, thereby creating a specific

legal relationship between itself and those who have paid for this special privilege. Because these national parks are outdoors and, therefore, subject to extreme and sometimes unexpected weather changes, structural failures such as the one at issue here, other flash floods, and major fires which occur, changes may be sudden and dramatic (because of acts of God or foibles of man). Therefore, the Government, in creating this relationship with citizens, also creates a duty for itself to develop orderly procedures for dealing with emergencies. It is imperative to have a plan in place because in such situations there is little time for reflection. Priorities should be established before an emergency arises; otherwise personnel are unprepared to deal with them.

Such appears to be the case here. Many well intentioned and otherwise well trained rangers were inadequately prepared to cope with the breach in Lawn Lake Dam. It appears, for example, that the rangers did not even know how much water was being held back by the dam. Elementary lapses, obvious with the clarity of hindsight, could have been avoided through the development of orderly procedures for warning and evacuating people in the park in the event a crisis arose. There was a duty to plan, the Government failed to develop a plan, and the Court here finds that that failure to have a plan in place was a proximate cause of the death of Terry Coates.

C. Failure to Warn

The events of July 15, 1982 were not atypical of emergencies. The situation was an evolving one and, particularly in the absence of a plan, there is a need to assimilate, correlate, and synthesize occurrences so as to make decisions in a reasonable fashion. Therefore, the Court's analysis must track as closely as possible decisions which would be reasonably made by persons at the time of the occurrence. The Court believes that Lawn Lake Dam had burst by 6:23 a.m. This resulted in the constant flow of a substantial stream of water down the side of the mountain. It would be unreasonable not to know that this water would hit the bottom with high velocity and would probably back up to the west of the trail area. The first major mistake made by the rangers was to assume that Horseshoe Park could absorb all of the water. At some point that belief became unreasonable and it should have become apparent that Cascade Dam was a focal point of the flood waters. In fact, poised as it was above the campground area, Cascade Dam was the key to what would occur downstream. It was absolutely critical that there be someone there to monitor what was happening at Cascade Dam and be prepared to act immediately.

There had been attempts earlier that morning to warn persons in the walk-in sites. Ranger Schultz had been dispatched to warn campers of the possibility of a flash flood. He was apparently given no indication of who was to be warned and with what sense of urgency the message was to be conveyed. It appears that, on his own, he also notified people in the island camp sites. The evidence shows that he told some that they had one to one and one-half hours to leave. It is clear that his tone was not one of urgency and that the information he conveyed was insufficient to alert them to the potential magnitude of the emergency. It seems obvious to the Court that the ranger's desire not to cause panic had to be balanced against the need to motivate the campers to clear the area in an expeditious manner. Moreover, Schultz did not warn those campers who were not presently in the immediate area, nor did he consider that others might have been hiking or exploring in these areas even though they did not actually have tents there. The exercise of reasonable care mandated, at a minimum, the issuance of careful and complete warnings to all of the people who were camped in or otherwise using areas of the park which were downstream from Lawn Lake Dam. The failure to issue these warnings constitutes negligence and, the Court here finds that that negligence was a proximate cause of the death of Terry Coates.

D. Damages

The first issue to be considered here is that of comparative negligence. Many of the same factors which the Court has just discussed with respect

to the Government also apply to Terry Coates. This is not a situation which happened all at once; it evolved. While it is true that he was never given direct notice of the potential flooding caused by the breach in the dam, Terry Coates and his wife did receive indirect notice from another camper concerning possible flooding. There was, as well, an increased level of activity within the campsites occasioned by people moving, albeit slowly, to evacuate the area. There was a general low level of information about the flood and at some point, the environment of the campground began to change. In fact, Terry Coates was sufficiently impressed with the awesomeness of the rising waters to take pictures. The pictures which he had taken establish clearly that, by reason of his own observation, Terry Coates had knowledge that the water was rushing in and that the water level was rising. He did not have much information, but the Court believes that it was enough to require him to take some reasonable steps for his own safety.

There are substantial discrepancies between the descriptions of the physical situation in the campground in the depositions of Wayne Haymon and Mrs. Thomas, the mother of Bridget Dorris, the other decedent. The deposition of Wayne Haymon described a situation of escalating seriousness, including an increase of noise, a rising level of churning, angry water, and a change in the nature and appearance of the water from clear to muddy. Mrs. Thomas, on the other hand, tendered a report which minimized the potential for danger and, seemingly, excused the decedents' failure to leave the area more quickly. The Court is inclined to give greater credence to Mr. Haymon's deposition because it has a ring of truth and, more importantly, much of what he says is corroborated elsewhere in the record. The Court fears that, under the circumstances, Mrs. Thomas may have suffered substantial residual guilt as a result of her daughter's tragic death. Wayne Haymon had no personal interest in the litigation (either directly or [**14] indirectly), and there is no indication of any motivation for subliminal and unintentional shading of the facts. The Court finds that Terry Coates had notice (albeit inadequate), had time to remove himself safely from the endangered area, and took inadequate steps to protect himself.

The Court has found that the actual pecuniary loss sustained by Rosemary Coates and her children was $800,000, and now holds that the negligence attributable to the United States Government is 60% of the total negligence which occurred in this case, and that attributable to Terry Coates is 40%. Therefore, the total liability of the Government to the Plaintiffs is Four Hundred Eighty Thousand Dollars ($480,000).

CONCLUSIONS OF LAW

The Government has argued that the United States is entitled to offset that portion of Social Security payments made to Rosemary Coates and the children which were contributed by the Government. They concede that the amount which was actually contributed by Terry Coates and his employers is a collateral source and, therefore, is not appropriately considered as offset. The Government cites the case of *Steckler v. United States*, 549 F.2d 1372 (10th Cir. 1977) which is apparently the only case applying Colorado law to the issue of offsetting a Federal Tort Claims Act judgment with Social Security benefits. In that case, the Court of Appeals upheld, as proper, the offsetting of that amount of Social Security benefits which was not attributable to the direct contributions of the employee and his employers.

The Plaintiffs counter by first noting that the Social Security Fund is a separate fund from the general treasury and they argue that all amounts contributed by all workers and all employers to the fund are collateral sources. As support for this position, they cite *Smith v. United States*, 587 F.2d 1013 (3rd Cir. 1978). In that case, the court performed rather extensive analysis of the approach used by other circuit courts of appeals in dealing with the collateral source doctrine and its application to the Federal Tort Claims Act. Generally, federal courts seem to hold that there is a distinction between those benefits which come from unfunded general revenues and those which come from a "special fund supplied in part by the beneficiary or a relative upon whom the beneficiary is dependent." Two of the cases considered were States, supra, and *United States v. Harue Hayashi*, 282 F.2d 599 (9th Cir. 1960), which dealt specifically with the application of the doctrine to Social Security benefits. In Harue Hayashi the court specifically held that Social Security survivor benefits are "collateral" to an FTCA recovery, finding that: "The money which goes into this fund is provided by a system of excise taxes on employers and income taxes on employees, designed to be actuarially sound and self-supporting."

Again, in Steckler, the court dealt with Social Security payments and their analysis concluded that, because the Government has supplemented the fund when necessary, some burden should rest on a plaintiff to establish, if possible, that portion contributed by the Government so as to permit offset of other portions as collateral sources.

The court in Smith then states its agreement with the Ninth Circuit in Harue Hayashi that:

> Where state law recognizes the 'collateral source' doctrine, Social Security benefits should not be deducted from a recovery under the Federal Tort Claims Act. FTCA recoveries come out of general revenues; Social Security benefits are funded almost entirely from employee and employer contributions.... The Government here did not argue that a FTCA recovery must be reduced by that portion of Social Security benefits attributable to the Government's contribution out of general revenues. Nevertheless, we are constrained to note our disagreement with the Tenth Circuit in Steckler, supra, and decline therefore to adopt its approach. We believe that the Government's payments are so minimal and so difficult to trace that such an approach would be impracticable.

The Court believes that this is a more realistic analysis of this issue and finds that, since it is virtually impossible to ascertain accurately the portion of the payments contributed by the Government, Steckler, considered in its entirety, is not fundamentally inconsistent with the holding in Smith. It is, therefore, held that the Social Security payments made to Rosemary Coates and her children should not offset the damage award made to them in this Order.

There were also several legal findings made during the pretrial stages which the Court now incorporates as a part of this Opinion.

Because the Defendant in this case is the United States, some special procedural determinations were necessary. Title 28 U.S.C.A. § 2674 subjects the Government to liability in the same manner and to the same extent as a private person under like circumstances. The choice of law decision is governed by 28 U.S.C.A. § 1346(b) which provides that the liability of the United States will be determined by the law of the place where the act or omission occurred. Therefore, the applicable standard to be used in assessing the Government's liability is that of a private landowner operating a campground in Colorado.

A landowner in the State of Colorado is charged with the obligation to discover dangers and take relevant precautions and to warn of those dangers which he knows or should know exist to the same extent that any reasonable man in view of the probability or foreseeability of injury to persons on the land would be required to do. *Mile High Fence Company v. Radovich*, 175 Colo. 537, 489 P.2d 308; *Hartzell v. United States*, 539 F.2d 65 (10th Cir. 1976). One factor in ascertaining the existence and extent of the landowner's duty and conduct is the plaintiff's status as trespasser, licensee or invitee. *Mile High Fence Company v. Radovich*, supra. Because the Plaintiffs paid a monetary fee for the privilege of camping in Rocky Mountain National Park, a special relationship was created which required the United States to treat them as licensees.

The Court concluded that the version of comparative negligence which has been adopted by Colorado and set out in Colorado Revised Statutes, § 13-21-111, permits recovery by a plaintiff whose negligence is less than that of the defendant, but requires that his damages be reduced by the percentage of negligence attributable to him.

Finally, the Court holds that the conduct of the United States in this case is not covered by the statutory exceptions to tort liability set out in the Tort Claims Act, 28 U.S.C.A. § 2680(a)(h). The Government's failure to develop an adequate emergency plan and failure to give adequate warning to all campers is not comprehended in the conduct which Congress intended to protect through enactment of that section of the Tort Claims Act. *United States v. S.A. Empresa De Viacao Aerea Rio Grandense* (Varig Airlines), 467 U.S. 797, 52 U.S.L.W. 4833, 81 L. Ed. 2d 660, 104 S. Ct. 2755; *Hylin v. United States*, 715 F.2d 1206 (7th Cir. 1983); *Universal Aviation Underwriters v. United States*, 496 F. Supp. 639 (D.C. Colo. 1980).

For all of the reasons set forth in this Memorandum Opinion, judgment is hereby entered on behalf of Plaintiffs in the amount of Four Hundred and Eighty Thousand Dollars ($480,000) on this 6th day of May, 1985.

QUESTIONS

1. Why did the government have a duty to inspect the dams in this case?
2. What was the basis of the government's duty to create a plan for evacuation in case the dams might fail?
3. What was the basis for the government's duty to warn campers of dam failure?
4. How should the government have monitored the dams' condition? Should there have been a ranger stationed nearby 24/7/365? What would you suggest, and why?
5. Explain how a warning should be created and spread to people who might be threatened by a dam failure.
6. How might a warning system's requirements change if a threatened population was a city of 25,000 instead of a relatively small number of campers?
7. Review the discussion in Chapter 15 "Limited English Proficient Populations and Emergency Management: Legal Requirements and Practical Issues." Assume that

the threatened city of 25,000 contains 5,000 residents (20%) who speak six languages other than English, including Spanish, Russian, mandarin Chinese, Vietnamese, Somali, and Korean. Explain how you would plan in advance how to ensure that the city's entire population could receive timely warnings of an imminent dam break or other natural or man-made disaster.

Leake v. Murphy

617 S.E.2d 575 (Ga. App. 2005)

OPINION

MIKELL, Judge.

This is an appeal from the grant of a motion to dismiss a negligence action brought by Alan and Sandy Leake, individually and as parents of Anna Elisabeth Leake ("Anna"), a child who was grievously injured in February 2002 when a deranged individual attacked her with a hammer at Mountain Park Elementary School in Gwinnett County. The Leakes contend that the defendants, including the individual members of the Gwinnett County Board of Education ("Board") and the Superintendent of the Gwinnett County School District, J. Alvin Wilbanks, failed to develop a safety plan for the school which addressed security issues, as required by OCGA § 20-2-1185. That Code section provides in pertinent part that "[e]very public school shall prepare a school safety plan to help curb the growing incidence of violence in schools [and] to respond effectively to such incidents. . . . School safety plans . . . shall address security issues." In an additional count, the complaint alleges that the defendants, including the school's principal, Debbie Allred, and her front office staff members, Connie Finn and Melissa Switzer, are liable for Anna's injuries based on their negligent failure to implement and enforce measures designed to control access to the school as well as to monitor people entering the school under a sign-in policy developed a year earlier. The defendants moved to dismiss the complaint on the basis of official immunity, and the trial court granted the motion. For the reasons that follow, we reverse as to the individual Board members and the Superintendent on the claim alleging failure to prepare a security plan pursuant to OCGA § 20-2-1185. The remainder of the judgment is affirmed.

Construed in its proper light, the complaint shows that on February 6, 2001, one year prior to the attack on Anna, a deranged convicted felon, William Cowart, walked into the school holding the picture of a young girl. Allred, the principal, confronted him; he left, and she called the police. Following this incident, the school instituted an access control policy which involved stationing an individual in the lobby to screen persons entering the school and ensure that they signed in at the principal's office, which was located adjacent to the lobby. The policy also required the principal and her office staff to monitor such persons through the office's floor-to-ceiling glass window.

The following year, on February 21, 2002, at approximately 2:50 p.m., Chad Brant Hagaman, a paranoid schizophrenic who heard voices telling him to kill people, walked through the school's front doors armed with a hammer. The complaint asserts that Hagaman was not confronted, screened, detained, or examined as to his purpose although he walked past the principal's office. Hagaman walked approximately 100–150 feet until he came upon a row of fourth-grade students lined up in a hallway. When he reached ten-year-old Anna, Hagaman swung the hammer and embedded the claw end of it in her skull. The metal claws penetrated her brain, leaving her with permanent neurological deficits as well as post-traumatic stress disorder.

1. We first address the Leakes' contention that the trial court erred in ruling that the defendants are protected by official immunity from any claim arising from their failure to prepare a school safety plan which addressed security issues.

Under Ga. Const. of 1983, Art. I, Sec. II, Par. IX (d), "public officials are immune from damages that result from their performance of discretionary functions, unless those functions were undertaken with malice or intent to cause injury." Conversely, official immunity does not apply to the performance of ministerial duties. Therefore, the threshold inquiry is whether the duty to prepare a safety plan pursuant to OCGA § 20-2-1185 is ministerial rather than discretionary.

As noted above, OCGA § 20-2-1185 states in part: "Every public school *shall* prepare a school safety plan to help curb the growing incidence of violence in schools. . . . School safety plans prepared by public schools *shall* address security issues in school safety zones as defined in [OCGA § 16-11-127.1 (a) (1)]." The word "'[s]hall' is generally construed as a word of command." Therefore, OCGA § 20-2-1185 mandates the preparation of a school safety plan which addresses security issues for every public school in this state. The duty is absolute, and, as a result, ministerial. Furthermore, we hold that the legislature has conferred this duty upon the county school superintendent and the county board of education through the passage of OCGA § 20-2-59, which provides that the superintendent and the board "shall make rules to govern the county schools of their county." Accordingly, we hold that the duty to prepare a school safety plan for the school at issue fell to Superintendent Wilbanks and the Board members. We further hold that the remaining defendants, the principal and her front office staff, are not vested by the legislature with rule-making authority and thus cannot be held liable for damages for failure to prepare such a plan.

The defendants contend that the use of the word "shall" in OCGA § 20-2-1185 does not create a ministerial duty, citing *Norris v. Emanuel County*. That case is distinguishable. The statute under consideration in *Norris*, OCGA § 32-6-50 (c) (1), states that "[c]ounties and municipalities shall place and maintain upon the public roads of their respective public road systems such traffic-control devices as are necessary to regulate, warn, or guide traffic." We held that the " 'as necessary' language connotes discretion," such that the county road superintendent and road crew supervisor had official immunity from an action alleging that they negligently failed to

warn of an eroded road shoulder. In contrast, OCGA § 20-2-1185 says that *every* public school shall prepare a school safety plan which shall address security issues. No exceptions are listed. Thus, the duty to prepare such a plan for a public school is ministerial, not discretionary.

In terms of immunity, however, the duty to prepare the plan must be distinguished from its contents. We do not agree with the Leakes' assertion that the manner in which a school safety plan is prepared, and its ultimate contents, require solely the performance of ministerial functions. In this regard, the relevant portions of OCGA § 20-2-1185 state:

> (a) . . . School safety plans of public schools shall be prepared with input from students enrolled in that school, parents or legal guardians of such students, teachers in that school, community leaders, other school employees and school district employees, and local law enforcement, fire service, public safety, and emergency management agencies.

Thus, the statute calls for the input of a variety of individuals in the development of a safety plan. And, although the plan must address "security issues," those issues are not defined. The legislature has not given specific direction on what elements to include in a safety plan; for example, whether to install electronic scanning devices at the entrance to the school, or to keep the doors locked such that visitors gain entrance by buzzer. These procedures would necessarily differ from school to school, and addressing these issues is left to the discretion of the school authorities. "A discretionary act . . . calls for the exercise of personal deliberation and judgment, which in turn entails examining the facts, reaching reasoned conclusions, and acting on them in a way not specifically directed." Based on this definition, we conclude that the development of the contents of a school safety plan calls for the exercise of discretion.

Contrary to the Leakes' assertion, *Meagher v. Quick* does not support their argument that the development of the contents of a school safety plan is a ministerial function. In *Meagher*, we held that police officers were not entitled to official immunity from a wrongful death action based on their failure to complete a Family Violence Report, as required by OCGA § 17-4-20.1 (c), after responding to a complaint of child abuse. The child was beaten to death after officers left the residence to which they had been summoned. OCGA § 17-4-20.1 (c) provides: "Whenever a law enforcement officer investigates an incident of family violence, whether or not an arrest is made, the officer shall prepare and submit…a written report of the incident entitled 'Family Violence Report.' "The statute lists 13 pieces of information that the report must contain. Construing the statute, we held that completion of the report was a ministerial function, requiring "merely the execution of a specific duty."

In the case sub judice, although the preparation of a school safety plan which addresses security issues is a mandatory ministerial duty under OCGA § 20-2-1185, the development of its contents and the manner of its enforcement have not been specifically directed by the legislature, unlike the Family Violence Report prescribed by OCGA § 17-4-20.1 (c). Thus, the development of the contents and manner of enforcement of a school safety plan are discretionary functions. If the present lawsuit alleged that the defendants had developed an inadequate plan and that the plan's inadequacies resulted in Anna's injury, the defendants would be entitled to official immunity. It is the total absence of any plan which precludes dismissal of the lawsuit.

Moreover, the absence of such a plan from the record in this case necessitates reversal of the trial court's order on this issue. A motion to dismiss should not be granted unless "the movant establishes that the claimant could not possibly introduce evidence within the framework of the complaint sufficient to warrant a grant of the relief sought." At this procedural juncture, there is no evidence of record that the Superintendent and the Board members have fulfilled their legislative mandate to prepare a safety plan that addressed security issues for the school at issue. "It is a well-established principle that a public official who fails to perform purely ministerial duties required by law is subject to an action for damages by one who is injured by his omission." It follows that the trial court erred in dismissing

that portion of the complaint which alleges that the Superintendent and the Board members failed to develop a safety plan for the school as required by OCGA § 20-2-1185. On the other hand, if a motion for summary judgment is filed and evidence of a plan that predates the attack on Anna and addresses security issues is presented, the defendants would be entitled to official immunity.

2. "[M]onitoring, supervising, and controlling the activities of students is a discretionary action protected by the doctrine of official immunity." The Leakes argue that the trial court erred in finding that the defendants' acts or omissions related to a duty to monitor, supervise, or control students. To the extent that the trial court's order can be interpreted as a ruling that the duty of Wilbanks and the Board members to prepare a school safety plan which addressed security issues was a discretionary function, the court erred, as we have held in Division 1. The applicability of this argument to the principal and her front office staff (defendants Allred, Finn and Switzer) will be discussed below.

3. In their third enumerated error, the Leakes argue that the trial court erred in finding that the failure of defendants Allred, Finn and Switzer to comply with school policies and procedures adopted after the 2001 incident was discretionary rather than ministerial. We disagree with the Leakes and agree with the trial court.

"Generally, the determination of whether an action is discretionary or ministerial depends on the character of the specific actions complained of…and is to be made on a case-by-case basis." "The Georgia courts have consistently held that making decisions regarding the means used to supervise school children is a discretionary function of the school principal." "Supervision of students is considered discretionary even where specific school policies designed to help control and monitor students have been violated." The rule that principals and teachers are immune from actions involving supervising and monitoring students has been applied uniformly in cases where students have been injured or killed.

Supervising and monitoring students includes, for purposes of deciding immunity questions, safeguarding the students from harm caused by intruders. A binding precedent is *Kelly v. Lewis* 21 Ga. App. 506 (471 SE2d 583) (1996). *Kelly* involved the fatal shooting of a student, apparently at or near the entrance to a public high school. The plaintiffs alleged that because a teacher failed to stand at the entrance in the morning, which was his assigned duty, the teacher could be held liable for negligent failure to perform a ministerial duty. The plaintiffs further contended that the principal could be held liable for failing to post a duty roster identifying the persons responsible for monitoring the entrance. We disagreed, stating "that the complete failure to perform a discre-

tionary act is the same as the negligent performance of that act for the purposes of determining whether such action was discretionary or ministerial." Similarly, the failure of defendants Allred, Finn and Switzer to monitor the entrance lobby to guard against assailants like Hagaman is a "complete failure to perform a discretionary act." Georgia's doctrine of official immunity means that they cannot be held accountable for their failure.

Judgment affirmed in part and reversed in part. Andrews, P. J., and Phipps, J., concur.

QUESTIONS

1. The Court states that "The word " '[s]hall' is generally construed as a word of command." Therefore, [Georgia law] "mandates the preparation of a school safety plan which addresses security issues for every public school in this state. The duty is absolute, and, as a result, ministerial." What implications does this ruling hold for local units of government where state law says (as in Indiana Code § 10-14-3-17 (2009):

   (**h**) Each local or interjurisdictional agency shall:

   (**1**) prepare; and

   (**2**) keep current; a local or interjurisdictional disaster emergency plan for its area.

2. The Court also states that "If the present lawsuit alleged that the defendants had developed an inadequate plan and that the plan's inadequacies resulted in Anna's injury, the defendants would be entitled to official immunity." What are the implications of this statement for emergency planners?

3. Why does the distinction between not planning at all and not planning competently exist? Divide into two groups and debate whether, as a matter of public policy, the distinction does or does not make sense.

4. Note that the Court cites: OCGA § 20-2-1185:

   (a) ...School safety plans of public schools shall be prepared with input from students enrolled in that school, parents or legal guardians of such students, teachers in that school, community leaders, other school employees and school district employees, and local law enforcement, fire service, public safety, and emergency management agencies.

   • What does "input" mean in this context?

   • What obligation does this law place on "local law enforcement, fire service, public safety, and emergency management agencies?"

   • Does the Court's ruling mean that these words in the statute have no meaning?

   • If the school presents the emergency management agency with a plan that calls for tornado protection to be in the glass-enclosed front hall of the school and the EM agency does not object, what are the potential results if a tornado comes and students are killed by flying glasss?

5. Happy County is a rural jurisdiction located in the Appalachian Mountains whose county seat is Appleton. The county has few resources, and a small tax base. Happy County's major employer is BugEx, a pesticide manufacturer. BugEx is losing business

due to foreign competition. The business has no insurance and no assets. It will close within the month.

Emergency management is the responsibility of Fred Taylor, the county sheriff, who has two part-time deputies. Although Sheriff Taylor knows that planning is one of his emergency management duties, he has not had time to create a plan.

The reason that no plan has been created is a lack of time. Recently, a number of meth labs have been setting up in rural abandoned buildings in Happy County. Sheriff Taylor has been working overtime to identify and eradicate the meth labs. The drugs they create as well as the extremely hazardous materials they use are, in the sheriff's mind, a "clear and present danger." The meth lab operators know that Taylor is after them, and they have armed themselves with heavy weaponry, including fully automatic submachine guns.

Joe Green and his wife Shawna enjoy driving along back roads and viewing rural scenery. Joe maintains that they are looking for the "real America." One day, they drive into Happy County. Shortly after they leave Appleton, Joe is delighted to find a narrow road leading into the hills. Shawna

is driving as they ascend the road, making several turns onto different roads on the way. Eventually, they decide that they are lost.

One morning, an employee at BugEx leaves the wrong valve open, releasing a cloud of poison pesticide into the air. An hour later, the cloud drifts over the road Joe and Shawna are on, and they drive into it. They are overcome by the fumes. Joe loses his sight, and he and Shawna suffer permanent damage to their lungs.

Joe and Shawna sue Happy County and Sheriff Taylor. They allege that their injuries are the result of 1.) Happy County's failure to create an emergency plan, and 2.) Sheriff Taylor's failure to warn them of the pesticide release. They do not sue BugEx, as it has no assets.

- Under the reasoning in these cases, what is the likely result of the lawsuit?
- Would the case's outcome be any different if Sheriff Taylor had made a plan that did not comply with the state law's requirement that it include an evacuation plan?
- What would happen if the state law merely said "create a plan" without further specific guidance?

Jonathan Paige et al. v. David Green et al.

Superior Court of Connecticut Judicial District of Waterbury

2007 Conn. Super. LEXIS 3342

OPINION BY: Judge Thomas F. Upson

FACTS

On June 13, 2006, the plaintiffs, Jonathan Paige, Eric Dency, Ronald Rolfe, Christopher Tardif and Patrick Tripodi, filed a fifteen-count complaint against the defendants, David Green (Green) and Phoenix Soil, LLC (Phoenix Soil) for injuries sustained during a training exercise held by the Connecticut urban search and rescue team. In the first, fifth, ninth, thirteenth and sev-

enteenth counts of their revised complaint, the plaintiffs allege the following: On June 6, 2004, they were part of a squad that participated in a confined space rescue training exercise held inside of a tower located at the Hop Brook Lake dam in Middlebury. At all times relevant hereto, Green was the director of operations of Connecticut task force one of the Connecticut urban search and rescue team. He was also the owner, operator and managing partner of Phoenix Soil at all relevant times.

The plaintiffs further allege that during the simulated rescue exercise, Green introduced into the exercise two military type Smith & Wesson smoke grenades. As the plaintiffs descended into the base of the tower, the two smoke grenades were set off, creating a thick, hazardous smoke and causing the plaintiffs' various injuries. The plaintiffs allege that Green was negligent in that he knew or should have known that the two smoke grenades were not intended for interior use and contained hazardous chemicals. The plaintiffs further assert that at the time he introduced the two grenades, he was not acting within the scope of his position as the director of operations of task force one, because such grenades were not part of the planned training exercise.

In the second, sixth, tenth, fourteenth and eighteenth counts of their revised complaint, the plaintiffs allege negligence on the part of Phoenix Soil in that it owned the two smoke grenades used and that its employee, Green, introduced the two grenades into the training exercise.

In the third, seventh, eleventh, fifteenth and nineteenth counts of their revised complaint, the plaintiffs allege that Green's conduct in supplying the grenades for the rescue exercise was an ultra hazardous activity that involved a risk of serious harm to the plaintiffs and as such, he is strictly liable for the plaintiffs' damages. The plaintiffs also assert claims of strict liability against Phoenix Soil for the same grounds in counts four, eight, twelve, sixteen and twenty of their complaint.

Green moved to dismiss the action as to him on the ground that the court lacks subject matter jurisdiction, asserting that he is immune from suit pursuant to General Statutes §28-13(a).[2] Furthermore, Green asserts that the plaintiffs' exclusive remedy is through the Workers' Compensation Act. Green submitted a memorandum of law in support of the motion. The plaintiffs filed a memorandum of law in opposition to Green's motion to dismiss. On March 26, 2007, this court denied Green's motion to dismiss. Subsequently, Green filed a motion to reargue on May 30, 2007, on the grounds that no memorandum of decision was ever issued with regard to the denial of his motion to dismiss. The motion to reargue was granted by this court on June 15, 2007 and oral argument was heard on August 6, 2007.

DISCUSSION

"Pursuant to the rules of practice, a motion to dismiss is the appropriate motion for raising a lack of subject matter jurisdiction." *St. George v. Gordon*, 264 Conn. 538, 545, 825 A.2d 90 (2003). "[T]he doctrine of [statutory] immunity implicates subject matter jurisdiction and is therefore a basis for granting a motion to dismiss." (Internal quotation marks omitted.) *Martin v. Brady*, 261 Conn. 372, 376, 802 A.2d 814 (2002). "[A] challenge to the jurisdiction of a court to render a judgment may be raised at any time, because the lack of subject matter jurisdiction cannot be waived." *DiBerardino v. DiBerardino*, 213 Conn. 373, 377, 568 A.2d 431 (1990). "A motion to dismiss admits all facts well pleaded and invokes any record that accompanies the motion, including supporting affidavits that contain undisputed facts." (Internal quotation marks omitted.) *Coughlin v. Waterbury*, 61 Conn.App. 310, 314, 763 A.2d 1058 (2001). "The burden rests with the party who seeks the exercise of jurisdiction in his favor... clearly to allege facts demonstrating that he is a proper party to invoke judicial resolution of the dispute." (Internal quotation marks omitted.) *Goodyear v. Discala*, 269 Conn. 507, 511, 849 A.2d 791 (2004).

Green moves to dismiss the negligence and strict liability counts as to him on the grounds that he was acting solely within the scope of his position as director of operations for the Connecticut urban search and rescue team at the time

---

[2] General Statutes §28-13(a) provides in relevant part: "Neither the state nor any political subdivision of the state nor, except in cases of wilful misconduct... any member of the civil preparedness forces of the state nor any person authorized by such civil preparedness forces or by any member of such civil preparedness forces complying with or attempting to comply with this chapter... shall be liable for the death of or injury to persons or for damage to property as a result of any such activity."

of the incident, and as such, he is entitled to statutory immunity. In support of the motion, Green points to §28-13(a), which provides immunity to all civil preparedness members, except in cases of "wilful misconduct," for injuries sustained to persons in the course of civil preparedness activities. Green argues that based on the definitions of "civil preparedness" and "civil preparedness forces" provided in General Statutes §28-1(4) and (5),[3] the actions he took in providing grenades for the training exercise were clearly within the ambit of protections provided by General Statutes §28-13(a). He further maintains that because the plaintiffs were members of the "civil preparedness forces" and because the use of smoke grenades was for the purposes of an official training exercise, he is statutorily immune from suit pursuant to §28-13(a).

Additionally, Green argues that as members of the civil preparedness forces, the plaintiffs are entitled to compensation through the Workers' Compensation Act, pursuant to General Statutes §28-14(a). It is Green's position that because the plaintiffs are entitled to workers' compensation, they are barred from bringing suit against fellow employees under General Statutes §31-293a.

In their memorandum of law in opposition, the plaintiffs do not dispute that the Connecticut urban search and rescue team is part of the "civil preparedness forces" as defined by §28-1(5). Nevertheless, the plaintiffs argue that §28-13(a) provides immunity to members of such civil preparedness forces only when said members are "complying with or attempting to comply with [chapter 517]" and Green's supplying of unapproved and hazardous smoke grenades to the

training exercise on June 6, 2004, without following the proper procedures for doing so, was a negligent act that deviated from the scope and authority of his duties as director of operations.[4] Therefore, the plaintiffs argue, at the time he supplied the grenades, Green was not complying with chapter 517 and as such, he should be deprived of statutory immunity. They further maintain that in procuring such grenades, Green engaged in an ultra hazardous activity and as such, he should be held strictly liable.

"Because this case comes to us on a threshold [statutory] immunity issue, pursuant to a motion to dismiss...we do not pass on whether the complaint was legally sufficient to state a cause of action...In the posture of this case, we examine the pleadings to decide if the plaintiff has alleged sufficient facts...with respect to personal immunity...to support a conclusion that the [defendant] [was] acting outside the scope of [their] employment or wilfully or maliciously." (Internal quotation marks omitted.) *Martin v. Brady, supra,* 261 Conn. 376. In looking at the standard set forth in *Martin* in conjunction with the text of §28-13(a), in order for the plaintiffs to overcome the statutory immunity provided by §28-13(a), the plaintiffs must have alleged sufficient facts in the complaint and supporting affidavits to demonstrate that Green committed "wilful misconduct" through his procurement of the smoke grenades. However, in arguing that no §28-13(a) immunity should be afforded to Green, the plaintiffs base their argument on the allegation that Green was not "complying or attempting to comply with this chapter." This text of the statute which the plaintiffs attempt to use as the test for

---

[3] General Statutes §28-1(4) defines "civil preparedness" as including "(I) measures to be taken in preparation for anticipated attack, major disaster or emergency, including the establishment of appropriate organizations, operational plans and supporting agreements; the recruitment and training of personnel..." General Statutes §28-1(5) includes within its definition of "civil preparedness forces" the "Connecticut Urban Search and Rescue Team, under the auspices of the Department of Emergency Management and Homeland Security."

[4] The complaint's allegations indicate an intention by the plaintiffs to bring suit against Green individually and not in his capacity as director of operations of task force one. However, while our courts have not addressed the issue with regards to §28-13(a) specifically, they have held in cases involving §4-165, which provides statutory immunity to state officers and employees, that certain requirements must be met in order for a suit to be considered one against an individual and not one against the state. Those requirements were incorporated into a four-step test set forth in *Spring v. Constantino*, 168 Conn. 563, 568, 362 A.2d 871 (1975), a test the court used in determining whether an action was one commenced against an employee in his individual capacity or was one against an employee in his official capacity. While it does not appear the plaintiffs have satisfied the *Spring v. Constantino* test in bringing suit against Green individually, the court need not fully address this issue, as dismissal is appropriate based on statutory immunity and a consequent lack of subject matter jurisdiction.

providing immunity is merely a descriptive term used to define those in the civil preparedness forces whom immunity is normally provided. The plain meaning of the statute, when read in its entirety, clearly uses a "wilful misconduct" standard as the sole exception to immunity. Even if the court were to consider whether Green should be denied immunity for a failure to comply with chapter 517, the plaintiffs have not explained which provision of the chapter with which Green allegedly failed to comply. Moreover, even if the allegations would support a claim that Green had failed to comply with chapter 517, under the plain language of the statute, Green would still be immune if he was "attempting to comply" with that chapter. The plaintiffs' allegations fail to support a claim that Green was not even attempting to comply with that chapter.

"Wilful misconduct has been defined as intentional conduct.... While we have attempted to draw definitional boundaries between the terms wilful, wanton or reckless, in practice the three terms have been treated as meaning the same thing. [In sum, such] conduct tends to take on the aspect of highly unreasonable conduct, involving an extreme departure from ordinary care, in a situation where a high degree of danger is apparent." (Internal quotation marks omitted.) *Bhinder v. Sun Co.*, 246 Conn. 223, 242 n.14, 717 A.2d 202 (1998). "Wilful and serious misconduct means something more than ordinary negligence. Ordinary negligence could never be even serious misconduct, much less wilful misconduct, and, although gross negligence might present an instance of serious misconduct, it could never present a case of wilful misconduct, as our definition of wilful misconduct clearly indicates and as the authorities so hold. No misconduct which is thoughtless, needless, inadvertent or of the moment, and none which arises from an error of judgment can be wilful and serious misconduct." *Greene v. Metals Selling Corporation*, 3 Conn.App. 40, 45, 484 A.2d 478 (1984).

Thus, in order to overcome the statutory immunity provided under General Statutes §28-13(a), the plaintiffs must have sufficiently alleged that he committed "wilful misconduct" by actually intending to injure the plaintiffs through the use of the smoke grenades, or that his conduct and the circumstances at the time constituted "highly unreasonable conduct." In reading the plaintiffs' complaint and allegations broadly in their favor, it is apparent that they have not alleged sufficient facts to warrant a finding that Green committed "wilful misconduct" in procuring the smoke grenades to the training exercise. In fact, the plaintiff's complaint, while directed at Green individually, sounds in nothing more than mere negligence and strict liability, and at no time have the plaintiffs even employed the term "wilful misconduct" in their pleadings or affidavits. While Green may have acted negligently in supplying the smoke grenades for the training exercise without taking the proper precautions or receiving the proper approval, such actions do not constitute a valid claim of "wilful misconduct." The facts as pleaded by the plaintiffs in no way establish that Green intended to injure the plaintiffs, nor do they establish that Green's actions went beyond gross negligence. Consequently, no sufficient basis has been provided by the plaintiffs warranting a finding that Green committed "wilful misconduct," and as such, there is no basis for abrogating Green's statutory immunity.

Additionally, Green is correct as to his second argument that the plaintiffs' sole remedy is through the Workers' Compensation Act, pursuant to General Statutes §§28-14(a) and 31-293a. Section 28-14(a) provides that all members of the civil preparedness forces, both part-time and full-time members, who are injured, killed or disabled while engaging in civil preparedness duties, shall "be construed to be employees of the state for the purposes of chapter 568 and section 5-142 and shall be compensated by the state in accordance with the provisions of said chapter 568 and section 5-142." Here, the plaintiffs have alleged that they were members of the civil preparedness forces and were engaged in a training exercise for the Connecticut urban search and rescue team at the time they sustained their alleged injuries. Accordingly, they are subject to the provisions of chapter 568, commonly known as the Workers' Compensation Act, which includes §31-293a. That statute states that if an employee has a right to benefits "under this chapter on account of injury or death from injury caused by the negligence or wrong of a fellow employee, such right shall be the exclusive remedy of such injured employee . . . and no action may be brought against such fellow employee unless such wrong was wilful or malicious . . ." Because the plaintiffs are entitled to such benefits and the plaintiffs have failed to plead sufficient facts to warrant a finding of "wilful or malicious" misconduct, they are bound to the exclusivity of remedies that §31-293a provides.

For the foregoing reasons, the court vacates its original ruling on this matter and grant the defendants' motion to dismiss for a lack of subject matter jurisdiction, as the plaintiffs have not alleged a sufficient claim of "wilful misconduct" to fall within the exception to statutory immunity provided to Green pursuant to §§28-13(a). Additionally, because no "wilful or malicious" misconduct has been alleged by the plaintiffs, their exclusive remedy is through the Workers' Compensation Act.

QUESTIONS

1. Under what authority did the Connecticut urban search and rescue team operate? Why was the authority under which they operated important in the context of this case?
2. What enabled Green, Director of the USAR team in this case, to avoid liability?
3. The plaintiffs in this case did not prevail. What would have been necessary for them to do so?
4. Perform research to find out under whose authority your local SAR team operates. Discuss the pros and cons of the situation under your state's law.
5. Perform research to find out whether your SAR team is federally sponsored. Whether your team is federally sponsored or not, discuss the legal implications of such sponsorship.
6. There are several types of state-level resources that have federal sponsorship similar to SAR teams. Perform research to find out what they are. Discuss the legal differences between state and federal activation of such resources. How does the process work when these resources are activated at the state level then later federalized?

Bay St. Louis, MS, August 21, 2010 – The renovated Hancock County Courthouse is open for after the 2005 hurricane flooded and destroyed its first floor and marred its exterior. Tim Burkitt/FEMA photo.

Chapter 19

EMERGENCY PREPAREDNESS AND THE COURT SYSTEM

An important element in Continuity of Government (COG) is ensuring that the court system continues to operate in the aftermath of a disaster or emergency. Fair and timely adjudication of both criminal and civil matters can be threatened if the courts do not take pro-active steps to implement emergency preparedness in a comprehensive manner.

In Re: Emergency Preparedness Planning

2006 Miss. Lexis 206 (Miss. S.Ct. 2006)

OPINION BY: JAMES W. SMITH, JR.

ORDER

This matter came before the Court, *en banc,* on the Court's own motion to adopt the Guidelines for Emergency Preparedness Planning attached hereto, which were developed by the Mississippi Supreme Court Emergency Preparedness Committee. The Guidelines for Emergency Preparedness Planning are adopted and incorporated herein as a part of this order. The guidelines were established to aid each court in Mississippi in the process of developing a unique, local emergency preparedness plan aimed at keeping the courthouse open during emergencies and/or threats, so long as the safety of the public, court officers and court personnel is not compromised.

The guidelines designate an Emergency Coordinating Officer (ECO) for each courthouse. Within sixty (60) days of the entry of this order, the circuit, chancery, and county court judges shall review the Guidelines for Emergency Preparedness Planning, confirm the ECO as designated by the guidelines, and the ECO shall identify himself or herself to the Administrative Office of Courts (AOC).

IT IS THEREFORE ORDERED that the Guidelines for Emergency Preparedness Planning are hereby adopted by this Court.

IT IS FURTHER ORDERED that the circuit, chancery, and county court judges shall review the Guidelines for Emergency Preparedness Planning, confirm the ECO as designated by the guidelines, and the ECO shall identify himself or herself with the AOC, all within sixty (60) days of the entry of this order.

SO ORDERED, this the 18th day of April, 2006.

/s/ James W. Smith, Jr.

JAMES W. SMITH, JR., CHIEF JUSTICE

Guidelines for Emergency Preparedness Planning

Make every effort to keep our courthouses open without compromising the safety of the public, judicial officers, and court personnel. . . .

STATEMENT FROM THE CHAIR

Planning ahead for emergencies is not a revolutionary idea. However, on August 29, 2005, we were reminded of the importance of emergency preparedness when Hurricane Katrina devastated the Gulf Coast region. Lives, homes and businesses were lost during that natural disaster, and the aftermath will continue indefinitely into the future to test the resolve of Mississippians to move ahead.

Of significant import, Mississippi's judicial system was also drastically affected that day. Although many of the courthouses throughout our state were able to quickly rebound from the storm, others, including our Supreme Court, were forced to completely shut down due to lost utilities or severe damage. In order to aid in the prevention of future interruptions in our court system, Chief Justice James W. Smith, Jr., established the Supreme Court of Mississippi Emergency Preparedness Committee, and charged this committee with developing guidelines that will aid Mississippi's courts in the development of unique, local level emergency preparedness plans. Thus, these guidelines have been developed with the focus of "keeping the doors of justice open during times of crisis, " unless the safety of the public, judicial officers and court personnel will is compromised.

The committee is comprised of Justice George C. Carlson, Jr. (Chair), and Justices James E. Graves, Jr., and Michael K. Randolph. On behalf of Mississippi's judiciary, the committee would like to recognize the Workgroup on Emergency Preparedness from the State of Florida for the work it has done in the area of emergency preparedness. Many of the ideas developed by that group have proved useful in assisting our committee to establish guidelines for emergency preparedness planning in Mississippi.

OVERVIEW

Developing a single, comprehensive emergency preparedness plan to be implemented by each courthouse in this state would not prove to be effective. Each county and or court facility may have different assets that can be utilized during emergencies, threats and natural disasters. Therefore, the guidelines established are merely recommendations to help judges and

local level planning committees to develop customized plans. However, the goal that each court must strive for is common throughout Mississippi's judiciary: "Make every effort to keep the courthouses open without compromising the safety of the public, judicial officers, and court personnel."

It is important for the people of this state to know that the doors of justice will not be closed and that access to the courts is always available. The justice system in Mississippi will not desert the people in times of crisis. Unfortunately, there are circumstances that may require a courthouse to be physically shut down or require evacuation. These guidelines are intended to provide guidance to the trial courts while developing local level plans for the curtailment of court proceedings and the closure of court facilities during threats and emergencies.

The key to protecting staff and keeping the doors of justice open during threats and emergencies is preparation and planning that focuses on *communication* and the *cooperation* of others. Because communication is so vital, the Supreme Court has decided to establish the channels of communication between it and the trial courts throughout the state. The section to follow entitled "Designation of Emergency Coordinating Officers" is not a recommendation. This section designates an Emergency Coordinating Officer (ECO) for each court facility and charges him or her with several duties, one of which is serving as the inter-court communications contact person in times of crisis. At the local level of planning, the courts are completely autonomous in deciding intra-court communication methods.

As will be further explained in the guidelines, each ECO should establish a Court Emergency Managing Group (CEMG). The CEMG should be responsible for the development of the Emergency Preparedness Plan (Plan). The Plan should include measures for the immediate protection of the public, judicial officers and court personnel in the face of inclement weather, fire, civil disorder, natural disasters, loss of utilities, bio hazards, violence in the courthouse and other threats. The Plan should also consider whether temporary or long-term closure of the court facility will occur.

Alternate facilities should be located, and a plan for moving essential staff and equipment should be well thought out in advance to allow for a smooth transition when facing a long-term closure of a court facility. This task will be discussed further in the guidelines.

One other vital aspect of the emergency plan is the protection of court records. Efforts are currently being made within Mississippi's judiciary to establish a statewide Electronic Case Management system. Such a system will ultimately prove to be essential in record and document preservation. In the interim, the CEMG should evaluate current methods of storage and plan for the protection of court files.

As mentioned above, the other key component in the process of emergency planning is the *cooperation* of others. It is recommended that the CEMG be comprised of the ECO as well as representatives of state and local governmental agencies, representatives of state and local emergency management agencies, judges, the district attorney, the public defender, the court administrator, the clerk of court, the sheriff, and any other representative of the judicial and executive branches of government, if appropriate. Of course, not all of these persons may be able to participate in some counties. Further, there may certainly be other individuals not mentioned who possess expertise that would greatly benefit the CEMG in its planning efforts. In some instances, it may be feasible for several ECOs to combine their efforts into one CEMG that can develop a Plan for more than one court facility within the county.

The Mississippi Supreme Court asks that the needed cooperation begin today, starting with the trial courts. Please do not look upon the task of emergency planning as a burden. Being prepared when facing a threat, emergency or disaster may save lives. We encourage each court to take these guidelines and improve upon them in any way possible. Remember that the threats and dangers that face us are forever evolving. It is important that each CEMG stay active to ensure that the emergency preparedness plan evolves too. The local level Plans and procedures that are developed should be practiced and understood by all members of the CEMG, judicial officers and court personnel. Because safety is of utmost importance, great effort should be made to train court staff in how to respond to threats and emergencies....

The Plan

15. The first stage of the Plan developed by the CEMG should outline emergency procedures, as well as administrative procedures, to prepare for, respond to, recover from, and mitigate from emergencies not involving the use of alternate facilities.

16. The Plan should consider and detail procedures for dealing with emergencies and threats including, but not limited to: catastrophic natural occurrences such as hurricanes, tornados, ice storms, floods, lightning fires and earthquakes; bomb, chemical, or biological threats; courthouse violence; civil disorder; terrorism/extremism; pandemic/quarantine and other heath issues; and intentional or accidental fires.

17. Likewise, the CEMG should develop a Continuity of Operations Plan (COOP) to prepare for, respond to, recover from and mitigate from emergencies requiring the use of alternate facilities...

18. Upon receipt of the recommendations from the CEMG, the ECO should review the recommendations, approve or revise the recommendations, and publish the Plan. Although no deadline has been established, it is likely that the Mississippi Supreme Court will require that a copy of each court facility's Plan be submitted to the Court and remain there on file.

19. As soon as a credible threat or emergency is known to exist, the ECO shall activate the appropriate emergency Plan.

20. Each ECO should designate, by written order, the judge or court officer who will serve as the alternate ECO and have the authority to decide if court proceedings will be suspended during the ECO's absence. The alternate ECO will be given authority to evacuate a court facility to preserve the safety of the public and court personnel in the event that an emergency or other imminent threat so requires.

21. In the event the ECO decides to suspend court proceedings or close a court facility, a written order of closing should be filed with the court's clerk and the DECO should be immediately notified. A copy of the order should be filed with the Clerk of the Mississippi Supreme Court.

22. If the CECO enters a closing order, an additional order should be entered to suspend, toll, or otherwise grant relief from time deadlines as may be appropriate, including, without limitation, those affecting speedy trial procedures in criminal and juvenile proceedings, all civil process and proceedings, and all appellate time limitations.

23. Upon the cessation of an immediate emergency or threat, the ECO shall consult with the CEMG to assess what actions may be necessary to resume court operations.

24. Training programs should be developed for judges, court administrators and court personnel, including bailiffs and deputy sheriffs, relating to emergency preparedness. The court facilities should take steps to ensure its officers and employees are trained in emergency procedures.

25. The Court Emergency Preparedness Plan should:

(A) Delineate what procedures will be followed and what agencies will be responsible for the decontamination of personnel [*20] and court facilities exposed to hazardous materials.

(B) Specify an evacuation procedure for each court facility.

(C) Require that each judge and all employees of the judicial branch rehearse and understand their personal evacuation route for any court facility to which they may be assigned.

(D) Require an annual or semi-annual unannounced test that is conducted during peak business hours.

26. If a court facility does not already have one, each court facility should have a Basic Life Safety (BLS) kit and portable Automated External Defibrillator (AED) in a readily accessible location.

27. An emergency medical team should be designated and its members should be trained in the administration of basic first aid, CPR, and use of an AED.

28. The Plan should address methodologies for responding to a variety of medical emergencies.

29. Each court should provide judges and employees with information to complete a Family Disaster Plan. Court personnel should be encouraged to prepare for emergencies while away from the court...

30. The ECO should define a process in which disaster assessment will be made.

31. The ECO should establish an Emergency Management Team that would be convened during times of emergency or impending emergencies. It may be necessary to call upon members of the CEMG to assist with the emergency or when activating the COOP. The ECO should also notify his or her DECO of the event.

32. The ECO should implement methods to physically secure paper records, monitor and control environmental factors at paper records storage sites, develop emergency procedures to protect paper records when disaster is imminent, and develop mitigating measures if paper records are lost or destroyed.

33. The ECO should determine which personnel are essential and nonessential when a court facility is closed due to emergency or threat. A method of communication should be established to communicate with essential court personnel. Additionally, a method of communicating with all nonessential personnel should be established for notification of the emergency and the court's closing. A telephone tree can prove to be a useful means of communication. Additionally, essential contact information for each [*22] employee should be maintained by the court and/or department supervisors. (See Exhibit "D" for a sample form.)

34. To ensure efficient and clear lines of communication during emergencies, the ECOs should immediately activate, upon knowledge of an emergency threat, the court's telephone tree.

35. The ECOs should develop an Employee Emergency Recovery Guide for use by employees following an emergency.

36. The Plan should identify, in advance, individuals who can serve on recovery teams to

assist employees with identifying and accessing the support services needed.

37. The Plan should be disseminated to the appropriate agencies affected by the Plan, and to judges and court personnel who may have responsibilities for the execution of the Plan....

(Note that the Supreme Court of Mississippi's full Opinion is a comprehensive, well thought out, and detailed approach to emergency preparedness planning. Readers desiring complete information should consult the full Opinion.)

QUESTIONS

1. What is the role of the court system in the aftermath of a disaster?
2. Why is emergency preparedness planning important for the court system?
3. Perform research into the emergency planning in your local court system. How does it compare with the Mississippi approach?

New Orleans, LA, August 20, 2005 – Neighborhoods are flooded and New Orleans is being evacuated as a result of floods caused by Hurricane Katrina. Photo by Jocelyn Augustino/FEMA.

Chapter 20

"WHEN THE WIND BLOWS": THE ROLE OF THE LOCAL GOVERNMENT ATTORNEY BEFORE, DURING, AND IN THE AFTERMATH OF A DISASTER

JOSEPH G. JARRET AND MICHELE L. LIEBERMAN

I beg you take courage; the brave soul can mend even disaster.[1]

I. INTRODUCTION

In the Ojibwa[2] tongue, the term for disaster is whangdepootenawah: "an unexpected affliction that strikes hard."[3] When disasters do strike, it immediately becomes obvious that the legal issues involved in local government disaster planning are some of the most misunderstood and confusing aspects of the entire process of disaster preparation and recovery. This makes the prudent involvement of the local government attorney essential. The local government attorney is encumbered with the responsibility of understanding and interpreting the seemingly conflicting and ever-changing body of federal, state, and local regulatory laws, rules, and guidelines. It is likewise the local government attorney who is called upon to be the bearer of bad news regarding the legal consequences of a public entity's failure or inability to execute an effective disaster-recovery plan properly. This Article will explore the evolving body of emergency management law, as well as the practical, and at times impractical, application of that law. The Article's analysis of state law will focus on Florida, but its discussion of the challenges that a disaster poses should resonate with all local government attorneys.

II. THE LAW

While the local government attorney clearly must be familiar with a wide variety of matters, the knowledge of federal, state, and local laws governing emergency preparedness and response is one of the most critical. Failure to follow the requirements of all applicable laws during each phase of a disaster may significantly impact a local government's ability to receive assistance in its most crucial time of need.

Note: From Joseph G. Jarret and Michelle L. Lieberman, "When the Wind Blows: The Role of the Local Government Attorney Before, During, and in the Aftermath of a Disaster," *Stetson Law Review, 36*(2):293, 2007. Reprinted by permission

A. Federal Emergency-Response Legislation

In recognition of the loss and disruption caused by disaster, Congress has declared that "special measures, designed to assist the efforts of the affected States in expediting the rendering of aid, assistance, and emergency services, and the reconstruction and rehabilitation of devastated areas, are necessary."[4] From disaster preparation to post-disaster recovery, it is Congress' intent, through its emergency-response legislation, "to provide an orderly and continuing means of assistance by the Federal Government to State and local governments in carrying out their responsibilities...."[5] This assistance consists of improving and expanding disaster relief, providing private and public federal assistance programs, encouraging development of comprehensive disaster plans, improving coordination and responsiveness, encouraging hazard mitigation through land use and construction controls, and encouraging the procurement of insurance to reduce reliance upon federal aid.[6] Under this approach, the federal role in emergency management is clearly more facilitative than directive.

Specifically, "the Federal Government shall provide necessary direction, coordination, and guidance, and shall provide necessary assistance ... so that a comprehensive emergency preparedness system exists for all hazards."[7] The responsibility for emergency preparedness "vests" in the federal government, the states, and their political subdivisions jointly.[8] During a disaster, however, a presidential declaration of emergency must initiate the federal response.[9]

A presidential declaration may be issued, at the president's discretion, upon the request of the governor of an affected state that meets certain criteria, and upon "a finding that the situation is of such severity and magnitude that effective response is beyond the capabilities of the State and the affected local governments and that Federal assistance is necessary."[10] Further, the president may initiate an emergency response without a governor's request in those circumstances for which the primary responsibility for response rests with the United States because the emergency involves a subject area for which, under the Constitution or laws of the United States, the United States exercises exclusive or preeminent responsibility and authority.[11]

The effect of a presidential declaration upon an affected state is extensive.[12] Under such authority, the president is authorized to do the following: direct the resources of any federal agency to assist state and local governments; coordinate disaster relief among federal, state, local, and private providers; provide technical and advisory assistance on a myriad of issues related to health and safety; assist with debris removal; assist in providing medicine, food, and other essential supplies to state and local governments; and provide any other assistance necessary to the protection of life and property.[13] The extent of financial assistance, however, is limited absent congressional oversight and approval.[14]

B. Florida's Emergency-Response Legislation

Federal law states that a governor, before requesting a federal emergency declaration, must act in accordance with governing state law and initiate the state's own emergency plan.[15] In contrast to federal restrictions on the president's authority, however, Florida law does not condition the governor's ability to declare a state of emergency on any specific prerequisite other than the existence of an actual or impending "emergency."[16] Any emergency declaration results in the classification of the situation as a disaster and makes state assistance available to respond to the disaster.[17] In addition, the declaration of a state of emergency activates local emergency management plans, allows for such assistance as the distribution of necessary supplies and equipment within the possession of the State,[18] and vests authority in the governor as commander in chief of the Florida National Guard and "other forces available for emergency duty."[19]

Further, a state of emergency provides the governor with additional authority not otherwise present, such as the ability to impose curfews, order evacuations, determine means of ingress and egress to and from affected areas, and "commandeer or utilize private property" subject to compensation.[20] It should be noted that a local government generally exercises similar authority in the execution of its adopted emergency management plan. While, in practical application, aspects of emergency preparedness, response, and recovery such as evacuations and curfews are more often exercised by the local government, such authority must necessarily be subordinate to any order that the governor issues.

Unfortunately, emergency management is an all too familiar exercise in Florida. If Tornado Alley is defined as the "geographic corridor in the United States...[that] receives more tornados than any other,"[21] then Florida certainly sits within Hurricane Alley. As a state historically within the sights of many a hurricane eye, Florida has a State Emergency Management Act that is both extensive and complex.[22] Like the president, who looks to the Federal Emergency Management Agency (FEMA) to implement federal emergency programs, the governor, while maintaining ultimate state authority during a disaster, relies upon the Division of Emergency Management (the Division) for the administration of the State Emergency Management Act.[23] The Division is "responsible for maintaining a comprehensive statewide program of emergency management," and for "coordination with efforts of the Federal Government[,] with other departments and agencies of state government, with county and municipal governments and school boards, and with private agencies that have a role in emergency management."[24] The Division's responsibility to carry out the provisions of the State Emergency Management Act also includes the duty to "create, implement, administer, adopt, amend, and rescind rules, programs, and plans needed to carry out the provisions of [Florida Statutes Sections] 252.31-252.90...."[25]

In this role, the Division is specifically responsible for such activities as the preparation and implementation of Florida's comprehensive emergency management plan under the legislature's direction, as well as the establishment of standards for local emergency management plans to ensure consistency with the state plan.[26]

The State Emergency Management Act provides the Division with additional authority to fund certain financial assistance programs,[27] to review local government comprehensive emergency management plans,[28] and to establish rules governing the operation of local emergency management agencies.[29] Moreover, the Division maintains oversight of, and rulemaking authority for, such mandates to local governments as maintaining a registry of persons with special needs;[30] providing for the evacuation and sheltering of their citizens; and adequately addressing "preparedness, response, recovery, and mitigation."[31] A local government's compliance with such standards is often linked to critical funding.[32] For example, receipt of funds from the Emergency Management, Preparedness, and Assistance Trust Fund requires that, at a minimum, a local emergency management agency located in a county with a population of more than 75,000 have a program director who works at least forty hours a week in that capacity.[33] It is, therefore, essential that local government attorneys familiarize themselves with these requirements, and ensure compliance, prior to the occurrence of any impending or actual disaster.

C. County Governments' Statutory Authority and Responsibility

Irrespective of the significant, and apparently concurrent, role that the State maintains in the local emergency management function, it is the local government that is on the front line of any disaster. Thus, the local government attorney should become intimately familiar with the local comprehensive emergency plan and with the extent and limitations of authority that a local government may exercise during an impending or realized disaster.

Since the adoption of the State Emergency Management Act in 1974, the Florida Legislature has expressly declared that "safeguarding the life and property of its citizens is an innate responsibility of the governing body of each political subdivision of the state."[34] Pursuant to this declaration, counties are specifically empowered to serve the emergency management function for both their incorporated and unincorporated areas.[35] Such authority requires the establishment of a county emergency management agency and the creation of an emergency management plan that is consistent with the state plan.[36] In addition, each county emergency management agency must appoint a director who meets the qualifications established in a job description approved by the county, and who is to serve under the county's direction.[37] The director is accountable for the "organization, administration, and operation of the county emergency management agency" and is responsible for the coordination of emergency management activities.[38] The director also serves as the county's liaison to other state and local entities.[39] Accordingly, a county must inform the Division of Emergency Management of its appointment of a director and all other personnel within the local emergency management agency.[40]

The director of a local emergency management agency, however, need not be a direct employee of the county. As the statute specifically allows, a county may choose to appoint as director either a county constitutional officer or an employee of a constitutional officer, as long as the county provides the Division with advance notification of its intent to make the appointment.[41] Notably, however, no distinction exists between the authority and responsibility of an appointed constitutional officer or employee thereof and any other individual who may be appointed to such position.[42] State law thus unambiguously provides for a level of oversight consistent with a county government's "innate responsibility" to safeguard its citizens.[43] It is, therefore, advisable for the local government attorney to recommend a clear delineation of responsibilities and oversight of this position should the county elect to appoint as director an individual who would not traditionally be subject to the county's control.

Such a delineation may appropriately be accomplished through an interlocal agreement in accordance with Florida Statutes Chapter 163. The Florida Interlocal Cooperation Act provides statutory authority for local governments to cooperate with other localities on a basis of mutual advantage and thereby to provide services and facilities in a manner and pursuant to forms of governmental organization that will accord best with geographic, economic, population, and other factors influencing the needs and development of local communities.[44]

This ability of local governmental entities to exercise power jointly or to agree to provide services to each other is not limitless, however. While public agencies may jointly exercise "any power, privilege, or authority which such agencies share in common and which each might exercise separately,"[45] an interlocal agreement cannot "extend the existing authority of the parties to the agreement . . . when not all the parties to the interlocal agreement possess independent authority" to act.[46]

A review of the State Emergency Management Act establishes that any local actions taken during an emergency are taken on the sole authority of the local government.[47] As noted earlier, a county, as a political subdivision of the State, may act independently to "make, amend, and rescind such orders and rules as are necessary for emergency management purposes and to supplement the carrying out of the provisions of [Sections] 252.31-252.90," provided that such orders and rules are not inconsistent with any orders or rules of the State.[48] In so acting, a political subdivision shall have the power and authority, among other acts, to do the following: "provide for the health and safety of persons and property, including emergency assistance to the victims of any emergency"; "appoint, employ, remove, or provide, with or without compensation, coordinators, rescue teams, fire and police personnel, and other emergency management workers"; "establish, as necessary, a primary and one or more secondary emergency operating centers to provide continuity of government and direction and control of emergency operations"; "assign and make available for duty the offices and agencies of the political subdivision, including the employees, property, or equipment thereof relating to firefighting, engineering, rescue, health, medical and related services, police, transportation, construction, and similar items or services for emergency operation purposes"; and "request state assistance or invoke emergency-related mutual-aid assistance by declaring a state of local emergency."[49]

No constitutional officer is provided such authority, so there can be no joint exercise of that authority.[50] Therefore, any interlocal agreement allowing a constitutional officer, or employee thereof, to serve as a county's emergency management director should not purport to delegate to the director any authority that belongs exclusively to the county. Further, the agreement should clearly require the approval and oversight of the county as the party ultimately liable for any noncompliance.

D. Municipalities' Statutory Authority and Responsibility

Unlike counties, municipalities are not required to create an emergency management agency, but rather are "encouraged to create municipal emergency management programs" in coordination with the county's emergency management agency.[51] Should the municipality elect to do so, its authority is not independent. Rather, a municipal emergency management plan must comply not only with those laws and rules applicable to counties, but also with its county's emergency management plan.[52] In the absence of a separate municipal plan, the municipality is subject to the county's plan.[53] Regardless of whether a municipal plan exists, a municipality must coordinate its requests for federal and state emergency response assistance with the county in which it is located.[54]

This is not to say, however, that a municipal attorney has less responsibility than any other local government lawyer in an emergency situation. To the contrary, while a municipal attorney does need to be constantly aware of the hierarchy of authority that the State Emergency Management Act creates, he or she will be called upon to make the same difficult decisions and to perform the same analyses in a crisis situation as a county attorney. Notwithstanding any limitations on its authority, a municipality is included within the definition of a "political subdivision"

for purposes of the Act[55] and thus possesses the same expansive emergency power and authority as a county.[56]

E. The Role of County and Municipal Law Enforcement

County and municipal law enforcement agencies have an indisputably critical and essential role in the preparation for, and implementation of, an emergency management response. Specifically, law enforcement is responsible for enforcing the rules and orders promulgated by state and local authorities pursuant to the State Emergency Management Act.[57] Law enforcement, however, has limited authority to act independently in a crisis situation.

Sheriffs or designated city officials are authorized to declare a state of emergency within their jurisdictional limits[58] independently of county or municipal action when there has been an act of violence or a flagrant and substantial defiance of, or resistance to, a lawful exercise of public authority and [when], on account thereof, there is reason to believe that there exists a clear and present danger of a riot or other general public disorder, widespread disobedience of the law, and substantial injury to persons or to property, all of which constitute an imminent threat to public peace or order and to the general welfare of the jurisdiction affected or a part or parts thereof. . . . [59]

Under such authority, a sheriff or designated city official "may order and promulgate" such emergency measures as imposing curfews and closing places of public assemblage, with any limitations and conditions the official may deem appropriate.[60] The state of emergency can continue for up to seventy-two hours without approval from any other government official.[61]

In comparison, the authority of a local government to impose emergency measures pursuant to its emergency management authority is not limited to an impending threat of riot, general public disorder, or widespread disobedience of the law.[62] To the contrary, a local government may impose a wide array of restrictions pursuant to its inherent responsibility to protect the life and property of its citizens during times of declared emergency.[63] Thus, communication between law enforcement and local government both during and after a disaster regarding the need for such measures is essential to the maintenance of an orderly recovery phase.

III. LESSONS LEARNED, OR FORGOTTEN?

In 2004, Florida suffered devastating losses from a series of four hurricanes that crossed the state in rapid succession – Charley, Frances, Ivan, and Jeanne.[64] Subsequently, during the 2005 legislative session, approximately eighteen bills were filed that would have had a direct impact on emergency management.[65] Of those eighteen bills, only one survived.[66] A review of the failed legislation shows that much of it was almost prophetic, given the subsequent 2005 hurricane season. For example, legislation mandating that nursing homes have a source of alternate electrical power capable of providing power for a specified period, and requiring priority restoration of electrical power to such facilities, died in committee.[67] The devastating impact of Hurricane Katrina during the 2005 hurricane season unequivocally demonstrated the necessity of such critical requirements.[68]

But while the unforgettable images of Florida property damage in 2004 apparently did not spur action, the inescapable images of human tragedy in 2005 appear to have had a much more profound impact. In comparison to the eighteen bills filed in the 2005 legislative session, approximately thirty-three bills were filed during the 2006 session.[69] Several laws were enacted as a result, and many provisions from those bills that failed individually were swept into House Bill 7121.[70] The new legislation that survived independently included the following: state facilities suitable for use as public shelter space must be made available to local governments, and a list of such facilities must be created;[71] amendments to a local government comprehensive plan covering areas subject to coastal high-hazard regulations must establish a specific level of service with regard to hurricane evacuation clearance times;[72] and the Division of Emergency Management is now directly answerable to the governor.[73] While the Department of Community Affairs will continue to provide support services and assistance in nonemergency matters, the Division is no longer under its control, supervision, or direction.[74] Similar to the manner in which a county government exercises authority over its local emergency management agency, the Division's director will now be appointed by the governor, to serve at his or her pleasure.[75]

But clearly the most significant legislation to be adopted in 2006 was House Bill 7121 (HB 7121), which became Chapter 2006-71 of the Laws of Florida. This sweeping legislation has provided both extensive mandates and significant opportunities for local governments that could, in and of themselves, be the subjects of an article. Thus, the local government lawyer should review these changes beyond their cursory treatment in this Article.

HB 7121 provides more than $150 million in appropriations to further disaster planning and response.[76] Building upon the lessons learned from prior storm seasons, the Legislature committed significant funding to enhance public education, improve local emergency operation centers, retrofit shelters, improve storage capacity for needed supplies, equip special-needs shelters with a permanent alternate power source, and assist in evacuation planning.[77] Each local government must equip its special-needs shelters with a permanent alternative power supply with the "capacity...to provide...for necessary medical equipment for persons housed in the shelter and for heating, ventilating, and air-conditioning the facility."[78] This must occur by June 1, 2007.[79]

A clear sign of lessons learned is that the need for alternate power supplies seems to be the theme throughout HB 7121.[80] One commodity found to be indispensable to disaster response and recovery is fuel. Because evacuations necessarily require fuel, certain existing retail fuel facilities within close proximity to an evacuation route must take steps to install an alternate power source "capable of operating all fuel pumps, dispensing equipment, life-safety systems, and payment-acceptance equipment" by the start of the 2007 hurricane season.[81] As of July 1, 2006, no "newly constructed or substantially renovated motor fuel retail outlet" may receive a certificate of occupancy unless such capability exists.[82]

After a disaster, communities depend on fuel for recovery activities including the operation of generators, government officials' assessment of damage, and the operation of emergency-response vehicles.[83] For that reason, all motor fuel terminal facilities and wholesalers must also be capable of providing fuel through an alternate generated power source by June 1, 2007.[84] This equipment must be available within thirty-six hours, and must remain available for at least seventy-two hours after a disaster.[85] However, facilities whose owners operate a fleet of motor vehicles or sell fuel exclusively to such a fleet are exempt.[86] The many local governments that own and operate their own fuel terminals will likely fall under this exemption.

Beyond ensuring an ability to dispense fuel is the need to ensure its availability for essential government functions. The Florida Disaster Motor Fuel Supplier Program addresses this issue.[87] Each county can elect to participate in this program, which allows a retail motor fuel outlet located in the county "to participate in a network of emergency responders to provide fuel supplies and services to government agencies, medical institutions and facilities, critical infrastructure, and other responders, as well as the general public," during a state of emergency declared by the governor.[88] These outlets must be capable of providing fuel within twenty-four hours after a disaster[89] and must give priority to State Emergency Response Team members.[90] Further, emergency management officials may permit these facilities to operate during any imposed curfew for the purpose of providing fuel to emergency personnel.[91]

Because each county has the option of deciding whether to participate in the fuel program,

once the decision is made, local emergency management agencies are responsible for administering the program.[92] This responsibility includes the establishment of criteria for evaluating a motor fuel outlet's readiness and ability to participate in the program.[93] In addition, a county may charge a fee for the cost of reviewing and accepting a motor retail facility into the program.[94] A local government, however, is expressly preempted from regulating the placement of the alternate power source at any fuel facility, or otherwise regulating any other type of retail facility that the State has approved to participate in emergency response.[95]

While ensuring the availability of motor fuel in a disaster is clearly essential, another lesson learned from the most recent bout of hurricanes is that a loss of power may prevent escape for, or assistance to, a certain segment of the population – specifically, those citizens who reside in multiple-story buildings.[96] Accordingly, HB 7121 addresses the necessity for an alternate power supply source to ensure vertical accessibility. Existing multifamily residential dwellings that are seventy-five feet or more in height and that contain a public elevator must now have at least one such elevator pre-wired for an alternate power source capable of operating after a disaster.[97] The power source must also be capable of operating the building's connected fire alarm system and providing emergency lighting in public areas.[98] The owner or manager of such a multifamily dwelling unit was required to submit engineering plans to the local building inspector by December 31, 2006, establishing this capability, and the county must verify the installation and operation of such improvements and report to the local emergency management agency no later than December 31, 2007.[99]

As of the date of enactment, newly constructed multifamily residential dwellings must be built to meet this requirement, and the local government inspector must provide verification of the installation and operation of the alternate power source to the local emergency management agency prior to occupancy of the building.[100] All such buildings installing alternate power supplies must also maintain a generator key in a lockbox for emergency access.[101]

As Hurricane Katrina demonstrated, however, it was not merely an inability to evacuate that kept people from heeding the warnings of emergency officials.[102] More emotional issues often play a role in the decision not to evacuate. For example, the thought of leaving a beloved pet behind to suffer during a time of disaster is unthinkable to many people.[103] Moreover, animals that are left behind not only suffer from lack of care and food, but can pose hazards for government personnel attempting to conduct recovery activities.[104] In recognition of these concerns, HB 7121 charges the Division of Emergency Management, in conjunction with the Department of Agriculture and Consumer Services, with the development of strategies within the shelter component of the State Emergency Management Plan for the evacuation of persons with pets.[105] Once established, these strategies must also be reflected in local governments' emergency management plans.[106]

Finally, HB 7121 includes a substantial rewrite of Florida Statutes Section 381.0303 regarding special-needs shelters.[107] The new provisions attempt to delineate more clearly the division of responsibilities for the operation and staffing of such facilities.[108] Undoubtedly, the most vulnerable members of our population, those with special needs, often pose a tremendous challenge for local governments attempting to provide adequate care for a diverse group of individuals and ailments. Consequently, home health agencies,[109] nurse registries,[110] and hospice providers[111] must now include within their comprehensive emergency management plans a method to provide to their patients, within a special-needs shelter, continued services of the same type and quantity as the services provided prior to evacuation.[112] Moreover, a link between these entities and the local emergency operations center may be established to provide information to assist the entity in reaching a client in a specific disaster area.[113] In addition to the above-noted medical providers, home medical equipment providers are now also required to have an emergency management plan.[114] Such providers are subject to the same requirements for establishment of continued service to their patients within a special-needs shelter and are likewise able to establish a link to the local emergency operations center for reaching clients after a disaster.[115]

Challenges for the future of emergency management will necessarily involve the continued implications and implementation of HB 7121. A local government's compliance with, and utilization of, such provisions will undoubtedly serve to ensure the welfare of its citizens. Additionally, those issues that failed to pass, such as regulations pertaining to the sheltering of sexual offenders, will certainly return in some form during the 2007 session.[116] Thus, the local government lawyer should remain vigilant and aware as he or she assumes the role of advocate and counselor in the eye of the storm.

IV. THE ROLE OF THE LOCAL GOVERNMENT ATTORNEY BEFORE, DURING, AND AFTER THE STORM

Something as seemingly perfunctory as defining the role of the local government attorney in a post-disaster world can prove to be an arduous task. Due to the diverse nature of Florida's political subdivisions in such terms as size, sophistication, and resources, it would not be prudent to assume that the roles local government attorneys may be asked to play are the same throughout the State. Consequently, it is incumbent upon the attorney to learn, in advance, what will be expected of him or her in a disaster scenario. It is not at all uncommon for attorneys to be thrust into roles or given responsibilities that are not only outside of their job descriptions, but consist of areas of the law that are alien to them. Further, attorneys operating in an emergency or crisis environment are routinely asked to render timely legal decisions with either incorrect or insufficient information at best. Aggravating this situation is the fact that the power outages that accompany most disasters deny attorneys the use of electronic research, the ability to confer with colleagues, access to computer databases, and the use of other helpful resources. Finally, elected and appointed officials, likely to be overwhelmed by operational decisions often cloaked in political overtones, will look to the local government attorney to assist them in making a myriad of difficult, emotionally charged decisions, many of which are not supported by venerable precedent, black-letter law, or established policies and procedures.

By way of anecdote, in the aftermath of Hurricane Katrina, lawyers in Louisiana marshaled their numbers in an effort to respond to the needs of clients and the general public. In response to this crisis, the Louisiana Bar Association, recognizing both the need for, and willingness of, attorneys to step out of their various areas of expertise and comfort zones, created the Emergency/Disaster Training Manual for Volunteer Lawyers Following Hurricane Katrina.[117] Of the panoply of issues these lawyers faced, the most prevalent were:

- Assistance with filing for emergency assistance;
- Assistance with insurance claims (life, property, medical, etc.);
- Counseling on lessor-lessee, homeowner, and other housing problems;
- Assistance with home repair contracts;
- Assisting in consumer protection matters, remedies, and procedures;
- Counseling on mortgage foreclosure problems;
- Replacement of important legal documents destroyed in the natural disaster, such as wills, green cards, and the like ... ; ... Drafting of powers of attorney;
- Estate administration (insolvent estates);
- Tax questions;
- Preparation of guardianships and conservatorships; [and]
- Referring individuals to local or state agencies which might be of further assistance (e.g. consumer affairs).[118]

Putting the Louisiana experience into a purely local government law perspective, the local government attorney will undoubtedly be called upon to:

- Ensure that his or her governmental entity complies with federal and state declarations of disaster;
- Draft and enforce local declarations of emergency, and ensure that these declarations are lawfully renewed as needed;
- Draft contracts for the utilization of private resources;
- Muddle through and become familiar with the myriad federal and state regulations and codes governing outside assistance;
- Clearly define the effect of the most current FEMA regulations related to debris removal and reimbursement; and
- Take whatever steps are necessary to protect both the body politic and the entity's employees and volunteers from liability.

Local government attorneys play another crucial role. They must assist their respective entities in navigating the issues that emergency preparedness and response present by playing "a key role in creating functional models for intergovernmental coordination and cooperation."[119] Alan D. Cohn, chair of the Legal Issues Working Group of the National Urban Search and Rescue Response System Advisory Committee, defined this role to include the following:

- Understanding the sources and limitations of state and local government authority to prepare for and respond to catastrophic incidents;
- Knowing federal, state, and local statutes, regulations, and ordinances governing emergency preparedness and response, including those involving intergovernmental cooperation, hazardous materials response, and terrorism;
- Assisting in the construction of lines of authority for emergency preparedness and response that are consistent with both applicable legal authorities and identified operational prerogatives;

- Mediating among competing factions, either within a single jurisdiction (police, fire and rescue, emergency medical services) or among different jurisdictions or different levels of government (federal, state, local,..., etc.);
- Understanding workers' compensation and tort liability issues, including the workers' compensation exclusive remedy provision, the boundaries and limits of sovereign immunity, and applicable tort claims statutes;
- Addressing interoperability issues, including interoperable communications and equipment issues that require interpretation of legal standards and rulings; and
- Making sure that legal issues do not hamstring emergency operations, while ensuring that arrangements for emergency operations minimize the legal risk to participating jurisdictions.[120]

By remaining cognizant of the above, the local government attorney is in a better position to diffuse interlocal governmental rivalries and to refute claims that the attorney is invariably a part of the problem and not the solution.

V. A COMMON CHALLENGE

Although each disaster brings with it its own unique challenges, certain significant commonalities exist. For example, California officials conducting wildfire after-action reports in the 1970s noted the following problems:

- Too many people reporting to one supervisor;
- Different emergency response organizational structures;
- Lack of reliable incident information;
- Inadequate and incompatible communications;
- Lack of a structure for coordinated planning between agencies;
- Unclear lines of authority;
- Terminology differences between agencies; [and]
- Unclear or unspecified incident objectives.[121]

These findings are telling indeed, as you need only review the report generated by the House Select Bipartisan Committee on Hurricane Katrina to note striking similarities to the California report, especially with regard to the bureaucratic maze that public servants, attorneys among them, are often required to negotiate.[122] Both of these reports make clear that pre-planning and prior coordination with neighboring public entities are key factors essential to emergency preparedness and response. Local government attorneys should take the time to review one another's ordinances and resolutions in an effort to eliminate both duplicative effort and language that serves to hamper rather than foster intergovernmental cooperation.

To help achieve this goal of cooperation, the National Emergency Management Association has identified some of the steps local governments should take:

- Identify potential hazards...using an identification system common to all participating jurisdictions;
- Conduct joint planning, intelligence sharing and threat assessment development...[and] joint training at least biennially;
- Identify and inventory the current services, equipment, supplies, personnel and other resources...; [and]
- Adopt and put into practice the standardized incident management system approved by the State Emergency Management Agency.[123]

In addition, something as seemingly basic as learning whether emergency radio systems utilized by various entities can "talk to one another" when traditional telecommunications systems break down[124] affords the attorney the opportunity to ensure continuity of communications or to initiate the steps necessary to create continuity.

VI. MUTUAL AID AGREEMENTS

The entire notion of intergovernmental cooperation most clearly manifests itself in the creation and execution of mutual aid agreements. Florida law permits political subdivisions to enter into mutual aid agreements designed to provide reciprocal emergency aid during those times when one entity's resources are insufficient for the job at hand.[125] Florida law likewise empowers the governor to enter into mutual aid agreements or compacts with other states during times of disaster or emergency.[126] Needless to say, mutual aid relationships are essential to fulfilling an entity's emergency preparedness obligations. Rare is the local government entity that is staffed, equipped, and in possession of all of the knowledge and resources necessary to meet every conceivable disaster, natural or man-made, that comes its way. This is especially true when the disaster is of the magnitude of Hurricane Katrina. To be sure, a disaster almost by definition includes conditions, events, or occurrences of such magnitude or severity that the normal governmental services are insufficient, ill-prepared, or inadequate to respond. Consequently, reliance on mutual aid, joint powers, and intergovernmental assistance agreements becomes expected and the norm. Help from another municipality may come in the form of equipment, supplies, information, technology, and personnel. The duration and severity of the disaster will likely dictate the extent of assistance required from other governmental entities, whether local, state, or federal.[127]

As you would imagine, the utility and effectiveness of a mutual aid agreement will be wholly contingent upon the manner in which it is drafted. Although "one size fits all" agreements seldom translate well from region to region, there are nevertheless some key components that should be included in every agreement. Mr. Cohn, of the National Urban Search and Rescue Response System Advisory Committee, identified the following critical considerations that any attorney should address when drafting a mutual aid agreement:

- Clarify the legal authorities under which the jurisdictions are entering into the agreement, taking into account any limitations those authorities impose on the jurisdiction;
- Set forth the procedures to be used for requesting and providing assistance;
- Clarify workers' compensation arrangements, including whether each jurisdiction will be responsible for providing workers' compensation coverage for its own employees or whether the requesting jurisdiction will provide such coverage, and whether employees of the responding jurisdiction are intended to become special employees of the requesting jurisdiction for the purposes of the response;
- Address liability and immunity issues, including how governmental immunities are intended to apply...;
- Identify whether reimbursement will be available for services provided, and if so, set forth procedures, authorities, and rules for payment, reimbursement, and allocation of costs;
- Require each jurisdiction to develop [and share] standard operating procedures describing how the mutual aid agreement will be implemented; and
- Require the use of a standardized incident command or management system....[128]

Mr. Cohn further identified the following important, but less critical, drafting considerations:

- Spell out notification procedures;
- Define relationships with other agreements among jurisdictions;
- Recognize qualifications and certifications (e.g., emergency medical technician, paramedic) across jurisdictional lines;
- Encourage participation by a broad range of emergency responders;
- Mandate joint planning, training, and exercises, with liability provisions operating as if an actual emergency had occurred;
- Set up protocols for interoperable communications;
- Develop forms, manuals, and other job aids to facilitate requests for aid, recordkeeping

regarding movement of equipment and personnel, and reimbursement;

- Include a provision requiring arbitration of disputes concerning reimbursement; and
- Keep agreements as short as possible, and use appendices and standard operating procedures where possible.[129]

Recognizing the importance of mutual aid agreements, the Florida Association of County Attorneys created a standardized Mutual Aid Agreement.[130] The Agreement recognizes that, "in terms of major or catastrophic disaster, . . . the needs of the residents of the local communities will most likely be greater than individual local resources . . . available to meet such needs."[131] Consequently, counties throughout Florida have entered into these agreements in the hopes of fostering greater intergovernmental cooperation, as well as encouraging a pooling of talent and resources.

VII. A WORD ABOUT VOLUNTEERS

Volunteers can be indispensable partners in disaster recovery operations. Unfortunately, they can also be the bane of emergency management officials' existence if not properly managed. Often, entities are put in the tenuous position of hosting volunteers who lack necessary skills and who, despite their good intentions, prove to be more of a liability than an asset. Often, the unsavory task of culling out those volunteers who benefit operations from those who unwittingly hamper them falls to the local government attorney. Volunteers must be managed much like other human resources, with the glaring distinction that they are in the area for wholly altruistic reasons. They offer their time, talents, abilities, and equipment, give of wallet and purse, and permit the use of their real property. Florida law extends workers' compensation benefits to volunteers, a fact that local governments should consider prior to giving out volunteer assignments.[132] Further, some volunteers donate items that are not needed by disaster victims or workers. It is therefore advisable to establish a central clearing area so that items can be sorted and distributed accordingly.

VIII. DONATIONS MANAGEMENT

Author Jean Cox suggests that public entities should have in place a donations management plan designed to account for items received and to ensure that items of most use to the response and recovery effort are distributed in a timely manner.[133] In discussing the operation of New York's Comprehensive Emergency Management Plan during the terrorist attacks of September 11, 2001, Ms. Cox provided three checklists based on the Plan for managing donations in all phases of a disaster.

Preparing for donations management. Annually update the donations management plan by canvassing all human service agencies that address human services needs before and after a disaster:

- Develop and maintain an inventory of warehouse facilities;

- Develop a toll-free number to handle inquiries;
- Establish procedures to accept cash;
- Develop and maintain a database for recording offers of donated monies, goods[,] and services;
- Establish a policy for distributing the goods remaining after the relief effort ends;
- Exercise the donations plan; [and]
- Designate and train staff on the donations management team.[134]

Responding to a Disaster with a Donations Management Plan. Activate the donations management plan:

- Alert local, state, and federal governments;
- Notify the volunteer network;

- Place the donations management team on stand-by, and determine staffing and support needs based on the incident;
- Identify warehouse space and staging areas and secure agreements for their use, if necessary;
- Search the database to identify goods or previous offers that may be useful in this event;
- Coordinate with the public information officer to encourage the media to request that goods and services be held locally until needed;
- Initiate a toll-free line and phone bank for donated goods and services;
- Maintain continuous contact with involved agencies through the donations management team, to ensure a smooth flow of goods and services to the disaster area;
- Monitor news accounts, to the extent practicable, to anticipate the number and type of goods that may arrive and try to divert them to the appropriate staging or warehouse areas;
- Prepare daily status reports that document any issues and track the goods going into and out of the warehouses or staging areas; [and]
- Maintain all records of purchases, rentals, loans, and agreements to facilitate potential reimbursement.[135]

Post-Emergency. Assess the continuing needs of the agencies involved in the recovery effort:

- Determine if the scale of the donations management team is still appropriate to the effort;
- Scale back the team if needed;
- Ensure the database has been updated;
- Assist with thank you letters; [and]
- Conduct a post-event evaluation.[136]

IX. FINAL THOUGHTS

Oscar Wilde once wrote, "To expect the unexpected shows a thoroughly modern intellect."[137] Lawyering for a local government entity during a crisis is rarely planned or anticipated. Disasters generate a wide array of crises, making it almost impossible to foresee even a substantial fraction of the situations that will mandate legal acumen and intervention. Alfred O. Bragg, III, who served as legal counsel to the Florida State Emergency Response Team during the devastating hurricanes of 2004, later observed that, "Like the potential universe of smaller emergencies bred by one large disaster, the potential universe of legal issues awaiting emergency management counsel has no boundaries; counsel may expect almost anything."[138] Howard D. Swanson, a Local Government Fellow with the International Municipal Lawyers Association, offered the following advice to local government attorneys, based on his experiences in the Office of the City Attorney for Grand Forks, North Dakota, during a period of severe flooding:[139]

- Become involved in emergency management planning. Make sure that the emergency manager is aware of what services you can provide and the benefits of allowing your involvement. Demonstrate that you will be a help, not an impediment.
- Participate in any emergency or disaster drills that might be held, [and don't hesitate to recommend those that aren't being held].
- Be available to perform more than one function in combating an emergency or disaster. This may require you to serve as a legal advisor in addition to performing other functions such as equipment and supply procurement, volunteer coordination, evacuation and housing coordination, sandbagging, directing traffic...whatever is required. Request to be located in the emergency operation center during times of emergency or disaster, if possible.
- Be available for drafting emergency or disaster declarations or proclamations. Pre-drafted forms may be of assistance[;] however, be

sure to include sufficient flexibility to address whatever circumstances confront the municipality. Maintain an expansive grant of authority with liberal interpretation in any proclamation or declaration. Emergencies and disasters are inherently unpredictable. Encourage the proper official issuing the proclamation or declaration to take full advantage of any length of time the declaration can be made effective. It can be rescinded earlier if appropriate. This will retain flexibility for the community. There may, however, be competing political considerations to be weighed.

• Be as familiar as possible with state statutes, local ordinances, and charters regarding emergency authority, process and procedure, and limitations. It is also helpful to have some familiarity with the Federal Disaster Relief Act and [the] Stafford Act.

• Review local ordinances relating to emergencies and disasters and ensure that the ordinances grant as much authority as permitted, with as much flexibility and durability as possible. Problems and solutions during a disaster are not easy to predict. Consequently, [greater] authority over a longer duration may be of significant assistance to your community. Make sure your local ordinances addressing emergency and disaster situations are current and apply to something more than simply the cold war or nuclear attack. The ordinance should be drafted to apply to the most unanticipated emergencies and disasters.

• Be able to identify what penalties may apply to any persons violating the provisions of an emergency declaration, proclamation, or any order issued as a result thereof. Is it a fine or might it include jail time? In this regard, also anticipate the possibility of correctional centers being inaccessible or inoperative as well as courthouses being closed.

• Ensure that the emergency ordinances provide abilities to waive and possibly modify ordinances during the effective time of the proclamation. In specific instances, you may wish to be able to waive or modify inspection fees, building codes, inspection processes, occupancy certificates, zoning requirements, procurement requirements, etc.

• Be familiar with what authority the municipality may have over consumer protection, particularly in the areas of scams (charitable scams and contractor scams), unscrupulous contractors, price gouging, and fraud. Consider the possibility of a "one stop shop" which might include the police department, local inspections department, local permitting offices, state consumer protection divisions, state attorney general's office, secretary of state's offices, and others providing consumer protection and assistant services.

• Assist in the preparation of all necessary contract and emergency plan documents to allow coordinated responses and intergovernmental activities.

• Remain flexible. No matter what you think will happen, something else likely will. Make sure you and your staff are available to respond on short notice to needs of key emergency management and response team members.

• Review and have available for consultation mutual aid, joint power, and intergovernmental assistance agreements. Have an idea of what is contained in those documents well before they are actually needed. Identify what governmental entities you have such agreements with and when the agreements may be invoked. You should also be familiar with who can activate or invoke the provisions of such agreements.

• Make yourself available for the public information officers' needs. Against virtually all of my preconceived notions, I found that providing concise and accurate information to the public information officer and, where appropriate, directly communicating with the media, became imperative. It reduced anxiety and provided helpful information for virtually everyone.

- Where possible, have access to a computer with CD and online capability. The ability to perform legal research under less than ideal circumstances (something other than your comfortable office) may become essential.
- Expect the unexpected.[140]

H. G. Wells once said, "Human history becomes more and more a race between education and catastrophe."[141] It can only be hoped that in the field of emergency management, the calming, steadying hand of the local government attorney will help to win the race.

Legal Topics:

For related research and practice materials, see the following legal topics:

Governments – Local Governments – Finance, Public Health & Welfare, Law – Social Services Emergency Services, Torts – Negligence Proof, Custom –Business Customs

NOTES

1. Catherine II of Russia, quoted at Thinkexist.com, Catherine II Quotes, http://thinkexist.com/quotes/catherine_ii/ (accessed Jan. 17, 2007).
2. The Ojibwa, also known as the Chippewa, were considered one of the largest and most powerful Great Lakes Native American tribes east of the Mississippi. Native Americans: Chippewa, http://www.nativeamericans.com/Chippewa.htm (accessed Jan. 17, 2007).
3. Ambrose Bierce, The Devil's Dictionary 204 (Am. H. 1983).
4. 42 U.S.C. § 5121(a)(2) (2006).
5. Id. at § 5121(b).
6. Id.
7. Id. at § 5195.
8. Id.
9. Id. at § 5189(b).
10. Id. at § 5191(a).
11. Id. at § 5191(b).
12. For a more detailed discussion of the role of the federal government in an emergency, see David G. Tucker & Alfred O. Bragg, III, Florida's Law of Storms: Emergency Management, Local Government, and the Police Power, 30 Stetson L. Rev. 837, 861-867 (2001).
13. 42 U.S.C. § 5192(a).
14. Id. at § 5193(b).
15. Id. at § 5191(a).
16. Fla. Stat. § 252.36(2) (2006). An "emergency" is defined as "any occurrence, or threat thereof, . . . which results or may result in substantial injury or harm to the population or substantial damage to or loss of property." Id. at § 252.34(3). A state of emergency may be terminated by the governor or by concurrent resolution of the legislature, but in no event shall a declaration last

longer than sixty days absent renewal by the governor. Id. at § 252.36(2).

17. Id. at §§ 252.34(1)(b), 252.36(3). Disasters can be classified as minor, major, or catastrophic. Id. at § 252.34(1). While minor disasters require only minimal state or federal assistance, major and catastrophic disasters will "likely exceed local capabilities" and require more extensive assistance. Id.
18. Id. at § 252.36(3)(b).
19. Id. at § 252.36(4).
20. Id. at § 252.36(5)(d), (e), (g), (k).
21. The Weather Channel, Weather Glossary, http://www.weather.com/glossary/t.html (accessed Nov. 16, 2006).
22. See generally Fla. Stat. §§ 252.31-252.60 (containing the provisions of the State Emergency Management Act).
23. Id. at § 252.35. While the governor has always maintained this ultimate authority, the Division was under the direct control of the Department of Community Affairs until legislative amendments passed in 2006. Infra nn. 73-74 and accompanying text.
24. Id. at § 252.35(1).
25. Id. at § 252.35(2)(x). Specific rules that the Division has promulgated in carrying out this directive may be found in Subtitle 9G of the Florida Administrative Code.
26. Fla. Stat. § 252.35(2)(a), (d).
27. Specifically, the Department of Community Affairs administers the "Emergency Management, Preparedness, and Assistance Trust Fund to local emergency management agencies and programs pursuant to criteria specified [by] rule." Id. at § 252.373(2).
28. Id. at § 252.35(2)(c).

29. Id. at § 252.35(2)(b).
30. Id. at § 252.355(1).
31. Fla. Admin. Code Ann. r. 9G-6.0023(2) (2006).
32. Id. at r. 9G-19.014(1) (providing for loss of current or future funding and requiring the repayment of funds already awarded when local governments fail to comply with any of the conditions of the funding).
33. Fla. Stat. § 252.373(2)(a); Fla. Admin. Code Ann. rr. 9G-11.004(1), 9G-19.004(1), 9G-19.005.
34. Fla. Stat. § 252.38.
35. Id. at § 252.38(1)(a).
36. Id.
37. Id. at § 252.38(1)(b).
38. Id.
39. Id.
40. Id.
41. Id. This provision requires the direct appointment of an individual by the county. Id. It does not permit the appointment of an agency, nor can it be read to provide authority for a constitutional officer to appoint an employee. Id.
42. See id. (giving all directors, however appointed, the same responsibilities and authority).
43. See supra nn. 34-36 and accompanying text (describing the inherent duties and powers of county governments under the State Emergency Management Act).
44. Fla. Stat. § 163.01(2) (2006).
45. Id. at § 163.01(4).
46. Fla. Atty. Gen. Op. 97-10 (Feb. 14, 1997) (available at 1997 WL 68127).
47. See generally Fla. Stat. § 252.38 (describing the authority of local political subdivisions in emergency situations).
48. Id. at § 252.46(1).
49. Id. at § 252.38(3)(a).
50. A narrow exception to this statement is that sheriffs have a limited authority to declare a state of emergency. Fla. Stat. § 870.042(1) (2006). The circumstances under which this may occur are clearly delineated in Section 870.043. For a further discussion of law enforcement's role in emergency management, see infra Part II(E).
51. Id. at § 252.38(2).
52. Id.
53. Id.
54. Id. Notwithstanding this provision, reimbursement requests made through a federal disaster assistance program do not require coordination with the county. Id.
55. Id. at § 252.34(8).

56. For a discussion of counties' emergency powers under the Act, see supra Part II(C).
57. Fla. Stat. § 252.47.
58. Id. at § 870.042(1).
59. Id. at § 870.043.
60. Id. at § 870.045(1), (4).
61. Id. at § 870.047. The state of emergency may, however, be terminated by act of the governor, county commission, or city council prior to the expiration of seventy-two hours. Id. An extension of this time period may occur "by request from the public official and the concurrence of the county commission or city council." Id.
62. See id. at § 252.38(3)(a)(5) (allowing a political subdivision to declare a state of local emergency whenever any emergency occurs that affects only that subdivision).
63. Id. at § 252.38.
64. Natl. Climatic Data Ctr., Climate of 2004 Atlantic Hurricane Season, http://www.ncdc.noaa.gov/oa/climate/research/2004/hurricanes04.html (last updated Dec. 13, 2004).
65. Fla. H. 1937, 2005 Reg. Sess. (Apr. 4, 2005); Fla. H. 1551, 2005 Reg. Sess. (Mar. 7, 2005); Fla. H. 1125, 2005 Reg. Sess. (Feb. 24, 2005); Fla. H. 751, 2005 Reg. Sess. (Feb. 8, 2005); Fla. H. 621, 2005 Reg. Sess. (Jan. 26, 2005); Fla. H. 347, 2005 Reg. Sess. (Jan. 14, 2005); Fla. H. 191, 2005 Reg. Sess. (Jan. 5, 2005); Fla. Sen. 2616, 2005 Reg. Sess. (Mar. 8, 2005); Fla. Sen. 2092, 2005 Reg. Sess. (Mar. 7, 2005); Fla. Sen. 1604, 2005 Reg. Sess. (Feb. 18, 2005); Fla. Sen. 1544, 2005 Reg. Sess. (Feb. 16, 2005); Fla. Sen. 1488, 2005 Reg. Sess. (Feb. 15, 2005); Fla. Sen. 1228, 2005 Reg. Sess. (Feb. 8, 2005); Fla. Sen. 662, 2005 Reg. Sess. (Jan. 12, 2005); Fla. Sen. 572, 2005 Reg. Sess. (Dec. 21, 2004); Fla. Sen. 442, 2005 Reg. Sess. (Dec. 8, 2004); Fla. Sen. 240, 2005 Reg. Sess. (Nov. 16, 2004); Fla. Sen. 232, 2005 Reg. Sess. (Nov. 16, 2004).
66. The surviving bill was Senate Bill 592. As adopted, it provided that the failure to possess an occupational license while offering goods and services for sale during a declared emergency is a misdemeanor; that the governor may authorize any business selling certain commodities to operate in violation of an established curfew; and that the governor may provide for the extension of solid waste disposal facilities' hours of operation by executive order. 2005 Fla. Laws ch. 283. All other bills were either vetoed or died in committee or on calendar.

67. Fla. Sen. 240, 2005 Reg. Sess.; Fla. Sen., Senate 0240: Relating to Nursing Homes/Electrical Service, http://www.flsenate.gov/session/index.cfm?Mode=Bills&SubMenu=1&BI_Mode=ViewBillInfo&BillNum=0240 (accessed Mar. 1, 2006).

68. See e.g. Dave Reynolds, Witnesses: Poor Planning and Follow-through Led to Hundreds of Katrina Deaths, Inclusion Daily Express, http://www.inclusiondaily.com/archives/06/02/02/020206/lakatrina.htm (Feb. 2, 2006) (reporting that almost twenty percent of those who died in Hurricane Katrina were in hospitals or nursing homes and that many of these facilities had improperly installed electrical switches and generators).

69. Fla. H. 7139, 2006 Reg. Sess. (Mar. 16, 2006); Fla. H. 7125, 2006 Reg. Sess. (Mar. 15, 2006); Fla. H. 1721, 2006 Reg. Sess. (Mar. 15, 2006); Fla. H. 1435, 2006 Reg. Sess. (Mar. 3, 2006); Fla. H. 1359, 2006 Reg. Sess. (Feb. 28, 2006); Fla. H. 1209, 2006 Reg. Sess. (Feb. 21, 2006); Fla. H. 911, 2006 Reg. Sess. (Jan. 27, 2006); Fla. H. 739, 2006 Reg. Sess. (Jan. 17, 2006); Fla. H. 707, 2006 Reg. Sess. (Jan. 13, 2006); Fla. H. 603, 2006 Reg. Sess. (Jan. 4, 2006); Fla. H. 545, 2006 Reg. Sess. (Dec. 19, 2005); Fla. H. 285, 2006 Reg. Sess. (Oct. 25, 2005); Fla. H. 249, 2006 Reg. Sess. (Oct. 14, 2005); Fla. H. 165, 2006 Reg. Sess. (Sept. 20, 2005); Fla. H. 89, 2006 Reg. Sess. (Aug. 23, 2005); Fla. Sen. 2488, 2006 Reg. Sess. (Mar. 2, 2006); Fla. Sen. 2486, 2006 Reg. Sess. (Mar. 2, 2006); Fla. Sen. 2386, 2006 Reg. Sess. (Mar. 1, 2006); Fla. Sen. 2256, 2006 Reg. Sess. (Feb. 22, 2006); Fla. Sen. 2154, 2006 Reg. Sess. (Feb. 16, 2006); Fla. Sen. 1888, 2006 Reg. Sess. (Feb. 9, 2006); Fla. Sen. 1708, 2006 Reg. Sess. (Jan. 27, 2006); Fla. Sen. 1588, 2006 Reg. Sess. (Jan. 25, 2006); Fla. Sen. 1484, 2006 Reg. Sess. (Jan. 23, 2006); Fla. Sen. 1058, 2006 Reg. Sess. (Dec. 6, 2006); Fla. Sen. 862, 2006 Reg. Sess. (Nov. 16, 2005); Fla. Sen. 678, 2006 Reg. Sess. (Nov. 9, 2005); Fla. Sen. 638, 2006 Reg. Sess. (Nov. 8, 2005); Fla. Sen. 590, 2006 Reg. Sess. (Nov. 4, 2005); Fla. Sen. 568, 2006 Reg. Sess. (Nov. 3, 2005); Fla. Sen. 528, 2006 Reg. Sess. (Oct. 28, 2005); Fla. Sen. 426, 2006 Reg. Sess. (Oct. 21, 2005); Fla. Sen. 156, 2006 Reg. Sess. (Sept. 8, 2005).

70. House Bill 7121 was enacted on June 1, 2006, and amends a wide variety of statutes in an attempt to improve Florida's emergency management capabilities. 2006 Fla. Laws ch. 71.

71. 2006 Fla. Laws ch. 67.

72. Id. at ch. 68, § 2.

73. Id. at ch. 70.

74. Id.

75. Id.

76. Id. at ch. 71, §§ 2-7.

77. Id. at ch. 71, § 1.

78. Id. at ch. 71, § 1(2).

79. Id.

80. E.g. id. at ch. 71, §§ 1(2), 2.

81. Fla. Stat. § 526.143(3)(a) (2006).

82. Id. at § 526.143(2).

83. See Carol Brzozowski, The Impact of Outage, 4 Distributed Energy (May/June 2006), http://www.gradingandexcavation.com/de_0605_impact.html (discussing the problems arising when power outages render gas stations inoperable); Jeff Ostrowski, Lawmakers Want Gas Stations to Have Generators, Palm Beach Post (Oct. 29, 2005) (discussing Floridians' frustrations with fuel shortages that resulted after hurricanes knocked out power, leaving otherwise undamaged gas stations unable to provide fuel).

84. Fla. Stat. § 526.143(1).

85. Id.

86. Id. at § 526.143(4)(b).

87. Id. at § 526.144.

88. Id. at § 526.144(1)(c).

89. Id. at § 526.144(1)(d).

90. Id. at § 526.144(3).

91. Id. at § 526.144(2).

92. Id. at § 526.144(1)(b).

93. Id. at § 526.144(1)(d).

94. Id. at § 526.144(1)(f).

95. Id. at § 526.144(4). The Division of Emergency Management was also charged with studying the feasibility of incorporating nongovernmental agencies and private entities in the State's emergency management plan to provide for the distribution of essential commodities. 2006 Fla. Laws ch. 71, § 8. As part of this review, the Division is to establish criteria by which local regulation of retail establishments would be preempted during a disaster. Id.

96. See Bob LaMendola, Care Needs Get Higher Priorities after Storms, S. Fla. Sun-Sentinel 1A (July 7, 2006) (noting that the sick and the elderly can become trapped in their homes when storms put elevators out of operation); Aaron Sharockman, Condo Is Evacuated for Lack of Utilities, St. Pete. Times 1 (Sept. 30, 2004) (describing the experiences of residents of a high-rise condominium after Hurricane Jeanne knocked out facilities

dependent on electricity, including elevators, which eventually caused local officials to declare the building uninhabitable).

97. Fla. Stat. § 553.509(2) (2006). Specifically, the elevator must provide service to building residents for a certain number of hours each day for five days following a disaster. Id.

98. Id. at § 553.509(2)(b).

99. Id.

100. Id. at § 553.509(2)(c).

101. Id. at § 553.509(2)(d).

102. See Quincy C. Collins, Should I Stay or Should I Go? Decision Is Personal, with Many Variables, Sun Herald (Biloxi, Miss.) A2 (Aug. 28, 2005) (describing various residents' reasons for deciding not to evacuate before Hurricane Katrina, including concern over pets, fatalistic attitudes about storms, and the expense of hurricane preparation); Maria L. La Ganga & Elizabeth Mehren, Gulf Coast Besieged, L.A. Times 1 (Sept. 24, 2005) (reporting that, for many elderly or disabled residents, the stress and physical hardships of evacuation can lead to an increased risk of injury or death); Meghan Gordon & James Varney, Katrina Evacuation Plan Remains the Model, New Orleans Times-Picayune 9 (May 28, 2006) (listing lack of personal transportation, the inconvenience of evacuation, and the survival of past storms as reasons that led one in five Gulf Coast residents to decide against evacuating during Hurricane Katrina).

103. See Emily Bazar, Post-Katrina Chaos Forces Painful Choice to Leave Pets, U.S.A. Today 4A (Sept. 23, 2005) (describing one New Orleans evacuee's heartbreak at leaving his pets behind, a choice he referred to as "the worst decision ever"); Legislation, Bucks Co. (Pa.) Courier Times 6A (Sept. 23, 2005) (reporting on federal legislation introduced by lawmakers concerned that many people who ultimately died in Hurricane Katrina remained in their homes because they could not take their pets if they evacuated).

104. See Ann M. Simmons, Owners to Sue over Katrina Pet Shootings, L.A. Times 11 (Sept. 9, 2006) (noting that sheriff's deputies had reported having to shoot vicious dogs found wandering the streets after Hurricane Katrina); Alison Vekshin, House Bill Aims to Assure Evacuation of Pets in Disaster, Las Vegas Rev.-J. 2B (May 23, 2006) (noting lawmakers' concerns that abandoned animals could pose a threat to rescuers and returning evacuees); CNN Sat. Morning News (CNN Sept. 17, 2005) (TV broad.; transcr. available from http://transcripts.com/TRAN SCRIPTS/0509/17/smn.02.html) (reporting on rescue workers' concerns that the large number of starving, abandoned animals in New Orleans after Hurricane Katrina was becoming a health hazard).

105. Fla. Stat. § 252.3568.

106. Id.

107. 2006 Fla. Laws ch. 71, § 20. HB 7121 also expands and improves the process for notifying people with special needs of the ability to register with local emergency management agencies for special assistance. 2006 Fla. Laws ch. 71, § 16 (amending Fla. Stat. § 252.355).

108. Fla. Stat. § 381.0303 (2006).

109. 2006 Fla. Laws ch. 71, § 21 (codified at Fla. Stat. § 400.492 (2006)).

110. Id. at ch. 71, § 23 (codified at Fla. Stat. § 400.506(16)).

111. Id. at ch. 71, § 24 (codified at Fla. Stat. § 400.610(1)(b)).

112. Prior versions of the statutes regulating these facilities simply required the facilities to have plans providing for continued services to special-needs patients in case of an emergency, without requiring that the services be of the same type and quantity as those provided before the emergency, and often without specifically referencing evacuations. Fla. Stat. §§ 400.492, 400.506(16), 400.610(1)(b) (2005).

113. 2006 Fla. Laws ch. 71, §§ 21, 23, 24 (codified at Fla. Stat. §§ 400.492(3), 400.506(16), 400.610(1)(b), respectively).

114. 2006 Fla. Laws ch. 71, § 26 (codified at Fla. Stat. § 400.934(20)(a)).

115. 2006 Fla. Laws ch. 71, § 26 (codified at Fla. Stat. § 400.934(21), (22)).

116. Such legislation has previously been introduced, but has failed. Fla. Sen. 638, 2006 Reg. Sess. (Nov. 8, 2005); Fla. H. 165, 2006 Reg. Sess. (Sept. 20, 2005). Many local governments, however, have already passed such legislation at the local level. E.g. Ord. Code Jacksonville (Fla.) § 674.501 (2005) (requiring registered sexual predators or offenders to notify shelter operators of their status immediately upon entering a shelter).

117. La. St. B. Assn., Emergency/Disaster Training Manual for Volunteer Lawyers Following Hurricane Katrina (Morrison & Foerster 2005) (available at http://www.lsba.org/home1/FinalManual .pdf).

118. Id. at 6 (numbers from original source have been omitted and replaced with bullets and punctuation has been modified for consistency).

119. Alan D. Cohn, Mutual Aid: Intergovernmental Agreements for Emergency Preparedness and Response, 37 Urb. Law. 1, 3 (2005).

120. Id.

121. Incident Command Sys., National Training Curriculum: History of ICS 1-2 (Natl. Wildfire Coordinating Group 1994) (available at http://www.nwcg.gov/pms/forms/compan/history.pdf) (cited in Cohn, supra n. 119, at 5) (punctuation has been modified for consistency).

122. Select Bipartisan Comm. to Investigate the Preparation for and Response to Hurricane Katrina, A Failure of Initiative, H.R. Comm. Rpt. 109-377 (Feb. 15, 2006) (available at http://www.gpoaccess.gov/serialset/creports/katrina.html). A review of the Committee's Executive Summary of Findings alone discloses striking parallels between the problems in California in the 1970s and those in New Orleans in 2005. Id. at 1-5.

123. Model Intrastate Mutual Aid Legislation art. III (Natl. Emerg. Mgt. Assn. 2004) (available at http://www.emacweb.org/?150) (numbers from original source have been omitted and replaced with bullets, and punctuation has been modified for consistency).

124. Public emergency management entities across the nation are implementing 800 megahertz radios that utilize a trunking radio system. A trunking radio system is a single system that can act as though it were several separate, larger systems. See Cohn, supra n. 119, at 39-40 (discussing the effectiveness of such a system during the attack on the Pentagon on September 11, 2001). A radio system using trunking can be organized to provide groups of users and access arrangements tailored to meet local needs when traditional telecommunications equipment fails. Id.

125. Fla. Stat. § 252.40(1). This Section, entitled "Mutual aid arrangements," reads: The governing body of each political subdivision of the state is authorized to develop and enter into mutual aid agreements within the state for reciprocal emergency aid and assistance in case of emergencies too extensive to be dealt with unassisted. Copies of such agreements shall be sent to the [Division of Emergency Management]. Such agreements shall be consistent with the state comprehensive emergency management plan and program, and in time of emergency it shall be the duty of each local emergency management agency to render assistance in accordance with the provisions of such mutual aid agreements to the fullest possible extent.

126. Fla. Stat. § 252.40(2). This Section reads: The Governor may enter into a compact with any state if she or he finds that joint action with that state is desirable in meeting common intergovernmental problems of emergency management planning or emergency prevention, mitigation, response, and recovery.

127. Howard W. Swanson, The Delicate Art of Practicing Municipal Law under Conditions of Hell and High Water, 76 N.D. L. Rev. 487, 497 (2000); see also Joseph Jarret, In Times of Disaster: Minimize Local Government Business Interruptions, 86 Pub. Mgt. 21, 22-23 (Dec. 2004) (providing suggestions for local governmental interaction with state and federal emergency management officials).

128. Cohn, supra n. 119, at 21. With regard to governmental immunities, Florida law provides: Neither the state nor any agency or subdivision of the state waives any defense of sovereign immunity, or increases the limits of its liability, upon entering into a contractual relationship with another agency or subdivision of the state. Such a contract must not contain any provision that requires one party to indemnify or insure the other party for the other party's negligence or to assume any liability for the other party's negligence. Fla. Stat. § 768.28(19) (2006).

129. Cohn, supra n. 119, at 21-22.

130. The Florida Association of County Attorneys (FACA) is a non-profit organization affiliated with the Florida Association of Counties whose mission is to enhance the professionalism and technical expertise of attorneys employed by Florida's counties. FACA, About Us, http://www.fl-counties.com/legal/faca.shtml (accessed Mar. 1, 2007). For more information or a copy of the Association's mutual aid agreement, contact Joseph G. Jarret, Esquire, President, 2006-2007, at josephjarret@polk-county.net.

131. Model Mutual Aid Agreement 1 (FACA 2005) (copy on file with Stetson Law Review).

132. See Fla. Stat. § 440.02 (2006) (excluding volunteers generally from the definition of employees entitled to receive workers' compensation, but creating an exclusion within the exclusion for volunteers for governmental entities).

133. Jean Cox, Managing Donated Resources Following Catastrophic Events, in A Legal Guide to Homeland Security and Emergency Management for State and Local Governments 201, 203 (Ernest B. Abbott & Otto J. Hetzel eds., ABA 2005).

134. Id. (numbers from original source have been omitted and replaced with bullets, and punctuation has been modified for consistency).

135. Id. at 203-204 (numbers from original source have been omitted and replaced with bullets, and punctuation has been modified for consistency).

136. Id. at 204 (numbers from original source have been omitted and replaced with bullets, and punctuation has been modified for consistency).

137. Oscar Wilde, An Ideal Husband, in The Complete Works of Oscar Wilde 482, 526 (Barnes & Noble, Inc. 1994).

138. Alfred O. Bragg, III, Experiencing the 2004 Florida Hurricanes, in A Legal Guide to Homeland Security and Emergency Management, supra n. 133, at 139, 143.

139. Interestingly, the challenges facing Mr. Swanson during the flooding that plagued North Dakota cities in 2000 and the experiences of co-author Joseph G. Jarret during the spate of hurricanes that plagued Polk County (Frances, Charley, and Jeanne) in 2004 are strikingly similar.

140. Swanson, supra n. 127, at 501-503. The Robert T. Stafford Disaster Relief and Emergency Assistance Act, which Swanson references, provides the president with the authority to fund the restoration of property damaged by a major disaster. 42 U.S.C. §§ 5172, 5174 (relating to the restoration of public or nonprofit private facilities and to private, owner-occupied residences, respectively). The Act also allows the president to provide additional funding to enhance the restored facility's ability to resist similar damage in future disasters. Id. at §§ 5172(c)(2)(B), 5174(c)(2)(A); see also 44 C.F.R. § 206.226(e) (2005) (providing that the cost of any requirements to mitigate against future hazards that FEMA places on restoration will be eligible for federal assistance). To demonstrate the truth of Swanson's final admonition - to expect the unexpected - the Authors point out that, in 2004, while Polk County was in the process of recovery operations necessitated by Hurricanes Charley and Frances, Hurricane Jeanne was making landfall.

141. H.G. Wells, The Outline of History Vol. II, 594 (MacMillan Co. 1921).

QUESTIONS

1. How has your understanding of the attorney's role in emergency management changed after reading this article?

2. This article is written by two Florida lawyers and it revolves around that state's law. How might the fact that the law discussed is from Florida make it different from other states' laws?

3. Compare and contrast county governments' statutory authority and responsibility with that of municipalities as described in the article.

4. What is the role of county and municipal law enforcement under Florida emergency management law?

5. What are the authors' opinions regarding lessons learned? Do you agree or not, and why?

6. What are the best ways to create and use mutual aid agreements, according to the authors?

7. What are the biggest challenges a lawyer faces before, during and after a disaster?

ABOUT THE AUTHORS

Joseph G. Jarret is a County Attorney, Polk County, Florida. B.S., Troy State University (West Germany Campus), 1981; M.A., Public Administration, Central Michigan University, 1983; J.D., Stetson University College of Law, 1989; Graduate Certificate in Public Management, University of South Florida, 1996. Mr. Jarret has been the Polk County Attorney since 2003 and is a Certified Circuit Civil Mediator & Arbitrator. He is the 2006–2007 president of the Florida Association of County Attorneys and a former United States Army Combat Arms Officer who has published over eighty-five articles in various professional journals, nine of which have appeared in the Florida Bar Journal. He is the past president of the Manatee and Hardee County Bar Associations.

Michele L. Lieberman is Chief Assistant County Attorney, Citrus County, Florida. B.A., University of South Florida, 1993; J.D., Stetson University College of Law, 1997. Ms. Lieberman is board certified in City, County, and Local Government Law and has been with the Citrus County Attorney's Office since 2002. She is past president of the Citrus County Bar Association and a 2006 appointee to the Fifth Circuit Judicial Nominating Commission.

New Orleans, LA, September 30, 2005 – Houses were destroyed by flood waters after Hurricane Katrina came through the area and the levees broke. Some homes floated off their foundations and bumped into others homes or came to rest on streets. Many New Orleans residents are in shelters and are relying on FEMA to help them though the disaster. Marvin Nauman/FEMA photo.

Chapter 21

RECOVERY CASES

Recovery involves restoring the situation to the "status quo ante." This term translates into the state things in were before the emergent event occurred. However, well-crafted zoning and building codes can help to ensure that the result is a great improvement over what preceded the disaster. This is an area where previous pro-active, close cooperation with legal counsel during mitigation can really pay off.

The subject of recovery is complex and many-faceted. This chapter can only hope to scratch the surface of the legal issues. FEMA's website has much to offer: http://www.fema.gov/rebuild/ index.shtm

For those interested in a wider perspective on recovery matters, I recommend two books by authors who are veritable founts of wisdom for the reader's consideration:

David A. McEntire, Ph.D., *Disaster Response and Recovery*, Wiley (2007)

Brenda C. Phillips, Ph.D., *Disaster Recovery*, CRC Press (2009)

Bernofsky v. The Road Home Corporation

741 F. Supp. 2d 773 (U.S.D.Ct. W.D.La. 2010)

MEMORANDUM RULING

I. BACKGROUND

A. Facts

On August 29, 2005, Hurricane Katrina struck New Orleans, Louisiana, causing catastrophic damage to the city and to the residential property of husband and wife, Carl and ShirleyBernofsky ("the Bernofskys"). At the time, the Bernofskys owned the property at 6478 General Diaz Street, New Orleans, Louisiana 70124, in the Lakeview subdivision. The hurricane and resulting flooding severely damaged the Bernofskys' property.

In response to Hurricanes Katrina and Rita, the Federal Government awarded the State of Louisiana an initial $6.4 billion dollars, followed by an additional $4.2 million dollars, in Community Development Block Grants ("CDBG") for disaster recovery and rebuilding efforts. In return, Louisiana was required to develop a plan for disbursement of the CDBG funds to be approved by the United States Department of Housing and Urban Development ("HUD"). Louisiana developed "The Road Home Program," which was approved by HUD on May 30, 2006.

In accordance with federal statute, the state created the Louisiana Recovery Authority ("LRA") to oversee the disbursement of federal funds. *See* La. R.S. 49:220.4-220.5. The state authorized the Office of Community Development ("OCD")

within the Division of Administration ("DOA") to administer The Road Home Program. *See* La. R.S. 40:600.62(2). It is undisputed that the LRA and the OCD are state agencies. *See* Record Document 1 at 2. Additionally, the state contracted with ICF Emergency Management Services, LLC ("ICF"), a limited liability company domiciled in Delaware, to serve as "Louisiana's Road Home Housing Manager" from June 12, 2006, to June 11, 2009. ICF's responsibilities included "accept[ing] and process[ing] applications for financial assistance," "verify[ing] applicants' eligibility" and "determin[ing] amounts of assistance in accordance with State guidelines."

In an attempt to claim benefits, the Bernofskys filled out a registration application online with Louisiana's "Road Home Registry" on May 7, 2006. The Road Home Registry was a project administered through Governor Kathleen Blanco's office, designed to assist Louisiana residents with obtaining funds to rebuild. *See* id. The Road Home Registry website encouraged residents to "Begin the Housing Registration Process Here," informed readers that "[t]his registration process is the first step for the State to identify your home address and information in determining your eligibility for funding," and thanked them for taking "this important first step." Id. Upon completing the online registration, the Bernofskys received a confirmation page with a registration number. *See* id. The Road Home Registry, through the confirmation page, thanked them for their "participation in the pre-application registry process" and informed them that someone would be in contact "as soon as the program begins." Id.

Between May of 2006 and August of 2008, the Bernofskys took no action with regard to The Road Home Program. While the Bernofskys were not contacted by The Road Home Program as promised by the confirmation page, they also never inquired as to the status of their registration or how to complete the application process. Relying on newspaper reports, the Bernofskys concluded they were not eligible for Road Home benefits because they were selling their home.

The Bernofskys sold their gutted home at a great loss on February 14, 2007. They received insurance proceeds, but this amount did not cover their total losses. The Bernofskys claim an uncompensated loss of $89,391.88.

Later, the Bernofskys discovered they might be eligible for benefits, again through newspaper reports. They began making inquiries on August 1, 2008. The Bernofskys wrote multiple letters to Road Home, the Road Home Appeals Department, the LRA, the OCD, and Senator Mary Landrieu seeking information regarding their Road Home eligibility.

By this time, The Road Home Program application deadline of July 31, 2007, had already passed. On June 1, 2009, the Bernofskys allegedly received a telephone call from Ms. Judy Johnson-White ("Ms. Johnson-White"), representing Road Home. Ms. Johnson-White informed them that Road Home had "no record of their registration" or "listing of their home in their records." The Bernofskys allege they received another call from Ms. Johnson-White on July 28, 2009, in which she told them that they were "not qualified for Road Home benefits because they had not registered." Finally, on October 20, 2009, the OCD notified the Bernofskys that they were not eligible for Road Home benefits, because they had not timely filed an application for The Road Home Program and no exception for late filing could be given.

B. Procedural History

The Bernofskys filed this suit against defendants, The Road Home Corporation, ICF, the LRA, and the OCD on November 13, 2009. The Bernofskys claim they are entitled to $89,391.88 from the defendants for damage to their property. Specifically the Bernofskys allege the following:

(1) Defendants were negligent in failing to contact the Bernofskys following their online registration and in refusing to allow them to file an application after the deadline;

(2) Defendants violated their duty to serve Louisiana residents by failing to respond to the Bernofskys' requests for information regarding Road Home eligibility;

(3) Defendants' "disparate treatment" of the Bernofskys violated their rights under 42 U.S.C. § 1983 of the Civil Rights Act; and

(4) Defendants' "disparate treatment" of the Bernofskys violated their right to Equal Protection under the Fourteenth Amendment and Article 1, Section 3 of the Louisiana Constitution.

See Record Document 1. Plaintiffs' jurisdictional basis for this suit is 28 U.S.C. § 1331.

Defendants the LRA and the OCD (hereinafter "the agencies") move to dismiss the Bernofskys' claims for lack of subject-matter jurisdiction pursuant to Federal Rule of Civil Procedure 12(b)(1). The LRA and the OCD argue that they are immune from suit in federal court as an "arm of the state" under the Eleventh Amendment. The agencies also move to dismiss for failure to state a claim upon which relief can be granted pursuant to Federal Rule of Civil Procedure 12(b)(6).

Defendant ICF moves to dismiss the Bernofskys' claims for failure to state a claim under Federal Rule of Civil Procedure 12(b)(6) and for failure to join a party under Rule 19 pursuant to Rule 12(b)(7). *See* Record Document 20.

II. LAW AND ANALYSIS

A. Motion To Dismiss By LRA And OCD

1. Eleventh Amendment Immunity

The Eleventh Amendment 2 prohibits an individual from suing a state in federal court, "unless the state consents to suit or Congress has clearly and validly abrogated the state's sovereign immunity." *Perez v. Region 20 Educ. Serv. Ctr.*, 307 F.3d 318, 326 (5th Cir. 2002). In addition to protecting states from suits brought by citizens of other states, it is well-established that the Eleventh Amendment bars a federal court from "entertain[ing] a suit brought by a citizen against his own State." *Pennhurst State Sch. & Hosp. v. Halderman*, 465 U.S. 89 (1984) (citing *Hans v. Louisiana*, 134 U.S. 1 (1890)... Eleventh Amendment protection also extends to state law claims brought under pendent jurisdiction. *See* Pennhurst, 465 U.S. at 120-21. Finally, "a state's Eleventh Amendment immunity extends to any state agency or entity deemed an alter ego or arm of the state." Perez, 307 F.3d at 326 (internal quotations omitted).

Plaintiffs concede that the LRA and the OCD are "agenc[ies] of the State of Louisiana." The parties do not dispute that the agencies qualify as the "state" for purposes of Eleventh Amendment immunity. Thus, both the LRA and the OCD are immune from suit in federal court absent a valid waiver by the state or unequivocal abrogation by Congress. The court finds that the LRA and the OCD are immune from suit in federal court and the motion to dismiss must be granted.

2. Waiver Of Immunity By The State

Louisiana has not waived its Eleventh Amendment immunity from federal court jurisdiction. "The test for determining whether a State has waived its immunity from federal-court jurisdiction is a stringent one." *Atascadero State Hosp. v. Scanlon*, 473 U.S. 234, 241 (1985). Waiver requires a "clear declaration" by the state. *AT&T Comm-c'ns v. BellSouth Telecomm.*, 238 F.3d 636, 644 (5th Cir. 2001) (quoting *College Sav. Bank v. Fla.* Prepaid Postsecondary Educ. Expense Bd., 527 U.S. 666, 675-76 (1999)). Courts will find waiver "only where stated by the most express language or by such overwhelming implications for the text as will leave no room for any other reasonable construction." Edelman, 415 U.S. at 673 (internal quotations omitted).

The Bernofskys can point to no express waiver by the State of Louisiana of its Eleventh Amendment immunity in federal court. 4 HN4 The state has made clear that "[n]o suit against the state or a state agency or political subdivision shall be instituted in any court other than a Louisiana state court." La. R.S. 13:5106(A).

The "mere fact that a State participates in a program through which the Federal Government provides assistance for the operation by the State of a system of public aid is not sufficient to establish consent on the part of the State to be sued in the federal courts." Edelman, 415 U.S. at 673... The federal statute must "clearly and unambiguously" provide that "the state's particular conduct or transaction will subject it to federal court suits brought by individuals." AT&T, 238 F.3d at 644.

The federal statutes at issue here, Pub. L. No. 109-148, 119 Stat. 2680 (Dec. 30, 2005) and Pub. L. No. 109-234, 120 Stat. 418 (June 15, 2006), do not include a provision even referencing federal court jurisdiction, much less expressly conditioning the receipt of federal funds on the state's waiver of Eleventh Amendment immunity. The court finds the State of Louisiana's receipt of community development funds was not a waiver of Eleventh Amendment immunity.

3. Abrogation Of Immunity By Congress

Congress has not unequivocally abrogated Louisiana's sovereign immunity with regard to the Bernofskys' claims under 42 U.S.C. § 1983 and the Fourteenth Amendment. To find that Congress abrogated a state's Eleventh Amendment immunity, there must be an "unequivocal expression of congressional intent to overturn the constitutionally guaranteed immunity of the several States." Atascadero, 473 U.S. 234 at 240.

Such unequivocal expression of intent to abrogate immunity is not present here. The text of 42 U.S.C. § 1983 does not expressly override the states' Eleventh Amendment immunity. The Supreme Court has held that Congress did not abrogate states' sovereign immunity in passing section 1983. See *Will v. Mich. Dep't. of State Police*, 491 U.S. 58, 66 (1989); *Quern v. Jordan*, 440 U.S. 332 (1979).

The Bernofskys' only other federal law claim is an alleged violation of the Equal Protection Clause of the Fourteenth Amendment. 5 The Bernofskys seem to allege a violation of the Fourteenth Amendment distinct from their section 1983 claim, as they do not cite any statute or common law doctrine in support of their claim. Absent a statute that clearly and expressly abrogates the state's immunity, the court will not presume such intent. See Atascadero, 473 U.S. 234 at 240. The court finds no basis for concluding that Congress abrogated Louisiana's Eleventh Amendment immunity in this case.

Based on this analysis, the court finds that the State of Louisiana has not waived its Eleventh Amendment immunity and Congress has not ab-

rogated the state's immunity with respect to the Bernofskys' claims. The court finds the LRA and the OCD, as arms of the state, are entitled to the protections of the Eleventh Amendment.

B. Motion To Dismiss By ICF

1. Federal Rule Of Civil Procedure 12(b)(6) Standard

"To survive a Rule 12(b)(6) motion to dismiss, a complaint 'does not need detailed factual allegations,' but must provide the plaintiffs grounds for entitlement to relief-including factual allegations that when assumed to be true 'raise a right to relief above the speculative level.'" *Cuvillier v. Taylor*, 503 F.3d 397, 401 (5th Cir. 2007)...The plaintiff must plead "enough facts to state a claim to relief that is plausible on its face." Twombly, 550 U.S. at 570. "When considering a motion to dismiss, the court accepts as true the well-pled factual allegations in the complaint, and construes them in the light most favorable to the plaintiff." *Taylor v. Books A Million, Inc.*, 296 F.3d 376, 378 (5th Cir. 2002) (citation omitted). The court generally must not consider any infor mation outside the pleadings. *See Sullivan v. Leor Energy, LLC*, 600 F.3d 542, 546 (5th Cir. 2010).

It is well-established that pro se complainants like the Bernofskys are held to less stringent standards than formal pleadings drafted by lawyers. *See* Taylor, 296 F.3d at 378 (citation omitted). However, regardless of whether the plaintiffs are preceding pro se or are represented by counsel, conclusory allegations or legal conclusions masquerading as factual conclusions will not suffice to prevent a motion to dismiss. See id. (citation and quotations omitted). "[E]ven when a plaintiff is proceeding pro se, the complaint must contain either direct allegations on every material point necessary to sustain a recovery...or contain allegations from which an inference fairly may be drawn that evidence on these material points will be introduced at trial." *Govea v. ATF*, 207 Fed. Appx 369, 372 (5th Cir. 2006)...This court need not "conjure up unpled allegations or construe elaborately arcane scripts to save a complaint." Id. (quotations and citation omitted).

2. 42 U.S.C. § 1983 Claim

The Bernofskys allege a violation of their rights under 42 U.S.C. § 1983. Section 1983 "provides a federal remedy against a defendant who under color of state law deprives a plaintiff of any federal constitutional or statutory right." *Hull v. City of Duncanville*, 678 F.2d 582, 583 (5th Cir. 1982); *Cornish v. Correctional Servs. Corp.*, 402 F.3d 545, 549 (5th Cir. 2005) (internal quotations omitted). Assuming arguendo that ICF acted "under color of state law," the court must determine whether plaintiffs allege a violation of a federal constitutional or statutory right.

The Bernofskys claim that ICF's "disparate treatment of plaintiffs constituted a violation of plaintiffs' civil rights." Record Document 1 at 8. The Bernofskys fail to identify the federal statutory or constitutional right of which ICF allegedly deprived them. They also provide no explanation of what actions or inactions by ICF comprised the alleged "disparate treatment."

Accepting the Bernofskys' factual allegations as true, the Bernofskys have at most a tort claim against ICF. The Bernofskys have not alleged any facts supporting a theory of arbitrary or discriminatory action by ICF. The only possible basis for the Bernofskys section 1983 claim is ICF's alleged failure to properly process their registration application or otherwise transfer it into the Road Home system, to contact them upon the initiation of the program as promised in the registry's online confirmation page, to respond to letters sent by the Bernofskys in August 2008, and to provide them with an opportunity to file their application past the deadline of July 31, 2007.

However, "no Section 1983 action arises merely because tortious injury results from action or inaction" by a state agent. Hull, 678 F.2d at 584. The conduct must be "sufficiently egregious as to be constitutionally tortious," or in other words, "the sort of abuse of government power that is necessary to raise an ordinary tort by a government agent to the stature of a violation of the Constitution." Id. Such abuse of power is not alleged here. The Bernofskys allege only mismanagement and incompetence, which does not rise to the level of a constitutional violation. As the Bernofskys' allegations do not implicate a federal right, they have failed to state a claim under section 1983.

3. Fourteenth Amendment Equal Protection Claim

The Bernofskys appear to allege a violation of the Equal Protection Clause of the Fourteenth Amendment independent from their section 1983 claim. They do not cite any statute or common law doctrine in support of their claim. *See id.* To the extent the Bernofskys are alleging a stand-alone violation of the Fourteenth Amendment, the Fifth Circuit has expressly rejected such a claim. *See Hearth, Inc. v. Dep't of Pub. Welfare*, 612 F.2d 981 (5th Cir. 1980) (holding that where plaintiffs had an alternative remedy, the court would not substitute its judgment for that of Congress by creating an implied right of action). Here, the Bernofskys have an alternative remedy, as they may sue ICF in state court. This court sees no reason to depart from established jurisprudence by implying a federal right of action in this case. Without a right of action, the Bernofskys' claim fails for a lack of jurisdictional basis.

The Bernofskys also fail to allege an equal protection claim. HN16 To properly plead a "class of one" equal protection claim, a plaintiff must demonstrate that "(1) he or she was treated differently from others similarly situated and (2) there was no rational basis for the disparate treatment." *Stotter v. Univ. of Tex.*, 508 F.3d 812, 824 (5th Cir. 2007); *Vill. of Willowbrook v. Olech*, 528 U.S. 562, 120 S. Ct. 1073, 145 L. Ed. 2d 1060 (2000). The Bernofskys have not alleged any facts suggesting they were treated differently from others similarly situated. Nor have they argued that ICF did not have a rational basis for its actions or inactions. The court finds the Bernofskys have failed to allege an equal protection claim.

4. State Law Claims

The Bernofskys allege various state law claims in addition to their federal law claims, including negligence and a violation of Article 1, Section 3 of the Louisiana Constitution. All of the Bernofskys' federal law claims have been dismissed. The court declines to exercise its supplemental jurisdiction over the remaining state law claims and will accordingly dismiss these claims without

prejudice. 8 *See* 28 U.S.C. § 1367(c)(3); *Certain Underwriters at Lloyd's, London v. Warrantech Corp.*, 461 F.3d 568, 578 (5th Cir. 2006) (explaining "it is our 'general rule' that courts should decline supplemental jurisdiction when all federal claims are dismissed or otherwise eliminated from a case").

III. CONCLUSION

Based on the foregoing, the defendants' motions to dismiss (Record Documents 19 and 20) are GRANTED. All of the Bernofskys claims against the LRA and the OCD are dismissed without prejudice. All of the federal claims asserted against ICF are dismissed with prejudice. The court will exercise its discretion under 28 U.S.C. § 1367(c), in accordance with the general rule of this circuit, and decline to exercise supplemental jurisdiction over the Bernofskys' state

law claims and will dismiss these claims without prejudice.

A judgment consistent with the terms of this Memorandum Ruling shall issue herewith.

THUS DONE AND SIGNED at Shreveport, Louisiana, this 30th day of September, 2010.

/s/ Tom Stagg

JUDGE TOM STAGG

QUESTIONS

1. From your reading of the case, did the Bernofskys do anything inconsistent with what was required of them under the program? What additional steps might they have taken to ensure that their claim was properly processed?
2. Is it unreasonable to expect the Bernofskys to have done more than they did to preserve their claim? Why or why not?
3. The Court states that "The Bernofskys allege only mismanagement and incompetence, which does not rise to the level of a constitutional violation." Should they access to have some lesser legal remedy other than litigation? If so, what should it be? Construct a remedy that would resist potential fraudulent applicants, be cost-effective as well as user-friendly for people in shock in the aftermath of a disaster.
4. What more might the government at various levels, as well as private sector actors, have done to assist the Bernovskys?

St. Tammany Parish v. Fema And Department Of Homeland Security

556 F.3d 307 (5th Cir. 2009)

OPINION

KING, Circuit Judge:

We are asked to determine whether the discretionary function exception of the Robert T. Stafford Disaster Relief and Emergency Assistance Act, 42 U.S.C. § 5148, bars a suit based on the federal government's decision not to approve

funding for debris removal in the aftermath of Hurricane Katrina. St. Tammany Parish brings this lawsuit against the Federal Emergency Management Agency and the Department of Homeland Security because they denied funding for the removal of sediment deposited in the canals within St. Tammany Parish's Coin du Lestin community. The district court dismissed the case for lack of subject matter jurisdiction after concluding that the United States has not waived sovereign immunity for its agencies' discretionary

funding decisions. We agree that the agencies' decision not to fund the removal of the sediment was discretionary and therefore affirm.

I. FACTUAL, REGULATORY, AND PROCEDURAL BACKGROUND

A. Factual History and Relevant Regulations

On August 29, 2005, Hurricane Katrina made landfall along Louisiana's Gulf of Mexico coast. As a result, President George W. Bush declared that a major disaster existed in the State of Louisiana and initiated the federal government's involvement in the hurricane recovery effort. *See* Notice of the Presidential Declaration of a Major Disaster for the State of Louisiana, 70 Fed. Reg. 53,803-01 (Sept. 12, 2005). The President exercised his authority to declare major disaster areas pursuant to the Robert T. Stafford Disaster Relief and Emergency Assistance Act ("Stafford Act"), 42 U.S.C. §§ 5121-5208. *See* 42 U.S.C. § 5122(2) (defining major disaster by reference to presidential determination); *see also* 44 C.F.R. § 206.38(a) ("The Governor's request for a major disaster declaration may result in either a Presidential declaration of a major disaster or an emergency, or denial of the Governor's request.").

After the President declares a major disaster, the Stafford Act states that "[f]ederal agencies may on the direction of the President, provide assistance essential to meeting immediate threats to life and property resulting from [the] major disaster." 42 U.S.C. § 5170b(a). It permits federal agencies to assist in "debris removal" where it is "essential to saving lives and protecting and preserving property or public health and safety." Id. § 5170b(a)(3)(A). Section 5173 of the Stafford Act likewise states that "[t]he President, whenever he determines it to be in the public interest, is authorized...to make grants to any State or local government...for the purpose of removing debris...resulting from a major disaster from publicly or privately owned lands and waters." Id. § 5173(a)(2).

The Stafford Act authorizes the President to delegate his authority under the Act to a federal agency. *See* id. § 5164. The President exercised

this option and delegated most of his authority to the Federal Emergency Management Agency ("FEMA"), which is now part of the Department of Homeland Security ("DHS") (collectively, "defendants" or the "government"). Exec. Order No. 12,673: Delegation of Disaster Relief and Emergency Assistance Functions, 54 Fed. Reg. 12,571, § 1 (Mar. 23, 1989). In turn, FEMA promulgated certain regulations pursuant to the Stafford Act establishing the qualifications and procedures related to federal assistance for debris removal. *See* 44 C.F.R. § 206.224. The implementing regulations establish that debris removal must be in the "public interest" in order to be eligible for funding. Id. § 206.224(a). Debris removal is in the public interest when it is, inter alia, necessary to "[e]liminate immediate threats to life, public health, and safety." Id. § 206.224(a)(1). "Upon determination that debris removal is in the public interest, the Regional Director [, a FEMA official,] may provide assistance for the removal of debris and wreckage from publicly and privately owned lands and waters." Id. § 206.224(a).

Under the regulations, a specific project must be documented in a Project Worksheet, FEMA Form 90-91 ("PW"). *See* id. § 206.202(d). The Regional Director, or a designee, is charged with "review[ing] and sign[ing] an approval of work and costs on a Project Worksheet," id. § 206.201(j), and will then "obligate funds to the Grantee based on the approved Project Worksheet," id. § 206.202(e)(1). *See also* id. § 206.201 (i)(1) (authorizing FEMA to "approve a scope of eligible work and an itemized cost estimate before funding a project").

The President's August 29, 2005 declaration that a major disaster existed in Louisiana as a result of the damage caused by Hurricane Katrina in certain areas authorized FEMA "to allocate from funds available for these purposes such amounts as you find necessary for Federal disaster assistance and administrative expenses." 70 Fed. Reg. 53,803-01, at 53,803. The declaration also authorized FEMA "to provide . . . assistance for debris removal." Id. The President specifically identified St. Tammany Parish ("plaintiff" or the "Parish") as a municipality eligible for such assistance. Id.

Pursuant to the Stafford Act and its accompanying regulations, FEMA issued Recovery Policy 9523.13 to help facilitate debris removal from private property after Hurricane Katrina. *See* FEMA, Debris Removal from Private Property: Recovery Policy 9523.13 (Oct. 23, 2005), amending and replacing Recovery Policy 9523.13 (Sept. 7, 2005). Recovery Policy 9523.13 provides that:

> Hurricanes Katrina and Rita in some areas created catastrophic, widespread destruction resulting in vast quantities of debris which may require state or local government to enter private property to remove it in order to prevent disease and other immediate public health and safety threats. In these situations, debris removal from private property may be in the public interest and thus may be eligible for reimbursement, when the unconditional authorization for debris removal and indemnification requirements established by Sections 403 and 407 of the Stafford Act are met.

Recovery Policy 9523.13, at § 6(C). It offers guidance for reimbursing "state, county, and municipal governments for costs incurred in debris removal from private property." Id. at § 7. Section 7 of the Policy provides that FEMA will work with local governments to determine areas in which such debris removal "is in the 'public interest' under 44 C.F.R. § 206.224 and thus is eligible for FEMA reimbursement," § 7(A); requires the local government to submit a written request seeking reimbursement, see § 7(C); and grants FEMA the authority to approve or disapprove each request, *see* § 7(H). To clarify Recovery Policy 9523.13's application, Nancy Ward, FEMA's Director of the Recovery Area Command, issued a memorandum to FEMA's field offices that declared that "it is in the public interest to remove debris from private property because an immediate threat to public health and safety exists in" the Parish, thus satisfying § 7(A) of the Policy. *See* Memorandum from Nancy

Ward, Recovery Area Command Director, to FEMA Joint Field Offices (Sept. 10, 2005) (the "Ward Memorandum").

On September 12, 2005, the Parish filed a "Request for Public Assistance" for debris removal from public and private property within its jurisdiction. Part of the request sought debris removal from private canals in Coin du Lestin. Coin du Lestin, a private community consisting of approximately 250 residential homes, sits in the eastern part of the Parish. Coin du Lestin utilizes an aboveground drainage system that consists of drainage ditches, drains, and culverts. The drainage system is connected to a number of canals, which, in turn, drain into Bayou Bonfouca and then Lake Pontchartrain. The Coin du Lestin canals were navigable prior to Hurricane Katrina, reaching a depth of at least ten feet. Hurricane-related flooding, however, deposited construction and demolition ("C&D") materials, a boat, a submerged vehicle, as well as silt, mud, and vegetative materials, into the canals. Citing a potential flood hazard due to clogging in the Coin du Lestin canals, the Parish requested funding for removal of C&D debris and for the dredging of the canals to a depth of eight feet from bank to bank. The proposed scope of work included the removal of approximately 500,000 cubic yards ("CY") of debris.

In response, FEMA issued PW 2981, authorizing some, but not all, of the Parish's requested funding for debris removal from the Coin du Lestin canals. See FEMA, Project Worksheet Report 2981 (Feb. 6, 2006). According to PW 2981, FEMA debris specialists conducted an assessment of the canals on February 2, 2006, "to estimate the amount of debris in the canals that posed an immediate threat to improved property, public health and safety." Id. at 14. FEMA specialists determined that:

> [H]igh winds and storm surge associated with Hurricane Katrina . . . caused an estimated 130 CY of C&D debris and one (1) recreational boat to be deposited in the St. Tammany Coin [du]Lestin canals. The canals serve as access to the surrounding area and are within close proximity to parish residences; therefore, the debris is considered an immediate threat to public health and safety. Also

deposited in the canals were large quantities of marsh grass.

Id. at 2. As a result of this assessment, FEMA authorized funding for removal of approximately 130 CY of C&D debris and the recreational boat. FEMA, however, determined that "[m]arsh grass removal from the canal is considered not eligible as its removal is considered to be dredging." Id. at 3; *see also* id. at 14 (noting that some debris may not be considered eligible for funding because it "is not considered to pose an immediate threat to improved property, public health and safety, such as marsh grass, soil, and debris that is not in close proximity to improved property"). The approved PW 2981 authorized funding for a total of $ 7350 for debris removal. Id. at 1.

Nearly a year later, FEMA amended PW 2981 with PW 2981-1. See FEMA, Project Worksheet Report 2981-1 (Jan. 10, 2007). The new PW altered the work description to include removal of a newly discovered submerged vehicle that FEMA determined was eligible for funding. In addition, FEMA amended the cost figure to cover the use of an amphibious crawler and a barge to clear the C&D debris, the boat, and the submerged vehicle, which could not be cleared by conventional means. PW 2981-01 thus documented the total costs to be $31,979 (the original $7350 plus new costs totaling $24,629).

In an email dated May 4, 2007, Patrick W. Ruland, FEMA's Public Assistance officer, proposed an extension of the work to include the removal of some marsh grass to ensure flow within the canals. *See* Email from Patrick W. Ruland, Public Assistance Officer, to Joe Shoemaker et al. (May 4, 2007). Ruland notified the Parish that FEMA was willing to reconsider funding the removal of some marsh grass from the Coin du Lestin canals. Id. He wrote:

> FEMA understands that the Parishes [sic] two primary concerns are the removal of debris to reduce the threat of future flooding and to return the canals to a functional capacity for the residences. But with these waters being tertiary, non-navigable waterways, our concern is primarily to reduce further damages from occurring due to flood waters. The amounts of marsh grass removal FEMA PA

> would consider reasonable to allow the waters to flow for drainage purposes at the Coin Du Lestin area will be limited to 2' deep × 10' wide.... The debris removal will only be eligible at those areas of the waterways in which water flow is severely restricted.

Id. The record does not reveal that either FEMA or the Coin du Lestin community pursued this proposal or completed a new PW before the Parish filed the present suit.

B. Procedural History

On June 22, 2007, the Parish filed a five-count complaint against FEMA and DHS based on FEMA's refusal to fully fund the Parish's Request for Public Assistance seeking removal of all debris and sediment from the Coin du Lestin canals to a depth of eight feet. The Parish's first count alleged that defendants violated the Stafford Act, 42 U.S.C. §§ 5170 and 5173, by wrongfully refusing to fund dredging of the Parish's Coin du Lestin canals. The second count alleged that defendants were liable under the Federal Tort Claims Act ("FTCA"), 28 U.S.C. § 2674, for the same conduct. The third count alleged that defendants violated the Administrative Procedures Act ("APA"), 5 U.S.C. § 553, and the Stafford Act, 42 U.S.C. § 5165c, because FEMA's refusal to approve funding constituted a substantive rule change about which FEMA never provided the public with notice and an opportunity to comment. The fourth count alleged that defendants deprived the Parish of its right under the Stafford Act, id. § 5189a, and corresponding regulations, to appeal FEMA's denial of the requested funding. Lastly, the fifth count alleged that defendants violated the Stafford Act, id. § 5151, by failing to treat the residents of the Coin du Lestin community in a fair and equitable manner because FEMA funded similar projects in other communities. Plaintiff sought an order requiring FEMA to cover the costs of debris removal, a declaratory judgment that the government's policies violated the Stafford Act, and a monetary judgment in tort for sums required for debris removal.

Defendants moved to dismiss the case for lack of subject matter jurisdiction on the grounds of sovereign immunity under 42 U.S.C. § 5148 and of failure to exhaust administrative remedies. In support of their motion, defendants submitted an affidavit from Eddie Williams, FEMA's Deputy Public Assistance Officer. Williams confirmed that while FEMA did not initially agree that "complete removal of sediment was necessary to reduce an immediate threat of flooding," it has since "determined that sufficient flood hazards exist to allow for the removal of sediment (and swamp grass) to clear a 10 ft. wide by 2 ft. deep channel at drainage culverts throughout the approximately 1.5 miles of canals in the community." He also stated that a new version of PW 2981 was being considered and that "additional information from the Parish was needed."

In response to defendants' motion, the Parish argued that the Stafford Act, the FTCA, and the APA waived the United States's sovereign immunity for purposes of its suit because the Stafford Act and its corresponding regulations mandated that FEMA provide funding for debris removal from the Coin du Lestin canals. Furthermore, the Parish contended that Williams's affidavit showed that FEMA determined that there was an immediate threat of flooding from the Coin du Lestin canals, thus mandating funding of the requested dredging to remove the immediate threat.

- The district court granted defendants' motion to dismiss for lack of subject matter jurisdiction. The court first noted that § 702 of the APA waives sovereign immunity for claims alleging that a person suffered a legal wrong because of agency action, within the meaning of a relevant statute. The court also noted, however, that the waiver of sovereign immunity is inapplicable when "statutes

preclude judicial review." 5 U.S.C. § 701 (a)(1). The court thus weighed whether the Stafford Act's discretionary function exception, which precludes liability "for any claim based upon the exercise or performance of or the failure to exercise or perform a discretionary function or duty on the part of a Federal agency," 42 U.S.C. § 5148, applied to the Parish's claims. Noting the similarities between § 5148 and the FTCA's similar discretionary function exception to its waiver of sovereign immunity for tort claims, see 28 U.S.C. § 2680(a), the district court concluded that the same test for discretionary function applied under both statutes. It, therefore, applied the two-part test that the Supreme Court first described in *Berkovitz v. United States*, 486 U.S. 531 (1988), which defined discretionary functions under § 2680 (a). Under this test, the district court held "that FEMA's decisions whether and to what extent to remove the debris from the Coin du Lestin canals is a matter of judgment or choice and this judgment is of the kind that the discretionary function exception was designed to shield." 2 Thus, the district court held that the United States has not waived sovereign immunity for plaintiff's claims because they arise from discretionary functions that preclude suit under § 5148 and accordingly dismissed the suit for lack of subject matter jurisdiction pursuant to Rule 12(b)(1) of the Federal Rules of Civil Procedure.

On December 21, 2007, the Parish filed a timely notice of appeal. We have jurisdiction under 28 U.S.C. § 1291.

II. DISCUSSION

A. The Stafford Act's Discretionary Function Exception

"We review a district court's dismissal for lack of subject matter jurisdiction de novo." *Stiles v. GTE Sw., Inc.*, 128 F.3d 904, 906 (5th Cir. 1997).

"In our de novo review..., we apply the same standard as does the district court...." *Wagstaff v. U.S. Dep't of Educ.*, 509 F.3d 661, 663 (5th Cir. 2007) (internal quotation marks and citation omitted). Under our traditional explication of the standard applied by the district court, the district court "has the power to dismiss for lack of subject matter jurisdiction on any one of three separate bases: (1) the complaint alone; (2) the complaint supplemented by undisputed facts evidenced in the record; or (3) the complaint supplemented by undisputed facts plus the court's resolution of disputed facts." *Williamson v. Tucker*, 645 F.2d 404, 413 (5th Cir. 1981). Here, the district court did not resolve any disputed facts, so we, as did the district court, "consider the allegations in the plaintiff's complaint as true." Id. at 412. "[O]ur review is limited to determining whether the district court's application of the law is correct" and, to the extent its "decision [was] based on undisputed facts, whether those facts are indeed undisputed." Id. at 413. We then ask if dismissal was appropriate. *See United States v. Gaubert*, 499 U.S. 315, 327 ("'accept[ing] all of the factual allegations in [the plaintiff's] complaint as true' and ask[ing] whether the allegations state a claim sufficient to survive a motion to dismiss" (quoting Berkovitz, 486 U.S. at 540)).

We are asked to determine whether the United States has waived sovereign immunity for its agencies' decision not to fund the Parish's requested debris removal. Plaintiff bears the burden of showing Congress's unequivocal waiver of sovereign immunity. *See Kokkonen v. Guardian Life Ins. Co.*, 511 U.S. 375, 377 (1994) (holding that the party asserting the jurisdiction of a federal court bears the burden); *Peoples Nat'l Bank v. Office of the Comptroller of the Currency of the U.S.*, 362 F.3d 333, 336 (5th Cir. 2004) ("The party claiming federal subject matter jurisdiction has the burden of proving it exists."); *Paterson v. Weinberger*, 644 F.2d 521, 523 (5th Cir. 1981) (holding that, when faced with factual attack on federal court's subject matter jurisdiction, plaintiff bears the burden "of proving by a preponderance of the evidence that the trial court does have subject matter jurisdiction"); accord *Cole v. United States*, 657 F.2d 107, 109 (7th Cir. 1981) ("A party who sues the United States has the burden of pointing to a congressional act that gives consent." (citing *Malone v. Bowdoin*, 369 U.S. 643 (1962). At the pleading stage, plaintiff must invoke the court's jurisdiction by alleging a claim that is facially outside of the discretionary function exception.

"The basic rule of federal sovereign immunity is that the United States cannot be sued at all without the consent of Congress." *Block v. North Dakota ex rel. Bd. of Univ. & Sch. Lands*, 461 U.S. 273, 287 (1983); *see also Williamson v. U.S. Dep't of Agric.*, 815 F.2d 368, 373 (5th Cir. 1987) ("The doctrine of sovereign immunity is inherent in our constitutional structure and...renders the United States [and] its departments...immune from suit except as the United States has consented to be sued"). Because "[s]overeign immunity is jurisdictional in nature," *F.D.I.C. v. Meyer*, 510 U.S. 471, 475 (1994), Congress's "waiver of [it] must be unequivocally expressed in statutory text and will not be implied," *Lane v. Pena*, 518 U.S. 187, 192 (1996) (internal citation omitted)....

Plaintiff argues that Congress waived sovereign immunity for its present claims in three statutes: the FTCA, the APA, and the Stafford Act. Plaintiff alleges a claim under the FTCA for wrongful denial of funding under the Stafford Act. HN10The FTCA authorizes suits against the United States for damages arising from:

> injury or loss of property, or personal injury or death caused by the negligent or wrongful act or omission of any employee of the Government while acting within the scope of his office or employment, under circumstances where the United States, if a private person, would be liable to the claimant in accordance with the law of the place where the act or omission occurred.

28 U.S.C. § 1346(b)(1). Thus, the FTCA waives sovereign immunity and permits suits against the United States sounding in state tort for money damages. In re Supreme Beef Processors, Inc., 468 F.3d 248, 252 (5th Cir. 2006). As long as state tort law creates the relevant duty, the FTCA permits suit for violations of federal statutes and regulations. *See Johnson v. Sawyer*, 47 F.3d 716, 728 (5th Cir. 1995) (en banc) ("[T]he violation of a federal statute or regulation does not give rise to FTCA liability unless the relationship between the offending federal employee or agency and the injured party is such that the former, if a private person or entity, would owe a duty under state law to the latter in a nonfederal context. If the requisite relationship and duty exist, then the statutory or regulatory violation may constitute

or be evidence of negligence in the performance of that state law duty"). The FTCA, however, excepts discretionary functions and duties from this waiver of sovereign immunity. *See* 28 U.S.C. § 2680(a). This "discretionary function exception" provides that the waiver of sovereign immunity in § 1346(b) does not apply to:

> Any claim . . . based upon the exercise or performance or the failure to exercise or perform a discretionary function or duty on the part of a federal agency or an employee of the Government, whether or not the discretion involved be abused.

(Id). The "discretionary function exception is thus a form of retained sovereign immunity." In re World Trade Ctr. Disaster Site Litig., 521 F.3d 169, 190 (2d Cir. 2008). It "'marks the boundary between Congress' willingness to impose tort liability upon the United States and its desire to protect certain governmental activities from exposure to suit by private individuals.'" Berkovitz, 486 U.S. at 536 (quoting *United States v. S.A. Empresa de Viacao Aerea Rio Grandense* (Varig Airlines), 467 U.S. 797, 808 (1984)). It is intended to "assure protection for the Government against tort liability for errors in administration or in the exercise of discretionary functions." *Dalehite v. United States*, 346 U.S. 15 (1953) (citation omitted). In this case, plaintiff alleges a claim under the FTCA for FEMA's purported wrongful denial of nondiscretionary funding of the Parish's requested debris removal.

Plaintiff also brings a claim pursuant to the APA for improper rulemaking under the Stafford Act. The APA is a broadly applicable statute that "undoubtedly evinces Congress' intention and understanding that judicial review should be widely available to challenge the actions of federal administrative officials." *Califano v. Sanders*, 430 U.S. 99, 104 (1977). Section 702 of the APA authorizes suits against the United States through a limited waiver of sovereign immunity for "relief other than money damages" related to an agency's regulatory action. *See* 5 U.S.C. § 702. The APA, moreover, permits judicial review of claims for specific relief that result in the payment of money – such as a claim seeking a declaratory judgment ordering that an agency comply with a

mandatory funding requirement –because such actions are not for "money damages," in the form of compensation for a loss that the plaintiff has suffered or will suffer but for specific relief related to agency action. See *Bowen v. Massachusetts*, 487 U.S. 879, 893, 901 (1988). However, the waiver does not apply "to the extent that – (1) statutes preclude judicial review; or (2) agency action is committed to agency discretion by law." 5 U.S.C. § 701(a). In this case, the Parish alleges that FEMA's denial of full funding for its requested debris removal violated the APA, *see* id. § 553, and the Stafford Act, *see* 42 U.S.C. § 5165c, by altering FEMA's policy without providing notice and an opportunity to be heard.

Thus, plaintiff has alleged claims under two generally applicable statutes, the FTCA and the APA, for violations of the Stafford Act and its corresponding regulations. HN14Although the Stafford Act does not contain a waiver of sovereign immunity, *see Graham v. Fed. Emergency Mgmt. Agency*, 149 F.3d 997, 1001 (9th Cir. 1998), it does contain a discretionary function exception to governmental liability nearly identical to the one contained in the FTCA. See 42 U.S.C. § 5148. The Stafford Act's discretionary function exception provides that the United States will not be liable for:

> [A]ny claim based upon the exercise or performance of or the failure to exercise or perform a discretionary function or duty on the part of a Federal agency or an employee of the Federal Government in carrying out the provisions of this chapter.

(Id). The Stafford Act's discretionary function exception exists, despite the lack of an express waiver of sovereign immunity, to protect the government from liability for claims based on its discretionary conduct brought pursuant to the FTCA, APA, or other statutes of general applicability. Nonetheless, this provision "preclude[s] judicial review of all disaster relief claims based upon the discretionary actions of federal employees." *Rosas v. Brock*, 826 F.2d 1004, 1008 (11th Cir. 1987).

The government argues that the Stafford Act's discretionary function exception applies to the funding decisions that form the basis of the present claims. The Parish counters that it has alleged facts that give rise to a nondiscretionary duty to provide funding i.e., one not sheltered by the discretionary function exception. The first contention between the parties thus is the meaning of the term "discretionary function or duty" within the Stafford Act's discretionary function exception. Neither the Supreme Court nor this court has considered this question.

Plaintiff argues that the Supreme Court's interpretation of the term "discretionary function or duty" under § 2680(a) of the FTCA applies to define the same term under § 5148 of the Stafford Act because of the near identical phrasing of the two provisions. For the first time on appeal and despite its success under the Parish's proposed definition in the district court, the government argues that the Stafford Act's structure and its extraordinary purpose dictate that its discretionary function exception applies more broadly. According to the government, "any activity of the United States undertaken to carry out the provisions of the Stafford Act will necessarily trigger § 5148 Stafford Act immunity." Thus, the government now recommends interpreting § 5148 to bar claims with a "propinquity" to a disaster and a "close substantive nexus" to disaster assistance.

We hold that "discretionary function or duty" has the same meaning in § 5148 as it does in § 2680(a). We reach this conclusion because the relevant language of the two provisions is identical, because anecdotal evidence shows that Congress intended to incorporate § 2680(a)'s standard into § 5148, and because our sister courts of appeals and most district courts have looked to § 2680(a) case law when applying § 5148. Whatever the arguable differences between the two statutes, those differences are not at play here, and we conclude that the meaning of "discretionary function or duty" is the same under § 5148 and § 2680(a).

"The starting point in statutory interpretation is 'the language [of the statute] itself.'" *United States v. James*, 478 U.S. 597, 604 (1986) (citation omitted, alteration in original), abrogated in non-relevant part by *Central Green Co. v. United States*, 531 U.S. 425 (2001). We are bound to apply "the plain language of the statute, especially where . . . there is nothing in the statute or its legislative history to indicate a contrary intent." In re DP Partners Ltd., 106 F.3d 667, 671 (5th Cir. 1997). Moreover, "as a matter of statutory interpretation, in determining the meaning of a particular statutory provision, it is helpful to consider the interpretation of other statutory provisions that employ the same or similar language." *Flowers v. S. Reg'l Physician Servs. Inc.*, 247 F.3d 229, 233 n.4 (5th Cir. 2001) (citations omitted). We must nonetheless respect differences in the texts and the overall statutory schemes to ensure that the purposes of both acts are served by interpreting them in the same manner. See *Smith v. City of Jackson*, 351 F.3d 183, 189 (5th Cir. 2003).

In this case, the language of § 5148 of the Stafford Act mirrors that of § 2680(a) of the FTCA. Compare § 5148 (exempting claims based on "the exercise or performance of or the failure to exercise or perform a discretionary function or duty"), with § 2680(a) (exempting claims based on "the exercise or performance or the failure to exercise or perform a discretionary function or duty"). Both provisions prevent subject matter jurisdiction arising from the same type of conduct – "a discretionary function or duty." Thus, we hold that the phrase "discretionary function or duty" has the same meaning under both statutes.

Ancillary evidence supports our conclusion . . .

The government's counterarguments do not show that Congress intended "a discretionary function or duty" to have a different meaning in § 5148 than in § 2680(a). The government first attempts to distinguish the two provisions by noting that § 2680(a) is "a passive exception" to a waiver of sovereign immunity while § 5148 is "an active prohibition" against the imposition of liability based on a statute that does not contain an express waiver of sovereign immunity. This structural distinction, while relevant to the question of whether Congress waived sovereign immunity in the Stafford Act, does not answer the question at hand – what constitutes a "discretionary function or duty."

The government also attempts to distinguish the two statutes by reference to the few differences in the statutes' wordings. Where relevant, however, these differences support our holding. First, § 5148 contains the phrase "in carrying out the provisions of this chapter"; whereas, § 2680(a) applies to all discretionary conduct from the myriad substantive sources of FTCA liability. This distinction narrows the applicability of § 5148 to discretionary functions under the Stafford Act but, again, does not clarify the meaning of "discretionary function or duty." Second, § 5148 bars "[a]ny claim." Of course, § 2680(a) also mentions "any claim," but it only applies to tort claims for monetary relief (the only available relief under the FTCA). Although § 5148 covers all types of claims brought with respect to the Stafford Act, it does not cover all conduct, and our goal here is to identify the discretionary conduct that it covers. This difference in statutory language is not inconsistent with our conclusion.

This distinction, however, undermines the government's next argument-that interpreting § 2680(a) and § 5148 consistently would render the latter superfluous. Under our construction and contrary to the government's assertion, because § 5148 applies to any claim, it has an independent operational effect beyond that of § 2680(a): Where § 5148 is applicable, the government retains sovereign immunity for claims that are based on discretionary functions or duties, whether alleged under the FTCA, see Dureiko I, 1996 U.S. Dist. LEXIS 22365, 1996 WL 825402, *2; cf. Gaubert, 499 U.S. at 324; the APA, see Rosas, 826 [*322] F.2d at 1008; or, ostensibly, the Contract Disputes Act, 41 U.S.C. § 601-613, cf. Dureiko II, 209 F.3d at 1353. In support of its argument that § 5148 would be superfluous under our reading, the government contends that because the § 2680(a) existed at the time of the passage of the Stafford Act, Congress must have intended to provide § 5148 with a distinct meaning. Our conclusion that § 5148 is not superfluous nullifies this argument, and in any case, the legislative history does not support the government's position. The government offers the statement of Representative Whittington, Chairman of the House of Representatives Committee on Public Works, during the House of Representative's floor debate on the Stafford Act:

> We have further provided that if the agencies of the Government make a mistake in the administration of the Disaster Relief Act that the Government

may not be sued. Strange as it may seem, there are many suits pending in the Court of Claims today against the Government because of alleged mistakes made in the administration of other relief acts, suits aggregating millions of dollars because citizens have averred that the agencies and employees of Government made mistakes. We have put a stipulation in here that there shall be no liability on the part of the Government.

96 Cong. Rec. 11895, 11912 (1950); see also H.R. Rep. No. 81-2727, at 3 (report of Rep. Whittington) (summarizing that the proposed legislation contained an amendment "to provide that the Federal Government shall not be liable for any claims based upon the proper exercise or performance of a function or duty on the part of any Federal agency or any employee of the Government in carrying out the provisions of the section"). Placed in proper context and in light of the clear language of § 5148, this statement is unpersuasive. Representative Whittington made his statement immediately after the text of § 5148, including the phrase "a discretionary function or duty," was read to the House of Representatives. 96 Cong. Rec. at 11912. Thus, Representative Whittington's statement was in the context of the proposed statute's discretionary language. In addition, earlier in the debate, Representative Whittington commented that the version presented for passage was amended to ensure "that the Government shall not be liable for any claim based upon the performance, or the failure to perform, a discretionary function on the part of an agency or employee of the government." Id. at 11897 (emphasis added). Thus, Representative Whittington verbalized § 5148's focus on discretionary conduct. Overall, Representative Whittington's comment and one sentence from the committee's report cannot fairly be construed to override the clear, unambiguous language of § 5148. Our construction of "discretionary function or duty" thus does not render § 5148 superfluous and promotes a consistent application of congressional language intended to the have the same meaning. We therefore reject the government's arguments and conclude that "a discretionary function or duty" has the same meaning within § 5148 as it does within § 2680(a). Thus,

we will rely on precedent interpreting that phrase under § 2680(a).

B. The Funding Decision

We now determine the applicability of the discretionary function exception under § 5148 of the Stafford Act to this case by turning to the well-established precedent defining discretionary conduct under § 2680(a) of the FTCA. We hold that under that precedent, FEMA's decision not to approve funding for sediment dredging in the Coin du Lestin canals is discretionary and that § 5148 therefore bars the Parish's current claims.

The Supreme Court has developed a two-part test for determining whether agency conduct qualifies as a discretionary function or duty under this exception. *See* Gaubert, 499 U.S. at 322-23 (citing Berkovitz, 486 U.S. at 536-37). First, the conduct must be a "matter of choice for the acting employee." Berkovitz, 486 U.S. at 536. "The exception covers only acts that are discretionary in nature, acts that 'involv[e] an element of judgment or choice.'" Gaubert, 499 U.S. at 322 (quoting Berkovitz, 486 U.S. at 536) (alteration in original). Thus, "'it is the nature of the conduct, rather than the status of the actor' that governs whether the exception applies." Id. (quoting Varig Airlines, 467 U.S. at 813). If a statute, regulation, or policy leaves it to a federal agency to determine when and how to take action, the agency is not bound to act in a particular manner and the exercise of its authority is discretionary. *See* id. at 329. On the other hand, "[t]he requirement of judgment or choice is not satisfied" and the discretionary function exception does not apply "if a 'federal statute, regulation, or policy specifically prescribes a course of action for an employee to follow,' because 'the employee has no rightful option but to adhere to the directive.'" Id. at 322 (quoting Berkovitz, 486 U.S. at 536).

Second, "even 'assuming the challenged conduct involves an element of judgment,'" we must still decide that the "'judgment is of the kind that the discretionary function exception was designed to shield.'" Id. at 322-23 (quoting Berkovitz, 486 U.S. at 536); see also Varig Airlines, 467 U.S. at 813. "Because the purpose of the exception is to 'prevent judicial "second-guessing" of legislative and administrative decisions grounded in social, economic, and political policy through the medium of an action in tort,' when properly construed, the exception 'protects only governmental actions and decisions based on considerations of public policy.'" Gaubert, 499 U.S. at 323 (quoting Berkovitz, 486 U.S. at 537). In this regard, "if a regulation allows the employee discretion, the very existence of the regulation creates a strong presumption that a discretionary act authorized by the regulation involves consideration of the same policies which led to the promulgation of the regulations." Id. at 324.

In this case, plaintiff argues that the Stafford Act, regulations promulgated pursuant to it, Recovery Policy 9523.13, the Ward Memorandum, and PW 2981.1 created a nondiscretionary duty on FEMA to fund its requested debris and sediment removal in the Coin du Lestin canals to a depth of eight feet. Under the first prong of the Berkovitz test, plaintiff asserts that these authorities prevented FEMA from exercising any choice in whether to fund the dredging. Plaintiff's claim is predominately based on two deductions – first, that once FEMA declared debris removal from private property in the Parish to be in the public interest, it had no discretion to deny funding for the Parish's request in this specific case; and second, that once FEMA engineers determined that some threat existed to improved property, public health, or safety, it lacked discretion to deny funding for the debris removal. We conclude that, in this case, the cited authorities do not create a nondiscretionary duty mandating that FEMA fund the Parish's requested dredging of the Coin du Lestin canals.

Sections 5170b and 5173 of the Stafford Act grant FEMA authority to fund debris removal from private property in designated disaster areas. Those provisions, however, are cast in discretionary terms. Section 5170b states that: "Federal agencies may on the direction of the President, provide assistance essential to meeting immediate threats to life and property resulting from a major disaster." § 5170b(a) (emphasis added). Section 5173, likewise, states that "[t]he

President, whenever he determines it to be in the public interest, is authorized... to make grants to any State or local government... for the purpose of removing debris... resulting from a major disaster from publicly or privately owned lands and waters." § 5173(a)(2) (emphasis added). The uses of the words "may" and "is authorized" indicate that government approval of assistance for debris removal is discretionary. *See*, e.g., *Neuwirth v. La. State Bd. of Dentistry*, 845 F.2d 553, 557 (5th Cir. 1998) (stating that the use of the word "may" shows a legislature's intention to bestow discretion on an agency).

Nor do the corresponding federal regulations create a mandatory duty. The regulations permit FEMA to provide assistance for removal of eligible debris if in the public interest, but they do not mandate assistance even where that eligibility criterion is met. *See* 44 C.F.R. § 206.224(a) ("Upon determination that debris removal is in the public interest, the Regional Director may provide assistance for the removal of debris and wreckage from publicly and privately owned lands and waters." (emphasis added)); *see also* Lockett, 836 F. Supp. at 854 (holding that "[e]ven where an applicant meets the requirements of eligibility, the language, by including the term 'may,' signifies that FEMA has the discretion to award that assistance" (citation omitted)).

Similarly, Recovery Policy 9523.13, the Ward Memorandum, and PW 2891.1 do not mandate that FEMA fund the Parish's request for dredging. Recovery Policy 9523.13 states that FEMA has "authority" to fund debris removal, § 6(A), and that "debris removal from private property may be in the public interest and thus may be eligible for reimbursement," § 6(C) (emphasis added). Thus, under Recovery Policy 9523.13, funding decisions remain discretionary. Through the Ward Memorandum, FEMA favorably exercised its authority to label the Parish eligible for public-interest-based debris removal funding; however, the Ward Memorandum does not create a nondiscretionary duty to fund any or every request for debris removal. In addition, PW 2891.1, even if viewed as an agency confirmation that some threat to improved property, public health, and safety existed, does not create

a nondiscretionary duty to fund the debris removal presently sought by the Parish. While PW 2891.1 concludes that funding to remove the C&D debris, the boat, and the submerged vehicle was necessary to eliminate the threats to improved property, public health, and safety from further flooding, it provides no basis to conclude that additional dredging of sediment and marsh grass was necessary to reduce those threats. Finally, we note that, viewed as a comprehensive regulatory scheme, the combined import of these sources does not give rise to a nondiscretionary funding mandate. Thus, we conclude that under the first prong of the Berkovitz test, FEMA was under no nondiscretionary duty to fund dredging.

Under the second prong of the Berkovitz test, we hold that funding decisions related to the extent of debris removal that is necessary to protect improved property, public health, and safety are exactly the type of public policy considerations that § 5148 shields from judicial scrutiny. See Sunrise Vill. Mobile Home Park, 960 F. Supp. at 286 (HN25"[T]he Government's decisions on when, where, and how to remove debris after a major disaster are exactly the sort of policy-imbued decisions that fall within the second prong of the discretionary function exception."). Here, FEMA engineers determined that removal of C&D debris, the boat, and the submerged vehicle was necessary to protect the community but that dredging was not necessary. Eligibility determinations, the distribution of limited funds, and other decisions regarding the funding of eligible projects are inherently discretionary and the exact types of policy decisions that are best left to the agencies without court interference.

Thus, we hold that in this case, the Stafford Act, its regulations, and related agency guidance do not give rise to a mandatory duty. They instead permit discretionary, policy-oriented choices that cannot be the basis for the court's subject matter jurisdiction. As a result, FEMA's decision not to approve funding for sediment dredging in the Coin du Lestin canals is discretionary, and § 5148 therefore bars the Parish's current claims.

III. CONCLUSION

For the above explained reasons, we AFFIRM the district court's dismissal for lack of subject matter jurisdiction under Rule 12(b)(1).

_____

QUESTIONS

1. In light of this case's outcome, does the restoration of the "status quo ante" as mentioned at the beginning of the chapter have continued validity?
2. Why did the Court hold that the decision not to fund the removal of the sediment was discretionary?
3. Why did Congress create discretionary immunity for FEMA's actions under the Stafford Act as described in the case?
4. Do you believe that discretionary immunity for FEMA's actions under the Stafford Act as described in this case is a good thing from a public policy perspective? Explain your answer, with reference to the discussion of the two points of view as discussed in the case as well as your own perspective.

Section III

TOWARD THE FUTURE

Let us not seek the Republican answer or the Democratic answer, but the right answer. Let us not seek to fix the blame for the past. Let us accept our own responsibility for the future.

John F. Kennedy

Students in the author's Ethics in Criminal Justice class at North Carolina Central University take part in a spirited discussion of the role of ethics in law enforcement. Author photo.

Chapter 22

THE ETHICAL IMPERATIVE

William C. Nicholson

Both emergency responders and emergency managers are public servants. Their work is performed not with the goal of personal financial gain but to assist their fellow humans in time of need. All public servants have an overriding obligation to perform their duties in an ethical manner. This requirement is a matter of personal morality as well as legal responsibility. Failure to comply may expose the violator to legal sanctions as well as to the emotion of having failed to do the right thing. That feeling may be the worst punishment for a person whose greatest goal is to help others.

So how does one determine what is the ethical course of action when faced with a choice? The safest choice is generally to act as you have been trained, in conformance with the standard operating procedures (SOPs) or standard operating guidelines (SOGs) of your service.

If one is a state employee, there are often state statutes that cover ethical issues, such as Indiana's 40 IAC 2-1-1 *et seq.* (2011), the Indiana Code of Ethics for the Conduct of State Business. The code is described as "aspirational," but its list of goals contains some items whose violation could lead to discipline. These include:

(1) Duties should be carried out impartially.
(2) Decisions and policy should not be made outside of proper channels of state government.

(3) Public office should not be used for private gain.
(4) Public confidence in the integrity of government is essential to the exercise of good government.
(5) Actions, transactions, or involvements should not be performed or engaged in which have the potential to become a conflict of interest.[1]

Accepting food, drink, and items above minimal value are forbidden, as is moonlighting (working for others in ways incompatible with government employment or in ways that impede government employment). No political activity is allowed on duty, conflicts of interest must be avoided, and the State Ethics Commission must be consulted for an advisory opinion if such a conflict is suspected. Violations or suspected violations of ethics laws may be complained of directly to the Ethics Commission, which may also make investigations on its own. The rules make provision for a hearing and appeals process for violations.

Sometimes, however, the issue involved will not be covered by state ethics law because an individual is not a state employee or the state involved does not have as well-developed an ethics law structure as Indiana's. There will be times, as well, when SOPs or SOGs do not cover the situation. There may also be times when a

---

[1] 40 IAC 2-1-3 (2011)

public employee's internal sense of what is right tells him or her that the SOP or SOG is wrong, which is a most difficult situation, for it may im- mediately expose one to sanctions for a viola- tion. Under those circumstances, the approach in the box is suggested.

STRUCTURE FOR SOLVING AN ETHICAL DILEMMA

1. **Identify whether you face an ethical dilemma.**
   a. **If you take action, will your action cause conflict?** Conflict will usually be with co- workers or another aspect of the emergency response and emergency management system.
   b. **Does the situation raise issues of rights and morality?** For example, for EMS, a patient's right to confidentiality of medical records or the morality of stealing patient medicines or sexually molesting patients. For emergency management, choosing a radio system after the salesman has taken the contracting officer on lavish trips or refusing to include persons who do not speak English in the planning group.
   c. **State the ethical dilemma clearly** – specify the conflict involved and the issues of rights and morality

2. **State the relevant information regarding your knowledge of the situation, including:**
   a. **Facts and circumstances of the situation**
   b. **How you learned of the situation**
   c. **Identities of those involved**

3. **State the values that may conflict and personal elements**
   a. **"Formal" values** – Usually these are SOPs and SOGs. They also include the laws regarding conflict of interest, kickbacks, and so on.
   b. **"Informal" values** – These are usually the how you and your peers operate on a daily basis. Formal rules don't cover everything.
   c. **Personal values** – What do you believe? Can you live with yourself if you see wrong being done and do not take action. The worse the effects if you do not act may be an im- portant issue as you consider what to do.
   d. **Personal position in hierarchy** – When it comes to conflict, will this influence the out- come?
   e. **Potential effects on personal life** – Could you be fired? What other work could you get? What effect might a bad recommendation have on your future? Consider the discussion of "easy wrong" versus "hard right" that follows below when you think about potential conse- quences on your personal life. Remember that your decision may have very significant long-term effects on your life.

4. **Identify ethical theories and principles to be applied to the dilemma.** At this point, you may think of formal ethical theories that apply if you are familiar with any of them. You may use Utilitarianism, which suggests that the best act is the one that creates the greatest good for the greatest number of people. Or you may wish to apply Aristotle's Virtue Ethics, which man- dates that a person should act in all situations in conformity with the traditional virtues that are part of a good character. Consequentialism, Stoicism, Justice Ethics and other options also are available for study.

5. **Identify available options to deal with dilemma using ethical theories and principles listed in step 4.**
   a. **This is where formal analysis may take place.** However, you may believe that you have sufficient information to make an educated decision after you have analyzed the material in steps 1–3 without the need of further formal ethical analysis.
   b. **You must apply the results of your analysis** to all options identified. Referring to the analysis in 5.a., EXPLAIN to yourself and support to your personal satisfaction which is best choice. Your options include:
      i. **Formal action through vertical hierarchy.** In other words, bringing the issue to the attention of those further up in the organization so that they can address it. This works if you believe that the ethical problem is a "bad apple" but that the organization as a whole is sound.
      ii. **Informal action through horizontal structure.** This is working with your peers to get them to "knock it off" and perform in an ethical manner.
      iii. **Inaction.** You have decided not to act
      iv. **Other action.** This may be any other kind of option. It generally means going completely outside of the formal or informal structure. This is often the "whistleblower" option. People are usually extremely very reluctant to take this road. It means that they believe that the organization they are affiliated with will not reform itself if they report unethical conduct through the vertical hierarchy. Sometimes they have previously attempted to report unethical action within the system and either been ignored or been threatened with reprisal for their attempts to reform the system. Taking this action is a brave and often desperate act.

6. **Make a decision based on your analysis** of the dilemma and application of ethical principles. Decide "This is the right thing for me to do."
   a. **State the ethical decision**
   b. **Act on the decision**

Note: This approach is adapted from: Cyndi Banks, *Criminal Justice Ethics Second Edition*, Sage Publications (2008).

A tale from the so-called "war on terror" provides an illustration of how quickly a well-established rule of law could be thrown aside. Without a person of conscience in a position to say this is wrong and make it stop, we saw how easily an established war crime, rejected by the entire civilized world, became the policy of this nation after 9/11.[2] After the allied victory in World War II, the practice of waterboarding was the basis of prosecution for a number of enemy soldiers, who received prison time for engaging in the practice. The United States has prosecuted its own soldiers for waterboarding prisoners in the past.[3] Despite clear and unambiguous law, both domestic and international, prohibiting the practice,[4] waterboarding was practiced under the Bush administration. This illegality continues to be endorsed by some who are current and former nominees for President in 2012. As of the writing of this book, these include: Rick

---

[2] Evan Wallach, "Drop by Drop: Forgetting the History of Water Torture in the United States," *45 Colum. J. Transnat'l Law 468* (2007).

[3] See id. at 494-501.

[4] Wilson R. Huhn, "Waterboarding is Illegal." *89 Washington University Law Review 1* (1988).

Santorum,[5] Herman Cain, Michelle Bachman, and Rick Perry.[6]

The checks and balances of the Constitution should operate to protect the legal status quo and prevent such unlawful practices. Yet if an administration calls something a "war" – (even if Congress has not declared war) the Courts become very reluctant to step in, especially if there is some wisp of legal argument for the government to hide behind. The Bush/Cheney administration took cover behind legal opinions authorizing it to engage in waterboarding. It defended the attorneys who submitted the opinions – an interesting contrast to Bush's determination to prosecute whistleblowers who exposed the patently illegal use of the National Security Agency to conduct warrantless domestic electronic surveillance.[7]

The pressures on lawyers asked to create the opinions authorizing waterboarding must have been very heavy. If one were an attorney in a cushy White House or Department of Justice political job, it would be very difficult to resist the pressure to create the opinion desired by one's masters. Comply, and the large firm, big money opportunities to which such positions are stepping stones continue to beckon. Fail to go along and those doors might no longer be open.[8] Viewed in the most favorable light, these lawyers were endorsing breaking domestic and international law for the "greater good" of national security. They must appreciate; however, no less

than the nonviolent protester who is jailed after a sit-down strike, than when one violates the law, one must be willing to accept the consequences. The ugly thing about these lawyers in their comfortable Washington offices is that they were giving advice that would result in others, some of them American service members, committing legal and ethical violations. The legal opinions authorizing waterboarding could have led to professional discipline for all attorneys involved in their creation. The opinions stating that it was lawful misrepresented the state of the law and seriously interfered with the administration of justice, both violations of the District of Columbia Bar's Rules of Professional Conduct.[9]

Contrast the decisions of those lawyers to go along with endorsing violations of law with that of John Dean, White House Counsel under Richard Nixon. Dean told all to the Watergate Committee, and was excoriated by many. He had a hard time of it for many years. Yet now, history can fairly be said to have vindicated Dean, and he is widely regarded as something of a hero for his actions, in contrast to the vast majority of the Nixon White House staff.

During the nine years since the first edition of this book was published, the author has had the honor to teach emergency management and homeland security classes as well as Ethics in Criminal Justice to many fine students. These young men and women listened attentively as he

---

[5] Santorum: McCain Doesn't Understand Interrogation, Fox News, May 16, 2011, found on line at http://www.foxnews.com/politics/2011/05/18/santorum-mccain-doesnt-understand-interrogation/

[6] "Have Republicans 'gone off the rails' by endorsing waterboarding?" This Week, November 15, 2011, found on line at: http://theweek.com/article/index/221435/have-republicans-gone-off-the-rails-by-endorsing-waterboarding

[7] "Justice Drops Probe Of Leaker Who Exposed Bush-Era Wiretapping" NPR, April 26, 2011, found on line at: http://www.npr.org/blogs/thetwo-way/2011/04/26/135735752/report-justice-drops-probe-of-leaker-who-exposed-bush-era-wiretapping

[8] Also, it would be easy for an inexperienced young attorney to listen to one point of view and become convinced that no other path to protection exists. This is why lawyers are trained to perform complete research before forming an opinion. It is difficult to imagine that any representatives from the "win them over by reason and friendship" school of interrogation were present at the White House to discuss that path as an alternative.

[9] District of Columbia Bar Rules of Professional Conduct found on line at: http://www.dcbar.org/for_lawyers/ethics/legal_ethics/rules_of_professional_conduct/amended_rules/rule_eight/rule08_04.cfm

The DC Bar's Rules of Professional Conduct state:

Rule 8.4-Misconduct

It is professional misconduct for a lawyer to:

(a) Violate or attempt to violate the Rules of Professional Conduct, knowingly assist or induce another to do so, or do so through the acts of another;

(b) Commit a criminal act that reflects adversely on the lawyer's honesty, trustworthiness, or fitness as a lawyer in other respects;

(c) Engage in conduct involving dishonesty, fraud, deceit, or misrepresentation;

(d) Engage in conduct that seriously interferes with the administration of justice;

advised them to take their careers and livelihoods in their hands and tell truth to power when they see wrong being done by those endowed with the public trust. This is never an easy thing to do.

The choice always boils down to this – do the easy wrong or the hard right. The easy wrong will let you get along with your peers today, tomorrow, and perhaps for years to come. But there will always be a reckoning ahead, even if it is only with your own conscience as you struggle to sleep at night, knowing that you chose to do the wrong thing, to take the path downwards, when you saw right path, the road up to the light. The hard right may be the choice that makes your life a hell for the short run – your friends at work could reject you, you may lose your job, people might call you a rat or traitor, and you could be out in the cold for a good while. But inside, where it counts, you will be warm. Your sleep will not be troubled by visions of the consequences of your failure to be true to yourself. You will find that the people whose opinion really matters – whether or not they wield power – will be those to whom your ethical stand is an important measure of your character.

When the author teaches about good and bad ethical choices for professionals in the criminal justice system, he includes cops who shake down citizens and lawyers who create opinions authorizing torture in the same category. For, although the scope of their transgressions against the justice system may differ and their intent also is not identical, the nature of their offenses is the same. Both are discarding the rule of law, whether for personal self-aggrandizement or for an "ends justifying the means" argument of national security. In both cases, these acts weaken public support of and belief in the legal system and, at the heart of that system, the United States Constitution that underlies the power of the law. As ethics is to the individual, so is the Constitution to the nation.

In the emergency response and emergency management world, there are sometimes opportunities to do the wrong thing. In his capacity as agency Ethics Officer and General Counsel, the author pursued ethics violations such as, for example, taking the medicines or possessions of a patient being transported to a medical facility or converting emergency management supplies and funds to one's personal use.

These prosecutions often resulted from fellow employees who reported the wrongdoers. When asked why they reported the ethical wrongdoing, the most frequent reply was that they were disappointed in their fellow public servants. Further, the reporting individuals felt a positive duty to keep the profession "clean" and respected by the citizens. Like the vast majority of public servants, these people saw their work as a calling whose rewards were not monetary. Their job satisfaction came from spending their shifts helping people. For those who keep their priorities straight, ethical issues rarely arise.

QUESTIONS

1. Assign each class member to create an ethical dilemma for an emergency response or emergency management organization member, then have class members exchange their hypothetical situations with one another. Have each student address how to solve the dilemma using the six step process outlined in the box with the text and present to the class his or her solution. Critique one another's solutions and suggest alternate approaches.

2. Perform research on waterboarding. Divide the class into two groups to discuss the issue. What is your opinion of waterboarding? Does it work? Is it effective? After the discussion of whether it works, address the matter of its morality.

3. Put yourself in the position of the ambitious young attorney in the White House who is requested to write an opinion on the legality of waterboarding. You are a committed member of the staff who believes strongly in

President Bush and the War on Terror. You do not believe that "soft" interrogation techniques work on committed terrorists. 9/11 is in the recent past, we have high value terrorists in hand, and you believe that more big attacks are coming soon – maybe with WMD – nuclear, bio, or chemical. What will you do? What is your opinion of the "easy wrong" versus the "hard right" in this situation? Address the following matters:

a. To whom or what do you owe duties?

b. How do you decide that a given activity is wrong?

c. Whose interests do you need to consider when deciding what to do?

d. To whom do you report the wrong activity?

4. Assign each class member a different state and have him/her report on that state's law regarding public employee ethics. Discuss similarities and differences between the laws and why differing approaches to common issues be preferable. Have the class construct an ideal hybrid state ethics law.

5. The author states, "As ethics is to the individual, so is the Constitution to the nation." What do you think he means by this statement?

6. Why did so many people turn in ethics violators to the author as discussed in the final paragraph? Would you do so? Why or why not?

Soldiers with 2nd Battalion, 327th Infantry Regiment, 101st Airborne Division, return fire during a firefight with Taliban forces in Barawala Kalay Valley in Kunar province, Afghanistan, March 31, 2011. Photo Credit: Pfc. Cameron Boyd.

Chapter 23

"ALL OF THAT CHANGED ON SEPTEMBER 11"

William C. Nicholson

This chapter takes a rather broad view of emergency management's history since the terrorist attacks of September 11, 2001. This writing discusses political and historic events, both foreign and domestic, and relates them to emergency response and emergency management issues. As with the entire history of the disciplines, law has set their meets and bounds in these contexts. The chapter also attempts to forecast future events. This is a normal part of the Risk Management process. Doing so on such a global basis is unusual and not part of the typical planning process, but helpful in trying to project future man-made risks. Attempts to draw lessons from history at a short distance and predict their future effects often go awry. Any prognostication can only be based on the grossly inadequate information available at the time the prediction is made. Unforeseen factors can intervene and render deductions irrelevant.

It is important when discussing the events since 9/11, or any events, to replace nostalgia with historical fact and work from there. The reader will notice that the author's political views color his interpretation of the events. He sincerely hopes that those who disagree with his assessments will favor class members with their perspectives and that the result will be well-informed and lively discussions.

A. Introduction

In a Meet the Press interview in 2003, Vice President Richard Cheney described the Bush administration's approach to foreign policy just prior to the beginning of the U.S. invasion of Iraq in the following way, "All of that changed on September 11."[1] In particular, no longer would we wait for another nation to attack us before responding, nor would we try to contain aggression. Instead, unlike during the 1991 Iraq war, we would be the aggressors, destroy that nation's infrastructure, unseat Saddam Hussein, and move to destroy the nuclear, bioweapon, and chemical warfare capabilities that Cheney believed him to possess.[2] Cheney refused any delay in the administration's schedule for invasion of Iraq to get greater international support. This was itself a change from just after the attacks of September 11, 2001, when the international community was almost universally sympathetic to the United States. The Bush administration's "my way or the highway" attitude squandered that support ("changing" that – as Cheney put it), leaving the nation almost alone to wage and pay for ruinously expensive wars in Iraq and Afghanistan.

The media reflected a national attitude of change following the attacks, though the conclusions it drew from the alteration of viewpoint

---

[1] http://www.mtholyoke.edu/acad/intrel/bush/cheneymeetthepress.htm

[2] As of this writing, 8 years after that interview, no such capabilities have been discovered. Doubtless Cheney if asked would say that they were so well hidden that all our searching has not uncovered them.

401

were less black and white than those of the administration.[3] The claims of domestic change were in fact somewhat hard to document. For, indeed, what had changed? The September 11 attacks were often compared to Pearl Harbor, and the nation prepared for calls to make sacrifices like those after that attack in response. After the December 7 attack in Hawaii, taxes were raised, production shifted to a wartime footing, many luxuries (such as new autos) became unavailable, gasoline and other basic commodities were rationed and so on. In contrast, after the 9/11 attacks, taxes went down (particularly for the wealthiest Americans), President Bush encouraged spending to keep the economy growing, there were no limitations on luxury goods, and the military remained a volunteer force.

While the ten years since 9/11 establish that terrorism is still a concern, this period also demonstrates that natural hazards continue to inflict many billions of dollars in damage and kill thousands of people. Indeed, when worldwide disasters are taken into account, it is clear that predictable natural disasters such as major earthquakes, tsunamis, and volcanic eruptions have the potential to do significantly more economic damage and kill more people than a terrorist nuclear device, for example. The problem is that natural hazards can be mitigated, their effects greatly diminished, while the nuclear device **might** be prevented. The bottom line is political – since no administration could ever survive a nuclear release, they must be avoided at all costs, whether doing so at the expense of other mitigation efforts makes sense from a rational cost/benefit analysis or not.

B. Homeland Security Expenditures

How has our national treasure been spent since the 9/11 attacks? America has spent $1 trillion on the Iraq war through 2011.[4] The fact is that in the United States, one's chances of being killed by a terrorist are 1 in 3.5 million, and state and federal governments are spending $75 billion yearly on homeland security.[5] How much should be spent to lower those odds to 1 in 4 or 4.5 million?

At the same time as the federal government has been ladling out these huge sums, the fiscal challenges facing state and local governments continue to grow even as the economy stagnates. Unfunded federal mandates like those authorized by the Homeland Security Act and embodied in NIMS and CPG 101 (Version 2) set noble goals but do not provide the tools to achieve them. Congress has in the past appropriated monies for training and equipment for emergency responders and emergency management, but with the slavish attention to deficits that is the current rage in Washington, the future of such funding is uncertain at best. Washington's unfunded mandates to state and local governments will continue, including that for emergency management to obtain competent legal advice on an ongoing basis. This expensive personnel requirement has never been supported by federal dollars. One hopes that the reader understands by this point the importance of competent legal advice for emergency management.

As these fiscal demands have steadily grown, in what areas of homeland security (which includes emergency response and emergency management) has Washington appropriated our ever-scarcer national treasure? As mentioned elsewhere, law enforcement has been the first priority domestically beginning with the Bush incorporation of emergency management into DHS. The Bush administration linked its two wars – in Iraq and Afghanistan – directly to our domestic safety. The funds it expended for those conflicts – over a trillion dollars for Iraq alone – therefore appropriately may be added to monies spent for domestic homeland security law enforcement to gain a true perspective for comparison with emergency response and emergency

---

[3] See, e.g., www.time.com/time/magazine/article/0,9171,100312,00.html

[4] Iraq War Facts, Results & Statistics at October 28, 2011. found online at http://usliberals.about.com/od/homelandsecurit1/a/Iraq Numbers.htm

[5] Kim Murphy, "9/11 A Decade Later: Is Homeland Security Spending Paying Off?" *Los Angeles Times* August 28, 2011 Found online at: http://www.latimes.com/news/nationworld/nation/september11/la-na-911-homeland-money-20110828,0,4574475,full.story

management unfunded mandates under the G.W. Bush regime.[6]

Emergency management continues to face major, well-recognized hazards that are either unfunded or underfunded. One example is the widely discussed group of levees around Sacramento, California. Originally created for agricultural purposes, these water barriers are not designed to protect human habitation from flooding, yet they now encompass thousands of homes rather than farmers' fields. Estimates for repair of this probable disaster-in-the-making range from the Bay Delta Conservation Plan's (the official federal/state body) estimate of $15.7 billion to the California Environmental Water Caucus's $60 billion estimate. Greater funding for programs with demonstrated track records would be a wise use of DHS appropriations. Mitigation has proven to be a wise and cost-effective investment[7] – property acquisition and relocation, raising homes on stilts, assisting in retrofitting structures in earthquake zones and similar steps significantly increase safety. These are not pop-bang, gee-whiz, photo-op kinds of expenditures. They do their job very well, but do not give politicians nearly as much media exposure as putting their thumbs in the dike when their lack of proper planning results in a disaster. Infrastructure restoration would be a worthwhile outlay with the bonus of increased domestic employment. In 2008, CNN estimated that fixing all our bridges would cost $140 billion.[8] Comparing the costs of these and similar palpable improvements to public safety with the costs of the Bush administration's war in Iraq reveals that, had we kept even a fourth of these funds at home (a quarter of a trillion dollars – $250,000,000,000), there could be actual vital

physical improvements to show for our investment as well as many thousands fewer dead.

Other important factors are the additional pressures from Washington lobbyists as well as businesses and workers in home Districts benefitting from production and distribution of whatever the expensive anti-terrorism gadget of the day may be. Examples of boondoggles in the name of homeland security abound: the unsuccessful and cancelled SBInet program to build a "virtual fence" of sensors, cameras and radar along the nation's border after paying more than $1.1 billion to prime contractor Boeing (the Government Accountability Office concluded that poor management and an overreliance on Boeing, had caused staggering delays and cost overruns while producing inadequate results), the failed and cancelled Advanced Spectroscopic Portal (ASP) – a Raytheon contract that wasted $230 million but never worked, the Risk Assessment Management Program – a computer application intended to help officials figure out which buildings are most vulnerable to attack – that contractor Booz Allen promised in one year for $21 million that was junked after three years and $35 million,[9] $42,000 for full dive gear and a Zodiac with side-scan sonar in a Nebraska lake in a county with a population of less than 9,000, over 300 Bearcat 9-ton armored vehicles at $205,000 apiece for police departments across the nation – many bought with DHS funds, $3,000 worth of lapel pins in West Virginia – range from a pinprick in the national budget to hugely expensive, but few of them appear to be the result of rational prioritization.[10]

Like the military-industrial complex that became a permanent and powerful part of the American landscape during the Cold War, the vast network

---

[6] Although there is a law that discourages unfunded mandates, the Unfunded Mandate Reform Act of 1995 (P.L. 104-4) does not actually ban them.

[7] Multihazard Mitigation Council, *Mitigation Saves: An Independent Study to Determine the Future Savings From Mitigation Activities* (2005). Found online at: http://www.floods.org/PDF/MMC_Volume1_FindingsConclusionsRecommendations.pdf

[8] "Report: Repairing U.S. bridges would cost $140 billion." *CNN*, July 28, 2008. Found online at: http://www.cnn.com/2008/US/07/28/bridge.report/index.html

[9] Dan Froomkin, "9/11 Attacks Led To Half-Trillion-Dollar Homeland Security Spending Binge." *Huffington Post*, September 9, 1911. Found online at: http://www.huffingtonpost.com/2011/09/09/september-11-homeland-security-spending_n_953288.html

[10] Kim Murphy, "9/11 A Decade Later: Is Homeland Security Spending Paying Off?" *Los Angeles Times*, August 28, 2011 Found online at: http://www.latimes.com/news/nationworld/nation/september11/la-na-911-homeland-money-20110828,0,4574475,full.story

of Homeland Security spyware, concrete barricades and high-tech identity screening is here to stay. The Department of Homeland Security, a collection of agencies ranging from border control to airport security sewn quickly together after Sept. 11, is the third-largest Cabinet department and – with almost no lawmaker willing to render the U.S. less prepared for a terrorist attack – one of those least to fall victim to budget cuts.[11]

So it appears that we will have DHS and all of the ways in which it has come to resemble the Department of Defense with us for the foreseeable future. A homeland security-industrial complex, if you will – a massive, difficult to control bureaucracy where huge sums are spent to please home constituencies rather than being allocated as the result of a rational risk analysis. The overawing shadow of terrorism – that 1 in 3.5 million chance of death that properly conducted risk analysis would consign to the "[vanishingly] low probability, high impact" category – dominates allocation of DHS funds to such an extent that the nation is, in the author's opinion, needlessly exposed to much more likely risks from natural hazards.

C. Policy and Legal Changes

What are other effects of saying that "everything has changed?" I submit that another potential benefit seen by Vice President Cheney was that it opened up the entire range of policy matters to be looked at under the "changed light" of post-9/11 America. This sea change was accomplished through legal enactments, including emergency response and emergency management policy. In plain English, the national aftershock was somewhat similar to a poker player shouting "new deck" in the hope that a re-ordering of the landscape will change his luck. In fact, however, this simile is insufficient, for what happened to the national agenda after 9/11 was not the randomness of a changed deck of cards, but rather a brilliantly conceived and coldly

executed hijacking of domestic and international policy at all levels.

The immediate political effect of 9/11 was to open a window for extremely conservative legal enactments (which had long been desired by the right, and which the left and center had successfully resisted) to be imposed on the people of the United States over the largely supine bodies of those who historically had been watchdogs for individual rights. The conservatives proceeded to drive a tank through the window of opportunity opened by the 9/11 attacks. The USA PATRIOT Act, which limits individual freedoms in significant ways, is an illustration of such tactics. It was introduced by the Bush administration and marked up in a traditional, collegial bi-partisan way. Then at the last minute, literally in the middle of the night, the Bush administration pulled the agreed-upon bill, substituting one of its own, and insisted upon an immediate vote the next day. This occurred before any member of Congress could read the new bill, which was never engrossed, meaning that no member of Congress ever had any idea what it meant in context. The right's victory with the USA PATRIOT Act may be viewed as an early example of strong arm tactics that they continue to employ with such great success in their efforts to reconfigure the political landscape.[12]

One must understand that, from the Washington perspective, the ultimate unthinkable cost is not mass casualties but rather the loss of political power. Once anti-terrorism structures are in place, therefore, it becomes impossible to lessen or remove them for fear that bad timing will result in a successful attack on one's watch with resulting political disaster for the party in power. So, no matter how expensive or illogical, once any step is taken that is touted as preparedness against terrorists, it is immobile unless replaced – such as, for example, the much-lampooned color code Homeland Security Advisory System which was never safer than yellow, which has been replaced by a system which is not too different.[13]

---

[11] Id.
[12] See generally Susan N. Herman, *The USA PATRIOT Act and the Submajoritarian Fourth Amendment*, Harvard Civil Rights-Civil Liberties Law Review Vol. 41, 67-132(2006).
[13] http://www.dhs.gov/files/programs/ntas.shtm

Looking into weapons of mass destruction's dead, red apocalyptic eyes reveals sights far more horrid than even the aftermath of an atomic explosion. Chemical weapons can kill and create lifelong debilitating injuries that make quick death seem like a mercy. Bioweapons may target people directly or destroy our foodstuffs and leave us to mass starvation and a breakdown of social structure. Small wonder that many portions of the USA PATRIOT Act regarded as most intrusive by civil libertarians have been renewed through their sunset dates by bi-partisan majorities of both houses of Congress and signed by Presidents Bush[14] and Obama.[15]

Emergency management is a vital function of government, and it becomes more so with each passing day. Consider the following: world population recently passed seven billion – when some estimate that 1.5 billion is the total sustainable population of the planet, resulting in soaring commodity prices;[16] the built environment encroaches on the natural environment in many extremely vulnerable locations – often occupied by the poorest populations; the oceans are rising far faster than in the recent past as the ice caps melt.[17] The number of Presidential Disaster Declarations for natural disasters rises steadily over time. Further, both the national and world economies are sickly, with high unemployment and widespread disillusionment, potential fertile ground for growth of potential terrorists. In such circumstances, the need for well-funded emergency management with clear legal authority could not be more evident.

Unfortunately, the discussion of the history and future of FEMA in Chapter 12 illustrate that, on the federal level post-9/11, emergency management is structurally a small player in a law enforcement-oriented world. FEMA's poor performance after Hurricane Katrina is largely the consequence of this structure. Although Congress enacted the Post-Katrina Emergency Management Reform Act, that law did not restore true independence to FEMA, leaving it buried in the bureaucracy of the Department of Homeland Security. It is a sad day indeed for federal emergency management that its future adequacy must depend on the personal relationship between the President and the FEMA Director, rather than its being a legally defined cabinet post.

D. Wars in Iraq and Afghanistan

Recalling the history of the two wars, we first went into Afghanistan, where the radical Islamist Taliban government had provided a base and refuge for al Qaeda and its leader, Osama bin Laden. This made sense – al Qaeda had attacked us on 9/11. However, just as we had him surrounded in the mountains of Tora Bora, Defense Secretary Rumsfield allowed bin Laden to slip away from our grasp.[18] In the wake of this event, President Bush stated in 2002 that he was "not concerned" about Osama Bin Laden, the head of al Qaeda.[19] By 2006, the CIA closed its unit focused on catching bin Laden.[20] It would take another nine years after Bush's statement of indifference to bin Laden, and election of a

---

[14] "Bush signs renewal of Patriot Act into law" *USA Today*, March 8, 2006. Found online at http://www.usatoday.com/news/washington/2006-03-09-bush-patriot-act_x.htm

[15] "Patriot Act: three controversial provisions that Congress voted to keep" *Christian Science Monitor*, May 27, 2011. Found online at: http://www.csmonitor.com/USA/Politics/2011/0527/Patriot-Act-three-controversial-provisions-that-Congress-voted-to-keep

[16] http://www.businessinsider.com/commodities-boom-2011-11?op=1

[17] http://news.discovery.com/earth/are-the-oceans-rising.html

[18] "Rumsfeld let Bin Laden escape in 2001, says Senate report" *The Guardian*, November 29, 2009. Found online at http://www.guardian.co.uk/world/2009/nov/29/osama-bin-laden-senate-report. It has been reported that the Pakistani ISI intelligence service may have actually been the group that spirited bin Laden away. 'ISI helped bin Laden escape from Tora Bora.' *Hindustani Times*, September 13, 2011. Found online at http://www.hindustantimes.com/ISI-helped-bin-Laden-escape-from-Tora-Bora/Article1-745215.aspx

[19] "Bush 'Not Concerned' About Bin Laden in '02." *Los Angeles Times*, October 14, 2004. found online at http://articles.latimes.com/2004/oct/14/nation/na-osama14

[20] "C.I.A. Closes Unit Focused on Capture of bin Laden." *New York Times*, July 4, 2006. Found online at: http://www.nytimes.com/2006/07/04/washington/04intel.html

different President (from a different party, with a different belief of who the "enemy" is), before the terrorist mastermind met his fate at the hands of U.S. Navy SEALS. They found bin Laden in the compound he built in Pakistan shortly after President Bush called off American pursuit of the al Qaeda leader.

The complexities of our relationship with "ally" Pakistan are too convoluted to detail here. They revolve around the variety of groups in that Islamic nation – some of which are radical. The Pakistanis possess nuclear weapons. The nation is a center for nuclear proliferation – the demonstrated past willingness of at least one Pakistani scientist (who was never punished) to share the technology behind the weapons with other nations (including allegedly North Korea and Iran). We are very concerned that individual atomic weapons might get into the hands of terrorist groups such as al Qaeda. In the immediate aftermath of 9/11, the United States spent a great deal of energy to ensure that Pakistan would be a strong ally. The United States has sent Pakistan a great deal of foreign aid over the years. The fact that bin Laden ended up safe in Pakistan illustrates how ambivalent our relationship remains. The mission to take out bin Laden on May 2, 2011was undertaken without notice to the Pakistanis.

John O. Brennan, President Obama's chief counterterrorism adviser, said it was "inconceivable that bin Laden did not have a support system" in Pakistan that allowed him to live comfortably with his family in a town north of the capital. He said U.S. officials are pursuing this with the Pakistanis, who were pointedly not informed about the raid before it took place.[21]

Tensions between the United States and Pakistan resulted, and have grown following an apparent friendly fire NATO strike on a Pakistani border post in early December, 2011.[22] China is using the opportunity to move closer to Pakistan, a worrisome development for both the United States and our long-time ally, India. In fact, some have speculated that the most important reason for our involvement in Afghanistan is to keep Islamic militants as far as possible from Pakistan's atomic weapons.

What do we have to show for the Authorization for Use of Force Against Iraq[23] passed in October, 2002 that resulted in "Operation Iraqi Freedom"? (This law incidentally was what Vice President Cheney was advocating in favor of in the quotation that begins this chapter.) The Iraq war statistics are from About.com,[24] and they paint a grim picture of the effects of our intervention in that nation.

U.S. SPENDING IN IRAQ[25]

Spent & Approved War-Spending – About $1 trillion of U.S. taxpayers' funds spent or approved for spending through 2011.

Lost & Unaccounted for in Iraq – $9 billion of U.S. taxpayers' money and $549.7 million in spare parts shipped in 2004 to U.S. contractors. Also, per ABC News, 190,000 guns, including 110,000 AK-47 rifles.

Lost and Reported Stolen – $6.6 billion of U.S. taxpayers' money earmarked for Iraq reconstruction, reported on June 14, 2011 by Special inspector general for Iraq reconstruction Stuart Bowen who called it

---

[21] "Osama bin Laden killed in U.S. raid, buried at sea." *Washington Post*, May 2, 2011. Found online at: http://www.washingtonpost.com/national/osama-bin-laden-killed-in-us-raid-buried-at-sea/2011/05/02/AFx0yAZF_story.html

[22] "Pakistan Says U.S. Gave Wrong Information on Hit." *Wall Street Journal*, December 3, 2011. Found online at: http://online.wsj.com/article/SB10001424052970204012004577074042444708750.html

[23] Public Law 107-243, 116 Stat. 1498 (2002).

[24] Iraq War Facts, Results & Statistics at October 28, 2011. Found online at http://usliberals.about.com/od/homelandsecurit1/a/IraqNumbers.htm

[25] Iraq War Facts, Results & Statistics at October 28, 2011. Found online at http://usliberals.about.com/od/homelandsecurit1/a/IraqNumbers.htm

"the largest theft of funds in national history." (Source – CBS News) Last known holder of the $6.6 billion lost: the U.S. government.

Missing – $1 billion in tractor trailers, tank recovery vehicles, machine guns, rocket-propelled grenades and other equipment and services provided to the Iraqi security forces. (Per CBS News on December 6, 2007.)

Mismanaged & Wasted in Iraq – $10 billion, per February 2007 Congressional hearings

Halliburton Overcharges Classified by the Pentagon as Unreasonable and Unsupported – $1.4 billion

Amount paid to KBR, a former Halliburton division, to supply U.S. military in Iraq with food, fuel, housing and other items – $20 billion

Portion of the $20 billion paid to KBR that Pentagon auditors deem "questionable or supportable" – $3.2 billion

U.S. Annual Air-Conditioning Cost in Iraq and Afghanistan – $20.2 billion (Source – NPR, June 25, 2011)

U.S. 2009 Monthly Spending in Iraq – $7.3 billion as of October, 2009

U.S. 2008 Monthly Spending in Iraq – $12 billion

U.S. Spending per Second – $5,000 in 2008 (per Senate Majority Leader Harry Reid on May 5, 2008)

Cost of deploying one U.S. soldier for one year in Iraq – $390,000

TROOPS IN IRAQ

Troops in Iraq – Total 39,000 U.S. troops. All other nations have withdrawn their troops.

U.S. Troop Casualties – 4,485 U.S. troops; 98% male. 91% non-officers; 82% active duty, 11% National Guard; 74% Caucasian, 9% African-American, 11% Latino. 19% killed by non-hostile causes. 54% of U.S. casualties were under 25 years old. 72% were from the U.S. Army

Non-U.S. Troop Casualties – Total 316, with 179 from the UK

U.S. Troops Wounded – 32,219, 20% of which are serious brain or spinal injuries. (Total excludes psychological injuries.)

U.S. Troops with Serious Mental Health Problems – 30% of U.S. troops develop serious mental health problems within 3 to 4 months of returning home

U.S. Military Helicopters Downed in Iraq – 75 total, at least 36 by enemy fire

IRAQI TROOPS, CIVILIANS AND OTHERS IN IRAQ

Private Contractors in Iraq, Working in Support of U.S. Army Troops – More than 180,000 in August 2007, per TheNation.com.

Journalists killed – 15,098 by murder and 52 by acts of war

Journalists killed by U.S. Forces – 14

Iraqi Police and Soldiers Killed – 10,125, as of July 31, 2011

Iraqi Civilians Killed, Estimated – On October 22, 2010, ABC News reported "a secret U.S. government tally that puts the Iraqi (civilian) death toll over 100,000," information that was included in more than 400,000 military documents released by Wikileaks.com.

A UN issued report dated September 20, 2006 stating that Iraqi civilian casualties have been significantly underreported. Casualties are reported at 50,000 to over 100,000, but may be much higher. Some informed estimates place Iraqi civilian casualties at over 600,000.

Iraqi Insurgents Killed, Roughly Estimated – 55,000

Non-Iraqi Contractors and Civilian Workers Killed – 572

Non-Iraqi Kidnapped – 306, including 57 killed, 147 released, 4 escaped, 6 rescued and 89 status unknown.

Daily Insurgent Attacks, February, 2004 – 14

Daily Insurgent Attacks, July, 2005 – 70

Daily Insurgent Attacks, May, 2007 – 163

Estimated Insurgency Strength, November, 2003 – 15,000

Estimated Insurgency Strength, October, 2006 – 20,000–30,000
Estimated Insurgency Strength, June, 2007 – 70,000

QUALITY OF LIFE INDICATORS

Iraqis Displaced Inside Iraq, by Iraq War, as of May, 2007 – 2,255,000
Iraqi Refugees in Syria & Jordan – 2.1 million to 2.25 million
Iraqi Unemployment Rate – 27 to 60%, where curfew not in effect
Consumer Price Inflation in 2006 – 50%
Iraqi Children Suffering from Chronic Malnutrition – 28% in June, 2007 (Per CNN.com, July 30, 2007)
Percent of professionals who have left Iraq since 2003 – 40%
Iraqi Physicians Before 2003 Invasion – 34,000
Iraqi Physicians Who Have Left Iraq Since 2005 Invasion – 12,000
Iraqi Physicians Murdered Since 2003 Invasion – 2,000
Average Daily Hours Iraqi Homes Have Electricity – 1 to 2 hours, per Ryan Crocker, U.S. Ambassador to Iraq (Per *Los Angeles Times*, July 27, 2007)
Average Daily Hours Iraqi Homes Have Electricity – 10.9 in May, 2007
Average Daily Hours Baghdad Homes Have Electricity – 5.6 in May, 2007
Pre-War Daily Hours Baghdad Homes Have Electricity – 16 to 24
Number of Iraqi Homes Connected to Sewer Systems – 37%
Iraqis without access to adequate water supplies – 70% (Per CNN.com, July 30, 2007)
Water Treatment Plants Rehabilitated – 22%

RESULTS OF POLL

Taken in Iraq in August, 2005 by the British Ministry of Defense (Source: Brookings Institute)
Iraqis "strongly opposed to presence of coalition troops – 82%
Iraqis who believe Coalition forces are responsible for any improvement in security – less than 1%
Iraqis who feel less secure because of the occupation – 67%
Iraqis who do not have confidence in multi-national forces – 72%

Apparently, we have neither won the hearts and minds of the Iraqi people nor made them safe from insurgency. Indeed, it could be argued that by removing Saddam Hussein, we have significantly destabilized the Middle East. After all, he was a secular ruler, not a radical Islamist. Indeed, he saw groups like al Qaeda as threats to his cult of personality and took strong measures against religious leaders he viewed as less than completely supportive. Keeping Islamic radicalism bottled up was the major goal of his long war against the Mullahs in Iran.

The war in Afghanistan had a more rational basis than did the war in Iraq. After all, leading up to the 9/11 attacks, al Qaeda was based in Afghanistan and found refuge there, thanks to the Taliban, which governed that backward state. After the 9/11 attacks, Congress passed the Authorization for the Use of Military Force[26] on September 14, 2001 that sanctioned the response against those who had attacked the nation. The United States put together "Operation Enduring Freedom" (OEF) to attack Afghanistan and related targets in response to this authority. American OEF casualties included 1,823 deaths and 14,793 wounded in action by November 12, 2011.[27] Both military and civilian casualties have grown as time has gone by.

---

[26] Public Law 107-40, 115 Stat. 224 (2001).
[27] http://www.defense.gov/news/casualty.pdf

Year	Anti-Govt	Pro-Govt	Other	Total
2006	699	230		929
2007	700	629	194	1523
2008	1160	828	130	2118
2009	1630	596	186	2412
2010	2080	440	257	2777
Total	6269	2723	767	9759

Afghan Civilian Deaths[28]

Difficulties in Afghanistan are nothing new for invaders. Alexander the Great, the British Empire, and the Evil Empire (aka the Soviet Union) learned to their dismay that great conventional forces can be broken on the high, dry mountains by hard men with primitive weapons. Americans have found the same to be true.[29] Afghans also have a tendency to unite eventually against outsiders, despite their internal differences.

E. Iraq and Afghan War Veterans: Expertise and Challenges

Historically, both emergency response and emergency management have benefitted greatly from the expertise that military veterans have brought to their ranks. Their excellent training in a variety of disciplines with direct application to emergency services recommends them highly as employees. Experience in leadership positions in times of great stress can make military veterans valuable when disaster strikes. Although their background is helpful, it does not translate completely. Most useful perhaps is the "never say die" attitude that the veteran brings to the job.

How have these brave Americans, who still have so much to offer to emergency response and emergency management, fared after their service? Veterans of the post-9/11 Iraq and Afghanistan wars have been terribly scarred by their experiences. Suicide rates are very high – 38 per 100,000 for those using Veterans Administration

health care (statistics for those not using those services are not available) compared to 11.5 per 100,000 deaths for the general population. Nearly half of those who have been deployed have done so more than once. There are 1,286 service members that re amputees from those wars. Over 58% of those currently serving have been deployed at least once.

Once these folks have made their sacrifices (put their lives on the line, often grievously wounded in body and mind, and then returned to a grateful nation), what is their reward? Virtually 12.1% of Iraq and Afghanistan veterans are unemployed, compared with 9.1% nationwide. The situation is worst for young veterans – the unemployment rate in 2010 for vets between the ages of 18–24 was 20.9 percent, up from a pre-recession rate of 11.7 percent in 2007.[30]

F. Legal "Mission Accomplished?"

During the ten years since we were attacked by al Qaeda on 9/11, we have been at war in Iraq and Afghanistan, expending vast amounts with little to show for our efforts. President Bush flew to the USS Abraham Lincoln and declared "Mission accomplished" in 2003. Bush stated in 2002 that he was not worried about Osama Bin Laden, the head of al Qaeda. From 2003 through 2011, Bin Laden continued to be the idea man behind al Qaeda many international acts of terror, according to documents found when he was finally terminated. In May 2011, U.S. Navy SEALS killed Osama Bin Laden, and our national payback for the 9/11 attacks could truly be said to be accomplished. President Obama, on whose watch Bin Laden met his fate, declared that American troops would leave Iraq by the end of 2011.[31] As of the close of December, 2011, the United States had departed Iraq. Unfortunately, internal violence there has escalated following

---

[28] Afghanistan civilian casualties: year by year, month by month. Found at http://www.guardian.co.uk/news/datablog/2010/aug/10/afghanistan-civilian-casualties-statistics

[29] David Loyn, *In Afghanistan: Two Hundred Years of British, Russian and American Occupation.* Palgrave Macmillan (2009).

[30] Statistics in the previous two paragraphs come from: U.S. Veterans: By the Numbers. Found online at http://abcnews.go.com/Politics/us-veterans-numbers/story?id=14928136#4

[31] "U.S. Troops to Leave Iraq by Year's End, Obama Says." *New York Times*, October 11, 2011. Found online at http://www.nytimes.com/2011/10/22/world/middleeast/president-obama-announces-end-of-war-in-iraq.html?pagewanted=all

our leaving, and that nation's future is murky at best.[32] The author believes the most likely future for Iraq is ongoing factional strife with ever-increasing violence. Many internal groups will be tempted by overtures of support – arms, training and money – from interested external forces. These would include Iran, al Qaeda, the Saudis, the Russians, unacknowledged western players (probably including the United States and Great Britain) and other players to be named at a future date. The Iranians continue to see Iraq as the road to Saudi Arabia's oil, control of the Persian Gulf, and leadership of all Muslims. Iran has long coveted Saudi oil assets, and with recent push back from the House of Saud on the Iranian nuclear program,[33] pushing through a weakened Iraq to the Saudi oil may be a more attractive goal than ever. Emergency managers should perhaps plan for increased likelihood of oil shortages and nuclear exchanges thanks to the Iraq war.

When making the case to attack Iraq in his 2003 State of the Union Address, President Bush cited evidence that Saddam Hussein was well on the way to creating atomic weapons. Bush stated that Saddam, through African sources, sought uranium and that he had obtained specialized aluminum tubes used in refining fissionable materials. As well, he cited Saddam's past capacity and willingness to use chemical and biological WMD to justify use of force.[34] Congress, and the nation as a whole, initially agreed with these arguments. In fact, Congress had already autho-

rized use of force against Iraq in 2002.[35] Later, President Bush stated that his goal for Iraq was the same sort of nation building for which he derided his predecessors. He told the nation that America's task in Iraq "is not only to defeat an enemy, it is to give strength to a friend – a free, representative government that serves its people and fights on their behalf."[36] The means to these goals are legal. To accomplish those goals would require establishing western ideas of representative democracy and respect for human rights in a country where those ideas had no tradition and no roots. Getting "buy in" from those in power or seeking power was not difficult – a few cynical years of sham democracy in a place where few understand or desire it and then a return to the same old ways might be expected. Corruption is a major problem, and in 2007 it led to the resignation of the chief investigator of the issue, who alleged that American's chosen Prime Minister Maliki protects corrupt ministers.[37] The United States has long backed a unity government in Iraq, but the reality is that sometimes we have not been wise in our selection of allies in that nation. The Iraqi tradition of sectarian loyalty – and violence to uphold that loyalty – is strong. Saddam's coterie from Tikrit has been replaced, and the in-fighting between Sunnis, Shi'ites, and Kurds goes on with a Constitutional crisis brewing in the background.[38] The goals ultimately sought by President Bush in Iraq were legal goals – a Constitutional democracy, the rule of law, and what we in the west view as the

---

[32] Lolita C. Baldor, "Iraq's future uncertain as U.S. closes door on 9-year war." AP, December 15, 2011. Found online at: http://www.katu.com/news/politics/national/Iraqs-future-uncertain-as-US-closes-door-on-9-year-war-135655848.html

[33] "Saudi Suggests 'Squeezing' Iran Over Nuclear Ambitions." *Wall Street Journal*, June 22, 2011. Found online at: http://online.wsj.com/article/SB10001424052702304887904576400083811644642.html

[34] "Bush's State of the Union Speech" January 28, 2003. Found online at: http://articles.cnn.com/2003-01-28/politics/sotu.transcript_1_tax-relief-corporate-scandals-and-stock-union-speech?_s=PM:ALLPOLITICS

[35] "Public Law 107 - 243 – Authorization for Use of Military Force Against Iraq Resolution of 2002." Found online at: http://www.gpo.gov/fdsys/pkg/PLAW-107publ243/content-detail.html

[36] "The Reach of War: The President; Bush Lays Out Goals for Iraq: Self-Rule and Stability." Elisabeth Bumiller, *New York Times*, May 25, 2004. Found online at: http://www.nytimes.com/2004/05/25/world/reach-war-president-bush-lays-goals-for-iraq-self-rule-stability.html?pagewanted=all&src=pm

[37] "Iraqi Official 'Corruption Has Crippled Iraq'" MSNBC News, September 9, 2007. Found online at: http://www.msnbc.msn.com/id/20040662/ns/nbcnightlynews-nbc_news_investigates/t/iraqi-official-corruption-has-crippled-iraq/

[38] "Turkish PM lashes out at Iraqi counterpart, denies meddling" Reuters, April 21, 2012. Found online at: http://uk.reuters.com/article/2012/04/21/uk-turkey-iraq-idUKBRE83K0J420120421

See also "How to Save Iraq From Civil War" Ayad Allawi, Osama Al-Nujaifi and Rafe Al-Essawi, *New York Times*, December 27, 2011. Found online at: http://www.nytimes.com/2011/12/28/opinion/how-to-save-iraq-from-civil-war.html?_r=1&pagewanted=all

blessings that flow from a government and a nation that adheres to such a legal structure. While there is still a chance of success, overall at this point the legal mission at the heart of our Iraqi adventure is in continuous peril. The United States continues to support this fragile effort at democracy in a region that has not had a democratic tradition, but opponents are many and supporters are few and often not whole-hearted in their endorsements. As the following material indicates, some responsibility for that failure rests on the shoulders of the highest leaders in the United States.

Around the world, repressive regimes have responded to the post-9/11 desire for suppression of terrorism as an excuse to label any dissenters as terrorists, often brutally torturing and murdering them.[39] The Bush administration routinely engaged in "extraordinary rendition" of suspected terrorists for "enhanced interrogation" to countries that allowed torture during questioning.[40] Anti-American forces around the world argue that this practice demonstrated that for all of our protestations about human rights, the United States was nothing more than a nation of hypocrites.

> The Council of Europe's rights commissioner, Thomas Hammarberg, said the 10-year anniversary of the Sept. 11 attacks was an occasion to analyze whether the official responses have been proper and effective. "In attempting to combat crimes attributed to terrorists, countless further crimes have been committed in the course of the U.S.-led 'global war on terror,'" he said in a statement. "Many of those crimes have been carefully and deliberately covered up."[41]

The author believes that the Bush administration did grave harm to our national image as the "shining city on the hill" by permitting us to be publicly associated with a variety of actions previously associated with totalitarian regimes. Actions like waterboarding and extraordinary rendition are now part of the American story. Unpunished torture by Americans and end runs around American laws designed to prevent torture are black marks that will be used against our nation in propaganda for years to come. Every American who held up his or her head because our country just doesn't do that kind of thing is now in the unprecedented state of being embarrassed by something our country has done. This was not the transgression of a few soldiers pushed beyond their limits like Lt. Calley and his men in My Lai, Vietnam,[42] but calculated policy, set by the Bush White House. These policies' clear immorality and their high-level endorsement mean that they will not simply fade into history. The result may be more fertile ground for terrorist recruitment. Emergency managers should perhaps plan for increased terrorist deeds in the future thanks to these activities.

The reader might wonder why a book on emergency response and emergency management law peruses with such detail the results of foreign policy commitments. That reader has not been paying attention to a major theme of this book – the consequences of the intertwining of emergency response and emergency management with homeland security, and the blurring of domestic and international security in the aftermath of the terrorist attacks of September 11, 2001. The Bush administration created, and propelled through Congress a series of laws that mixed all of these matters – and the funding for them – together in such a way that to oppose any of them would leave a Representative or

---

[39] "Rightly or Wrongly, Thousands Convicted of Terrorism Post 9/11" September 24, 2011, MSNBC News. Found on line at: http://www.msnbc.msn.com/id/44389156/ns/us_news-9_11_ten_years_later/t/rightly-or-wrongly-thousands-convicted-terrorism-post-/

[40] Jane Mayer, "ANNALS OF JUSTICE: Outsourcing Torture: The Secret History of the America's Extraordinary Rendition Program." *The New Yorker*, February 14, 2005. Found online at: http://www.newyorker.com/archive/2005/02/14/050214fa_fact6

[41] "Official Blasts Antiterror Actions" *New York Times*. Found online at: http://www.nytimes.com/2011/09/02/world/europe/02briefs-Europe.html?ref=extraordinaryrendition

[42] Calley was convicted by Court Marshal of murder in the My Lai massacre, and later apologized – something Bush and Cheney will never do. Robert Mackey, "An Apology for My Lai, Four Decades Later." *New York Times*, August 24, 2009. Found online at: http://thelede.blogs.nytimes.com/2009/08/24/an-apology-for-my-lai-four-decades-later/

Senator open to charges of aiding and abetting terrorism. One major result of this heavy-handed political pressure is a nation whose preparedness for emergencies and disasters is significantly decreased due directly to the structural changes wrought by law. Although people of good will spoke up on both sides of the issue, history will show that subordinating FEMA to DHS and failing to restore its independent status after the Hurricane Katrina debacle was a grave mistake. Given the uncertainty of the future relationship between the President and the FEMA Director, emergency managers should prepare for delayed federal response to future events due to the ungainly structure of which FEMA is a part.

In DHS under the Bush regime, leadership became steadily dominated by law enforcement, a matter of some concern to those in the emergency management arena as discussed above. Senior management in DHS understandably drew from the ranks of constituent agencies, with many top slots going to senior Coast Guard personnel. Following Michael Brown's departure from New Orleans after Hurricane Katrina, Coast Guard Vice Admiral Thad Allen replaced him.[43] Even the liberal media loved the "can-do" attitude of the military men who contrast with FEMA leadership seen as incompetent – the *Washington Post* called LTG Russell L. Honore "The Catagory 5 General" when he led the "boots on the ground" effort to pick up the pieces in New Orleans after Katrina.[44]

As time went by following the 9/11 attacks, a more disturbing trend developed within the domestic security arena. DHS became steadily more subject to control by military personnel. Neither Congress nor the media nor the national think tanks engaged in thoughtful debate or discourse over the matter of whether domestic security should be dominated by military minds. It is the lack of thoughtful consideration of this development that makes it so troubling. In other

nations, military control of domestic security organs is often the first step to crushing dissent and imposing dictatorship. In the United States, we have probably the strongest tradition in the world of civilian control of the military, a tradition that is embraced by all involved. Still, to simply have it taken for granted that this will continue, rather than to appreciate that this is a new thing with potentially dangerous future effects is troubling in the extreme. Given the lack of deep thought on this issue, it may be a possibility that our national foundation – citizen election and control of a civilian leadership that governs our national destiny – will continue to erode.

G. Conclusion

The law – and more – has changed since 9/11. As Vice President Cheney accurately observed, the way in which our nation engaged in foreign conflicts transformed under his and President Bush's administration. We would begin the post-9/11 period as a deeply wronged nation with united international support, and then alienate the world through a combination of *hubris* and disregarding at a whim universally accepted international law with which we had already agreed. The United States would attack in Afghanistan to root out the Taliban leaders who gave aid and comfort to Osama Bin Laden and his al Qaeda organization who had attacked us, and then become distracted by an irrelevancy in Iraq. We began a major war in Iraq based on miniscule evidence. America would spew forth its national treasure to huge conglomerates (largely Halliburton, Vice President Cheney's former employer) on no-bid contracts and then fail to hold the companies accountable when they profit from fraud and waste.[45] Thousands of Americans died. Thousands – perhaps hundreds of thousands – of Iraqis died. More Americans continue

---

[43] "Coast Guard admiral assumes Katrina relief responsibilities." *USA Today*, September 9, 2005. Found online at: http://www.usatoday.com/news/nation/2005-09-09-allen_x.htm

[44] "The Category 5 General." *Washington Post*, September 12, 2005. Found online at: http://www.washingtonpost.com/wp-dyn/content/article/2005/09/11/AR2005091101484.html

[45] "Ethics at Halliburton & KBR" *Ethics in Business*. Found online at: http://www.ethicsinbusiness.net/case-studies/halliburton-kbr/

to die in Afghanistan, as do Afghans. From a risk management perspective, are we safer because of these wars?

Since the attacks of September 11, 2001, emergency response and emergency management law have evolved at a rate more rapid than ever before in their history. Whatever great ventures may be pursued by the world's leaders, emergency responders and emergency managers will be responsible for dealing with the consequences. They must do so within the meets and bounds determined by the law. The all-embracing nature of NIMS means that all responders can work more easily together no matter where they are from and no matter how far they must go to respond to a scene. Indeed, the ongoing efforts to unite national resources through NFPA 1600, NIMS, NRP, NRF, CPG 101 and other tools put the nation in a much better place than we were prior to 9/11, by allowing our resources to be utilized much more effectively. On 9/11, FEMA was an independent agency. Thanks to changes in law, FEMA ended as a very minor player in DHS, which is a huge criminal justice conglomerate focused on terrorism. DHS works so closely with our national defense structure that they are in many ways a united structure. This is a good thing when one considers the resources the Department of Defense (DoD) brings to counter the potential effects of major terrorist attacks and natural disasters (although few complaints were ever heard regarding DoD's performance under the old Federal Response Plan, where it was a well-oiled support agency). Close involvement with DoD is a mixed blessing from other perspectives — FEMA's small size becomes even more evident when measured against the mammoth DoD.

Moving from the large-scale emphasis of this chapter to the more local focus of the rest of the book, clearly many challenges persist. On the state and local levels, in particular, much remain to be done. Perhaps the most frustrating thing is that the knowledge of what to do for better mitigation and preparedness has long been present at these levels of government guided by FEMA's professional cadre — the problem has always been a lack of money. Unfortunately, following the 9/11 attacks, when funding finally became available to address long-standing problems, including the lack of legal advice, it was frittered away at the federal level, thanks to the homeland security/industrial complex's pernicious influence and the trillion dollar boondoggle of the Iraq war. One fears that this opportunity will never come again, and that the federal government has blown the opportunity to create a truly resilient structure of safety for the nation by wasting the funds that could have done so.

So once again, emergency management is left where it was before the 9/11 attacks, with unfunded mandates in many areas, with none perhaps more glaring than legal advice. The federal government creates complex rules and expects obedience to them, yet it fails to provide funding for independent legal counsel to read and interpret those rules. For example, CPG 101 requires legal advice in many areas of planning. The Guide also mandates a lawyer's input on a continuing basis, yet Washington provides no money for these major expenses. This is a situation in which our federal leaders are setting up state and local governments to violate the law. All levels of government must be funded for ongoing competent advice by properly trained legal counsel on emergency response and emergency management issues. Only when this takes place can it truly be said that our leaders have taken the steps necessary to properly manage the liability risks that face the populations whose protection is their greatest duty.

QUESTIONS

1. Read Vice President Cheney's Meet the Press interview using the web reference attached as a footnote. Assign class members to investigate each allegation made by Mr. Cheney and investigate the truth of his statements about Iraq and why we

should invade that country. Has the Iraq war proven to be a worthwhile event? Why or why not?

2. Perform research to examine how veterans have been treated by the United States and other nations following conflicts in the past. Discuss whether such treatment is correct, why it occurs, and explain how you would change it in the aftermath of future conflicts. Be specific regarding where funding would come from if you propose increased services.

3. Discuss how the rest of the world reacted to 9/11 immediately after the event and how its attitude toward the United States changed in the following years. What factors were behind the changes in attitude?

4. The author sees emergency management as linked with both domestic and international security. Do you agree or not? What are the effects of such a linkage?

5. Those who argue in favor of extraordinary rendition and waterboarding argue that they are effective ways of getting information quickly from hardened terrorists. Divide the class in half. Have each half perform research on either side to determine the effectiveness of extraordinary rendition and waterboarding and debate the matter.

6. The author believes that extraordinary rendition and waterboarding will do great long-term damage to the reputation of the United States. Do you agree or disagree, and why?

7. The author is concerned that civilian control over emergency management lessened after Hurricane Katrina. Do you agree? What structures exist to prevent or encourage such a development? What would you suggest to make such a development more or less likely?

8. Referring to the wars in Iraq and Afghanistan, the author asks "From a risk management perspective, are we safer because of these wars?" Divide the class and discuss this question.

9. A. What did the war in Iraq achieve for the United States in the short run? In the long run?

   B. What did the war in Iraq achieve for Iraq in the short run? In the long run?

10. In your opinion, what changed on 9/11 regarding emergency response and emergency management law? What are the things that have remained the same?

11. What are the things that we can rely upon to remain as the foundation for the future of emergency response and emergency management law?

12. How can state and local governments ensure that the federal government does not create unfunded mandates for them?

13. The author states (regarding upgrading mitigation and preparedness) that "One fears that this opportunity will never come again, and that the federal government has blown the opportunity to create a truly resilient structure of safety for the nation by wasting the funds that could have done so." Discuss.

Pleasant Grove, Alabama, May 21, 2011 – A disaster survivor writes his thoughts on the car from his home after a tornado destroyed the structure. As cleanup efforts move forward, state health officials said the greatest impact is emotional, especially for the youngest survivors; simply driving past the enormous destruction each day, and hearing stories of survival has taken a great toll on the children. Adam DuBrowa/FEMA.

Chapter 24

CONCLUSION

Emergency response and emergency management are complex topics. Common themes emerge, however, from examination of the law underlying these disciplines. All too often, preparedness for legal hazards is seen as a matter for concern after the emergency has occurred. Legal counsel may be called on only after an injured party files a lawsuit. Virtually every dispute covered by cases and other materials in this book could have been avoided by pro-active involvement of knowledgeable attorneys during all phases of operations.

Legal challenges overlie all areas of emergency response and emergency management. Unfortunately, this ever-present hazard is often invisible to the eye that is not trained to be alert to its existence. Dealing with legal issues requires more than mere awareness of them. Specialized knowledge is necessary, which means that an attorney must be a part of the team.

Additional pressures add to these difficulties. Preparedness costs money, a commodity always in short supply for local governments. This is particularly so in 2012, when calls for balanced budgets come at all levels of government even as tax receipts fall precipitously. Yet the old saw "pay me now or pay me later" is particularly pertinent when applied to emergency preparedness. There is perhaps no other domain where it rings more true that in the legal arena, where failure to take proper reasonable steps can result in multi-million dollar verdicts against units of government. The question of "all hazards" vs. "terrorism" funding will become ever more tense as pressures to cut all federal spending, balance budgets, and compromise vital programs in all arenas in order to raise the debt ceiling will put unprecedented pressure on every area of federal, state, and local effort. Funding will always be a huge and ongoing issue, as indicated in the discussion of the history of FEMA.

Further, some hazards may be even more pressing that anticipated. The March 2011 tsunami in Japan and the subsequent Fukushima nuclear disaster point up huge vulnerabilities in this country. These apply not only to a nuclear plant such as Diablo Canyon that is on an earthquake fault line and subject to a tsunami, but also to aging plants close to major population centers like New York's Indian Point. Fukushima's major radiation leaks have come from stored waste in cooling ponds, an issue the industry has not yet dealt with at any plant in the nation. Further, a plant need not be near the ocean to be threatened by rising waters. In June 2011, rising floodwaters in the Missouri River threatened two nuclear plants in Nebraska, about as far as one could get from the ocean. www.nytimes.com/ 2011/06/21/us/21flood.html

Some people who were either misinformed or misguidedly attempting to reassure the American public in the aftermath of the tsunami and Fukushima disasters stated that such events could never occur here because we are much better prepared that the Japanese.[1] In fact, the opposite

---

[1] It is true that our plants are newer designs and are presumably more resilient, but a similar disaster likely would have bad effects on a more modern design that was similarly situated. Also, the storage pools for spent fuel (where most of the radiation leakage reportedly originated) are essentially identical in the U.S. and Japan.

may be true. The Japanese are probably the best prepared nation of earth for a variety of disasters, because they have so many of them. Their nation is on the volcanic "ring of fire" that circles the Pacific rim, causing so much seismic activity. They have strong building and fire codes. Their populace engages in frequent emergency preparedness exercises. Yet, when a force 9 earthquake hit just off the coast and sent a tsunami ashore, they were overwhelmed. Honesty must force admission that any nation confronted with such an extreme event would be locally devastated with a veritable shower of consequential events. The United States is fortunate to be a much larger nation and thus more resilient. It has much greater resources and would be able to respond more rapidly given leadership and the will to do so, but the immediate effects of such an event would be similarly devastating to the area affected.

The threshold legal issue for emergency responders is when and how the duty to act arises. Not every government employee must respond to or even notify authorities of an emergency situation. Those whose duties include such response have higher obligations than are those whose duties do not. Clearly, a responder who has been dispatched to a scene has the duty to act in compliance with the dispatcher's instructions. Upon arrival, he or she must make an effort to ascertain whether the event to which the dispatcher has directed him or her actually exists. This effort must include not only inquiry of people in the surrounding area but also efforts to investigate the scene at which the event was reported. Only after such inquiry may the duty to respond be fairly said to be extinguished.

Members of law enforcement, the fire service, search and rescue, or emergency medical services must undergo extensive training prior to being allowed to respond to emergency situations in a professional capacity. Accidents that occur during that training expose sponsoring entities to potential liability. Workers compensation usually covers such liabilities. Failure to carefully legally craft the sponsoring entity, however, may result in uncertainty regarding who exactly is liable for training accidents. The attorney

crafting an agreement to create an interagency response force must be particularly precise in drafting the underlying documents creating the entity. Every state has different laws that must be carefully followed to assure predictable coverage for deaths or injuries incurred during training activity.

Vehicle issues frequently arise on the way to or from a response. For law enforcement, hot pursuit of fleeing criminals is an occupational hazard. That activity also poses significant legal dangers. The innocent third party injured as the result of hot pursuit, whether by the vehicle of the pursued person or by a police vehicle, may have a cause of action against law enforcement for damages. Failure to follow the law enforcement organization's policy on hot pursuits poses a significant risk of liability. It may be advisable to instruct pursuers to evaluate the nature of the danger posed by the person being pursued. In the event that a person is pursued for a minor traffic violation alone, continuing the pursuit, particularly into a crowded area, may result in liability. No action will lie against law enforcement, however, for violation of civil rights under § 1983 unless the behavior of law enforcement in the pursuit "shocks the conscience."

Other emergency response groups also run risks in the event of vehicle accidents. In general, emergency responders have the duty to drive with "all due regard" for other users of the highway. The exact parameters of "due regard" vary from state to state, but failure to exercise it will expose responders to liability. One may not be protected by immunities that otherwise might apply from the moment one leaves one's abode and begins to drive on the highway on the way to a scene or when one comes upon a scene as a volunteer who is not dispatched. Possible criminal action may be taken against an emergency responder whose vehicle accident is found by a court to be the result of criminal recklessness. Confinement in prison is one possible outcome of such a case. Emergency responders, their managers, and the attorneys who advise them must understand the particular state laws regarding vehicle use in order to craft and obey appropriate driving guidelines.

Dispatch is a key part of emergency response. Indeed, although the cases involved may have been decided on another basis, dispatch issues exist in many of the cases discussed in the emergency response portion of the book. All too often, dispatch personnel do not receive the training, the pay, or the respect due their vital work. The result can be liability for the dispatching entity. Striving to reassure sometimes-distraught callers, dispatch may promise more than emergency responders can deliver. Reliance on those promises may create a duty in the response organization, and a corresponding legal cause of action for the disappointed caller. Careful integration of a dispatch system with the responders the system serves is needed to avoid "holes" in service that may result in legal liability. The media portray dispatch as universally competent, well trained, and able to provide over-the-telephone medical instruction. The public expects dispatch to perform as well in the real world as they do on television. Failure to do so may expose dispatch and those dispatched to potential liability.

Emergency medical services (EMS) face a bevy of unique legal issues due to the nature of the service they provide. EMTs and medics in the field must make decisions every day that have significant legal ramifications. The best protection for EMS lies in a good understanding of the importance of complete documentation. The threshold matter of whether a patient consents to medical treatment often raises thorny issues. In an emergency, patient consent may be presumed if the patient is incapable of consenting or refusing treatment. Whether a patient is indeed capable of consent must first be ascertained, a process that involves evaluation of the patient's mental capabilities. Further, the patient's age must be considered. Minors generally can neither consent to nor refuse treatment. Persons, sometimes including EMS personnel, who provide assistance to the injured may be protected by "Good Samaritan" laws from negligence liability for their actions. While Good Samaritan Acts vary from state to state, they generally do not provide protection to one with a previously existing duty to the patient. Cases involving delayed response by EMS are an evolving trend. While the delays

may be a combined effect of dispatch problems and EMS timing issues, either entity may end up facing liability.

Complex emergency incidents demand well-trained and knowledgeable responders. No responder needs more knowledge and training than the commander of the response. The incident management system (IMS), sometimes known as the incident command system (ICS), has evolved into the preferred method for leadership during a response. A variety of laws mandate its use, and failure to properly use IMS may result in liability. More important, failure to properly use IMS will surely endanger responders' lives. Large incidents often require resources beyond those available to a single jurisdiction. Well-drafted mutual aid agreements (MAAs) are the vehicle by which emergency response organizations plan ahead for such situations. One characteristic of a good MAA is its insistence on use of IMS. Individual emergency response organizations utilize standard operating procedures (SOPs) to help plan for reliable reactions by individual responders in emergency situations. Working together, IMS, MAAs, and SOPs create a structure of safety. They reinforce one another, resulting in a system of checks and balances that assure responder safety. Keeping responders as safe as possible eliminates an important potential source of liability. The *Buttram* case is the intersection of multiple failures in all three areas that resulted in the deaths of two young firefighters. Their sacrifice will not be in vain if others can learn from the mistakes made by their superiors.

The National Incident Management System (NIMS) and the National Response Framework (NRF) have come into being in the aftermath of the September 11, 2001 attacks. In the wake of adoption of Homeland Security Presidential Directive 5 (HSPD) by President George W. Bush, IMS became the national standard for emergency response. IMS provides common approaches to common issues and increases safety while reducing cost. All emergency response organizations at all levels of government must comply with HSPD 5 standards or lose Federal grant monies. As incorporated through the NIMS and the NRF, these standards are indeed the law of the land. The NIMS Integration Center has created and disseminated detailed mandatory standards with the goal of making sure that all can work together when called to large interstate events.

Hazardous materials (HAZMAT) incidents require particular skills and training on the part of responders and those who supervise them. HAZMAT responses may be undertaken by the private sector as well as by governmentally sponsored emergency response organizations such as fire departments. Many HAZMAT response requirements are creatures of law. OSHA standards detail how HAZMAT response must be handled. Their violation may result in both fines and liability under tort law.

Volunteer resources enable emergency planners and responders to do more with fewer resources. Their contributions are widely recognized as extremely valuable. It is this perceived value that has led legislatures to enact various laws immunizing volunteers from some or all liability for their actions. The most prominent of these acts is the federal Volunteer Protection Act of 1997 (VPA). Like many such acts, the VPA offers limited immunity only. Indeed, some might argue that the immunities provided by the VPA are illusory, at best. Some states extend immunities more broadly in their statutes, and their courts may interpret those enactments more broadly yet. The result is a system of immunities that varies widely. The attorney advising a volunteer agency or emergency response group that utilizes the services of volunteers must closely consider local law.

Public policy considerations limit the remedies available to responders injured or killed in the line of duty. The rescue doctrine and exceptions to the "fireman's rule" have developed through case law or by act of state legislatures. These developments open narrow avenues for redress of injuries to responders. Post Traumatic Stress Disorder (PTSD) is a widely recognized challenge faced by emergency responders that is treated differently depending in which state one may be located.

The aftermath of the World Trade Center attacks left many who worked the Pile with severe physical ailments. Important legal and safety lessons for emergency responders and the attorneys who advise them arise from the resultant case law. Management mistakes in the response and clean up after the attacks directly resulted in significantly more debilitating injuries and early deaths among emergency responders than would have occurred had management properly performed its duties. The underlying decisions were made or not made, not in the heat of response to an unprecedented event, but as matters of standard operating procedures. The result was much finger pointing that reflects well on none of the agencies involved. Comparison of the *Buttram* case and the World Trade Center litigation reveals all too many similarities.

Emergency management is an all-hazards planning and coordinating discipline that allows leaders to mitigate potential losses in emergency events. During the response, emergency management helps bring together the resources needed by incident command to stabilize an emergency scene. After the immediate response, emergency management allows recovery to a state at least as safe as that prior to the event.

The emergency powers of governors vary considerably from state to state. A legally "strong" governor possesses the ability to control the response to events to a much greater degree than a legally "weaker" governor. The latter must operate much more by consensus, obtaining legislative support for ongoing emergency measures. While strength has its advantages in the short run, consensus may be a better way to achieve long-term success.

State and local governments must cooperate in order to create viable emergency management structures. Every state has laws that require it to create a state-wide plan. Local units of government are responsible for plans that apply to their own jurisdictions. Both must realize that assets and vulnerabilities are in a state of constant flux. Both must recognize that planning is an organic process, that a "current" plan must also be a living, evolving plan. Their plans must mesh together to form a strong structure. Their plans must incorporate many elements from the emergency response portion of this text, including use of volunteers, the incident management system, mutual aid agreements, and good dispatch systems. Those with responsibilities under the plan must train to the plan. The plan must, by law and by prudence, be exercised, and then revised to reflect lessons learned during the exercise. By law, local units of government must allow local emergency management and the Local Emergency Planning Committee to have sway over their respective spheres. Local government must also ensure that their plans work together to assure protection from all hazards.

On the federal level, emergency management is a continually evolving governmental responsibility. From its roots in civil defense, federal emergency management has grown to embrace natural hazards as well as more recent man-made concerns, such as terrorism. The Federal Emergency Management Agency (FEMA) coordinates the response of the federal government to the broad panoply of potential disasters. While FEMA has significant responsibilities for assuring that federal resources are put to work in the wake of disasters, its major role is perhaps as a grantor organization. FEMA issues, administers, and manages grants through states as grantees to persons and/or local units of government within the states. Such grants must be carefully controlled and monitored to assure that they are properly utilized. A large body of literature and great agency expertise is available from FEMA to assist in this task. FEMA ensures that grants are lawfully utilized, through enforcement if necessary. The Pre-Disaster Mitigation Program is an example of the latest trend in FEMA grants. Some funds are available under that Program for mitigation of natural hazards, but since the September 11 attacks, the trend has been toward anti-terrorism grants.

Federal emergency management spent much of the decade from 2001–2011 in a state of flux. FEMA was incorporated entirely into the cabinet-level Department of Homeland Security. Along with other emergency preparedness assets, the goal has been for FEMA to provide assurance that all hazards emergency management would be under a single federal roof. Unfortunately, this experiment in conglomeration

proved to be less than an unqualified success. In 2005, Hurricane Katrina tested the new structure, which Congress found to have contributed to problems in the New Orleans response. As a result, Congress enacted the Post-Katrina Reform Act of 2006, which addressed many issues, but kept FEMA within DHS. Submerged as it is within DHS, FEMA's future effectiveness will depend on the personal relationships that the agency's head has with senior members of the executive branch of government.

Ideally, the attorney should be an important partner for emergency management. Great liability exposure may result if this does not occur. Unfortunately however, many obstacles, both structural and societal, combine to block this partnership in far too many cases. All too often the two do not see themselves as team members, and funding does not exist for them to proactively work together. Both will benefit greatly if the attorney takes care to learn about emergency response and emergency management law prior to encountering an emergency event. Indeed, one might argue that a failure to do so comes perilously close to malpractice. The old argument that "it won't happen here" has lost any credence it might once have had in the wake of the wide variety of disasters that have befallen the nation. One is hard pressed to think of any location that has not suffered at least one of the following: flood, mass casualty event, tornado, earthquake, wildfire, hurricane, hazardous materials release, terrorism attack, drought, civil unrest, or oil spill. Additionally, during the decade after the terrorist attacks of September 11, 2001, the number and severity of disasters has not decreased. The government attorney in particular who does not understand at least the basics of this complicated area of law is himself or herself a hazard requiring mitigation.

The number of disasters has grown steadily over the last hundred years as human population has increased, with the poor expanding into ever more vulnerable regions – which are also the least expensive areas in which to live. Given this reality, the number of natural disasters will continue to grow. Further, high value growth thrives in locations where land and water intersect such as beaches and other coastal regions – places where flooding and erosion are probable. Despite efforts to reduce human development in the most disaster-prone regions through such tools as flood plain mapping, owner buyouts, zoning changes and the like, the force of population growth is simply stronger, particularly in the less developed regions of the globe. So, for the future, we can anticipate more disasters where they can least be afforded and continuing pressure to permit development and re-building in the United States in places where we know that there will be repetitive losses, such as the Gulf of Mexico coastal region. The attorney working for mitigation of litigation may face opposition from many quarters as he or she attempts to lessen the likelihood of lawsuits arising from repetitive losses and other causes.

Potential negligence liability in emergency management may arise from a variety of causes. Statutory protection for emergency response is widespread. Courts are loath to second-guess executive decisions made during planning or response. Immunity as a "governmental function" or "discretionary action" may also apply. Clear variance from proper procedure or generally accepted practice may, however, result in liability. If planning or other preparedness steps are required by law, failure to fulfill the statutory mandate may be the basis for liability. The best plans will assist responders rather than limiting their actions unreasonably.

Emergency management is responsible for assuring that the various emergency response agencies have plans that fit together. Assuring training to the plan's requirements is also an emergency management function. Emergency management is responsible for ensuring that the overall emergency plan is exercised, with all response agencies taking their parts. Emergency management must ensure that the plan is revised to reflect lessons learned during plan exercises. For emergency management to fulfill all of its legal responsibilities, however, it must be strongly supported by the head of a unit of government. When individual agencies within a unit of government refuse to fulfill their legal duties, whether in preparedness or in response, the responsibility for fixing the situation lies squarely on the shoulders of the unit of government's leader.

All the planning in the world is useless if the emergency plans and directives formulated under them cannot be communicated to the public. In order to utilize any government services, communication in the language understood by the person who is the target of the message is vital. For this reason, the federal government requires emergency managers to include people from all language groups in the planning process. This allows partnering ahead of time with people who will be links to various groups who are trusted by all parties during an emergency event.

Each state and local jurisdiction has its own set of laws that must be kept in mind when answering any legal question. Florida, as a state that has suffered more than its share of disasters, possesses evolved law as well as experienced attorneys in the area of emergency management. Therefore, including an article on that state's approach to this area of law is particularly appropriate. Those from around the nation may profit from the Sunshine State's learning experiences without going through the pain that led to the lessons recounted for their edification.

The recovery phase imposes its own set of challenges. Among the most difficult are ensuring that both individuals and units of government receive all relief funds to which they may be rightfully entitled. Those not entitled to funds must not be permitted to commit fraud and obtain monies sorely needed by real victims. Others will come into the ravaged area and try to gouge victims out of money for shoddy goods and services or simply take their money, promising relief that they never deliver. Lawyers can help to protect against such schemes and, where appropriate, prosecute their practitioners.

All government employees are bound by codes of ethics as well as legal requirements that limit actions that they may take. In the wake of some unethical practices that occurred after the 9/11 attacks in the procurement area, both emergency managers and the attorneys who advise them need to keep these limitations in mind. The overwhelming majority of public servants enter their careers with the focus on service, but the misdeeds of a few can reflect on many. One must remember that the employer is the public at large, and if one becomes aware of persons who are not properly serving that employer, there is a duty to report them to the proper authorities. The duty of ethical service can be a heavy one.

In the aftermath of the September 11, 2001 attacks, emergency response and emergency management changed in important and basic ways. They became, to an unprecedented extent, nationally directed and controlled forces. As part of the Department of Homeland Security, FEMA became a very small player in a gargantuan organization. Despite all claims to the contrary, FEMA's future effectiveness will continue to depend very much on the closeness of its head to the President. FEMA's budget is tiny compared to that of DHS, and under the Bush/Cheney administration, its responsibilities were steadily nibbled away. DHS is a law enforcement organization, and FEMA is the odd person out within its ambit. DHS and the Department of Defense cooperate so closely on matters of national security, that for many purposes, they may be seen as a single entity. Under Bush/Cheney, that entity spent $1 trillion on the war in Iraq, and one is hard pressed to identify any return that the nation has received for that investment. Far easier is sight of the costs, which go well beyond mere money to include many lives, wounded and maimed service members and civilians, as well as strategic positions in the middle east. From an emergency management viewpoint, the funds could have been used to rejuvenate our tired infrastructure as well as to fund and educate attorneys and local emergency managers in **specific state and local emergency management law**.

Looking to a future that will ensue if our leaders do not address the structural issues underlying the ongoing increase in disasters, emergency managers may be facing a "perfect storm" that could potentially overwhelm even the best prepared of jurisdictions – local, state and even national. The number of vulnerable people likely to need more help during a disaster continues to rise. Reasonable mitigation measures face resistance, sometimes from well-connected and well-financed opponents. Limited funds may be expended on projects that are of questionable worth, while pressing dangers go unaddressed. The earth is populated well beyond its carrying capacity, and population continues to rise. Arable land is being overrun by expanding population centers. Changing weather patterns pose multiple daunting challenges, including the possibility of crop failures.[2] The world does not have food reserves sufficient to see its residents through a single year without harvests – famine is a possibility. Shortages of clean drinking water are endemic worldwide, especially in the poorest nations. The globe possesses dwindling natural resources – the low-hanging fruit has already been harvested.

Examine, for example, the issue of energy, which is in ever-increasing demand by the continually growing world population, all of which aspires to live at the level of the people of the United States. Petroleum reserves are finite, and the search for new ones leads steadily deeper into the earth and into ever more fragile ecosystems. Other energy sources like coal may be more abundant options, but they too have limits, and carry their own costs. These include significant impact on the land from which coal is extracted and the release of particulates into the air and pollutants into the water. Some observers blame the particulates in part for increases atmospheric heating (and demand for energy for air conditioning – a downward cycle). These observers believe that heating melts ice caps, raising sea levels, resulting in floods. Those in favor of nuclear power tout its long life and call it clean, but the storage of long-lived poisonous waste has never been adequately addressed (Note the previous discussion of the 2011 Japanese tsunami). All of the above approaches to energy needs have a significant problem – they just "kick the can" down the road for future generations to deal

---

[2] For a broad and provocative look at weather issues resulting from climate change as well as policy challenges, *See* Jane Lubchenco and Thomas R. Karl, "Predicting and Managing Extreme Weather Events." 65 *Physics Today No. 3*: 31 (March 2012). Found online at: http://www.physicstoday.org/resource/1/phtoad/v65/i3/p31_s1?bypassSSO=1

with, and do not directly deal with the underlying issues. Tidal, solar and wind power have promise, but to date their performance has not lived up to the starry eyed visions of promoters. All too often, politicians continue to have a horizon that looks no further than the next election, and business leaders look no further than the end of the next quarter. Meanwhile, these potential structural disasters continue to stare the world in its unseeing face. In the not-too-distant future, emergency management lawyers may find that their mitigation activities consist of writing contracts to ensure that their jurisdiction has access to a limited pool of resources at the expense of others to ensure short-term survival – a far cry from the cooperation and multiplication of effort encouraged by NIMS and the NRF of today.

This book will serve as a starting point for attorneys and policy makers who want to play their part in safeguarding emergency response and emergency management from legal liability. Emergency responders and emergency managers themselves will also find the text accessible and of assistance in their daily activities. These groups need to take another step if they wish to provide the fullest protection from liability. They must become partners in protection. The best attorneys know both the law and their clients' business. Lawyers should spend time in the field with emergency responders and emergency managers to learn about their concerns and perspectives. Only by sharing information on legal concerns and listening respectfully to one another's views can emergency responders, emergency managers, and their attorney advisors enter into a comprehensive alliance. Only through a truly equal partnership can appropriate steps be taken to mitigate and prevent legal liability.

QUESTIONS

1. What was the most interesting chapter of the book to you? Explain your answer.
2. What do you plan to do with the knowledge you have gained in this course?
3. In your opinion, where in the local government table of organization should emergency management be located? Explain your answer. Be prepared to defend your point of view in class.
4. What is your opinion regarding NIMS and the NRF? Are their goals worthwhile? To what extent are they achieving their goals? How could they do a better job? What downsides do you see to their "one size fits all" approach?
5. What are the potential liabilities for the utilities that operate nuclear plants? Why is this the case? Do you believe that the current system of liability provides sufficient protection to the public? Why or why not?
6. The author discusses "structural issues underlying the ongoing increase in disasters" that if not dealt with may cause a "perfect storm" that could potentially overwhelm even the best prepared of nations. Discuss the issues he raises. What additional energy sources does he not mention? In what ways does he underestimate politicians and business leaders? What ways do you see of avoiding the calamities he foresees for future emergency managers and the lawyers who advise them? What can students of emergency management, emergency managers and lawyers do to help avoid structural disasters?

Appendix A

LOCAL DISASTER OR EMERGENCY ORDINANCE*

Section 7-0701. Intent.

It is the intent of this article to provide the necessary organization, powers, and authority to enable the timely and effective use of all available City resources to prepare for, respond to, and recover from emergencies and/or disasters, whether natural or manmade, likely to affect the health, security, safety, or property of the inhabitants of the City. It is intended to grant power as broad as permitted.

Section 7-0702. Definitions.

(1) Civil emergency: Conditions of unrest including, but not limited to riot, civil disturbance, unlawful assembly, hostile military or paramilitary action, war, terrorism, or sabotage.

(2) Disaster: The occurrence of widespread or severe damage, injury, or loss of life or property resulting from any natural or manmade cause including but not limited to flood, fire, cyclone, tornado, earthquake, severe high or low temperatures, blizzard, landslide, mudslide, hurricane, building or structural collapse, high water table, water pollution, air pollution, epidemic, riot, drought, utility emergency, sudden and severe energy shortages, volcano, earthquake, snow, ice, windstorm, waves, hazardous substance spills, chemical or petroleum spills, biological material release or spill, radiological release or spill, structural failure, public health emergency, or accidents.

(3) Emergency: Any occurrence or threat of natural or manmade disaster of a major proportion in which the safety and welfare of the inhabitants of the City or their property are jeopardized or placed at extreme peril that timely action may avert or minimize.

(4) Utility emergency: Any condition which endangers or threatens to endanger the safety, potability, availability, transmission, distribution, treatment, or storage of water, natural gas, fuel, or electricity through their respective systems.

Section 7-0703. Authorization to issue declaration of emergency or disaster.

The Mayor is authorized to declare a local emergency or disaster if the Mayor finds that the City or any part thereof is suffering from or is in eminent danger of suffering a natural or manmade emergency or disaster.

Section 7-0704. Filing and posting of declaration.

Any declaration of an emergency or disaster by the mayor or person acting in place of the Mayor under authority of this Ordinance shall be promptly filed with the City Auditor and the public shall be notified through posting the declaration in a public place and general publicity of said declaration. The declaration shall be effective when filed or, when filing is not practicable due to emergent circumstances, when posted

---

* This material is based on a form ordinance found in: Howard D. Swanson, "The Delicate Art of Practicing Municipal Law Under Conditions of Hell and High Water." *North Dakota Law Review* 76: 487 (2000).

and given general publicity. The declaration shall be filed with the City Auditor as soon as practicable if emergent circumstances prevent its immediate filing.

Section 7-0705. Term of declaration.

The declaration of a local emergency or disaster shall be in effect as determined by the Mayor or person acting in place of the Mayor under authority of this Ordinance for a period of up to thirty (30) days. This period may be extended only upon the approval of the City Council.

Section 1-0106. Succession of authority.

If the Mayor is unavailable, the President of the City Council shall have the same authority as is granted to the Mayor hereunder, followed by the Vice President of the City Council, the Chair of the Public Safety Committee, the Chair of the Public Service Committee, the Chair of the Finance Committee, the Chair of the Urban Development Committee, and then followed by the next most senior member of the City Council. In the event that none of these are available, the same authority shall be granted to the City Emergency Management Director, followed by the City Chief of Police, City Fire Chief, and the Head of the Department of Public Works. When a person higher in the line of authority specified herein becomes available, that person shall assume the powers granted under this Emergency Ordinance during the term of the Declaration.

Section 1-0107 Powers.

Upon the issuance of an emergency or disaster declaration, the Mayor (or the person acting in his or her stead under Section 1-0106) may exercise the following powers, including, but not limited to:

(1) an order establishing a curfew during such hours of the days or nights and affecting such categories of persons as may be designated.

(2) an order to direct and compel the evacuation of all or a part of the population from any stricken or threatened areas within the City if the Mayor deems this action is necessary for the preservation of life, property or other disaster or emergency mitigation, response

or recovery activities, and to prescribe routes, modes of transportation and destination in connection with an evacuation.

(3) an order controlling, restricting, allocating, or regulating the use, sale, production, or distribution of food, water, fuel, clothing, and other commodities, materials, goods, services, and resources.

(4) an order requiring the closing of businesses deemed nonessential by the Mayor.

(5) an order suspending or limiting the sale, distribution, dispensing, or transportation of alcoholic beverages, firearms, explosives, and/or combustible products and requiring the closing of those businesses or parts of businesses insofar as the sale, distribution, dispensing, or transportation of these items are concerned.

(6) an order prohibiting the sale or distribution within the City of any products which, the mayor determines, could be employed in a manner which would constitute a danger to public health or safety.

(7) an order closing any streets, alleys, sidewalks, public parks, public ways, or other public places.

(8) an order closing the access to any buildings, streets, alleys, sidewalks or other public or private places.

(9) establish and control routes of transportation ingress or egress.

(10) control ingress and egress from a disaster or emergency area.

(11) subject to requirements for compensation, commandeer, or utilize private property if necessary to cope with emergency or disaster conditions.

(12) appropriate and expend funds, exclude contracts, authorize the obtaining and acquisition of property, equipment, services, supplies, and materials without the strict compliance with procurement regulations or procedures.

(13) transfer the direction, personnel, or functions of City departments and agencies for the purposes of performing or facilitating emergency or disaster services.

(14) utilize all available resources of the City as may be reasonably necessary to cope with

the emergency or disaster whether in preparation for, response to, or recovery from an emergency or disaster.

(15) suspend or modify the provisions of any ordinance if strict compliance thereof would in any way prevent, hinder, or delay necessary action in coping with any emergency or disaster.

(16) accept services, gifts, grants and loans, equipment, supplies and materials whether from private, nonprofit, or governmental sources.

(17) temporarily suspend, limit, cancel, postpone, convene, schedule, or continue all meetings of the City Council, and any City committee, commission, board, authority, or other City body as deemed appropriate by the Mayor.

(18) suspend or limit the use of the City's water resources.

(19) suspend or limit the burning of any items of property within the City limits and up to two (2) miles outside the corporate City limits.

(20) requiring emergency services of any City officer or employee. If regular City forces are determined to be inadequate, then to acquire the services of such other personnel as are available, including citizen volunteers. All duly authorized persons rendering emergency services shall be entitled to the privileges and immunities as provided by state or local law.

(21) hire or contract for construction, engineering, architectural, building, electrical, plumbing, and/or other professional or construction services essential to continue the activities of the City without the advertising of bids or compliance with procurement requirements.

(22) make application for local, state, or federal assistance.

(23) terminate or suspend any process, operation, machine, device, or event that is or may negatively impact the health, safety, and welfare of persons or property within the City.

(24) delegate authority to such City officials as the Mayor determines reasonably necessary or expedient.

(25) requiring the continuation, termination, disconnection, or suspension of natural gas, electric power, water, sewer, or other public utilities.

(26) close or cancel the use of any municipally owned or operated building or other public facility.

(27) prescribe routes, modes of transportation and destination in connection with any evacuation.

(28) exercise such powers and functions in light of the exigencies of emergency or disaster including the waiving of compliance with any time consuming procedures and formalities, including notices, as may be prescribed by law pertaining thereto.

(29) issue any and all such other orders or undertake such other functions and activities as the Mayor reasonably believes is required to protect the health, safety, welfare of persons or property within the City or to otherwise preserve the public peace or abate, clean up, or mitigate the effects of any emergency or disaster.

Section 7-0708. Enforcement of orders.

A. The members of the police department and such other law enforcement and peace officers as may be authorized by the Mayor are further authorized and directed to enforce the orders, rules, and regulations made or issued pursuant to this chapter.

B. During the period of a declared emergency or disaster, a person shall not:

(1) enter or remain upon the premises of any establishment not open for business to the general public, unless such person is the owner or authorized agent of the establishment.

(2) violate any of the orders duly issued by the Mayor or authorized personnel.

(3) willfully obstruct, hinder, or delay any duly authorized City officer, employee, or volunteer in the enforcement or exercise of the provisions of this chapter, or of the undertaking of any activity pursuant to this chapter.

Section 1-0109. Authority to enter a property. During the period of a declared emergency or disaster, a City employee or authorized agent may enter onto or upon private property if the employee or authorized agent has reasonable grounds to believe that there is a true emergency and an immediate need for assistance for the protection of life or property, and that entering onto the private land will allow the person to take such steps to alleviate or minimize the emergency or disaster or to prevent or minimize danger to lives or property from the declared emergency or disaster.

Section 1-0110. Location of governing body meetings and departments.
(1) Whenever an emergency or disaster makes it imprudent or impossible to conduct the affairs of the City at its regular locations, the governing body may meet at any place, inside or outside the city limits. Any temporary disaster meeting location for the governing body shall continue until a new location is established or until the emergency or disaster is terminated and the governing body is able to return to its normal location.
(2) Whenever a disaster makes it imprudent or impossible to conduct the affairs of any department of the City at its regular location, such department may conduct its business at any place, inside or outside the city limits, and may remain at the temporary location until the emergency or disaster is declared ended or until the department is able to return to its normal location.
(3) Any official act or meeting required to be performed at any regular location of the governing body or of its departments is valid when performed at any temporary location under this section.

(4) The provisions of this section shall apply to all executive, legislative, and judicial branches, powers, and functions conferred upon the city and its offices, employees, and authorized agents.

Section 1-0111. Mutual aid agreements.
(1) The Mayor may, on behalf of the City, enter into such reciprocal aid, mutual aid, joint powers agreements, intergovernmental assistance agreements, or other compacts or plans with other governmental entities for the protection of life and property. Such agreements may include the furnishing or exchange of supplies, equipment, facilities, personnel, and/or services.
(2) The governing body or any of its committees, commissions, or authorities may exercise such powers and functions in light of the exigencies of the emergency or disaster and may waive compliance with time consuming procedures and formalities prescribed by law pertaining thereto.
(3) The foregoing shall apply to all executive, legislative, and judicial powers and functions conferred upon the City and its officers, employees, and authorized agents.

1-0112. Severability.
The provisions of this chapter are declared to be severable, and if any section, sentence, clause, or phrase of this chapter shall for any reason be held to be invalid or unconstitutional or if the application of this chapter to any person or circumstance is held to be invalid or unconstitutional, such holding shall not affect the validity of the remaining sections, sentences, clauses, and/or phrases of this ordinance.

Appendix B

WEB SITES OF INTEREST TO EMERGENCY RESPONDERS AND EMERGENCY MANAGERS

American College of Emergency Physicians
acep.org
The American College of Emergency Physicians (ACEP) exists to support quality emergency medical care, and to promote the interests of emergency physicians.

American Police Beat
http://www.apbweb.com
The online voice on the Nation's police community.

British Intelligence
www.mi5.gov.uk/
The official website of the Security Service (MI5).

Center for Defense Information
http://www.cdi.org/
The Center for Defense Information is a nonpartisan, nonprofit organization committed to independent research on the social, economic, environmental, political and military components of global security.

Centers for Disease Control and Prevention
http://www.cdc.gov/
CDC website informs about bioterrorism issues, including anthrax.

Chemical, Biological, Radiological, and Nuclear Defense Information Analysis Center
https://www.cbrniac.apgea.army.mil/Pages/default.aspx

A Department of Defense Information Analysis Center – The Premier Resource for Authoritative CBRN Defense and Homeland Security Scientific and Technical Information.

EPA Emergency Management Homepage
http://www.epa.gov/emergencies/index.htm
To ensure that this nation is better prepared for environmental emergencies, EPA is working with other federal partners to prevent accidents as well as to maintain superior response capabilities. One of our roles is to provide information about response efforts, regulations, tools, and research that will help the regulated community, government entities, and concerned citizens prevent, prepare for, and respond to emergencies.

CNN News
http://www.cnn.com/
Provides up-to-date breaking news stories.

Cops on Line
http://www.copsonline.com/
Cop site by real cops.

Counterrorism Office: U.S. Department of State
http://www.state.gov/s/ct/
Works to support U.S. Counterterrorism Policy.

Department of Homeland Security
http://www.dhs.gov/
Umbrella agency responsible for most domestic terrorism prevention. Houses many sub-agencies, including FEMA.

Federal Bureau of Investigation
http://www.fbi.gov/
Offers many resources.

Federal Emergency Management Agency
http://www.fema.gov/
The FEMA website is loaded with information
on mitigation, preparedness for all hazards, response, and recovery.

Federation of American Scientists
http://www.fas.org/
The Federation of American Scientists conducts
analysis and advocacy on science, technology
and public policy.

Indiana Department of Homeland Security
http://www.in.gov/dhs/
Contains a variety of information on state level
emergency and terrorism preparedness.

Institute for Homeland Security
http://www.homelandsecurity.org/
Provides executive education and public awareness of the challenges to homeland security in
the 21st century.

International Association of Emergency Managers
http://www.iaem.com/
Professional emergency managers from around
the world.

International Code Council (ICC)
http://www.iccsafe.org/
The International Code Council is a member-fo-
cused association dedicated to helping the building safety community and construction industry
provide safe, sustainable and affordable construction through the development of codes and
standards used in the design, build and compliance process. Most U.S. communities and many
global markets choose the International Codes.

International Policy Center for Counter-Terrorism
http://www.ict.org.il/
ICT is a research institute and think tank dedicated to developing innovative public policy solutions to international terrorism.

Moxie Safety Resources
http://www.moxietraining.com/oshalinks/bioter
rorism_state.htm
Web Site connecting to multiple other sites.

National Association of Amusement Ride Safety
Officials (NAARSO)
http://naarso.com/
The National Association of Amusement Ride
Safety Officials is composed of Amusement Ride
Inspectors representing jurisdictional agencies,
insurance companies, private consultants, safety
professionals, and federal government agencies.

National Association of EMS Physicians
http://www.naemsp.org/
The National Association of EMS Physicians is an
organization of physicians and other professionals who provide leadership and foster excellence
in out-of-hospital emergency medical services.

National Association of State EMS Officials
http://www.nasemso.org/
The National Association of State EMS Officials
is the lead national organization for EMS.

National Association of State Fire Marshals
(NASFM)
http://www.firemarshals.org/
The membership of National Association of State
Fire Marshals (NASFM) comprises the most
senior fire officials in the United States. State Fire
Marshals' responsibilities vary from state to state,
but Marshals tend to be responsible for fire safety
code adoption and enforcement, fire and arson
investigation, fire incident data reporting and
analysis, public education and advising Governors
and State Legislatures on fire protection. Some
State Fire Marshals are responsible for fire fighter
training, hazardous materials incident responses,
wildland fires and the regulation of natural gas
and other pipelines.

National Conference of State Legislatures
http://www.ncsl.org/
A resource for state legislators and others interested in developments on the state side. Contains
model legislation.

National Emergency Management Association
http://www.nemaweb.org/
NEMA is the professional association of state, Pacific and Caribbean insular state emergency management directors.

National Institute for Occupational Safety and Health (NIOSH)
http://www.cdc.gov/niosh/
The National Institute for Occupational Safety and Health (NIOSH) is the Federal agency responsible for conducting research and making recommendations for the prevention of work-related disease and injury. The Institute is part of the Centers for Disease Control and Prevention (CDC). NIOSH conducts the Fire Fighter Fatality Investigation and Prevention Program. The reports arc an invaluable safety and liability prevention source.
http://www.cdc.gov/niosh/fire home.html

National Fire Protection Association
http://www.nfpa.org/
The mission of the international nonprofit NFPA is to reduce the worldwide burden of fire and other hazards on the quality of life by providing and advocating scientifically based consensus codes and standards, research, training and education.

National Highway Traffic Safety Administration
http://www.nhtsa.dot.gov/
NHTSA promotes highway safety and keeps statistics on motor vehicle crashes.

National Guard
http://www.nationalguard.mil
Official website of The National Guard Bureau. Provides vital domestic military support capabilities in public safety / disaster preparedness.

National Response Center
http://www.nrc.uscg.mil/index.htm
The NRC is the sole federal point of contact for reporting oil and chemical spills.

National Safety Council
http://www.nsc.org
The National Safety Council is the nation's leading advocate for safety and health.

National Weather Service
http://www.nws.noaa.gov/
A user friendly interface to the weather.

NIMS Resource Center
http://www.fema.gov/emergency/nims/

Office of Homeland Security
http://www.whitehouse.gov/homeland
A link to White House positions on Homeland Security.

Oklahoma City National Memorial Institute for the Prevention of Terrorism
http://www.mipt.org/
The Oklahoma City National Memorial Institute for the Prevention of Terrorism is dedicated to preventing and reducing terrorism and mitigating its effects. Serving the line officer through police training.

Page, Wolfberg & Wirth
www.pwwemslaw.com
Page, Wolfberg & Wirth is a National EMS, Ambulance, & Medical Transportation Industry Law Firm.

Public Health Emergency Preparedness – Dept. of HHS
http://www.phe.gov/preparedness/Pages/default.aspx
Provides resources for public health emergency preparedness on national, state and local levels.

Terrorism Questions and Answers
http://www.cfr.org/issue/135/
From the Council on Foreign Relations.

Transportation Security Administration
http://www.tsa.gov/
A part of DHS, the Transportation Security Administration protects the Nation's transportation systems to ensure freedom of movement for people and commerce.

U.S. Army Research, Development and Engineering Command
http://www.army.mil/rdecom

The U.S. Army Research, Development and Engineering Command is the Army's technology leader and largest technology developer. RDECOM ensures the dominance of Army capabilities by creating, integrating and delivering technology-enabled solutions to our Soldiers. To meet this commitment to the Army, RDECOM develops technologies in its eight major laboratories and research, development and engineering centers. It also integrates technologies developed in partnership with an extensive network of academic, industry, and international partners.

U.S. Fire Administration
http://www.usfa.fema.gov/
As an entity of the Federal Emergency Management Agency, the mission of the USFA is to reduce life and economic losses due to fire and related emergencies, through leadership, advocacy, coordination, and support.

U.S. Army Corps of Engineers
http://www.usace.army.mil/
The mission of the US Army Corps of Engineers is to provide quality, responsive engineering services to the nation.

U.S. Department of Energy
http://www.energy.gov/
The Department of Energy's overarching mission is enhancing national security.

U.S. Geological Survey
http://www.usgs.gov/
USGS provides accurate scientific information to help prevent and recover from natuiral disasters.

U.S. Nuclear Regulatory Commission
http://www.nrc.gov/
The NRC regulates US commercial nuclear power plants and regulates civilian use of nuclear materials.

Appendix C

EMERGENCY PLANNING WEB RESOURCES

Emergency Plans

Biological and Chemical Terrorism: Strategic Plan for Preparedness and Response
http://www.cdc.gov/mmwr/preview/mmwrhtml/rr4904al.htm

National Center for HIV/AIDS, Viral Hepatitis, STD, and TB Prevention Strategic Plan 2010–2015
http://www.cdc.gov/nchhstp/docs/10_NCHHSTP%20strategic%20plan%20Book_semi%20final508.pdf

National Response Framework
http://www.fema.gov/pdf/emergency/nrf/nrf-core.pdf

National Oil and Hazardous Substances Pollution Contingency Plan (commonly referred to as the National Contingency Plan or NCP)
http://www.epa.gov/OEM/content/lawsregs/ncpover.htm

Other Resources

Developing and Maintaining Emergency Operations Plans: Comprehensive Preparedness Guide 101 (CPG 101 Version 2, November 2010)
http://www.fema.gov/pdf/about/divisions/npd/CPG_101_V2.pdf

National Flood Insurance Program
http://www.fema.gov/about/programs/nfip/index.shtm

Society for College and University Planning
http://www.scup.org/

Understanding Your Risks: identifying hazards and estimating losses (FEMA 386-2)
http://www.fema.gov/library/viewRecord.do?id=1880

INDEX

William C. Nicholson, Photo by Chip Devine.

ABOUT THE AUTHOR

William Charles Nicholson is an internationally known expert in emergency response and emergency management law and policy. He assists businesses and units of government as they seek innovative, proactive techniques to mitigate liability risks. Nicholson formerly served on the faculty of North Carolina Central University as an Assistant Professor in the Department of Criminal Justice, where he taught Emergency Management and Recovery, Homeland Security Law and Policy, Ethics in Criminal Justice and Criminal Justice Management Theory. While at Widener University School of Law, Nicholson created the nation's first Terrorism and Emergency Law course. At the University of Delaware, he produced his pioneering Homeland Security Law and Policy course. Prior to his academic career, Nicholson served as General Counsel to the Indiana State Emergency Management Agency, Indiana Department of Fire and Building Services, and Public Safety Training Institute. He also provided legal advice to the following affiliated Boards and Commissions for which the Agencies provide staff support: Indiana Public Safety Training Institute Board of Directors; Emergency Management, Fire and Building Safety and Public Safety Training Foundation; Fire Protection and Building Safety Commission; Emergency Medical Services Commission; Indiana Emergency Response Commission; Regulated Amusement Device Safety Board; and Boiler and Pressure Vessel Safety Board. He served as the agencies' Ethics Officer and supervised the Human Resources area. Previously, he was part of a complex litigation practice in Washington, DC. He is a Member of the Editorial Boards for *Best Practices in Emergency Services: Today's Tips for Tomorrow's Success and Journal of Emergency Management.*

Notable Publications

Books

Emergency Response and Emergency Management Law, (2nd Ed.). Charles C Thomas Publisher, Ltd. (2012).

Homeland Security Law and Policy, Charles C Thomas Publisher, Ltd. (2005).

Emergency Response and Emergency Management Law, Charles C Thomas Publisher, Ltd. (2003).

Law Review Articles

Obtaining Competent Legal Advice: Challenges for Local Emergency Managers and Attorneys. Natural Disaster Law issue, *California Western Law Review, 46*, No. 2, 343–368 (2010).

Seeking Consensus on Homeland Security Standards: Adopting the National Response Plan and the National Incident Management System. *Widener Law Review, 12*, No. 2, 491–559 (2006).

Legal Issues in Emergency Response to Terrorism Incidents Involving Hazardous Materials: The Hazardous Waste Operations and Emergency Response ("HAZWOPER") Standard, Standard Operating Procedures, Mutual Aid and the Incident Command System. *Widener Symposium Law Journal, 9*, Number 2, 295 (2003).

Other Selected Articles

"Point/Counterpoint: Should FEMA Stay in DHS?" *Best Practices in Emergency Services* Vol. 12, No. 1 (February 2009); "An Essential Team: Local Emergency Managers and Legal Counsel" *The Public Manager*, Vol. 37, No. 4 (Winter 2008-2009); "Courts Address When Dual Function Fire Services Employees Qualify for Overtime Pay" *Best Practices in Emergency Services* Vol. 11, No. 10 (October 2008); "Emergency Planning and Potential

Liabilities for State and Local Governments" *State and Local Government Review* Vol. 39, No. 1, 44–56 (2008); "Emergency Management Legal Issues" invited chapter for revision of the International City Management Association (ICMA) publication *Emergency Management: Principles and Practice for Local Government*, Kathleen Tierney and William Waugh, editors (2007); "Emergency Management and Law" invited chapter in *Disciplines, Disasters and Emergency Management: The Convergence and Divergence of Concepts, Issues and Trends in the Research Literature*, David A. McEntire, Ph.D. (Editor), Charles C Thomas Publisher, Ltd., Springfield, IL. (2007); invited chapter in *McGraw-Hill Handbook of Homeland Security* entitled "Legal Issues for Businesses in Responding to Terrorist Events" David G. Kamien, editor (2006); "Legal Issues in Public Health" invited chapter in American Bar Association State and Local Government Law Section's book Legal Issues in Homeland Security and Emergency Management Errnest B. Abbott & Otto J. Hetzel, Editors (2005).

Education

| Reed College | Portland, OR | BA |
| Washington and Lee University School of Law | Lexington, VA | Juris Doctor |

Contact: emergency_management_law@yahoo.com